OPTIMAL RADAR TRACKING SYSTEMS

Altair Radar Antenna on Kwajalein Atoll

OPTIMAL RADAR TRACKING SYSTEMS

George Biernson
Communication Systems Division
GTE Government Systems

A WILEY-INTERSCIENCE PUBLICATION
John Wiley & Sons, Inc.
NEW YORK / CHICHESTER / BRISBANE / TORONTO / SINGAPORE

Library of Congress Cataloging in Publication Data:

Biernson, George.
  Optimal radar tracking systems/George Biernson.
     p.   cm.
  "A Wiley-Interscience publication."
  Bibliography: p.
  ISBN 0-471-50673-7
  1. Tracking radar.   I. Title.
  TK6580.B44 1989   89-32555
  621.3848—dc20   CIP

Printed in the United States of America

10 9 8 7 6 5 4 3 2 1

Contents

Preface

This book is based on the study of a sophisticated radar tracking system, the Altair radar, which uses Kalman filtering to perform optimal long-range tracking of ballistic missile warheads. This engineering example provides a means for explaining Kalman filter theory in a relatively simple manner as well as many other technical issues that are critical to the design of a modern optimal radar tracking system.

To give the background needed to understand the study of Altair, the book presents simplified explanations of the following:

Feedback control.
Modulation and demodulation of signals.
Digital sampled-data systems.
Digital computer simulation.
Statistical analysis of random signals.
Detection and tracking processes in a radar system.

These discussions start at an elementary level and develop the concepts in a concrete manner.

The book is directed to all who are interested in target tracking, including mathematicians. The reader is assumed to have a basic knowledge of mathematics (including the Laplace transform), physics, and AC circuit analysis.

For security reasons, certain details of the Altair radar system are replaced with a comparable radar model. The calculations applied to the model are the same as those that have been used to analyze the performance of the Altair radar, and so the reader is presented with the equivalent of a direct study of Altair. The Kalman filter tracking equations that are given are implemented in Altair, but the radar parameters and some of the tracking coordinate transformation equations are different.

The Altair radar antenna, shown in the frontispiece, is physically very impressive. It has a huge steerable parabolic dish, 150 ft in diameter, which operates simultaneously at VHF and UHF. This radar is located in the Kwajalein Atoll of the Marshall Islands, 2100 miles west and south of Hawaii. It performs long-range tracking of warheads from ballistic missiles launched at Vandenburg Air Force Base in California and dropped into the lagoon of the atoll.

Although this book deals directly with radar tracking technology, the basic principles apply to any target-tracking system. Hence, the book should also be useful to individuals involved in other tracking applications that use optical signals, sonar signals, RF telemetry signals, etc.

This study of Altair includes a considerable amount of detail concerning the operation of a complex electronic system and thereby presents a study that is unusual in the unclassified literature. It gives a practical introduction to the system analysis of modern sophisticated electronic equipment by illustrating the multitude of analytical issues involved in its design.

ACKNOWLEDGMENT

I am grateful for the support provided by GTE Government Systems in preparing this book.

The material presented in Chapter 8, on which this book is based, evolved from work sponsored by the Ballistic Missile Defense Systems Command of the U.S. Army, under technical direction of the Massachusetts Institute of Technology, Lincoln Laboratory.

The cooperation of the U.S. Army in reviewing and approving this material for publication is greatly appreciated. I am indebted to Mr. George Fangmann of the GTE Altair Program Office for his assistance in achieving this approval.

I want to thank Analog Devices, Inc. for graciously providing the illustrations for describing noise signals given in Figs 6.2-7 and 6.2-8.

GEORGE BIERNSON

Concord, Massachusetts
January 1990

Chapter 1

Introduction

Target tracking in modern radar systems is often very sophisticated, involving several engineering disciplines. An effective system engineering approach to this field requires a broad understanding of its many aspects. This should include Kalman filter theory [8.1, 8.2], which has become an important element in the design of tracking systems. However, the esoteric manner in which this theory has been presented has placed it beyond the understanding of many engineers engaged in target tracking.

One of the best ways to learn engineering is to study its application in a practical setting. This book presents a design study of a sophisticated radar tracking system, the Altair radar, which uses Kalman filtering to perform optimal long-range tracking of ballistic missile warheads. A study of this system provides an effective means of explaining Kalman filtering, along with many other engineering issues of a complex radar tracking system.

Although this book deals directly with radar tracking technology, the basic principles apply to any target-tracking system. Hence, the book should also be useful to individuals involved in other tracking applications that use optical signals, sonar signals, radio-frequency (RF) telemetry signals, etc.

1.1 OVERVIEW OF BOOK

The Altair radar, shown in the frontispiece, has a huge steerable parabolic antenna, 150 ft in diameter, which operates simultaneously at very high frequency (VHF) and ultrahigh frequency (UHF). This radar is located in the Kwajalein Atoll of the Marshall Islands, 2100 miles west and south of Hawaii. It tracks warheads from ballistic missiles launched at Vanderburg Air Force Base in California and dropped into the lagoon of the atoll.

The study of Altair, which is presented in Chapter 8, gives a practical explanation of the many engineering aspects of its tracking system. This includes Kalman filter theory, feedback control, digital sampled-data theory, and radar signal processing. It shows how these different principles are integrated into a complex modern radar tracking system.

In order that the reader can comprehend this study of the Altair system, background material is presented in Chapters 2–7. The book is directed to all who are interested in target tracking, including the mathematician. The discussions in Chapters 2–7 start at an elementary level and develop the concepts in a concrete manner. The reader is assumed to have a basic knowledge of mathematics (including the Laplace transform), physics, and AC circuit analysis.

In Chapters 2–7, the reader is led by the hand through the following material needed to comprehend the complex issues of a sophisticated optimal radar-tracking system:

a. Chapter 2 summarizes the principles of feedback control, and Chapter 3 extends this with techniques for calculating the error of a tracking radar due to target motion.

b. Chapter 4 discusses modulation, carrier-frequency filtering, and demodulation of signals.

c. Chapter 5 presents the theory of digital sampled-data systems. This includes the design of algorithms for digital filtering and digital computer simulation. A simulation method is presented that allows a very complex dynamic system to be modeled using BASIC computer language on a personal computer. The state variable, state equation, and transition matrix are explained in a simple manner.

d. Chapter 6 describes the statistical analysis of system response to random inputs, and applies this to receiver noise and wind gusts. Basic noise theory is explained, including noise bandwidth, the Gaussian distribution, and correlation functions.

e. Chapter 7 analyzes the detection and tracking processes of a tracking radar system. Radar performance is calculated for acquiring a target signal, and for tracking the target in range and angle.

For security reasons, certain details of the Altair radar system are replaced with a comparable radar model. The calculations applied to the model are equivalent to those that have been used to analyze the performance of the Altair radar. Hence, the reader is presented with the equivalent of a direct study of Altair. The Kalman filter tracking equations that are given are the same as those actually implemented in Altair, but the radar parameters and some of the tracking coordinate transformation equations are different.

The following is a summary of the discussion of Altair given in Chapter 8:

a. In Section 8.1, a radar model is presented, which differs quantitatively from Altair, but explains the radar aspects of the Altair tracking system. Radar parameters are examined along with the equations from which they are derived. This model shows how Altair senses radar signal strength to determine the measurement-noise covariance matrix used in the Kalman filter equations.

b. Section 8.2 describes the basic structure of the target-tracking equations. Radar tracking data are converted to rectangular coordinates, and Kalman filtering is performed in terms of these converted signals. Computations are included in the filters to account for predicted accelerations of the target, caused by gravity, coriolis effects, and atmospheric drag.

c. In Sections 8.3 and 8.4, the principle of the Kalman filter is explained in physical terms. From the Kalman-filter matrix equations, algorithms are derived that optimally adjust the smoothing parameters of the Altair Kalman-filtering tracking system. This Altair example yields a simple yet in-depth understanding of Kalman filter theory.

d. In Section 8.5, the tracking-filter equations of Altair are analyzed and reduced to signal-flow diagram form by applying the sampled-data and feedback-control principles developed earlier. This derivation shows that the Kalman filter configuration is equivalent to a conventional tracking system that filters the target information in terms of the coordinates of the target. It is the equations for calculating the optimal smoothing coefficients of the tracking filters that are unique to Kalman filter theory.

e. In Section 8.6, Altair Kalman-filter responses are simulated, yielding plots that show how this optimal tracking system performs while tracking a ballistic missile warhead. The plots characterize the acquisition phase, the steady-state conditions during exoatmospheric flight, and the strong transient that occurs when the warhead reenters the atmosphere.

Section 1.2 of this Introduction summarizes the background material presented in Chapters 2–7. Section 1.3 summarizes the discussion of the Altair system given in Chapter 8.

1.2 BACKGROUND MATERIAL FOR UNDERSTANDING THE ALTAIR DISCUSSION

1.2.1 Feedback Control Theory

Tracking is a feedback control process, and the Altair tracking system incorporates many feedback control loops. Feedback control is employed in the servos that position the radar antenna, in the range-tracking circuitry that keeps the range gate centered over the target pulse, and in the tracking-filter equations implemented in the tracking computer, which smooth the radar data and determine the trajectory of the target. Thus, feedback is the foundation of target tracking, and so the book begins in Chapter 2 with a summary of the principles of feedback control.

The summary of Chapter 2 is based on feedback control concepts that have been developed by the author, which are presented in *Principles of Feedback Control*, volumes 1 and 2 [1.1, 1.2].

The discussion of feedback control begins by examining a second-order lowpass filter. This introduces the transfer function concept, magnitude and phase plots of the frequency response, the transient response to a step input, and the natural frequency and damping ratio of a pair of complex poles. This and subsequent material are presented in a concrete manner using specific examples, so that a mathematician unfamiliar with the frequency and transient response tools of the electrical engineer is given the essential background needed to understand the book.

Feedback control is defined along with the basic variables and transfer functions of a feedback loop. Approximations are presented that relate frequency response parameters of a feedback loop to the transient responses for step and ramp inputs.

The primary dynamic parameter of a feedback loop is its *gain crossover frequency* ω_{gc}, which is the frequency where the magnitude of the open-loop transfer function (the *loop gain*) drops below unity. This is the "control bandwidth" of a feedback loop, because a feedback loop has effective control action only at frequencies where the loop gain is greater than unity. As will be explained, this parameter is quite distinct from the *half-power frequency* of a feedback loop, which is the frequency where the magnitude of the closed-loop frequency response is $1/\sqrt{2}$. The ratio of the gain crossover frequency ω_{gc} to the half-power frequency can vary more than $5:1$ for different practical feedback control loops.

The book describes the magnitude asymptote plot of the loop transfer function, which is a simple and powerful tool for feedback control analysis. Based on this plot, the *asymptotic gain crossover frequency* ω_c is defined. In practical feedback control loops, ω_c is approximately equal to the true gain crossover frequency ω_{gc}.

A method is developed for expressing the open-loop transfer function of a feedback loop so that the gain parameter of the transfer function is the asymptotic gain crossover frequency ω_c. This approach greatly simplifies feedback control design, because it yields a gain parameter that has fundamental significance, and thereby provides a means of approximating the dynamics of complicated multi-loop feedback control systems.

Based on the asymptotic gain crossover frequency ω_c, methods are presented in Section 2.6 for approximating the closed-loop transfer functions of a feedback control loop, from input to feedback signal and from input to error signal. These approximations are applied in Section 2.8 to analyze a multi-loop servomechanism having a current feedback loop, a velocity feedback loop, and a position feedback loop. This control system is used later in Chapter 5 as a model for digital computer simulation.

Section 2.7 presents a general discussion of the dynamic response of a feedback control system to a low-frequency input, which serves as an introduction to more detailed analyses in Chapter 3. This section shows how integral networks are used in control systems to reduce the error for a low-frequency input. Adding an integral network to a basic single-integration feedback loop produces a double integration at low frequency, which gives the loop *velocity memory*. Adding a second integral network produces a triple integration at low frequency, which gives the loop *acceleration memory*. An integral network is characterized primarily by its upper break frequency, denoted ω_i.

The feedback control principles developed in this chapter are applied later in Chapter 8 to analyze the responses of the tracking filters used in the Altair radar. The individual Altair tracking filters are feedback loops having velocity memory and sometimes acceleration memory. The smoothing parameters of the Altair Kalman filter are expressed in terms of the basic feedback loop parameters of the tracking filters: the gain crossover frequency ω_c and the values of the integral-network break frequency ω_i.

Chapter 3 presents a method for calculating the response of a feedback control system to an arbitrary input, using error-coefficient terms that are proportional to the derivatives of the input. This provides a general means of characterizing the error of a tracking system caused by target motion. The traditional error-coefficient approach is based on an infinite-series expansion of the system response. The theory of Chapter 3 employs a truncated error-coefficient expansion that includes transients along with error-coefficient terms. As will be shown, this provides a much clearer understanding of the nature of the response of a control system to a low-frequency input.

A general theory of the response of a system to an arbitrary input is presented in Section 3.3. A much simpler approach is given in Section 3.2, which employs the approximate error response of a feedback loop developed in Chapter 2, Section 2.6.

The method of Section 3.3 applies straight-line approximations to the input derivative curves. Bounds are derived for the inaccuracy in the calculated response associated with these approximations. The bounds are expressed in terms of parameters for each derivative called *remainder coefficients*. Associated with each remainder coefficient is a *remainder transient*, which describes the time over which a remainder-coefficient bound applies.

The remainder-coefficient principle is used in Chapter 8 to develop a simple optimal control theory applicable to the Altair tracking system. This simple optimal theory provides a frame of reference for evaluating and explaining the performance of the Kalman-filter tracking algorithms used in Altair.

1.2.2 Digital Signal Processing

Modern radar tracking systems usually implement tracking by means of digital signal processing, in which signals are conveyed in terms of sampled data. The principles of sampled-data theory are presented in Chapter 5. The sampling of a continuous signal, and the conversion of the sampled data back to a continuous signal, are analyzed. A method is developed for designing a sampled-data algorithm to implement a specified frequency response. This principle is extended to develop a simulation technique that allows the time response of a complex dynamic system to be simulated easily.

In a computer program, a dynamic transfer function is produced by a difference equation, which is a weighted sum of present and past values of the input and output signals for a computation step. The Laplace transform of a difference equation can be taken by expressing each time delay of one computation cycle as the transfer function $\exp[-sT]$, where T is the sampled period. The equivalent signal-frequency transfer function implemented by a difference equation can be obtained by replacing each $\exp[-sT]$ factor with the expression

$$\exp[-sT] = \frac{1 - pT/2}{1 + pT/2} \tag{1.2-1}$$

The variable p is a pseudo complex-frequency, which is approximately equal to the true complex-frequency s, for frequencies up to $1/4$ of the sampling frequency. (Note that a sampled signal can only convey information at frequencies up to half the sampling frequency.) The relations between the pseudo complex-frequency p and the true complex-frequency s are explained in Chapter 5. This discussion provides a rigorous theory for designing a sampled-data difference equation to satisfy a desired signal transfer function.

Chapter 5 describes a very convenient method of digital computer simulation called *serial simulation* that is based on the following principles:

1. The relation of Eq 1.2-1 is applied to obtain the computer algorithm that corresponds to each transfer function being simulated.
2. A feedback loop in a computation process must include a time delay of one computation cycle, because the feedback information is derived from a value computed in the previous cycle. The phase lag caused by this time delay is compensated for by inserting the lead factor $(1 + pT)$ into the computation loop.

The serial-simulation process allows a very complex dynamic system to be simulated easily on a personal computer using no more than BASIC computer language.

Chapter 5 explains the concepts of the state variable, the state equation, and the transition matrix, which are important elements in Kalman filter theory. The principles of Runge–Kutta integration are described, which is the conventional method for solving the differential equations of a dynamic process on a digital computer.

1.2.3 Analysis of Random Signals

Kalman filter theory assumes that the signals of the process are random and are characterized statistically. Chapter 6 describes the basic principles of statistical signal analysis, including the concepts of noise bandwidth, the Gaussian (or normal) distribution, correlation functions, and spectral densities.

These statistical principles are applied in Section 6.4 to analyze the effect of wind gust forces on a large satellite communication antenna. A wind gust spectrum is presented that evolved from measurements of wind torques on a large antenna. This spectrum has been used extensively as a wind gust standard in the specification of antenna tracking requirements in high-wind conditions. The dynamic responses of the antenna tracking servos to this wind gust spectrum are analyzed to determine the loss of communication signal caused by wind-induced tracking errors.

In the Kalman filter described in Chapter 8, variations of target motion are defined in terms of statistically characterized disturbances applied to a target model. The statistical wind model discussed in Section 6.4 can serve as a physically meaningful example of a typical disturbance spectrum in a Kalman filter.

Receiver noise is another variable that is characterized statistically. Thermal noise and shot noise are explained in Chapter 6. This noise theory is applied in Chapter 7 to analyze the target detection and tracking processes of a radar system.

1.2.4 Radar Signal Processing

An optimal tracking system is only as good as the data it uses for tracking, and so optimal tracking of a radar system must include optimal processing of the radar signal itself. Hence, Chapter 7 presents a detailed explanation of the principles of radar signal processing in a tracking radar.

As stated previously, Chapter 6 acts as an introduction to Chapter 7, by explaining the statistical analysis of random variables. A further introduction to radar systems is given in Chapter 4, which discusses amplitude modulation and demodulation of signals. It shows how a filter that operates on a modulated waveform affects the envelope of the waveform. This concept is applied in Chapter 7 to analyze the response of an intermediate-frequency (IF) bandpass filter in a radar receiver. The effect of the filter on the shape of an IF radar pulse is calculated.

The modulation of an amplitude-modulated waveform can be recovered by an envelope detector, which follows the peaks of the signal, ignoring its phase. Chapter 4 describes another type of detector, called the phase-sensitive detector, which includes the effect of phase in the demodulation process. As explained in Chapter 7, an envelope detector provides noncoherent detection, whereas a phase-sensitive detector provides coherent detection. A coherent detector yields a much better signal than a noncoherent detector when the signal-to-noise ratio at the input to the detector is less than unity.

Chapter 7 builds on the concepts presented in Chapters 4 and 6 to provide a detailed explanation of the radar signal processing that is associated with target detection and target tracking in range and angle. The material is presented by calculating the performance of an assumed radar tracking system during its acquisition and tracking phases.

The Rayleigh distribution and the Gaussian distribution are applied to calculate the probabilities of target detection and false alarm. The matched filter is explained, along with the signal loss due to noncoherent detection. Techniques of range and angle discrimination are analyzed, consisting of the early-late range gate, and conical-scan and monopulse angle-error detectors. As shown in the discussions, to optimize the characteristics of the range and angle tracking loops, and the bandpass filter response of the IF receiver amplifier, a compromise is required between transmission of receiver noise and degradation of information contained in the target signal.

The parameters of the example are chosen for convenience to provide a system that is realistic but not necessarily practical. (Instead of an X-band radar at 10 GHz for the assumed application, one would probably use a radar at a lower frequency.) Nevertheless, the radar parameters for the example are adequate to illustrate the principles of radar signal processing in a practical system.

This explanation of radar in Chapter 7 is not intended as a substitute for more rigorous material, such as that presented by Skolnick [7.1, 7.2], Barton [7.4], and Barton and Ward [7.5]. It is merely a convenient introduction to radar signal processing that provides the background needed to understand the radar analysis of the Altair system in Chapter 8. The material in Chapter 7 may also serve as an introduction to the rigorous radar system discussions in the preceding references.

There are a multitude of engineering issues associated with the development of a radar system. A practical and modern introduction to the overall field of radar system engineering is given by Brookner in *Radar Technology* [1.3], which contains contributions from many experts in the field.

1.3 THE ALTAIR RADAR

1.3.1 General Discussion of the Altair Radar System

The Altair radar system is described in Chapter 8. An essential part of the task of achieving optimal radar tracking is the radar signal processing that is implemented in the radar receiver. For security reasons, the actual radar parameters associated with Altair are not given. Instead, a model is presented that is based on a commercial high-power klystron transmitter tube operating at about the same frequency band as Altair. Although the radar performance parameters derived for this model are not the same as those for Altair, the type of calculations used to compute these parameters are the same as those that have been used to characterize Altair. Hence, the reader is presented with the equivalent of a direct study of Altair radar signal processing.

This radar analysis yields equations for the root-mean-square (RMS) angle error, and the RMS range error, per radar pulse. These are expressed as functions of target range and target radar cross-section, and are primary inputs to the Kalman-filter covariance matrix computations.

A key element of the Altair radar is a calibrated monopulse tracking feed, which allows the angle tracking error of the target to be measured very accurately. With this feed, the Altair antenna need not be aimed directly at the target to achieve accurate tracking. This allows the target tracking equations to operate independently of the servos that drive the monstrous 150-ft Altair antenna. The antenna servos must merely keep the antenna boresight reasonably close to the target.

Target tracking in Altair is implemented by means of digital signal processing in a computer. The Altair tracker has two separate tracking functions:

1. The *angle tracker*, which provides three-dimensional target tracking in terms of rectangular target coordinates.
2. The *range tracker*, which provides direct tracking of target range.

The reason for a separate range tracker is that the positional error corresponding to angular tracking error is very much larger than the range-tracking error (as much as 1000 times greater). The angle tracker uses both range and angle tracking data, but is called an "angle tracker" because its accuracy is limited primarily by angular tracking error.

The angle tracker operates in the following manner. The system measures the antenna gimbal angles, the angular tracking errors derived from the monopulse tracking feed, and the range derived from the range tracker. This information is transformed to rectangular axes (east, north, vertical) fixed at the radar site, to obtain the *observed* coordinates of the target. Three separate tracking-filter feedback loops (operating in the east, north, vertical axes) filter the *observed* target data to obtain the *estimated* target coordinates. Each of these filters has an integral network to provide velocity memory. Based on the position and velocity estimates of the target derived from the tracking filters, the predicted accelerations of the target are computed that are due to gravity, coriolis effects, and (when the target is within the atmosphere) atmospheric drag. These predicted accelerations are fed into the three tracking-filter feedback loops as command acceleration inputs. The predicted accelerations effectively allow the tracking feedback loops to filter the tracking data as if those loops operated in terms of the coordinates of the target.

The estimated position and velocity signals, derived from the three tracking feedback loops, are converted to antenna coordinates to obtain position and velocity commands for pointing the antenna.

The range tracker is a conventional range-tracking feedback loop, except that it receives a range-acceleration command derived from the estimated target data of the angle tracker. The range tracker loop has a triple integration, which provides acceleration memory as well as velocity memory.

1.3.2 Summary of the Kalman Filter

The Altair radar uses a Kalman filter to perform target tracking. The following is a brief overview of Kalman filter theory that introduces the basic functions and elements of the seven Kalman filter equations. This should help the reader to follow the detailed explanation of these equations given in Chapter 8.

The Kalman filter equations are as follows:

Models for Target and Measurement Processes:

$$\underline{x} = \underline{\Phi}\underline{x}_p + \underline{G}\underline{u} \quad \text{(target model)} \tag{1.3-1}$$

$$\underline{z} = \underline{H}\underline{x} + \underline{n} \quad \text{(measurement model)} \tag{1.3-2}$$

Tracking-Filter Difference Equations:

$$\hat{\underline{x}}[-] = \underline{\Phi}\hat{\underline{x}}_p[+] \tag{1.3-3}$$

$$\hat{\underline{x}}[+] = \hat{\underline{x}}[-] + \underline{K}(\underline{z} - \underline{H}\hat{\underline{x}}[-]) \tag{1.3-4}$$

Calculation of the Tracking-Filter Smoothing Parameters:

$$\underline{K} = \underline{P}[+]\underline{H}^T\underline{R}^{-1} \tag{1.3-5}$$

Calculation of the Error Covariance Matrix $\underline{P}$:

$$\underline{P}[-] = \underline{\Phi}\underline{P}_p[+]\underline{\Phi}^T + \underline{G}\underline{Q}\underline{G}^T \tag{1.3-6}$$

$$\underline{P}[+]^{-1} = \underline{P}[-]^{-1} + \underline{H}^T\underline{R}^{-1}\underline{H} \tag{1.3-7}$$

In these equations, an underline indicates that the symbol is a vector and a tilde (~) under a symbol indicates that it is a matrix. The subscript p indicates "past value of." Thus, the elements of the vectors $\underline{x}_p$, $\hat{\underline{x}}_p[+]$ are the past values (computed in the preceding cycle) of the elements of the vectors $\underline{x}$, $\hat{\underline{x}}[+]$. The elements of the matrix $\underline{P}_p[+]$ are the past values of the elements of the matrix $\underline{P}[+]$.

Equations 1.3-1, -2 define the target and measurement models, and so are not actually implemented in the Kalman filter. Equation 1.3-1 gives the assumed dynamic equations of the target, and Eq 1.3-2 describes the measurement process of the radar. These equations define the matrices Φ, G, and H, which are used in the computations that are actually implemented in the tracking-filter process.

In Eq 1.3-1, x is the state vector of the target. Its elements are usually the three-dimensional components of position, velocity, and acceleration of the target. The matrix Φ is the transition matrix of the target model, which characterizes the dynamic equations of the target. The variable u is a vector representing the disturbances that cause the target trajectory to change in an unpredictable manner, and G is the matrix that relates these disturbances to the state vector x.

The formidable-looking relation in Eq 1.3-1 is explained in a simple manner in Chapter 5, Section 5.5, which develops an equivalent set of equations for an electrical network. When applied to a practical example, matrix equations are easy to understand.

The radar provides only an indirect measure of the target state vector x. The measurement model in Eq 1.3-2 relates the components of the state vector x of the target to the measurement vector z observed by the tracking radar. In a typical pulse radar, the coordinates of the measurement vector z are the range to the target, and the elevation and azimuth angles of the target, as observed by the radar. These measurements are corrupted by radar noise, which is represented by the vector n. The matrix H relates the state variables of the target x to the radar measurement variables z.

The target disturbance vector u and the radar noise vector n are characterized statistically in the Kalman filter by the matrices Q and R. The matrix R and usually the matrix Q have elements only along the diagonal of the matrix, the other elements being zero. These diagonal elements are the mean-square values of the components of the noise vector n and the disturbance vector u. Thus, matrix R contains the mean-square values of the radar noise errors in the range, elevation, and azimuth measurements of the radar. Matrix Q contains the mean-square values of the components of the disturbance vector u that alter the target trajectory in a random manner.

Equations 1.3-3, -4 are the tracking-filter equations, which process the radar measurement data, described by the vector z, to compute the estimate state vector of the target, denoted $\hat{x}$. The variables $\hat{x}[+]$ and $\hat{x}[-]$ are two related versions of the estimated target state vector. The subscript p indicates that $\hat{x}_p[+]$ is the past value of $\hat{x}[+]$ computed in the previous cycle. The equations include the transition matrix Φ of the target model and the measurement matrix H of the radar, defined by Eqs. 1.3-1, -2. They also include the matrix K, which contains the smoothing parameters of the tracking filters.

The tracking-filter equations described by Eqs. 1.3-3, -4 represent a set of feedback control loops. The smoothing parameters of the K matrix are dynamic parameters of those loops. Chapter 8 shows how the elements of the K matrix are related to the basic feedback loop parameters ω_c, ω_i discussed in Chapter 2.

The tracking-filter smoothing parameters of the K matrix are calculated by Eq 1.3-5 from the matrix P, which is called the error covariance matrix. This equation also includes the matrix H defined in Eq 1.3-2, which characterizes the measurement model, and the matrix R, which contains the mean-square noise values of the radar measurements of target range and angle. The values of the R-matrix elements are determined by measuring the strength of the received target signal in the radar. The lower the target signal, the greater are the mean-square noise values in the R matrix.

The superscript T indicates that H^T is the transpose of the matrix H. The rules for manipulating matrices are summarized in Appendix C. The appendix shows that the transpose of a matrix is formed simply by interchanging the rows and columns of the matrix. The appendix also gives the rules for multiplying and inverting matrices. The superscript -1 indicates that R^{-1} is the inverse of the matrix R. As shown in the appendix, matrix inversion is a very complicated process when the order of the matrix is high.

The error covariance matrix P gives statistical parameters of the expected errors in the estimates $\hat{x}$ of the target state-vector components. The diagonal of this matrix contains the expected mean-square errors of the estimated state-vector components. The other elements of the matrix are the covariance values obtained by cross-correlating the expected error components.

The error covariance matrix P is computed in an iterative manner from the sampled-data difference equations given in Eqs. 1.3-6, -7. There are two related forms of the P matrix, denoted $P[+]$ and $P[-]$. Matrix $P_p[+]$ contains the elements of $P[+]$ computed in the previous computation cycle.

In Eqs. 1.3-6, -7, $P[-]^{-1}$ is the inverse of the matrix $P[-]$, and $P_p[+]$ is the inverse of the matrix $P[+]^{-1}$ computed in the previous cycle. Hence, this pair of equations requires two inversions of the error covariance matrix for each computation cycle. This indicates that the equations are extremely difficult to solve when the order of the P matrix is high.

To avoid this problem, Eqs. 1.3-5 to -7 can be replaced by the following:

$$K = P[-]H^T\left(HP[-]H^T + R\right)^{-1} \qquad (1.3\text{-}8)$$

$$P[+] = (1 - KH)P[-] \qquad (1.3\text{-}9)$$

The relation between the two forms of the Kalman filter is explained by Gelb [8.5] (pp. 110–112). This alternate set of equations requires one inversion of the following matrix:

$$\left(HP[-]H^T + R\right) \qquad (1.3\text{-}10)$$

The measurement matrix H relates the state variables of the target x to the radar-measurement variables z. The combined matrix $HP[-]H^T$ is a square matrix, the order of which is equal to the number of variables z measured by the radar. Generally, the radar measures no more than four variables: range, range rate, and two angular coordinates. Hence, this square matrix is generally no more than fourth order. The matrix $\underset{\sim}{R}$ gives the mean-square noise components of the radar signal, and so is a square matrix of the same order. Hence, with the alternate form of the Kalman filter given in Eqs. 1.3-8, -9, the Kalman filter computations generally require the inversion of a matrix that is no more than fourth order.

1.3.3 Implementation of the Kalman Filter in Altair

In the Altair angle tracker, the state vector includes the position and velocity components of the target in three dimensions, which results in a sixth-order tracking filter. The range tracker also includes the range acceleration in its state vector and so is of third order. Hence, the total Altair tracker is of ninth order.

On the other hand, the error covariance matrices $\underset{\sim}{P}[+]$ and $\underset{\sim}{P}[-]$ in Eqs. 1.3-6, -7 are reduced to second order in the Altair tracking system. They have only a single position component and a single velocity component, and so are only 2×2 matrices. Nevertheless, these two matrix expressions of Eqs. 1.3-6, -7 result in six nonlinear equations for the angle tracker (and six more for the range tracker), which are solved iteratively to obtain the coefficients of the P-matrix. Two more nonlinear equations are solved for the angle tracker (and similarly for the range tracker) to obtain the smoothing coefficients of the tracking filters, represented by the K-matrix.

The Altair Kalman-filter tracking equations have achieved very effective tracking of ballistic missile warheads. The primary reason for this is that tracking is performed in terms of the coordinates of the target. During exoatmospheric flight, the warhead is moving in a highly predictable ballistic trajectory. By taking advantage of trajectory prediction, a very long smoothing period can be allowed, and hence very accurate tracking can be achieved. On the other hand, this aspect of the Kalman filter (which is implemented by Eqs. 1.3-3, -4) is not unique to Kalman filter theory. The same approach would evolve from the application of conventional tracking principles.

The aspect of Kalman filtering that is unique to Kalman filter theory is the set of computations given in Eqs 1.3-5 to -7, which determine the optimum smoothing coefficients of the tracking filters. To evaluate the effectiveness of these optimal computations, the corresponding equations of the Altair system were simulated, and the results are shown in Section 8.6. Three different phases of Altair tracking of ballistic missile warheads were studied: (1) the acquisition phase, which occurs immediately after radar lock-on; (2) the steady-state tracking phase during exoatmospheric flight; and (3) the strong transient that occurs when the warhead reenters the atmosphere.

To help explain these transient responses, a simpler optimal tracking theory is presented, based on the author's analysis of the response of a system to an arbitrary input, given in Chapter 3. A comparison of the filter smoothing coefficients derived from these two optimal theories helps to provide physical understanding of the Kalman filter calculations.

1.3.4   Time-Varying Filter Parameters

Since the parameters of the Kalman filter are time varying, it has been claimed that a Kalman filter cannot be characterized by steady-state parameters, such as noise bandwidth. The answer to this claim is that the parameters of practical Kalman filters must vary quite slowly—so slowly that the steady-state filter parameters still apply to a good approximation.

This issue is analyzed in Section 8.6.2. It shows that the smoothing gain of a simple tracking filter can be varied so rapidly its filter action cannot be characterized by noise bandwidth with reasonable accuracy. However, essentially all practical tracking filters are more complicated than this, and have at least velocity (or input rate) memory. In order for a filter with velocity memory to achieve an accurate, stable solution, its filter parameters must be varied slowly. Steady-state parameters, such as noise bandwidth, apply with good accuracy to any practical time-varying tracking filter having velocity memory.

1.4   NOTES ON SYMBOLISM AND TERMINOLOGY

In this book, functional relationships are expressed exclusively with square [] brackets. These brackets are used to represent multiplication only in cases where they cannot be confused with functional relationships. For example, $F[x + y]$ always indicates that F is a function of $(x + y)$ and cannot represent F times $(x + y)$; whereas $F(x + y)$ always means F times $(x + y)$.

Although angular frequencies, denoted by the symbol ω, have the units of radian/second, the term *radian* is generally omitted in this book in order to simplify the calculation of units. For example, the expression $\omega_1 = 4$ rad/sec is usually written $\omega_1 = 4$ sec^{-1}.

The sampled-data literature uses the symbol z to represent $\exp[sT]$, the Laplace transform for a time shift in the future of one sample period. In contrast, this book deals with $\exp[-sT]$, the Laplace transform for a one-period delay, which is represented as $\bar{z}$. This variable $\bar{z}$ is therefore equal to $1/z$. The conventional symbolism has the disadvantage that z normally occurs only in the form z^{-n} because time shifts in the future are not realizable.

The logarithm to base 10 of x is expressed as $\log[x]$, while the logarithm to base e (the natural logarithm) is expressed as $\ln[x]$. To represent a logarithmic signal-amplitude ratio, the book generally uses the decilog (abbreviated dg) instead of the decibel (abbreviated dB). The decilog, which is 10 times the logarithm of the signal amplitude ratio, is discussed in Section 2.3.4. An important advantage of the decilog is that a signal gain in decilogs can be easily interpreted in terms of the corresponding signal amplitude ratio, whereas a signal gain in decibels cannot.

Chapter 2

Summary of Feedback Control Principles

This chapter presents a summary of feedback control principles. The discussion is based on the author's approach to feedback control design given in *Principles of Feedback Control*, volumes 1 and 2 [1.1, 1.2].

This approach emphasizes the gain crossover frequency ω_{gc} of a feedback control loop, which is the frequency where the open-loop gain crosses unity. This is the effective control bandwidth of a feedback control loop and so is its primary dynamic parameter. Feedback control analysis is greatly simplified when the loop transfer function is expressed such that its gain parameter is the asymptotic gain crossover frequency ω_c, which is approximately equal to the true gain crossover frequency ω_{gc}.

2.1 SUMMARY OF CHAPTER

To introduce the reader to feedback control dynamics, the chapter starts in Section 2.2 with an analysis of a second-order lowpass filter. This illustrates the Laplace transfer function, the magnitude and phase plots of the frequency response, and the transient response to a step input. The damping ratio ζ, natural frequency ω_n, and oscillation frequency ω_o for a pair of complex poles are defined.

Section 2.3 introduces the variables and transfer functions of a feedback loop. A feedback loop is reduced to the basic form shown in Fig 2.1-1, which has the variables

$$X_i = \text{loop input}$$

$$X_e = \text{loop error}$$

$$X_b = \text{loop feedback signal}$$

These variables are the Laplace transforms of the actual time-varying signals.

15

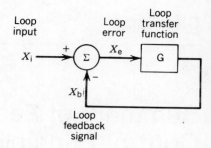

Figure 2.1-1   Basic feedback loop.

The loop error is defined as the loop input minus the loop feedback signal:

$$X_e = X_i - X_b \qquad (2.1\text{-}1)$$

The transfer function between the loop error X_e and the loop feedback signal X_b is called the loop transfer function, denoted G:

$$G = \frac{X_b}{X_e} = \text{loop transfer function} \qquad (2.1\text{-}2)$$

This is the total transfer function around a feedback loop and is often called the *open-loop transfer function*. The transfer function between the loop input X_i and the feedback signal X_b is called the feedback transfer function, denoted G_{ib}:

$$G_{ib} = \frac{X_b}{X_i} = \text{feedback transfer function} \qquad (2.1\text{-}3)$$

This is often called the *closed-loop transfer function*. The transfer function between the loop input X_i and the error signal X_e is called the error transfer function, denoted G_{ie}:

$$G_{ie} = \frac{X_e}{X_i} = \text{Error transfer function} \qquad (2.1\text{-}4)$$

Combining Eqs 2.1-1 to -4 gives the following equations relating G_{ib} and G_{ie} to G:

$$G_{ib} = \frac{G}{1 + G} \qquad (2.1\text{-}5)$$

$$G_{ie} = \frac{1}{1 + G} \qquad (2.1\text{-}6)$$

The magnitude of the loop transfer function G, denoted $|G|$, is called the loop gain. Feedback control action occurs only at frequencies where the loop gain is greater than unity. Hence, the frequency where the loop gain drops below unity is the limit to the region of feedback control, and so is the effective control bandwidth of a feedback loop. This is called the gain crossover frequency of the loop, denoted ω_{gc},

$$\omega_{gc} = \text{gain crossover frequency;} \quad \text{the frequency } |G| = 1.$$

Another common bandwidth parameter is the frequency where the magnitude of the feedback (or "closed-loop") transfer function G_{ib} is $1/\sqrt{2}$. This is called the half-power frequency ω_{hp}:

$$\omega_{hp} = \text{half-power frequency;} \quad \text{the frequency } |G_{ib}| = 1/\sqrt{2}.$$

At the half-power frequency, the power gain, which is proportional to the square of the signal-amplitude ratio, is $1/2$. The half-power frequency ω_{hp} cannot be used to estimate the gain crossover frequency ω_{gc}, because the frequency ratio ω_{hp}/ω_{gc} can vary by more than $5:1$ for different practical feedback control loops.

Section 2.3 gives general approximations that relate these and other frequency response parameters to the transient responses of a feedback control loop for step and ramp inputs. It discusses stability and shows that the peak overshoot of the step response can be derived approximately from the amount of peaking of the $|G_{ib}|$ frequency response.

Section 2.4 explains the use of the magnitude asymptote plot of a loop transfer function in feedback control analysis. This is an approximate plot of $|G|$ versus frequency, on log-log coordinates. It consists of straight-line segments called asymptotes that are proportional to $\omega^{-1}, \omega^{-2}, \omega^{-3}, \dots$. Magnitude asymptote plots are derived for different types of feedback control loops to demonstrate, among other things, how integral compensation is used to increase the loop gain at low frequencies.

In a practical feedback control loop, the magnitude asymptote plot of a loop transfer function must have a broad frequency region, at or near gain crossover, where the magnitude asymptote is proportional to ω^{-1} (or $1/\omega$). The frequency where this asymptote (extended if necessary) crosses unity gain is the asymptotic gain crossover frequency, denoted ω_c. Hence, the asymptote is a plot of the function ω_c/ω.

Section 2.5 shows how the loop transfer function of a feedback control loop is written in terms of this ω_c/ω asymptote, so that the gain constant of the transfer function is ω_c. This approach greatly simplifies feedback control analysis because it provides a gain parameter that is of fundamental significance. For practical feedback control loops, ω_c is approximately equal to the true gain crossover frequency ω_{gc}, which is the primary dynamic parameter of a feedback control loop.

Based on the asymptotic gain crossover frequency ω_c and the magnitude asymptote plot, Section 2.6 develops simple techniques for approximating the closed-loop transfer functions G_{ib} and G_{ie}. These G_{ib} and G_{ie} approximations are used in analyses presented in Sections 2.7 and 2.8, and in Chapter 3, Section 3.2.

Section 2.7 describes the response of a feedback control loop to a low-frequency input. The error response of a loop with a single integration below gain crossover is approximately proportional to the input velocity. To reduce this error, an integral network can be added to the loop to produce a double integration at low frequencies. The integral network gives the loop *velocity memory*. This compensates for the error component that is proportional to input velocity, and so the resultant error is approximately proportional to input acceleration.

The discussion in Section 2.7 serves to introduce Chapter 3, which presents general techniques for calculating the response of a feedback control loop to an arbitrary input.

Section 2.8 analyzes a multiloop feedback control system, which uses a DC motor to position an optical instrument stage. The servo has feedback of motor current, tachometer feedback of motor velocity, and position feedback of linear stage displacement. The position loop has an integral network to reduce the low-frequency components of the servo error.

The transfer functions of the servo elements are derived, to characterize the motor with its mechanical load, the tachometer, the position sensor, and the operational amplifier circuits of the servo. The transfer functions of the various feedback loops are calculated, and the corresponding magnitude asymptotes are plotted. This analysis applies the principles for approximating the G_{ib} transfer function presented in Section 2.6.

The servo studied in Section 2.8 illustrates the principle that a complex feedback control system consists of different feedback control loops, each characterized by its own gain crossover frequency. This servo is used in Chapter 5, Section 5.5, as a model to illustrate the "serial" method of digital computer simulation developed in that chapter.

Thus, Chapter 2 gives a quick introduction to feedback control principles, which should be adequate for understanding subsequent feedback control material in this book. For more information, the reader is referred to the books written by the author in Refs [1.1, 1.2].

2.2   DYNAMIC RESPONSE OF SECOND-ORDER LOWPASS FILTER

Frequency response analysis is a key element in the design of feedback control systems. To apply this tool effectively, one must understand the relations between frequency response and transient response. As an introduction to this issue, let us examine the transient and frequency responses of the lowpass filter

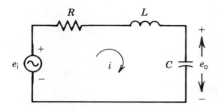

Figure 2.2-1   Circuit for second-order low-pass filter.

shown in Fig 2.2-1. The differential equations of the filter are

$$e_i = Ri + L\frac{di}{dt} + \frac{1}{C}\int i\,dt \qquad (2.2\text{-}1)$$

$$e_o = \frac{1}{C}\int i\,dt \qquad (2.2\text{-}2)$$

where i is the current, e_i is the input voltage to the filter, and e_o is the voltage across the capacitor, which is the output voltage of the filter. *If initial conditions of the circuit are zero*, the Laplace transforms of these equations are obtained by making the following substitutions:

$$e_i \rightarrow E_i \qquad (2.2\text{-}3)$$

$$e_o \rightarrow E_o \qquad (2.2\text{-}4)$$

$$i \rightarrow I \qquad (2.2\text{-}5)$$

$$\frac{d}{dt} \rightarrow s \qquad (2.2\text{-}6)$$

$$\int dt \rightarrow \frac{1}{s} \qquad (2.2\text{-}7)$$

This gives the following Laplace transforms of Eqs 2.2-1, -2:

$$E_i = RI + LsI + \frac{1}{Cs}I \qquad (2.2\text{-}8)$$

$$E_o = \frac{1}{Cs}I \qquad (2.2\text{-}9)$$

The variables E_i, E_o, I are the Laplace transforms of the voltage and current variables e_i, e_o, i. Each differentiation of the differential equation is replaced by s, and each integration by $1/s$. A double differentiation (d^2/dt^2) would be replaced by s^2, a double integration by $1/s^2$, etc.

This simple rule for deriving the Laplace transform equations from the differential equations applies only when the initial conditions of the dynamic system are zero. However, an initial condition can always be forced to be zero by replacing it with an equivalent signal source. The responses to these initial-condition signal sources are calculated separately and added to the response to the input forcing signal (in this case e_i). This book assumes that initial conditions are always set to zero in this manner.

Solving Eqs 2.2-8, -9 simultaneously gives the following transfer function, which relates the input voltage E_i to the output voltage E_o:

$$\frac{E_o}{E_i} = \frac{1/LC}{s^2 + s(R/L) + (1/LC)} \tag{2.2-10}$$

Strictly speaking, E_i and E_o should be called the Laplace transforms of the input and output voltages. However, the discussion is much more convenient when the expression "the Laplace transform of" is omitted with reference to such variables. Accordingly, this terminology is used commonly in this book. The reader should be able to tell from the context whether the variable being discussed is an actual time-varying signal or a transform of that signal.

Analysis using the transfer function of Eq 2.2-10 is greatly simplified if the denominator quadratic expression is replaced by the following general form:

$$\left(s^2 + 2\zeta\omega_n s + \omega_n^2\right) = \left[s^2 + s(R/L) + (1/LC)\right] \tag{2.2-11}$$

where the parameters ω_n and ζ are called

ω_n = natural frequency
ζ = damping ratio

The damping ratio ζ describes the degree of damping of the response. At low values of damping ratio ζ, the natural frequency ω_n is approximately the frequency of oscillation. Equating corresponding terms of Eqs 2.2-11 gives

$$\omega_n^2 = \frac{1}{LC} \tag{2.2-12}$$

$$2\zeta\omega_n = \frac{R}{L} \tag{2.2-13}$$

Solving these for ω_n and ζ gives the following values of natural frequency and damping ratio for this circuit:

$$\omega_n = \frac{1}{\sqrt{LC}} \tag{2.2-14}$$

$$\zeta = \frac{1}{2\omega_n}\frac{R}{L} = \frac{R}{2}\sqrt{\frac{C}{L}} \tag{2.2-15}$$

Substituting Eqs 2.2-12, -13 into Eq 2.2-10 gives the following expression for the transfer function of the filter in terms of the general parameters:

$$\frac{E_o}{E_i} = \frac{\omega_n^2}{s^2 + 2\zeta\omega_n s + \omega_n^2} \tag{2.2-16}$$

Since the denominator is a second-order polynomial in s, this is called the *basic second-order lowpass filter*.

Let us assume that the input voltage $e_i[t]$ is a unit step. According to Laplace transform theory, the Laplace transform of this input voltage is

$$E_i = \frac{1}{s} \tag{2.2-17}$$

Solve Eq 2.2-16 for E_o and substitute Eq 2.2-17 for E_i. This gives

$$E_o = \frac{E_i \omega_n^2}{s^2 + 2\zeta\omega_n s + \omega_n^2} = \frac{\omega_n^2}{s(s^2 + 2\zeta\omega_n s + \omega_n^2)} \tag{2.2-18}$$

To obtain the inverse transform of this equation, the roots of the denominator expression (which are the poles of E_o) must be found. This yields $s = 0$ and the roots of the following equation:

$$s^2 + 2\zeta\omega_n s + \omega_n^2 = 0 \tag{2.2-19}$$

The quadratic formula gives the roots

$$\begin{aligned} s &= -\zeta\omega_n \pm \sqrt{(\zeta\omega_n)^2 - \omega_n^2} \\ &= -\zeta\omega_n \pm \omega_n\sqrt{\zeta^2 - 1} \end{aligned} \tag{2.2-20}$$

When s is equal to one of these roots, the transfer function E_o/E_i is infinite. Hence these roots are called the poles of E_o/E_i.

If the damping ratio ζ is greater than unity, Eq 2.2-20 yields two real poles. If ζ is less than unity, it yields a pair of complex poles. For $\zeta = 1$, there is a double-order pole at $s = -\omega_n$. These three cases are defined as

$$\begin{array}{lll} \text{Overdamped case:} & \zeta > 1 & \text{(2.2-21)} \\ \text{Critically damped case:} & \zeta = 1 & \text{(2.2-22)} \\ \text{Underdamped case:} & \zeta < 1 & \text{(2.2-23)} \end{array}$$

When $\zeta < 1$, Eq 2.2-20 is expressed in the form

$$\begin{aligned} s &= -\zeta\omega_n \pm \omega_n\sqrt{(-1)(1 - \zeta^2)} \\ &= -\zeta\omega_n \pm j\omega_n\sqrt{1 - \zeta^2} \end{aligned} \tag{2.2-24}$$

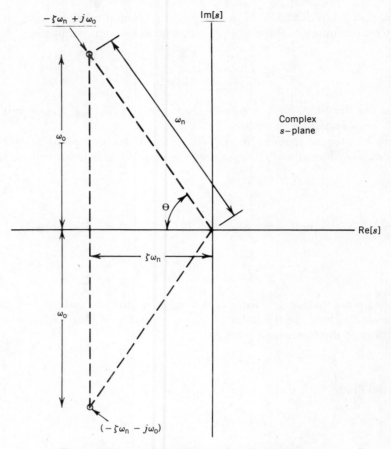

Figure 2.2-2 s-plane plot of a pair of complex poles.

where $\sqrt{-1}$ was replaced by j. The oscillation frequency ω_0 is defined as

$$\omega_o = \omega_n\sqrt{1 - \zeta^2} \quad \text{(Oscillation frequency)} \qquad (2.2\text{-}25)$$

Hence, the poles of Eq 2.2-24 become

$$s = -\zeta\omega_n \pm j\omega_o \qquad (2.2\text{-}26)$$

An s-plane plot of these complex poles is shown in Fig 2.2-2. Applying Eq 2.2-26 to Eq 2.2-18 allows the denominator to be factored as follows when $\zeta < 1$:

$$E_o = \frac{\omega_n^2}{s(s + \zeta\omega_n + j\omega_o)(s + \zeta\omega_n - j\omega_o)} \qquad (2.2\text{-}27)$$

This can be solved by expanding the expression into partial fractions. It can be shown that the inverse Laplace transform of the result is

$$e_o = 1 - e^{-\zeta\omega_n t}\left(\cos[\omega_o t] + \frac{\zeta}{\sqrt{1 + \zeta^2}}\sin[\omega_o t]\right) \tag{2.2-28}$$

For $\zeta > 1$, the poles are real, and the response becomes

$$e_o = 1 + \frac{1}{(\omega_2 - \omega_1)}\left(\omega_2 e^{-\omega_1 t} - \omega_1 e^{-\omega_2 t}\right) \tag{2.2-29}$$

where

$$\omega_1 = \omega_n\left(\zeta - \sqrt{\zeta^2 - 1}\right) \tag{2.2-30}$$

$$\omega_2 = \omega_n\left(\zeta + \sqrt{\zeta^2 - 1}\right) \tag{2.2-31}$$

For $\zeta = 1$, there is a double-order pole at $s = -\omega_n$, and the response becomes

$$e_o = 1 - e^{-\omega_n t}(1 + \omega_n t) \tag{2.2-32}$$

Figure 2.2-3 shows plots of this step response for various values of the damping ratio ζ. The time scale is expressed as $\omega_n t$ to normalize it in terms of the natural frequency ω_n.

Another way to characterize the dynamics of this lowpass filter is to consider its response to a sinusoid of varying frequency. A sinusoidal input voltage can be expressed as

$$e_i = |E_i|\cos[\omega t + \phi_i] \tag{2.2-33}$$

The parameter ω is the frequency of the sinusoid in radians per second, (rad/sec), and $|E_i|$ and ϕ_i are the magnitude and phase (in radians) of the cosine wave. This sinusoid can be represented by the following rotating complex vector:

$$E_i = |E_i|e^{j(\omega t + \phi_i)}$$
$$= |E_i|\cos[\omega t + \phi_i] + j|E_i|\sin[\omega t + \phi_i] \tag{2.2-34}$$

The variable E_i is a vector in the complex plane that rotates counterclockwise at an angular velocity of ω rad/sec. The actual signal e_i is the real part of this rotating vector:

$$e_i = \text{Re}[E_i] = |E_i|\cos[\omega t + \phi_i] \tag{2.2-35}$$

Note that $\text{Re}[X]$ denotes the real part of the complex quantity X, and $\text{Im}[X]$ denotes the imaginary part of X. One could instead assume that the time

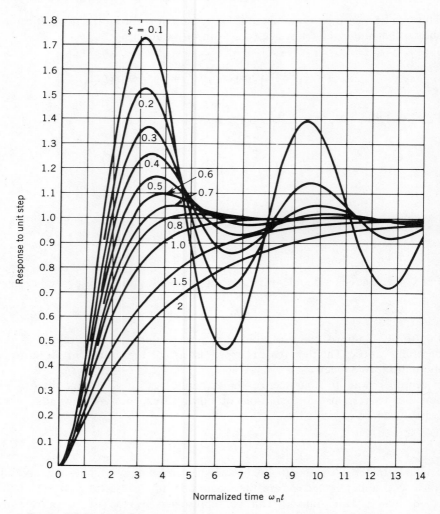

Figure 2.2-3   Transient response of second-order lowpass filter to unit step.

signal is the imaginary part of the rotating vector. For that convention, the time-varying voltage e_i corresponding to the vector E_i of Eq 2.2-34 would be

$$e_i = \text{Im}[E_i] = |E_i|\sin[\omega t + \phi_i] \qquad (2.2\text{-}36)$$

In AC circuit analysis, the magnitude of a complex vector (or *phasor*) is often defined as the RMS value of the sine wave. However, this book sets the vector amplitude equal to the peak value. Either convention is acceptable provided it is applied consistently.

The frequency response of the lowpass filter can be obtained from the Laplace transfer function of Eq 2.2-16 by replacing s by $j\omega$. This gives

$$\frac{E_o}{E_i} = \frac{\omega_n^2}{(j\omega)^2 + 2\zeta\omega_n(j\omega) + \omega_n^2}$$

$$= \frac{\omega_n^2}{(\omega_n^2 - \omega^2) + j2\zeta\omega_n\omega} \tag{2.2-37}$$

The variables E_i, E_o are now the complex vector representations of the time-varying voltages e_i, e_o, rather than the Laplace transforms of those voltages.

Thus, one can convert a Laplace transfer function into a frequency response transfer function, and vice versa, by interchanging s and $j\omega$. During analysis, it is often convenient to express a transfer function in terms of s rather than $j\omega$, even when frequency response is being considered. Hence, s and $j\omega$ are often interchangeable in this book.

To simplify the plotting of this frequency response, the frequency variable ω is normalized as follows:

$$u = \frac{\omega}{\omega_n} \tag{2.2-38}$$

Equation 2.2-37 becomes

$$\frac{E_o}{E_i} = \frac{1}{(1 - u^2) + j2\zeta u} \tag{2.2-39}$$

The magnitude and phase of this are

$$\frac{|E_o|}{|E_i|} = \frac{1}{\sqrt{(1 + u^2)^2 + (2\zeta u)^2}} \tag{2.2-40}$$

$$\text{Ang}\left[\frac{E_o}{E_i}\right] = -\arctan\left[\frac{2\zeta u}{1 - u^2}\right] \tag{2.2-41}$$

The magnitude and phase given in Eqs 2.2-40, -41 are plotted in Fig 2.2-4 versus the normalized frequency $u = \omega/\omega_n$, for various values of the damping ratio ζ.

As will be shown in Section 2.4.3, the transfer function of this basic second-order lowpass filter is the same as the closed-loop, or *feedback*, transfer function of a feedback loop that commonly serves as a model in feedback control design. The transient and frequency responses of this lowpass filter in Figs 2.2-3, -4 are often used as a basis for estimating relations between

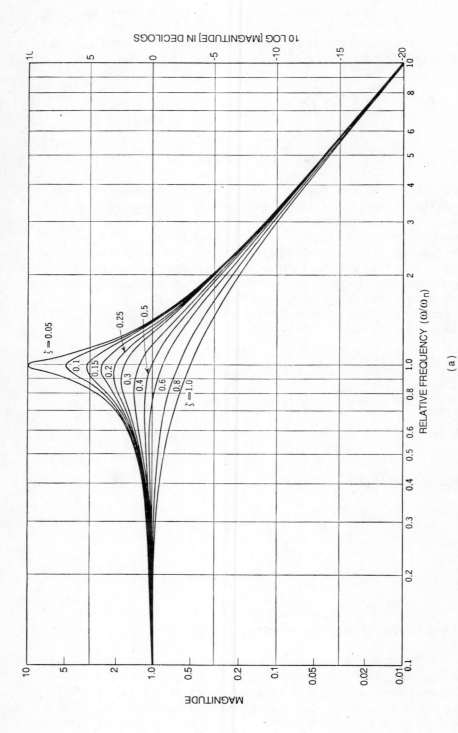

Figure 2.2-4 Frequency response of second-order lowpass filter: (a) magnitude; (b) phase. Reprinted with permission from Chestnut and Mayer [2.2] Figures 11.3-6 and 11.3-7.

(a)

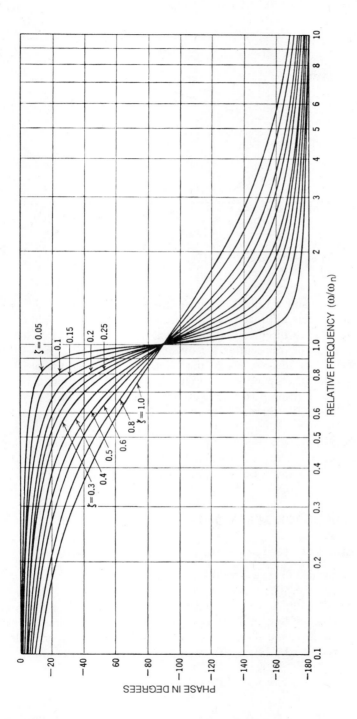

(b)

Figure 2.2-4 (*Continued*)

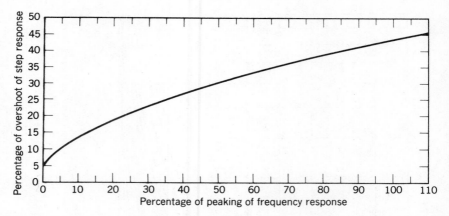

Figure 2.2-5 Plot of peak overshoot of step response versus peaking of frequency response for second-order lowpass filter.

frequency response and transient response in the design of feedback control systems. Although this approach has advantages, as will be explained it has been overdone and has often caused confusion.

A comparison of the transient and frequency response plots of Figs 2.2-3, -4 shows that the higher the peaking of the frequency response, the greater the peak overshoot of the step response. Figure 2.2-5 shows a plot of the peak overshoot of the step response (in percent) versus the peaking of the frequency response magnitude (in percent) for the basic second-order lowpass filter. Although this plot was derived for a particular lowpass filter, it applies approximately to any reasonable lowpass filter. As will be shown in Section 2.3.2, the plot also applies approximately to the closed-loop (or feedback) frequency response of any reasonable feedback control system.

2.3 BASIC FEEDBACK RELATIONS

2.3.1 The Effect of Feedback Control

The principle on which feedback control is based is illustrated in the signal-flow diagram of Fig 2.3-1a. The controlled variable X_c is measured to form the feedback signal X_b, which is subtracted from the input signal X_i to produce the error signal X_e. The amplified error signal X_e changes the controlled variable X_c in the direction that reduces the error. When the error is close to zero, the feedback signal X_b is approximately equal to the input X_i:

$$X_b \cong X_i \qquad (2.3\text{-}1)$$

The feedback gain K_f is the sensitivity of the circuit that measures the

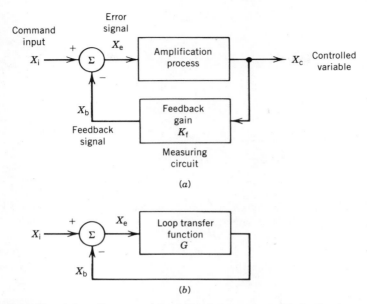

Figure 2.3-1 Signal-flow diagrams: (*a*) feedback loop to provide feedback control, (*b*) basic form of feedback loop.

controlled variable to form the feedback signal. The feedback signal X_b is related to the controlled variable X_c by

$$X_b = K_f X_c \qquad (2.3\text{-}2)$$

Combining Eqs 2.3-1, -2 gives the following approximation relating the controlled variable X_c to the input signal X_i:

$$X_c = \frac{1}{K_f} X_b \cong \frac{1}{K_f} X_i \qquad (2.3\text{-}3)$$

Thus, the feedback loop works to keep the controlled variable X_c proportional to the input X_i.

The feedback loop can be expressed in a more general form by combining the amplification process and the measuring circuit into a single block, called the *loop transfer function*, as shown in Fig 2.3-1*b*. The controlled variable X_c is ignored, and the basic feedback loop is concerned only with the three general loop variables X_i, X_e, X_b. If one knows how accurately the feedback signal X_b follows the input X_i, one can readily determine the response of the controlled variable X_c by setting X_c equal to $(1/K_f)X_b$, in accordance with Eq 2.3-3. Thus, the controlled variable X_c need not be considered in the basic analysis of the feedback loop response.

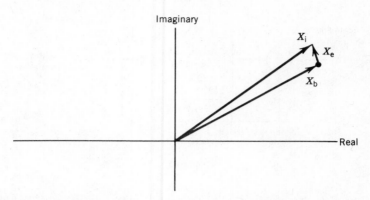

Figure 2.3-2 Typical vectors (or phasors) representing input (X_i), error (X_e), and feedback X_b signals for sinusoidal input X_i.

Figure 2.3-1 shows that the loop error X_e is equal to the loop input X_i minus the loop feedback signal X_b:

$$X_e = X_i - X_b \tag{2.3-4}$$

The relationships among the three basic loop variables X_i, X_b, X_e can be expressed in a simple fashion by assuming that the input X_i is a sinusoidal signal of varying frequency. At any frequency, these three signals can be represented by complex vectors (or *phasors*), as shown in Fig 2.3-2. In accordance with Eq 2.3-4, the error vector X_e is the difference between the input vector X_i and the feedback vector X_b.

The loop gain, denoted $|G|$, is the ratio of the magnitude of the feedback vector X_b to the magnitude of the error vector X_e:

$$\text{Loop gain} \doteq |G| = \frac{|X_b|}{|X_e|} \tag{2.3-5}$$

The quantity $|X_e|$ is the length of the X_e vector and $|X_b|$ is the length of the X_b vector. If the loop gain $|G|$ is much greater than unity, the feedback vector X_b in Fig 2.3-2 is much longer than the error vector X_e. For this condition, the feedback vector X_b must nearly coincide with the input vector X_i. Thus, at a frequency where the loop gain $|G|$ is much greater than unity, the feedback signal X_b is nearly equal to the input X_i. The exact value of the loop gain is not important, provided that the loop gain is much greater than unity and the loop has good stability.

The loop gain varies with frequency in the general manner indicated in Fig 2.3-3. Since the loop gain is zero at infinite frequency, it must decrease with increasing frequency. The frequency at which the loop gain drops below unity

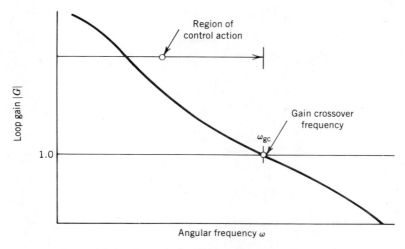

Figure 2.3-3   Typical plot of loop gain versus frequency.

is called the gain crossover frequency, denoted ω_{gc}. The feedback loop has effective feedback-control action only at frequencies below ω_{gc}, where the loop gain is greater than unity. Hence, the gain crossover frequency ω_{gc} is the upper limit to the region of feedback control action, and so is a good criterion for the "control bandwidth" of the feedback loop.

The transfer function relating the loop error X_e to the loop feedback signal X_b is called the loop transfer function, denoted G:

$$G = \frac{X_b}{X_e} \quad \text{(loop transfer function)} \tag{2.3-6}$$

The loop transfer function G can be characterized by frequency response plots of its magnitude and phase. The magnitude of G is called the *loop gain* $|G|$, and the phase of G is called the *loop phase*, denoted $\text{Ang}[G]$. (In this book, the phase angle of a complex variable F is denoted $\text{Ang}[F]$.)

Figure 2.3-4 shows a general signal-flow diagram of a feedback loop. The transfer functions relating the loop input X_i to the feedback signal X_b and to the error signal X_e are called the feedback and error transfer functions of the loop:

$$G_{ib} = \frac{X_b}{X_i} \quad \text{(feedback transfer function)} \tag{2.3-7}$$

$$G_{ie} = \frac{X_e}{X_i} \quad \text{(error transfer function)} \tag{2.3-8}$$

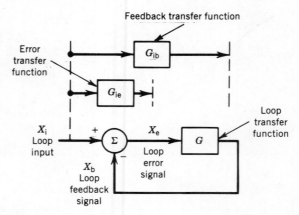

Figure 2.3-4 Definition of general loop variables and transfer functions.

The loop transfer function G is commonly called the *open-loop* transfer function, and the feedback transfer function G_{ib} is commonly called the *closed-loop* transfer function.

The transfer functions G_{ib} and G_{ie} can be related to the loop transfer function G by combining Eqs 2.3-4, -6, -7, and -8 to obtain

$$G_{ib} = \frac{G}{1 + G} \tag{2.3-9}$$

$$G_{ie} = \frac{1}{1 + G} \tag{2.3-10}$$

A fundamental problem with using feedback to achieve control is that feedback can also produce an oscillator. To illustrate this point, suppose that G is equal to -1 at a particular frequency. The expression $(1 + G)$ would be zero, and so G_{ib} and G_{ie} in Eqs 2.3-9, -10 would be infinite. This indicates that the loop would oscillate at that frequency. This oscillatory condition, $G = -1$, corresponds to the following values of magnitude and phase of G:

$$|G| = 1 \qquad \text{Ang}[G] = -180° \tag{2.3-11}$$

This shows that a feedback loop exhibits a stable oscillation if the loop phase lag is 180° at the gain crossover frequency. It can also be proven that a feedback loop exhibits a growing oscillation if the phase lag is greater than 180° at gain crossover.

Thus, to be stable, a feedback loop must have less than 180° of phase lag at gain crossover. On the other hand, a well-designed feedback control system must be more than absolutely stable; it must have good stability. To achieve good stability, a feedback loop should have no more than 135° of phase lag at gain crossover.

This requirement is a necessary but not sufficient condition for good stability. There are certain conditions, based on the Nyquist stability theorem, that also must be satisfied in order to assure that the loop is stable in an absolute sense. When these are satisfied, the primary criterion for good stability is that the maximum value of $|G_{ib}|$, denoted $\text{Max}|G_{ib}|$, should not exceed 1.3. When $\text{Max}|G_{ib}| = 1.3$, the peak overshoot of the step response is about 25%.

2.3.2 Relations between Frequency Response and Transient Response

Figure 2.3-5 shows general frequency response plots of the loop (or "open-loop") transfer function G and the feedback (or "closed-loop") transfer function G_{ib}, which illustrate the major characteristics of these frequency responses. Figure 2.3-6 shows the corresponding transient responses of the feedback and error signals of the loop to step and ramp inputs. These responses illustrate approximations that relate transient and frequency responses.

In Fig 2.3-5, there are three distinct "bandwidth" parameters derived from the frequency response plots of a feedback control loop. The ω symbols indicate that these are angular frequency parameters, which are measured in radian/second (*not* hertz). These parameters are

ω_{gc} = gain crossover frequency; the frequency (rad/sec) where the magnitude of the loop transfer function G (the loop gain) drops below unity

ω_{hp} = half-power frequency; the frequency (rad/sec) where the magnitude of the feedback transfer function G_{ib} drops below $1/\sqrt{2} = 0.707$

ω_{ϕ} = frequency (rad/sec) of one radian of phase lag of the feedback transfer function G_{ib}

Note that ω_{gc} is derived from the open-loop transfer function G; whereas ω_{hp} and ω_{ϕ} are derived from the closed-loop transfer function G_{ib}.

If the feedback loop has good stability and reasonably high loop gain at low frequency (more than a factor of 6), the frequencies ω_{gc} and ω_{ϕ} are approximately equal to one another:

$$\omega_{gc} \cong \omega_{\phi} \tag{2.3-12}$$

On the other hand, the ratios ω_{hp}/ω_{gc} and $\omega_{hp}/\omega_{\phi}$ can vary by a factor of 5 or more in practical feedback control systems.

The major parameters of the step response are illustrated in Fig 2.3-6a. There are two *response-time* parameters, the rise time T_R and the delay time

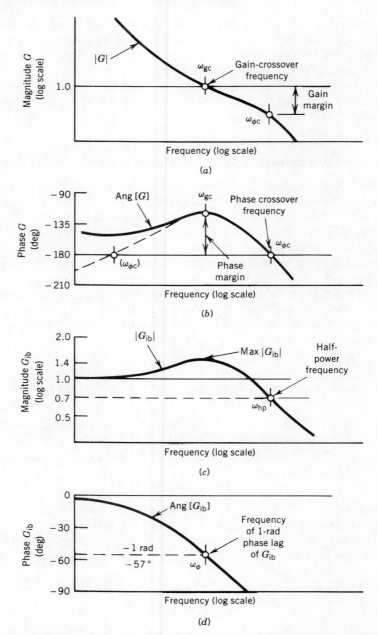

Figure 2.3-5 Definition of feedback-loop frequency-response parameters: (*a*) magnitude plot of *G*; (*b*) phase plot of *G*; (*c*) magnitude plot of G_{ib}; (*d*) phase plot of G_{ib}.

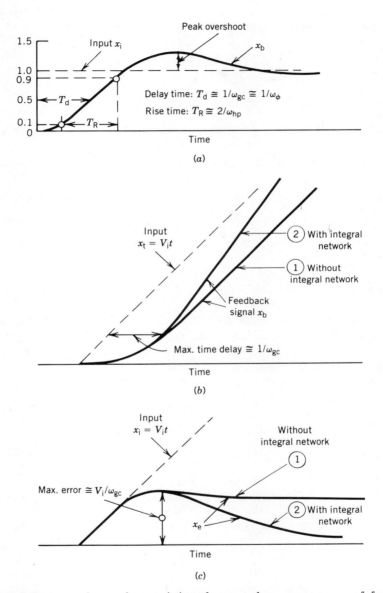

Figure 2.3-6 Approximate characteristics of step and ramp responses of feedback loop: (a) response of feedback signal x_b to unit step; (b) response of feedback signal x_b to ramp of slope V_i; (c) response of error signal x_e to ramp of slope V_i.

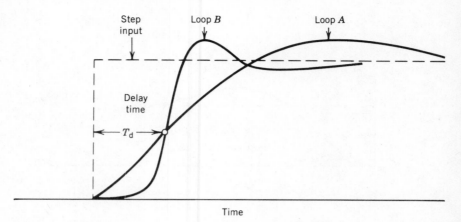

Figure 2.3-7 Typical step responses of two feedback control loops having same delay time, but 5 : 1 difference in rise time.

T_d, which are defined as

T_R = rise time; the time for the step response to rise from 10% to 90% of the *final* (not peak) value

T_d = delay time; the time after initiation of the input step for the step response to reach 50% of the *final* (not peak) value

These response-time parameters are related approximately as follows to the bandwidth parameters (expressed in rad/sec):

$$T_R \cong \frac{2}{\omega_{hp}} \tag{2.3-13}$$

$$T_d \cong \frac{1}{\omega_\phi} \cong \frac{1}{\omega_{gc}} \tag{2.3-14}$$

Since the bandwidth ratio ω_{hp}/ω_ϕ can vary widely, the response-time ratio T_R/T_d can also vary widely. Figure 2.3-7 shows two step responses (A, B) with the same delay time, where one has 5 times the rise time of the other. The corresponding frequency responses have nearly the same values for the gain crossover frequency ω_{gc} and the frequency ω_ϕ of one radian of phase lag of G_{ib}. However, the half-power frequency ω_{hp} for loop B is about 5 times greater than that for loop A.

The half-power frequency ω_{hp} is recognized as a good practical bandwidth criterion for describing the responses of amplifiers and filters in communication systems. The greater the half-power frequency ω_{hp} is, the steeper is the rise in the step response, and so (if the response has good stability) the more accurately the step response waveform conforms to the shape of the input step. This relationship applies also to an input signal of complex arbitrary shape. The greater the half-power "bandwidth" ω_{hp} of the filter; the more closely the output waveform duplicates the shape of the input waveform.

On the other hand, the fact that the input and output waveforms have nearly the same shape does not necessarily mean that they are nearly equal to one another, because there can be an appreciable time delay between the two waveforms. The time delay between a step input and its response is character-ized by the *delay time* T_d for the step response to reach 50% of the final value.

In terms of frequency response, time delay between input and output sinusoids represents phase lag. Therefore, delay time in the step response is directly related to the phase curve of the closed-loop G_{ib} frequency response. As was shown in Eq 2.3-14, the delay time T_d is approximately equal to the reciprocal of the frequency, denoted ω_ϕ, where the phase of G_{ib} is -1 rad (or $-57°$).

Time delay between the input and output is generally of little concern to the communications engineer, as long as the output waveform has essentially the same shape as the input waveform. However, to the control engineer, time delay is crucially important. He is generally concerned with minimizing the error between the input and output signals, and time delay between these signals represents dynamic error. Waveform fidelity is usually not important in a control application. In fact, (as will be shown in Fig 2.7-14) control loops often have low-frequency compensation networks (called integral networks) that purposely distort the waveform of the output signal in order to minimize the low-frequency components of the error.

Thus, the goals of the communications and control engineers are quite different: The communications engineer is generally concerned with duplicat-ing waveform shape; while the control engineer is generally concerned with minimizing dynamic error. Consequently, the communications engineer should measure bandwidth from the magnitude of the closed-loop frequency re-sponse, because this curve characterizes the ability of the amplifier to duplicate waveform shape. In contrast, the control engineer should measure bandwidth from the phase of the closed-loop frequency response, because this curve characterizes the time delay between the input and the output, which is the cause of dynamic error. Thus ω_ϕ and ω_{gc} are good measures of control system bandwidth, but ω_{hp} is not.

Chapter 6 will explain statistical techniques for analyzing the response of systems to random inputs, including receiver noise. An important parameter used in this analysis is *noise bandwidth*. For a feedback control loop, the noise bandwidth is derived from the magnitude of the closed-loop G_{ib} frequency response. If the loop has good stability, the noise bandwidth is roughly equal to the half-power frequency ω_{hp}.

Diagram *b* of Fig 2.3-6 shows responses of the feedback signal of a feedback loop to a ramp input. Two types of responses are illustrated, one of which has a low-frequency integral network. (The use of integral-network compensation will be discussed in Section 2.4.4.) An important characteristic to note from these responses is that the maximum time delay between the input ramp x_i and the x_b feedback signal is approximately equal to $1/\omega_{gc}$. This holds regardless of whether the loop has integral compensation. Thus,

$$\text{Maximum time delay for ramp input} \cong \frac{1}{\omega_{gc}} \qquad (2.3\text{-}15)$$

Diagram c of Fig 2.3-6 shows the x_e error responses of a feedback loop to a ramp input. This is equal to the difference in diagram b between the input x_i and the feedback signal x_b. As shown, the low-frequency integral-network compensation gradually reduces the final value of the error for a ramp input to zero. However, the integral network does not appreciably affect the maximum value of the error. For any feedback loop with good stability, the maximum error for a ramp input is approximately given by:

$$\text{Max. Error for ramp input} \cong \frac{V_i}{\omega_{gc}} \qquad (2.3\text{-}16)$$

The parameter V_i is the slope (or velocity) of the ramp input.

The stability of the response of the feedback loop is characterized by overshoot and ringing of the step response and by magnitude peaking of the feedback frequency response G_{ib}. Figure 2.2-5 showed a plot of the percentage overshoot of the step response versus the percentage peaking of the frequency response magnitude for a second-order lowpass filter. Although that plot was derived for a particular lowpass filter, it applies approximately to lowpass filters in general and to the G_{ib} response of a feedback control loop.

Figure 2.3-8 gives a general set of plots of peak overshoot of the step response versus the maximum magnitude of the G_{ib} frequency response for a feedback control loop. Step response peak overshoot is defined in Fig 2.3-6a, and the maximum magnitude of the G_{ib} frequency response, denoted $\text{Max}|G_{ib}|$, is defined in Fig 2.3-5a. As shown in Ref [1.1] (Section 2.4), the information in Fig 2.3-8 was derived by comparing these parameters for a large variety of different types of practical feedback control loops. All of the cases fell within the limits of the dashed maximum and minimum curves in Fig 2.3-8. The solid median plot provides a convenient standard for estimation.

A good practical stability criterion for feedback control design is that the maximum value of the magnitude of G_{ib} should not exceed 1.3:

$$\text{Practical stability criterion:} \ \text{Max}|G_{ib}| \leq 1.30 \qquad (2.3\text{-}17)$$

As shown by the dashed line in Fig 2.3-8, this practical criterion, which represents 30% peaking of $|G_{ib}|$, corresponds to a median peak overshoot of 25% for the step response. The deviation from the median is $\pm 3\%$, and so the overshoot can vary from 22% to 28%.

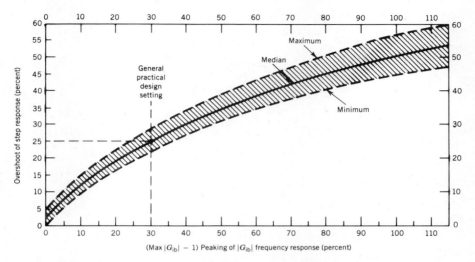

Figure 2.3-8 Practical range of variation of peak overshoot of step response, versus peaking of the $|G_{ib}|$ frequency response.

Sometimes the step response has high-frequency ringing, as shown in Fig 2.3-9. This response has poor stability even though the peak overshoot is not excessive. A response of this type is encountered frequently in the velocity feedback loop of a radar antenna servo, and is due to mechanical resonance in the antenna structure. When the step response has this characteristic, the magnitude of the G_{ib} frequency response has two peaks, as shown in Fig 2.3-10. Detailed discussions of secondary resonance and the effects of mechanical structural resonance are given in Ref [1.2] (Chapter 10).

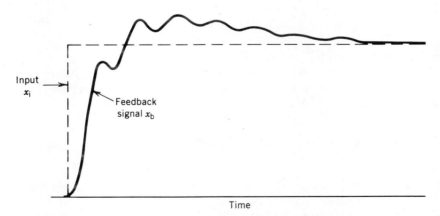

Figure 2.3-9 Step response of feedback loop having poorly damped secondary oscillation.

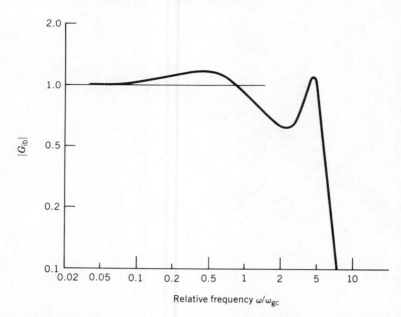

Figure 2.3-10 Frequency response of feedback loop having poorly damped secondary oscillation.

A feedback loop is best designed in terms of the magnitude and phase plots of its ("open") loop transfer function G. The corresponding plots of the feedback (or "closed-loop") transfer function G_{ib} can be obtained by plotting the magnitude and phase of G on the Nichols chart, which graphically solves the equation

$$G_{ib} = \frac{G}{1 + G} \qquad (2.3\text{-}18)$$

The variables G and G_{ib} are complex quantities, specified in terms of magnitude and phase or real and imaginary parts, which vary with frequency.

To illustrate the use of the Nichols chart, Fig 2.3-11 shows on the Nichols chart a plot of the G-locus for the following loop transfer function:

$$G = \frac{\omega_c(1 + \omega_1/s)}{s(1 + s/\omega_2)} \qquad (2.3\text{-}19)$$

where $\omega_c = 1.0 \text{ sec}^{-1}$, $\omega_1 = 0.203 \text{ sec}^{-1}$, and $\omega_2 = 1.79 \text{ sec}^{-1}$. (The value for ω_c was selected to minimize $\text{Max}|G_{ib}|$ for given values of ω_1, ω_2.) The magnitude of G (the loop gain) in decibels* is plotted on the vertical axis (the ordinate), and the phase of G (the loop phase) in degrees is plotted on the

* The decibel (dB) is explained in Section 2.4.1.

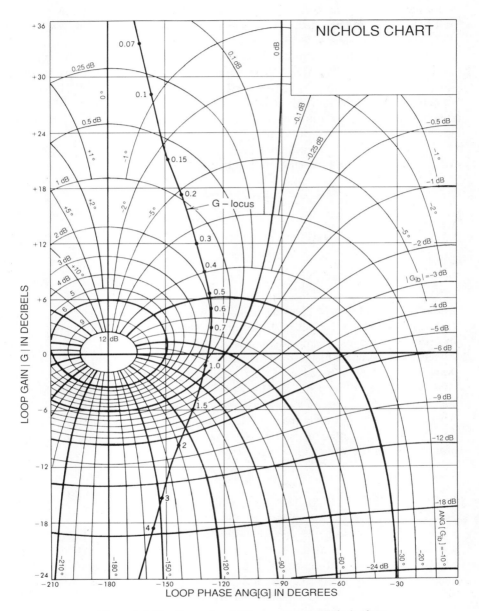

Figure 2.3-11 Example of *G*-locus plot on Nichols chart.

horizontal axis (the abscissa). Each point on the G-locus gives the magnitude and phase of G for the particular angular frequency ω labeled on the plot. For example, consider the G-locus point labeled 0.3. The ordinate of this point is $+12$ decibels, and the abscissa is $-135°$. Hence, at $\omega = 0.3$ rad/sec, $|G| = +12$ dB, and Ang$[G] = -135°$.

The curved contours on the Nichols chart are plots of constant magnitude and phase of G_{ib}. The point $\omega = 0.3$ rad/sec on the G-locus lies halfway between the contours for $|G_{ib}| = +1.0$ dB and $|G_{ib}| = +2.0$ dB. Hence, the magnitude of G_{ib} at 0.3 rad/sec is $+1.5$ dB. This G-locus point lies 30% of the distance between the contours for Ang$[G_{ib}] = -10°$ and Ang$[G_{ib}] = -20°$. Hence, the phase of G_{ib} at 0.3 rad/sec is $-13°$.

To summarize, the coordinates corresponding to the point on the G-locus labeled 0.3 yield the following information:

$$\omega = 0.3 \text{ rad/sec} \tag{2.3-20}$$

$$|G| = +12 \text{ dB} = +6 \text{ dg} = 4.0 \tag{2.3-21}$$

$$\text{Ang}[G] = -135° \tag{2.3-22}$$

$$|G_{ib}| = +1.5 \text{ dB} = +0.75 \text{ dg} = 1.189 \tag{2.3-23}$$

$$\text{Ang}[G_{ib}] = -13° \tag{2.3-24}$$

Equations 2.3-21 to -24 give the magnitude and phase of G and G_{ib} at the frequency indicated in Eq 2.3-20. The reader should verify each of these values in order to achieve a clear understanding of the Nichols chart.

The Nichols chart has left–right symmetry about the vertical axis at which Ang$[G_{ib}] = -180°$. Because of this symmetry, the Nichols chart need not cover a full 360° of loop phase. The abscissa of the particular Nichols chart of Fig 2.3-11 gives values of loop phase, Ang$[G]$, only from 0 to $-210°$. When Ang$[G]$ is more negative than $-210°$, the locus for the conjugate of G, denoted G^*, is plotted instead of G. The magnitude and phase of G^* are obtained from the values for G as follows:

$$|G^*| = |G| \tag{2.3-25}$$

$$\text{Ang}[G^*] = -360° - \text{Ang}[G] \tag{2.3-26}$$

Since Ang$[G]$ is normally negative, the term, $-\text{Ang}[G]$, in Eq 2.3-26 is normally positive. When the locus of G^* is plotted on the Nichols chart, the curved contours yield data for G_{ib}^*, which is the conjugate of G_{ib}. The magnitude and phase of G_{ib} are derived as follows from the magnitude and phase values of G_{ib}^* read from the Nichols chart:

$$|G_{ib}| = |G_{ib}^*| \tag{2.3-27}$$

$$\text{Ang}[G_{ib}] = -360° - \text{Ang}[G_{ib}^*] \tag{2.3-28}$$

Thus, one can derive from the Nichols chart of Fig 2.3-11 the magnitude and phase plots of G_{ib} for any value of loop phase, provided that the loop gain lies within the magnitude range of the chart (-24 to $+36$ dB).

One can evaluate the stability of a feedback loop by examining the frequency-response plot of $|G_{ib}|$, to determine $\text{Max}|G_{ib}|$ and the characteristics of any secondary resonance peak. However, it is much more convenient to evaluate stability from the open-loop frequency response G. The degree of stability can be estimated by applying criteria based on gain margin and phase margin. As shown previously, a loop would oscillate if the loop phase ($\text{Ang}[G]$) were $-180°$ at a frequency where the loop gain $|G|$ is unity. Phase margin and gain margin describe the margins in loop gain and loop phase relative to this oscillatory condition.

As shown in Fig 2.3-5b, the phase crossover frequency, denoted $\omega_{\phi c}$, is defined as

> *Phase Crossover Frequency*, $\omega_{\phi c}$: the frequency where the phase lag of the loop transfer function G is $180°$.

All practical feedback control loops have a phase crossover frequency where the loop gain is less than unity. A few feedback loops (with a low-frequency triple integration) have a second phase crossover frequency at low frequency where the loop gain is greater than unity. This special case, which is illustrated by the dashed curve in Fig 2.3-5b, is ignored here. The reader is referred to Ref [1.1] (Section 3.1) for further discussion of this issue.

The phase margin is measured at the gain crossover frequency ω_{gc}, and the gain margin is measured at the phase crossover frequency $\omega_{\phi c}$. These parameters are derived from the phase and magnitude plots of the loop transfer function G, and are defined as:

> *Phase Margin*: $180°$ minus the phase lag of the loop transfer function G at the gain crossover frequency ω_{gc};

> *Gain Margin*: the reciprocal of the magnitude of the loop transfer function G at the phase crossover frequency $\omega_{\phi c}$.

The gain margin and phase margin are illustrated on the plots of $|G|$ and $\text{Ang}[G]$ in diagrams a, b of Fig 2.3-5. Since the magnitude plot is expressed on a logarithmic scale, the difference represented as gain margin in Fig 2.3-5a corresponds to a ratio of the two values of gain.

Practical design criteria for the minimum allowable values of phase margin and gain margin are as follows:

$$\text{Phase margin} \geq 45° \qquad (2.3\text{-}29)$$

$$\text{Gain margin} \geq \text{factor of } 2.5 \qquad (2.3\text{-}30)$$

If these gain margin and phase margin criteria are satisfied, the loop usually has reasonably good stability. The value of $\text{Max}|G_{ib}|$ may exceed 1.3 somewhat, but not by a large factor. (A phase margin of 45° is necessary if $\text{Max}|G_{ib}|$ is not to exceed 1.30, but is not sufficient to assure this condition.)

The gain margin requirement of Eq 2.3-30 is not based directly on the $\text{Max}|G_{ib}|$ specification. The gain margin can be as low as 1.8 without $\text{Max}|G_{ib}|$ exceeding 1.30. However, the gain margin should be at least a factor of 2.5 for the following reasons:

1. The stability of a feedback loop with a gain margin below 2.5 is very sensitive to loop gain variations, and degrades drastically with a small increase of loop gain;
2. A feedback loop with a gain margin below 2.5 tends to have a poorly damped secondary oscillation in its transient response, similar to that shown in the step response of Fig 2.3-9.

Thus, the degree of stability of a feedback loop can be estimated from the values of gain margin and phase margin, derived from the "open" loop transfer function G. However, more accurate stability information can be obtained from the magnitude plot of the feedback (or "closed-loop") transfer function G_{ib}.

2.4 MAGNITUDE ASYMPTOTE PLOTS IN FEEDBACK CONTROL ANALYSIS

2.4.1 Logarithmic Units: The Decibel and Decilog

The magnitude of a frequency response is generally plotted on a logarithmic scale. It is often convenient to represent the magnitude in terms of a logarithmic unit, the decibel, which allows the magnitude scale markings to be linear. However, the decibel is basically defined in terms of a power ratio, and so is being misused when power ratio has no meaning. For such cases, this book uses an alternate term, the decilog. The decibel (dB) and the decilog (dg) are explained as follows.

Consider an amplifier with an input power P_i and an output power P_o. The power gain of the amplifier in decibels is defined as

$$\text{Power gain } \frac{P_o}{P_i} \text{ in decibels (dB)} = 10 \log\left[\frac{P_o}{P_i}\right] \qquad (2.4\text{-}1)$$

When the input and output impedances of the amplifier are the same, the power ratio in proportional to the square of the voltage amplitude ratio:

$$\frac{P_o}{P_i} = \left(\frac{|E_o|}{|E_i|}\right)^2 \qquad (2.4\text{-}2)$$

where $|E_i|$ is the amplitude of the amplifier input voltage and $|E_o|$ is the amplitude of the output voltage. When Eq 2.4-2 holds, the power gain of the amplifier in decibels can be expressed as follows directly in terms of the voltage ratio:

$$\text{Power gain } \frac{P_o}{P_i} \text{ in decibels (dB)} = 10 \log \left[\left(\frac{|E_o|}{|E_i|} \right)^2 \right]$$

$$= 20 \log \left[\frac{|E_o|}{|E_i|} \right] \qquad (2.4\text{-}3)$$

As this definition was used, it became convenient to refer directly to the amplifier voltage gain in decibels. Thus, Eq 2.4-3 was reinterpreted as

$$\text{Voltage gain } \frac{|E_o|}{|E_i|} \text{ in decibels (dB)} = 20 \log \left[\frac{|E_o|}{|E_i|} \right] \qquad (2.4\text{-}4)$$

The decibel concept became such a powerful tool it was generalized to apply to situations where power ratio has no significance, such as transfer function gains in feedback control systems. A signal amplitude ratio is given in decibels by applying the voltage-gain definition of Eq 2.4-4. Thus, any signal amplitude ratio $|X_o|/|X_i|$ can be expressed in decibels as follows:

$$\text{Signal ratio } \frac{|X_o|}{|X_i|} \text{ in decibels (db)} = 20 \log \left[\frac{|X_o|}{|X_i|} \right] \qquad (2.4\text{-}5)$$

Instead of the decibel (dB), this book uses the decilog (abbreviated dg) when power ratio has no significance. The decilog was introduced by E. I. Green [2.1]. For any nondimensional ratio A, the corresponding value in decilogs is

$$A \text{ in decilogs (dg)} = 10 \log [A] \qquad (2.4\text{-}6)$$

A signal amplitude ratio expressed in decilogs is

$$\text{Signal ratio } \frac{|X_o|}{|X_i|} \text{ in decilogs (dg)} = 10 \log \left[\frac{|X_o|}{|X_i|} \right] \qquad (2.4\text{-}7)$$

Hence, for a signal ratio, the decilog and the decibel are related by

$$2 \text{ decibels (dB)} = 1 \text{ decilog (dg) (Signal ratio)} \qquad (2.4\text{-}8)$$

Although the decibel is widely used for describing signal amplitude ratios for feedback control systems, the author uses the decilog instead for the following reasons:

1. It is much easier to translate mentally a decilog value to the corresponding signal ratio than it is to translate a decibel value.
2. The decibel is basically defined in terms of a power ratio and is being misused to describe a signal ratio when power ratio has no significance.

Point 1 is discussed further in Ref [1.1] (Section 2.3), which shows how a decilog value can be conveniently related to the corresponding signal amplitude ratio.

2.4.2 Principle of Magnitude Asymptote Plot

The magnitude asymptote plot is an important tool in feedback control analysis. To understand it, consider the magnitude of the transfer function $(1 + s/\omega_x)$. The frequency response is obtained by setting $s = j\omega$, which yields $(1 + j\omega/\omega_x)$. The magnitude of this is

$$\left| 1 + \frac{j\omega}{\omega_x} \right| = \sqrt{1 + \left(\frac{\omega}{\omega_x} \right)^2} \tag{2.4-9}$$

This magnitude is shown by the solid curve in Fig 2.4-1.

At a low frequency, where $\omega \ll \omega_x$, the magnitude of Eq 2.4-9 is approximately equal to unity; while at a high frequency, where $\omega \gg \omega_x$, the magnitude is approximately equal to ω/ω_x. Hence, at low frequencies the magnitude

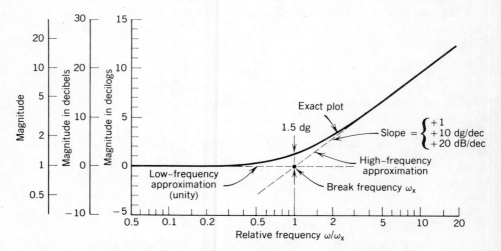

Figure 2.4-1 Plot of magnitude of $(1 + j\omega, \omega_x)$ versus frequency.

is approximated by unity, and at high frequencies it is approximated by ω/ω_x. These approximations are shown by the dashed lines in Fig 2.4-1, which are called magnitude asymptotes.

The magnitude asymptotes of Fig 2.4-1 intersect at the frequency $\omega = \omega_x$, which is called the *break frequency*. The magnitude plot is approximated by assuming that it follows the low-frequency magnitude asymptote at frequencies below the break frequency, and the high-frequency magnitude asymptote at frequencies above the break frequency. The maximum logarithmic deviation of this asymptote approximation from the exact magnitude curve occurs at the break frequency. As the plot shows, the maximum deviation is a factor of $\sqrt{2}$, which corresponds to 3 decibels (dB) or 1.5 decilogs (dg).

The high-frequency magnitude asymptote is a plot of ω/ω_x and so is proportional to the frequency ω. This asymptote has a slope of $+10$ decilogs/decade (dg/dec), or $+20$ decibels/decade (dB/dec), and is defined as having a logarithmic slope of $+1$. In like manner, a magnitude asymptote that is proportional to ω^2 has a logarithmic slope of $+2$, which corresponds to $+20$ dg/dec or $+40$ dB/dec. Similarly, a magnitude asymptote that is proportional to $1/\omega^2 = \omega^{-2}$ has a logarithmic slope of -2, which corresponds to -20 dg/dec or -40 dB/dec.

In many applications, the deviation of the magnitude curve from the magnitude asymptote approximation can be neglected, at least under certain conditions, and so the magnitude asymptote plot can be used directly in analysis.

2.4.3 Simple Feedback Loops

Let us examine the magnitude asymptote plots of the loop transfer functions of some simple feedback loops. Consider the signal-flow diagram for loop A in Fig 2.4-2a. The loop transfer function of this loop is

$$G_A = \frac{\omega_c}{s} = \frac{\omega_c}{j\omega} \qquad (2.4\text{-}10)$$

In the diagram, the $1/s$ integration factor is separated from the ω_c gain factor. Since the output from the integrator is the feedback signal X_b, the input to the integrator is the derivative of the feedback signal, denoted $\dot{X}_b$.

This loop could represent, at least ideally, a servo that controls the angle of a shaft. The variable X_b would be the shaft angle, and $\dot{X}_b$ would be the angular velocity of the shaft, which is proportional to the motor angular velocity. In this servo, the angular velocity $\dot{X}_b$ is proportional to the servo error signal X_e. Hence, the servo could be implemented by a motor, with drive amplifier, in which the motor angular velocity is proportional to the voltage applied to the drive amplifier, which in turn is proportional to the error signal X_e. For such a servo, the integration transfer function $1/s$ in the signal-flow diagram of Fig 2.4-2a represents the integration between angular velocity and

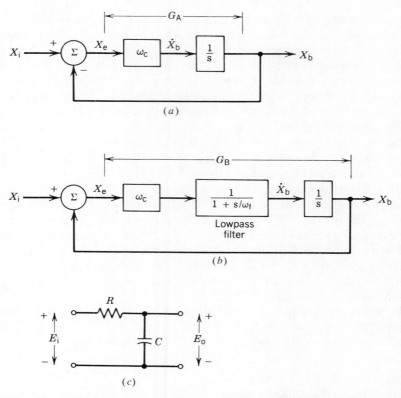

Figure 2.4-2   Signal-flow diagrams for (a) loop A and (b) loop B; (c) circuit diagram for lowpass filter in loop B.

angle. It occurs because the loop feeds back a signal proportional to angle, which is the integral of the shaft angular velocity.

The magnitude of G_A in Eq 2.4-10 is ω_c/ω. A plot of this magnitude is shown in Fig 2.4-3. Since ω_c/ω is equal to unity at $\omega = \omega_c$, this magnitude plot passes through unity at the frequency ω_c, which is the gain crossover frequency of the loop. As will be shown, in more complicated transfer functions the parameter ω_c will become the *asymptotic gain crossover frequency*, which is only approximately equal to the true gain crossover frequency ω_{gc}.

The plot of Fig 2.4-3 is proportional to ω^{-1} and so has a logarithmic slope of -1. In terms of decilogs, the plot has a slope of -10 decilogs/decade (-10 dg/dec). The G_{ib} transfer function for loop A is

$$G_{ib(A)} = \frac{G_A}{1 + G_A} = \frac{\omega_c/s}{1 + \omega_c/s} = \frac{1}{1 + s/\omega_c} \qquad (2.4\text{-}11)$$

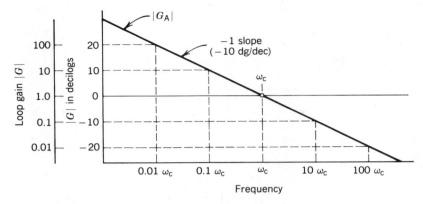

Figure 2.4-3 Magnitude plot of loop transfer function for loop A.

Let us assume that the simple RC lowpass filter shown in Fig 2.4-2c is placed in series with the motor of the G_A servo. This lowpass filter might simulate the dynamic lag caused by motor inductance. The transfer function of the lowpass filter is

$$\frac{E_o}{E_i} = \frac{1}{1 + sRC} = \frac{1}{1 + s\tau} = \frac{1}{1 + s/\omega_f} \tag{2.4-12}$$

The parameter τ is the time constant of the filter, which is equal to the product RC. The reciprocal of this is the break frequency of the filter, denoted ω_f, which is equal to

$$\omega_f = \frac{1}{\tau} = \frac{1}{RC} \tag{2.4-13}$$

A magnitude asymptote plot of this filter transfer function is shown in Fig 2.4-4a. This plot is unity at frequencies below ω_f, and has a logarithmic slope of -1 above ω_f. The filter provides attenuation at frequencies above the filter break frequency ω_f.

The transfer function E_o/E_i in Eq 2.4-12 for the lowpass filter is the same as that for $G_{ib(A)}$ in Eq 2.4-11, if ω_c is replaced by ω_f. Hence, Fig 2.4-4a is also the G_{ib} magnitude-asymptote plot of loop A for $\omega_f = \omega_c$.

When this lowpass filter is inserted into feedback loop A, the signal-flow diagram changes to that of Fig 2.4-2b. This loop, called loop B, has the loop transfer function

$$G_B = \frac{\omega_c}{s(1 + s/\omega_f)} = \frac{\omega_c \omega_f}{s(s + \omega_f)} = \frac{N}{D} \tag{2.4-14}$$

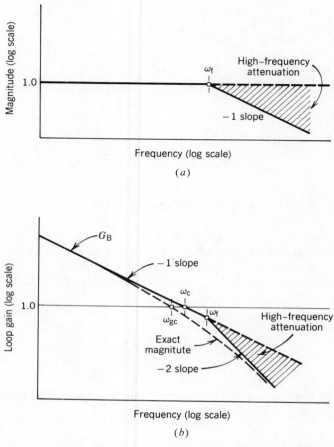

Figure 2.4-4 (*a*) Magnitude asymptote plot for lowpass filter. (*b*) Loop transfer function for loop *B*.

As shown, the numerator is denoted N and the denominator D. The corresponding feedback transfer function is

$$
G_{\text{ib(B)}} = \frac{G}{1 + G} = \frac{N}{D + N}
$$

$$
= \frac{\omega_c \omega_f}{s(s + \omega_f) + \omega_c \omega_f} = \frac{\omega_c \omega_f}{s^2 + \omega_f s + \omega_c \omega_f} \tag{2.4-15}
$$

A magnitude asymptote plot of the loop transfer function of Eq 2.4-14 is shown by the solid curve in Fig 2.4-4*b*. This is obtained by adding the logarithmic plots of Figs 2.4-3 and -4*a*, which is equivalent to multiplying the magnitude values of the corresponding transfer functions. At high frequencies, the magnitude asymptote is proportional to $1/\omega^2$ and so has a logarithmic slope of -2, which corresponds to -20 dg/dec. The dashed curve shows the exact magnitude plot. The frequency where the dashed plot crosses unity gain is the true gain crossover frequency ω_{gc}. The parameter ω_c is the asymptotic gain crossover frequency. In any practical feedback loop, the asymptotic gain crossover frequency ω_c is a reasonable approximation of ω_{gc}.

The loop transfer function of Eq 2.4-15 has the following general form:

$$G_{ib(B)} = \frac{\omega_n^2}{s^2 + 2\zeta\omega_n s + \omega_n^2} \tag{2.4-16}$$

Setting Eqs 2.4-15, -16 equal gives

$$2\zeta\omega_n = \omega_f \tag{2.4-17}$$

$$\omega_n^2 = \omega_c\omega_f \tag{2.4-18}$$

Hence, the natural frequency ω_n and damping ratio ζ are equal to

$$\omega_n = \sqrt{\omega_c\omega_f} \tag{2.4-19}$$

$$\zeta = \frac{\omega_f}{2\omega_n} = \frac{1}{2}\sqrt{\frac{\omega_f}{\omega_c}} \tag{2.4-20}$$

The loop-B feedback transfer function $G_{ib(B)}$ has the same form as the E_o/E_i transfer function of the second-order lowpass filter shown in Section 2.2, Eq 2.2-15. Hence the transient responses in Fig 2.2-3 and the frequency responses in Fig 2.2-4 also apply to the G_{ib} feedback response of loop B. This loop has been extensively studied in the control literature, and is commonly called the basic second-order feedback loop. Its dynamic responses are often used as models for those of much more complicated feedback transfer functions. This approach has merit. However, it has been overdone and has resulted in some important misconceptions, as will be explained in Section 2.4.5.

Figure 2.4-5 shows general plots of a number of parameters of the transient and frequency responses in Figs 2.2-3, -4, expressed as functions of the damping ratio ζ. The plots show:

1. Normalized rise time $T_R\omega_{hp}/2$.
2. Normalized delay time $T_d\omega_\phi$.
3. Frequency ratio ω_ϕ/ω_n.
4. Frequency ratio $\omega_{hp}/2\omega_\phi$.
5. Peak overshoot of the unit step response $(\text{Max}[x_b] - 1)$.
6. Peaking of the frequency response magnitude $(\text{Max}|G_{ib}| - 1)$.
7. Frequency ratio ω_{gc}/ω_n.
8. Normalized delay time $T_d\omega_{gc}$.

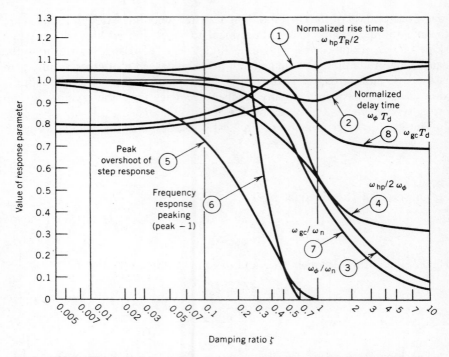

Figure 2.4-5 General plots of response parameters of second-order lowpass filter.

The parameters ω_{hp} and ω_ϕ are measured from the magnitude and phase plots of G_{ib} shown in Fig 2.2-4. The parameter ω_{gc} is obtained from the magnitude plot of G illustrated qualitatively in Fig 2.4-4b. According to Eqs 2.3-13, -14 the expressions $\omega_\phi T_d$, $\omega_{gc} T_d$, and $\omega_{hp} T_R/2$ should all approximate unity. Figure 2.4-5 shows that the largest variation from unity for these expressions occurs with $\omega_{gc} T_d$, which has a value of 0.7 for large ζ.

2.4.4   Effect of Integral Network

Many feedback control loops have compensation networks, called *integral networks*, that increase the loop gain at low frequencies in order to decrease the error from low-frequency inputs. Section 2.7 and Chapter 3 will show how an integral network affects the time response of a feedback loop. Let us first examine the effect of an integral network on the loop gain plot.

Compensation networks are usually implemented by operational amplifier (or opamp) circuits. An opamp is indicated schematically in Fig 2.4-6a. Signals $E_{(+)}$, $E_{(-)}$ are the voltages on the $(+)$ and $(-)$ input terminals of the opamp, and E_o is the output voltage. At low frequencies, the opamp response can be represented as

$$E_o = K_A\big(E_{(+)} - E_{(-)}\big) \qquad (2.4\text{-}21)$$

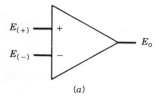

(a)

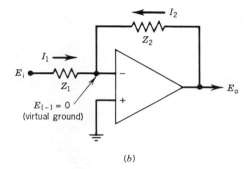

(b)

Figure 2.4-6 (a) Circuit diagram of opamp. (b) General inverting opamp circuit.

where K_A is the low-frequency gain of the opamp. An opamp is always used with negative feedback, in which the output voltage E_o is fed back to the $(-)$ input terminal. The output voltage changes until the difference between the voltages on the $(+)$ and $(-)$ input terminals is small. At low frequencies, the gain K_A is very high, and so this input voltage difference is essentially zero. Hence, the opamp controls the voltage $E_{(-)}$ on the negative input terminal to be essentially equal to the voltage $E_{(+)}$ on the positive input terminal.

Let us apply this principle to analyze the basic inverting opamp circuit shown in Fig 2.4-6b. The negative feedback to the $(-)$ input terminal continually changes the voltage on that terminal until it is equal to the voltage on the grounded $(+)$ input terminal, which is zero. Hence, the opamp keeps the voltage on the $(-)$ input terminal equal to zero. Since this $(-)$ input terminal is held at ground potential, it is called a *virtual ground*.

Because the voltage on the $(-)$ input terminal is zero, the currents I_1, I_2 flowing through impedances Z_1, Z_2 are

$$I_1 = \frac{E_i}{Z_1} \tag{2.4-22}$$

$$I_2 = \frac{E_o}{Z_2} \tag{2.4-23}$$

Since the input impedance to the opamp is very high, essentially no current flows into the opamp. Hence, the sum of currents I_1, I_2 is zero:

$$I_1 + I_2 = 0 \tag{2.4-24}$$

When Eqs 2.4-22 to -24 are solved simultaneously, they yield the following transfer function for the opamp circuit:

$$-\frac{E_o}{E_i} = \frac{Z_2}{Z_1} \tag{2.4-25}$$

The circuit has signal inversion: When the input voltage E_i is positive, the output voltage E_o is negative. It is convenient to separate this inversion from the transfer function achieved by the circuit. If the inversion is not wanted, it can be compensated for by changing the sign of another element in the loop or by adding another inverting opamp stage. The sign of the transfer function for a motor, a tachometer, or a position transducer can usually be reversed simply by swapping the windings of the device. For these reasons, the $(-)$ sign in Eq 2.4-25 is placed on the left side of the equation, and the transfer function provided by the circuit is considered to be Z_2/Z_1.

Figure 2.4-7a applies this principle to achieve a pure integration transfer function. The impedances Z_1, Z_2 are

$$Z_1 = R \tag{2.4-26}$$

$$Z_2 = \frac{1}{sC} \tag{2.4-27}$$

Hence, by Eq 2.4-25, the transfer function achieved by the opamp circuit is

$$-\frac{E_o}{E_i} = \frac{Z_2}{Z_1} = \frac{1/sC}{R} = \frac{1}{sCR} \tag{2.4-28}$$

This is a pure integration. This circuit could provide the $1/s$ integration shown in the signal-flow diagrams for loops A and B in Fig 2.4-2. In Section 2.8, Fig 2.8-5 shows the circuit diagram of a servo system, which uses this integrator circuit in a current feedback loop.

Figure 2-4.7b shows an opamp circuit that provides increased loop gain at low frequencies. Comparing this with Fig 2.4-6b gives the opamp impedances

$$Z_1 = R \tag{2.4-29}$$

$$Z_2 = R + \frac{1}{sC} \tag{2.4-30}$$

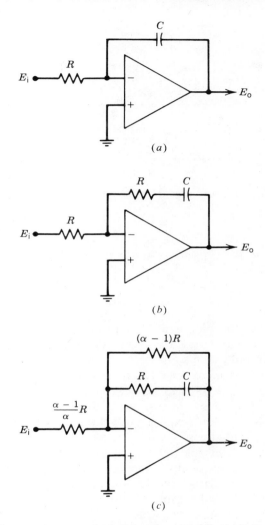

Figure 2.4-7 Opamp compensation networks providing: (a) pure integration; (b) integral network; (c) integral network with finite gain α at zero frequency.

Applying these to Eq 2.4-25 gives the transfer function of the circuit:

$$-\frac{E_o}{E_i} = \frac{Z_2}{Z_1} = \frac{R + 1/sC}{R} = 1 + \frac{1}{sCR}$$

$$= 1 + \frac{\omega_i}{s} = 1 + \frac{\omega_i}{j\omega} \qquad (2.4\text{-}31)$$

This circuit is called an *integral network*, and the parameter ω_i is called the

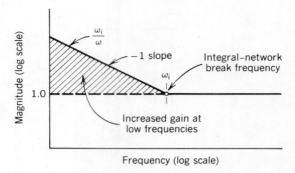

Figure 2.4-8 Magnitude asymptote plot for integral network transfer function $(1 + \omega_i, j\omega)$.

integral-network break frequency, which is equal to

$$\omega_i = \frac{1}{RC} \tag{2.4-32}$$

A magnitude asymptote plot of E_o/E_i for Eq 2.4-31 is shown in Fig 2.4-8. At high frequencies (for $\omega \gg \omega_i$), the magnitude of the term $\omega_i/j\omega$ in Eq 2.4-31 is much less than unity, and so the term can be neglected. At low frequencies (for $\omega \ll \omega_i$), the magnitude of this term is much greater than unity, and so the unity term in Eq 2.4-31 can be neglected. Hence, the transfer function approximates $\omega_i/j\omega$ at low frequencies and unity at high frequencies. As shown in Fig 2.4-8, the low-frequency magnitude asymptote is a plot of ω_i/ω for $\omega < \omega_i$ and is unity for $\omega > \omega_i$.

Let us assume that this integral network is inserted into loop A, which was shown in Fig 2.4-2a. This results in the signal-flow diagram of Fig 2.4-9a. The transfer function of the dashed block is the sum of two paths: one with a gain of unity, and the other with a gain of ω_i/s. Hence, the sum of these two paths is $(1 + \omega_i/s)$. This feedback loop, which is called loop C, has the loop transfer function

$$G_C = \frac{\omega_c(1 + \omega_1/s)}{s} = \frac{\omega_c(s + \omega_1)}{s^2} = \frac{N}{D} \tag{2.4-33}$$

The feedback transfer function is

$$G_{ib(C)} = \frac{G_C}{1 + G_C} = \frac{N}{D + N}$$

$$= \frac{\omega_c(s + \omega_i)}{s^2 + \omega_c(s + \omega_i)} = \frac{\omega_c(s + \omega_i)}{s^2 + \omega_c s + \omega_c \omega_i} \tag{2.4-34}$$

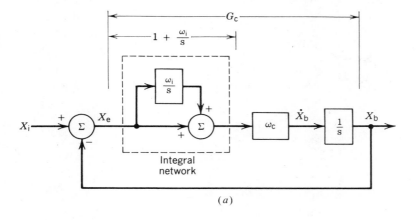

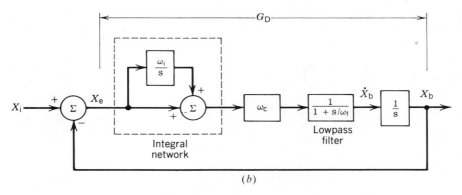

Figure 2.4-9 Signal-flow diagrams for: (a) loop C; (b) loop D.

The magnitude asymptote plot of the loop transfer function for loop C is shown in Fig 2.4-10a. This is formed by adding the logarithmic magnitude asymptote plot for the integral network in Fig 2.4-8 to that for G_A in Fig 2.4-3. As the figure shows, the integral network increases the loop gain at low frequencies, which decreases the loop error for a low-frequency input. The effect of this network on the general time response of the loop is explained in Section 2.7 and in Chapter 3.

Let us now assume that the integral network is applied to loop B, in Fig 2.4-2c. The signal-flow diagram for the resultant loop is shown in Fig 2.4-9b. This is called loop D and has the loop transfer function

$$G_D = \frac{\omega_c(1 + \omega_i/s)}{s(1 + s/\omega_f)} = \frac{\omega_c\omega_f(s + \omega_i)}{s^2(s + \omega_f)} = \frac{N}{D} \qquad (2.4\text{-}35)$$

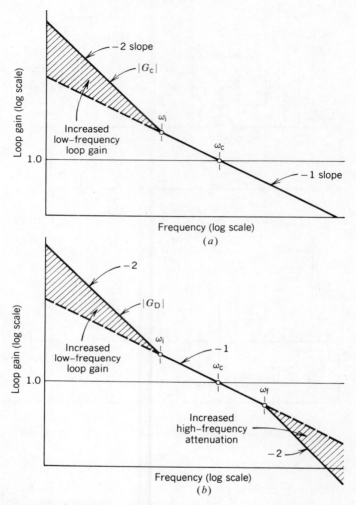

Figure 2.4-10 Magnitude asymptote plots of loop transfer functions for: (a) loop C; (b) loop D.

The feedback transfer function is

$$G_{ib(D)} = \frac{G_D}{1 + G_D} = \frac{N}{D + N}$$

$$= \frac{\omega_c \omega_f (s + \omega_i)}{s^2 (s + \omega_f) + \omega_c \omega_f (s + \omega_i)}$$

$$= \frac{\omega_c \omega_f (s + \omega_i)}{s^3 + \omega_f s^2 + \omega_c \omega_f s + \omega_c \omega_f \omega_i} \qquad (2.4\text{-}36)$$

The magnitude asymptote plot of the loop transfer function for loop D is shown in Fig 2.4-10b. This is formed by adding the logarithmic magnitude asymptote plot for the integral network in Fig 2.4-8 to that for G_B in Fig 2.4-4b. The figure shows that the integral network increases the loop gain at low frequencies, and the lowpass filter provides attenuation (relative to loop A) at high frequencies.

The integral-network opamp circuit in Fig 2.4-7b has infinite gain at zero frequency. If the feedback loop is driven into saturation, a large charge can build up on the capacitor C, which results in a long transient when the loop recovers from saturation. Figure 2.4-7c shows an integral network with finite gain at zero frequency, which minimizes this problem. The input resistance is chosen to achieve unity gain in the circuit at high frequencies. It can be shown that the transfer function of this circuit is

$$\frac{E_o}{E_i} = \frac{1 + \omega_i/s}{1 + \omega_i/\alpha s} \tag{2.4-37}$$

where the break frequency ω_i is equal to

$$\omega_i = \frac{1}{RC} \tag{2.4-38}$$

A magnitude asymptote plot of this transfer function is shown in Fig 2.4-11a. This network has a gain of α at low frequency, and a gain of unity at high frequency. Thus, the network increases the low-frequency gain by the factor α. This factor α is usually at least 10.

When this integral network is inserted into loop B, the resultant loop, which is called loop E, has the loop transfer function

$$G_E = \frac{\omega_c(1 + \omega_i/s)}{s(1 + \omega_i/\alpha s)(1 + s/\omega_f)} \tag{2.4-39}$$

A magnitude asymptote plot of this loop transfer function is shown in Fig 2.4-11b. This is similar to that for loop D in Fig 2.4-10b, except that the gain at low frequency is increased only by the factor α relative to the loop gain for loop B.

In the magnitude asymptote plots of the loop transfer functions for loops A to E in Figs 2.4-3, -4b, -10a, -10b, and -11b, the magnitude asymptote has a logarithmic slope of -1 at the gain crossover frequency. This condition must be satisfied, or at least approximated, in order for the loop to have good stability. Essentially all practical feedback control loops have a broad region of -1 magnitude asymptote slope at gain crossover, or close to gain crossover.

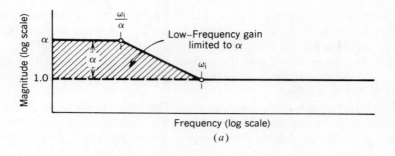

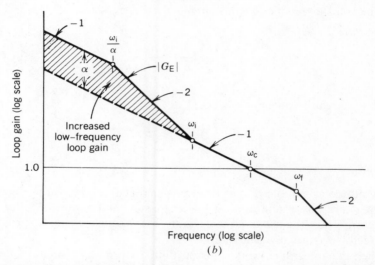

Figure 2.4-11 Magnitude asymptote plots of: (*a*) integral network with finite gain α at zero frequency; (*b*) loop transfer function of loop E with this integral network.

By *broad* we mean that this asymptote covers at least a factor of 7 in frequency. If this condition is not satisfied, the phase lag at gain crossover is so great the loop cannot have good stability.

For example, loop A, which has a magnitude asymptote slope of -1 over all frequencies, has the loop transfer function

$$G_A = \frac{\omega_c}{s} = \frac{\omega_c}{j\omega} \qquad (2.4\text{-}40)$$

The factor j in the denominator represents a phase lag of $90°$. Hence, loop A has a phase lag of $90°$ over all frequencies. Consider the following loop transfer function, which has a magnitude asymptote slope of -2 over all

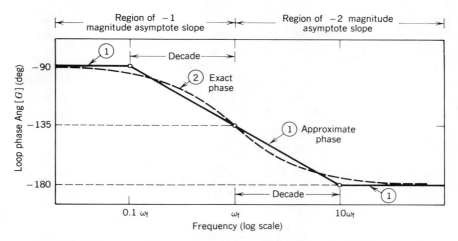

Figure 2.4-12   Approximate and exact plots of loop phase Ang[G] for loop B.

frequencies:

$$G = \frac{K}{s^2} = \frac{K}{(j\omega)^2} = -\frac{K}{\omega^2} \qquad (2.4\text{-}41)$$

This loop has 180° of phase lag at all frequencies and so is unstable.

As shown in Fig 2.4-4b, the magnitude asymptote slope for loop B is -1 at low frequencies and -2 at high frequencies. Hence, the phase varies from $-90°$ at low frequencies to $-180°$ at high frequencies. The equation for the phase of the loop transfer function G_B is

$$\text{Ang}[G_B] = -90° - \arctan[\omega/\omega_f] \qquad (2.4\text{-}42)$$

This phase is shown by the dashed curve (2) in Fig 2.4-12. The solid curve (1) is a straight-line approximation of the phase, which is within $\pm 5°$ of the exact phase.

The phase approximation shows that the phase corresponds directly to the magnitude asymptote slope for frequencies a decade or more from the break frequency. Thus, for $\omega < 0.1\omega_f$ (where the magnitude asymptote slope is -1), the phase is approximately $-90°$; and for $\omega > 10\omega_f$ (where the magnitude asymptote slope is -2), the phase is approximately $-180°$. Between $0.1\omega_f$ and $10\omega_f$ the phase makes a transition from $-90°$ to $-180°$, reaching $-135°$ at the break frequency ω_f.

As was stated earlier, the phase lag at gain crossover should not exceed 135° if the loop is to have good stability. Figure 2.4-12 indicates that $-135°$ of phase occurs at the break frequency ω_f. Hence, to achieve good stability, the gain crossover frequency ω_{gc} of this loop should not exceed ω_f.

Loop B has only a single break frequency (at ω_f). The phase plots for loops D and E are more complicated than that for loop B in Fig 2.4-12. Nevertheless, by applying the principles used in this figure, one can readily see why the loop transfer functions for loops D and E (or for any other reasonable feedback control loop) must have a broad region of -1 magnitude asymptote slope at or near gain crossover to achieve good stability. If this condition is not satisfied, the phase lag at gain crossover is much greater than $135°$, except for artificial loop transfer functions that represent poor designs. This point is discussed further in Ref [1.1] (Section 3.2).

2.4.5   Dominant Closed-Loop Poles

The feedback transfer function for loop D in Eq 2.4-36 can be simplified if the asymptotic gain crossover frequency ω_c is set equal to the geometric mean of the break frequencies ω_i and ω_f:

$$\omega_c = \sqrt{\omega_i \omega_f} \tag{2.4-43}$$

This is a reasonable gain setting for this loop. Let us define the ratio ω_c / ω_i as β. By Eq 2.4-43, this is equal to

$$\beta = \frac{\omega_c}{\omega_i} = \frac{\omega_f}{\omega_c} \tag{2.4-44}$$

When this is substituted into Eq 2.4-36, the expression $(s + \omega_c)$ can be factored from the denominator, and the feedback transfer function for loop D simplifies to

$$G_{ib(D)} = \frac{\beta \omega_c^2 (s + \omega_c / \beta)}{(s + \omega_c)[s^2 + (\beta - 1)\omega_c s + \omega_c^2]} \tag{2.4-45}$$

This feedback transfer function for loop D, and the feedback transfer function for loop C in Eq 2.4-34, both have a quadratic factor in the denominator. When these quadratic factors are expressed in terms of the natural frequency ω_n and damping ratio ζ, Eqs 2.4-34, -45 become

$$G_{ib(C)} = \frac{2\zeta \omega_n (s + \omega_n / 2\zeta)}{s^2 + 2\zeta \omega_n s + \omega_n^2} \tag{2.4-46}$$

$$G_{ib(D)} = \frac{(1 + 2\zeta)\omega_n^2 [s + \omega_n / (1 + 2\zeta)]}{(s + \omega_n)(s^2 + 2\zeta \omega_n s + \omega_n^2)} \tag{2.4-47}$$

The values of ω_n and ζ for these loops are

Loop C:

$$\omega_n = \sqrt{\omega_1 \omega_c} \tag{2.4-48}$$

$$\zeta = \tfrac{1}{2}(\omega_c \omega_n) = \tfrac{1}{2}\sqrt{\omega_c / \omega_1} \tag{2.4-49}$$

Loop D:

$$\omega_n = \omega_c \tag{2.4-50}$$

$$\zeta = \tfrac{1}{2}(\beta - 1) \tag{2.4-51}$$

Let us compare these transfer functions with that for loop B, which was given in Eq 2.4-16 and is repeated as follows:

$$G_{ib(B)} = \frac{\omega_n^2}{s^2 + 2\zeta\omega_n s + \omega_n^2} \tag{2.4-52}$$

The feedback transfer functions for each of the loops B, C, and D have a denominator factor of the form $(s^2 + 2\zeta\omega_n s + \omega_n^2)$. The roots of these factors are commonly called the *dominant closed-loop poles* of the feedback loop. The root-locus technique, which is often used to design feedback control loops, is based on the assumption that the transient response of a feedback loop can be characterized reasonably well by setting the damping ratio and natural frequency of the dominant closed-loop poles. The damping ratio of this dominant pole pair is assumed to be a good criterion for the stability of the loop. Let us test this hypothesis by comparing the damping ratio of the closed-loop poles with the peak overshoot of the step response for loops B, C, and D.

This issue is studied in Ref [1.1] (Section 2.4). The results are shown in Fig 2.4-13, which gives plots of the percent overshoot of the step response versus the damping ratio of the dominant closed-loop poles for loops B, C, and D.

A peak overshoot of 25% is a good practical stability criterion for most feedback control loops. Figure 2.4-13 shows that for 25% overshoot the damping ratio of the dominant closed-loop poles is 0.4 for loop B, 0.6 for loop C, and 1.0 for loop D. (For 25% overshoot, loop D has a triple-order pole at $s = -\omega_c$.) This indicates that the damping ratio of the closed-loop poles is a poor criterion for stability.

Loop B, the basic second-order loop, is generally used as a design model when the root-locus technique is applied. If one desires a peak overshoot of 25% for the step response, an examination of the step response for this basic second-order feedback loop would suggest that the damping ratio of the dominant closed-loop poles should be 0.4. However, if the feedback loop that is being designed is like loop D, this criterion would result in a peak overshoot of nearly 50% rather than 25%.

Thus, the dynamic performance of practical feedback control loops cannot be constrained effectively by specifying the dominant closed-loop pole pair. Loop C differs from the basic second-order loop B in that it also has a zero near gain crossover, and loop D also has a real pole and a real zero near gain

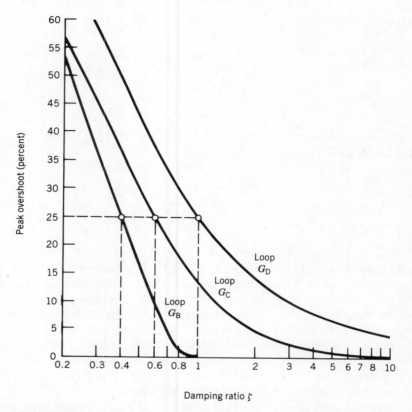

Figure 2.4-13 Plots of peak overshoot of step response versus damping ratio of dominant closed-loop poles, for loops G_B, G_C, and G_D.

crossover. All of these poles and zeros must be considered in order to characterize the response adequately.

The way out of this problem is to base the design on frequency response rather than on poles and zeros. It is easy to obtain the frequency response plot of the magnitude of G_{ib} for any feedback loop. The maximum value of the G_{ib} magnitude plot is an effective stability criterion.

Figure 2.3-8 gave general plots of percent step-response peaking versus percent peaking of the $|G_{ib}|$ frequency response for a feedback control loop. All of the plots for practical feedback control loops lie within the cross-hatched region of the figure. For 25% peak overshoot of the step response, the value of Max$|G_{ib}|$ varies approximately from 1.25 to 1.35 for practical feedback control loops. *Therefore, frequency response data, when appropriately applied, provides a much better basis for feedback control design than information based on the dominant closed-loop poles.*

2.5 EXPRESSING A LOOP TRANSFER FUNCTION IN TERMS OF ITS ASYMPTOTIC GAIN CROSSOVER FREQUENCY ω_c

As was indicated in Section 2.4.4, essentially all practical feedback control loops have a broad frequency region, at or near gain crossover, where the magnitude asymptote slope is -1. The frequency where this asymptote of -1 slope (extended if necessary) crosses unity loop gain is the asymptotic gain crossover frequency, denoted ω_c.

A magnitude plot of the loop transfer function for loop B was shown in Fig 2.4-4b for the condition $\omega_c < \omega_f$. If ω_c is increased to be greater than ω_f, the plot has the form of Fig 2.5-1. In this case, the magnitude asymptote of -1 slope must be extended (as shown by the dashed line) to reach unity loop gain. The frequency where this extension crosses unity gain is the asymptotic gain crossover frequency ω_c.

In Fig 2.5-1, the magnitude asymptote plot has a slope of -2 where it crosses unity loop gain. The frequency where this magnitude asymptote of -2 slope crosses unity gain should *not* be confused with the asymptotic gain crossover frequency.

In a loop such as this, where the -1 magnitude asymptote must be extended to reach unity loop gain, the value of ω_c differs significantly from the true gain crossover frequency ω_{gc}. Nevertheless, even for such cases these two parameters rarely differ by more than a factor of 1.5. Thus, one can generally

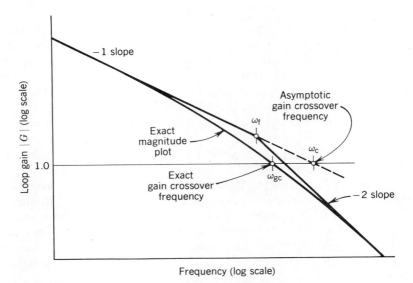

Figure 2.5-1 Definition of asymptotic gain crossover frequency when asymptote slope at gain crossover is greater than -1.

assume that ω_{gc} is within the limits

$$\omega_c/1.5 < \omega_{gc} < 1.5\omega_c \tag{2.5-1}$$

This magnitude asymptote of -1 logarithmic slope is proportional to ω^{-1} and, by definition, is equal to unity at $\omega = \omega_c$. Hence it is a plot of the function ω_c/ω. In terms of the transfer function, this asymptote corresponds to $\omega_c/j\omega$ or ω_c/s.

In all of the loop transfer functions considered in Section 2.4, the expression for G is written so that the gain constant is ω_c. This approach greatly simplifies system design, because the asymptotic gain crossover frequency ω_c is always approximately equal to the true gain crossover frequency ω_{gc}, which is the primary dynamic parameter of a feedback control loop.

To understand how a loop transfer function is written so that ω_c is the gain constant, consider again loop E, which has the loop transfer function

$$G_E = \frac{\omega_c\left\{1 + \dfrac{\omega_i}{j\omega}\right\}}{j\omega\left\{1 + \dfrac{\omega_i/\alpha}{j\omega}\right\}\left\{1 + \dfrac{j\omega}{\omega_f}\right\}} \tag{2.5-2}$$

This was given earlier (in terms of s) in Eq 2.4-39. The factors in this transfer function are written in two different forms: high-frequency factors, with break frequencies above the frequency region of the ω_c/ω asymptote, are written in the form $(1 + j\omega/\omega_x)$; and low-frequency factors, with break frequencies below the frequency region of the ω_c/ω asymptote, are written in the inverted form $(1 + \omega_x/j\omega)$. In Eq 2.5-2, the break frequencies of loop G_E at ω_i and ω_i/α apply to low-frequency factors, and the break frequency ω_f applies to a high-frequency factor.

A high-frequency factor $(1 + j\omega/\omega_x)$ approximates unity at low frequency, and so its magnitude asymptote is unity at frequencies below the break frequency ω_x. A low-frequency factor $(1 + \omega_x/j\omega)$ approximates unity at high frequency, and so its magnitude asymptote is unity at frequencies above the break frequency ω_x. Hence, in the mid-frequency region near gain crossover, the magnitude asymptotes of low-frequency as well as high-frequency factors are all unity. For G_E of Eq 2.5-2, this means that near ω_c the magnitude asymptotes of all of the factors in braces are unity, and so the magnitude asymptote of G_E near ω_c is determined by the transfer function $\omega_c/j\omega$, the magnitude of which is ω_c/ω.

The general principle for expressing a loop transfer function in terms of ω_c is as follows. For a feedback loop with real left-half plane poles and zeros, the loop transfer function is written as

$$G = \frac{\omega_c}{s}\left[\frac{(1 + \omega_{Lfz}/s) \cdots}{(1 + \omega_{Lfp}/s) \cdots}\right]\left[\frac{(1 + s/\omega_{hfz}) \cdots}{(1 + s/\omega_{hfp}) \cdots}\right]$$ (2.5-3)

The parameters represent

ω_c = asymptotic gain crossover frequency

ω_{Lfz} = break frequency for a low-frequency zero

ω_{Lfp} = break frequency for a low-frequency pole

ω_{hfz} = break frequency for a high-frequency zero

ω_{hfp} = break frequency for a high-frequency pole

The multiplying factor ω_c/s in Eq 2.5-3 is the transfer function that corresponds to the ω_c/ω asymptote. If the feedback loop has complex left-half plane poles and/or zeros, the corresponding factors are written in the forms:

High-frequency pole or zero;

$$\left[1 + 2\zeta(s/\omega_n) + (s/\omega_n)^2\right]$$ (2.5-4)

Low-frequency pole or zero;

$$\left[1 + 2\zeta(\omega_n/s) + (\omega_n/s)^2\right]$$ (2.5-5)

The break frequencies for low-frequency poles and zeros are, by definition, below the frequency region of the ω_c/ω asymptote, and those for high-frequency poles and zeros are above the frequency region of that asymptote. Hence, the magnitude asymptotes of the factors for all of these poles and zeros are unity in the frequency region of the ω_c/ω asymptote.

In Fig 2.5-2, magnitude asymptote plots are shown for a low-frequency zero, a low-frequency pole, a high-frequency zero, and a high-frequency pole. The corresponding transfer functions are also given. ·

By applying these plots of Fig 2.5-2, the magnitude asymptotes for G_E of Eq 2.5-2 are constructed in Fig 2.5-3. Curve (1) is the plot of ω_c/ω. This is constructed by locating point (A) at the frequency ω_c at a loop gain of 0 dg, and point (B) at the frequency $0.1\omega_c$ at a loop gain of 10 dg. The line (1) drawn through points (A) and (B) is the ω_c/ω asymptote, which has a slope of -10 dg/decade and is unity at $\omega = \omega_c$. Curve (2) is the magnitude asymptote plot of the low-frequency pole factor $1/[1 + (\omega_i/\alpha)/s]$, which follows diagram b in Fig 2.5-2. Curve (3) is the magnitude asymptote plot of the low-frequency zero factor $(1 + \omega_i/s)$, which follows diagram a in Fig 2.5-2. Curve (4) is the magnitude asymptote plot of the high-frequency pole

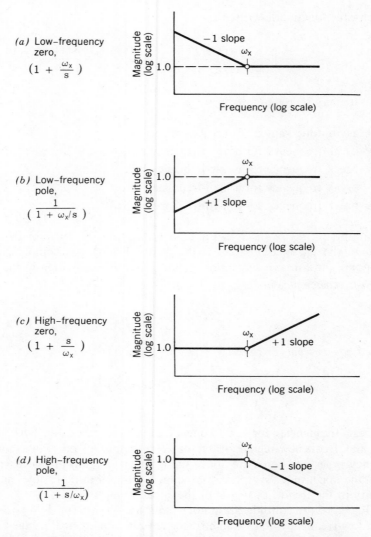

Figure 2.5-2 Magnitude asymptote plots of loop transfer function factors for low-frequency and high-frequency zeros and poles.

factor $1/(1 + s/\omega_f)$, which follows diagram d in Fig 2.5-2. Curves (1) to (4) are summed to form the dashed curve (5), which is the magnitude asymptote plot for the loop transfer function G_E. This is the same as the plot shown earlier in Fig 2.4-11b.

The conventional method for writing a loop transfer function is to express all of the factors in the form $(1 + j\omega/\omega_x)$, so that they all approximate unity at low frequency. To convert from the expressions given above to the conventional form, the factors for low-frequency poles and zeros are modified as follows:

$$\left(1 + \frac{\omega_x}{s}\right) = \frac{\omega_x}{s}\left(\frac{s}{\omega_x} + 1\right) = \frac{\omega_x}{s}\left(1 + \frac{s}{\omega_x}\right) \qquad (2.5\text{-}6)$$

The loop transfer function for loop G_E given in Eq 2.5-2 is as follows when written in terms of s:

$$G_E = \frac{\omega_c\left(1 + \dfrac{\omega_i}{s}\right)}{s\left(1 + \dfrac{\omega_i/\alpha}{s}\right)\left(1 + \dfrac{s}{\omega_f}\right)} \qquad (2.5\text{-}7)$$

Applying Eq 2.5-6 to the low-frequency zero and pole of G_E gives

$$G_E = \frac{\omega_c\left(\dfrac{\omega_i}{s}\right)\left(1 + \dfrac{s}{\omega_i}\right)}{s\left(\dfrac{\omega_i/\alpha}{s}\right)\left(1 + \dfrac{s}{\omega_i/\alpha}\right)\left(1 + \dfrac{s}{\omega_f}\right)}$$

$$= \frac{K_1\left\{1 + \dfrac{s}{\omega_i}\right\}}{s\left\{1 + \dfrac{s}{\omega_i/\alpha}\right\}\left\{1 + \dfrac{s}{\omega_f}\right\}} \qquad (2.5\text{-}8)$$

The parameter K_1 is called the velocity constant, which is equal to

$$K_1 = \alpha\omega_c \qquad (2.5\text{-}9)$$

Based on the conventional form of Eq 2.5-8, the magnitude asymptote plot of G_E is constructed in Fig 2.5-4. At very low frequencies, all of the factors of Eq 2.5-8 in braces approximate unity, and so G_E approximates K_1/s, and the loop gain approximates K_1/ω. Hence, the low-frequency magnitude asymptote of G_E is a plot of K_1/ω. This asymptote is constructed by locating point (A) at zero dg gain at the frequency $\omega = K_1 = \alpha\omega_c$, and point ($B$) at 10 dg, at the frequency $\omega = 0.1K_1$. Curve (1), which is drawn through points (A) and (B), has a slope of -10 dg/dec and is equal to unity at $\omega = K_1$. Hence, curve (1) is a plot of K_1/ω, which is the low-frequency asymptote of G_E. Curves (2) and (4) are the magnitude asymptote plots for the poles of break frequency ω_i/α and ω_f, which follow the plot for a high-frequency pole in Fig 2.5-2d. Curve (3) is the magnitude asymptote plot for the zero of break frequency ω_i, which follows the plot for a high-frequency zero in Fig 2.5-2c. Curves (1) to (4)

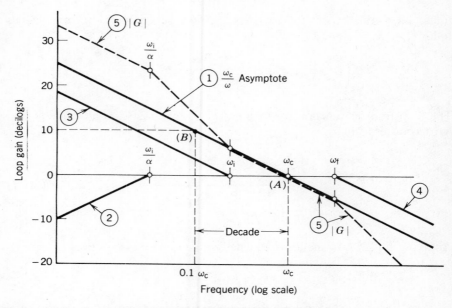

Figure 2.5-3   Construction of magnitude asymptote plot for loop E, based on ω_c/ω asymptote.

are summed to form the dashed curve (5), which is the magnitude asymptote plot of the loop transfer function G_E. This is the same as the plots shown earlier in Figs 2.4-11b and 2.5-3.

The general form of a loop transfer function written in the conventional manner is as follows, for a loop that has real left-half-plane poles and zeros:

$$G = \frac{K_n\{(1 + s/\omega_z) \cdots \}}{s^n\{(1 + s/\omega_p) \cdots \}} \qquad (2.5\text{-}10)$$

where ω_z and ω_p are the break frequencies for a zero and a pole. When G has complex poles or zeros, the corresponding factors are expressed in the form of Eq 2.5-4, which is unity at zero frequency. The parameter n is the number of integrations in the loop at zero frequency (or the number of poles at the origin). The low-frequency asymptote of G has a slope of $-n$ at low frequencies. The constant K_n is commonly called by the following names for $n = 0, 1, 2$:

$$n = 0: \quad K_0 = \text{position constant}$$

$$n = 1: \quad K_1 = \text{velocity constant (sec}^{-1})$$

$$n = 2: \quad K_2 = \text{acceleration constant (sec}^{-2})$$

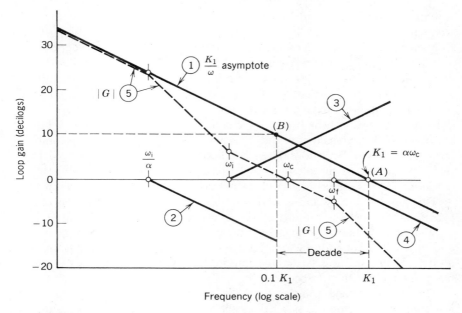

Figure 2.5-4 Construction of magnitude asymptote plot for loop E, based on low-frequency asymptote K_1/ω.

As shown, the velocity constant has units of $\sec^{-1}$, and the acceleration constant has units of $\sec^{-2}$. The position constant is nondimensional.

As was shown in Eq 2.4-35, the loop transfer function for loop D is expressed as follows in terms of ω_c:

$$G_D = \frac{\omega_c\left(1 + \dfrac{\omega_i}{s}\right)}{s\left(1 + \dfrac{s}{\omega_f}\right)} \tag{2.5-11}$$

When this is expressed in the conventional form, it becomes

$$G_D = \frac{K_2\left(1 + \dfrac{s}{\omega_i}\right)}{s^2\left(1 + \dfrac{s}{\omega_f}\right)} \tag{2.5-12}$$

where K_2 is the acceleration constant, which is equal to

$$K_2 = \omega_i\omega_c \tag{2.5-13}$$

Equation 2.5-12 is formed by applying Eq 2.5-6 to the factor for the break frequency ω_i in Eq 2.5-11. Loop G_D has a double integration at zero frequency, and so its low-frequency asymptote has a logarithmic slope of -2.

When the loop transfer functions for loops E and D are expressed in the conventional form, as shown in Eqs 2.5-8 and -12, the equations are radically different. The gain constants of the equations, which are K_1 for loop E and K_2 for loop D, do not even have the same units. This would suggest that the dynamic responses of the two loops should also be radically different, but this is not so. If the values for ω_c and ω_i are the same for the two loops, and α is large, it is very difficult to differentiate between the two loops using dynamic tests. Magnitude asymptote plots of G for these two loops were shown in Figs 2.4-10b, -11b. These are the same except at very low frequencies below ω_i/α.

This similarity between loops D and E is obvious when the loop transfer functions are written in terms of ω_c, as was done in Eqs 2.5-7 and -11. The only difference between the two transfer functions is that loop E has the additional denominator factor $[1 + (\omega_i/\alpha)/s]$. The gain constants for both loop transfer functions is the asymptotic gain crossover frequency ω_c, which has fundamental significance.

The fact that loop transfer functions are usually written in terms of the low-frequency asymptote has led to the mistaken belief that the associated constants K_0, K_1, K_2, etc. have fundamental significance. As shown below, these constants characterize the "steady-state" error responses of the loops to inputs of "very low frequency". However, as explained in Ref [1.1] (Section 3.2), an input of very low frequency often has little physical significance. Chapter 3 presents a general method for describing the response of a feedback loop to an arbitrary input, which avoids the confusing concept of the very low-frequency input.

At very low frequencies, G of Eq 2.5-10 approximates K_n/s^n. Hence, the approximate error transfer function at very low frequencies is

$$G_{ie} = \frac{1}{1 + G} \cong \frac{1}{1 + K_n/s^n} = \frac{s^n}{s^n + K_n} \tag{2.5-14}$$

For $n > 0$, the term s^n in the denominator of Eq 2.5-11 can be neglected at very low frequencies. For $n = 0$, the term s^n is unity in the numerator and denominator. Hence, at very low frequencies the approximation of Eq 2.5-14 becomes

$$G_{ie} \cong \frac{1}{1 + K_0} \quad \text{for } n = 0 \tag{2.5-15}$$

$$G_{ie} \cong \frac{s^n}{K_n} \quad \text{for } n > 0 \tag{2.5-16}$$

Since $G_{ie} X_e/X_i$, the approximate transform of the error X_e at very low frequencies is

$$X_e = G_{ie}X_i \cong \frac{X_i}{1 + K_0} \quad \text{for } n = 0 \quad (2.5\text{-}17)$$

$$X_e = G_{ie}X_i \cong \frac{s^n X_i}{K_n} \quad \text{for } n > 0 \quad (2.5\text{-}18)$$

The inverse transforms of Eqs 2.5-17, -18 at very low frequencies are

$$x_e \cong \frac{x_i}{1 + K_0} \quad \text{for } n = 0 \quad (2.5\text{-}19)$$

$$x_e \cong \frac{1}{K_n} \frac{d^n x_i}{dt^n} \quad \text{for } n > 0 \quad (2.5\text{-}20)$$

Apply to Eq 2.5-20 the values for constants K_1, K_2 for loops E and D in Eqs 2.5-9, -13. This gives the following approximations that hold at very low frequencies:

$$\text{Loop } E: \quad x_e \cong \frac{1}{K_1} \frac{dx_i}{dt} \cong \frac{1}{\alpha\omega_c} \frac{dx_i}{dt} \quad (2.5\text{-}21)$$

$$\text{Loop } D: \quad x_e \cong \frac{1}{K_2} \frac{d^2 x_i}{dt^2} \cong \frac{1}{\omega_i\omega_c} \frac{d^2 x_i}{dt^2} \quad (2.5\text{-}22)$$

Thus, the error response of loop D to an input of very low frequency is proportional to the first derivative of the input dx_i/dt (the input velocity), and the error for loop E is proportional to the second derivative $d^2 x_i/dt^2$ (the input acceleration). However, this result often has little practical significance because an input of such low frequency may rarely, if ever, occur.

As will be shown in Chapter 3, for loops D and E, the approximate error responses to an input that is low in frequency relative to the gain crossover frequency are:

$$\text{Loop } E: \quad x_e \cong \frac{1}{\alpha\omega_c} \frac{dx_i}{dt} + \frac{1}{\omega_i\omega_c} \frac{d^2 x_i}{dt^2} \quad (2.5\text{-}23)$$

$$\text{Loop } D: \quad x_e \cong \frac{1}{\omega_i\omega_c} \frac{d^2 x_i}{dt^2} \quad (2.5\text{-}24)$$

The error response for loop D in Eq 2.5-24 is the same as the very low-frequency error of Eq 2.5-22, which is proportional to input velocity. However, Eq 2.5-23 shows that loop E has two error terms: the very low-frequency error term in Eq 2.5-21, which is proportional to velocity, plus a term proportional to acceleration, which is equal to the error for loop D.

If the frequency of the input is low enough, the error term for loop E that is proportional to the input velocity dx_i/dt predominates over the term proportional to the input acceleration d^2x_i/dt^2. Hence, the error response for such an input approximately reduces to the expression of Eq 2.5-21.

However (if α is reasonably large), the control system often saturates in velocity before this very low-frequency condition is reached, except for input accelerations that are so small the response is clouded by noise and nonlinearities. Thus, the very low-frequency conditions characterized by the steady-state error constants K_0, K_1, K_2, etc. may never be experienced in practice. The discussion presented in Chapter 3 will clarify these issues and provide a firm foundation for specifying the low-frequency response of a feedback control system.

2.6   APPROXIMATION OF FEEDBACK AND ERROR TRANSFER FUNCTIONS G_{ib} AND G_{ie}

2.6.1   Approximation Method

Exact calculations of the feedback and error transfer functions G_{ib} and G_{ie} from the loop transfer function G are quite complicated. However, good approximations of these transfer functions can be obtained quite simply if the loop has good stability. The function $(1 + G)$ can be approximated by

$$(1 + G) \cong G \quad \text{for } |G| \gg 1 \tag{2.6-1}$$

$$(1 + G) \cong 1 \quad \text{for } |G| \ll 1 \tag{2.6-2}$$

This yields the following approximations for G_{ib} and G_{ie}:

$$G_{ib} = \frac{G}{1 + G} \cong \begin{cases} G/G = 1 & \text{for } |G| \gg 1 \\ G/1 = G & \text{for } |G| \ll 1 \end{cases} \tag{2.6-3}$$

$$G_{ie} = \frac{1}{1 + G} \cong \begin{cases} 1/G & \text{for } |G| \gg 1 \\ 1/1 = 1 & \text{for } |G| \ll 1 \end{cases} \tag{2.6-4}$$

If the feedback loop has good stability, these approximations can be extended to the frequency where $|G| = 1$ to give

$$G_{ib} = \frac{G}{1 + G} \cong \begin{cases} 1 & \text{for } |G| > 1 \\ G & \text{for } |G| < 1 \end{cases} \tag{2.6-5}$$

$$G_{ie} = \frac{1}{1 + G} \cong \begin{cases} 1/G & \text{for } |G| > 1 \\ 1 & \text{for } |G| < 1 \end{cases} \tag{2.6-6}$$

These approximations have significant error in the region near gain crossover (where $|G|$ is close to unity). However, if the loop has good stability, the error

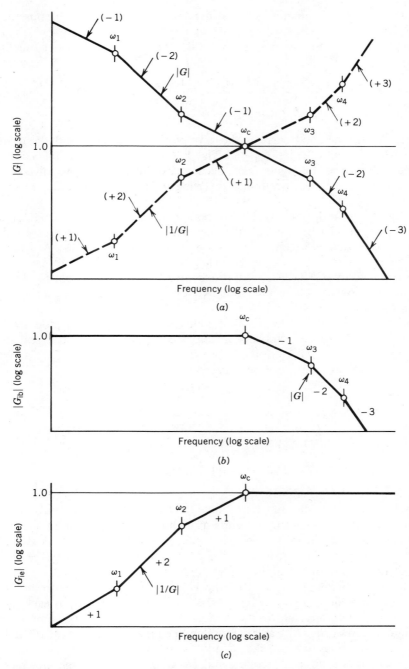

Figure 2.6-1 Development of approximate magnitude asymptote plots of G_{ib} and G_{ie}: (a) plots of G and 1/G; (b) plot of approximate G_{ib}; (c) plot of approximate G_{ie}.

is within acceptable bounds for most dynamic analyses when the approximations are used properly.

Let us apply these approximations to the following loop transfer function:

$$G = \frac{\omega_c(1 + \omega_2/s)}{s(1 + \omega_1/s)(1 + s/\omega_3)(1 + s/\omega_4)} \qquad (2.6\text{-}7)$$

In Fig 2.6-1a the solid curve is the asymptotic plot of $|G|$, and the dashed curve is the asymptote plot of $|1/G|$. The $|1/G|$ plot is formed by reflecting the $|G|$ plot about the unity-gain (zero decilog) axis to form the negative of the plot. Since the scale is logarithmic, inverting the plot generates the reciprocal of $|G|$.

Diagrams b and c in Fig 2.6-1 show the approximate plots of $|G_{ib}|$ and $|G_{ie}|$. At frequencies less than ω_c, the loop gain $|G|$ is greater than unity, and so $|G_{ib}|$ is approximated by unity and $|G_{ie}|$ is approximated by $|1/G|$. At frequencies greater than ω_c, the loop gain $|G|$ is less than unity, and so $|G_{ib}|$ is approximated by $|G|$ and $|G_{ie}|$ is approximated by unity. Note that G_{ib} has the frequency response of a lowpass filter, while G_{ie} has the frequency response of a highpass filter.

Assume that all of the break frequencies shown in the plots of the approximate G_{ib} and G_{ie} transfer functions correspond to negative real zeros and poles. The following are the approximate equations for the G_{ib} and G_{ie} transfer functions, which can be written by inspection:

$$G_{ib} \cong \frac{1}{(1 + s/\omega_c)(1 + s/\omega_3)(1 + s/\omega_4)} \qquad (2.6\text{-}8)$$

$$G_{ie} \cong \frac{(1 + \omega_1/s)}{(1 + \omega_2/s)(1 + \omega_c/2)} \qquad (2.6\text{-}9)$$

To write these transfer functions, note that G_{ib} approaches unity at low frequency, while G_{ie} approaches unity at high frequency. Therefore, all of the factors of G_{ib} are put in the form $(1 + s/\omega_x)$ so that they approximate unity at low frequency. All of the factors of G_{ie} are expressed in the form $(1 + \omega_x/s)$ so that they approximate unity at high frequency.

All of the break frequencies of the G_{ib} transfer function correspond to poles, because the G_{ib} plot in Fig 2.6-1b is concave downward at each break frequency. Therefore, all of the factors of Eq 2.6-8 for G_{ib} are in the denominator. For the G_{ie} transfer function, the break frequencies at ω_2 and ω_c correspond to poles, because the G_{ie} plot is concave downward at ω_2 and ω_c; the break frequency at ω_1 corresponds to a zero, because the G_{ie} plot is concave upward at ω_1. The factors of G_{ie} for the poles at ω_2 and ω_c are in the denominator; whereas the factor of G_{ie} for the zero at ω_1 is in the numerator.

2.6.2 Closed-Loop Poles of G_{ib} and G_{ie}

The approximate equations for G_{ib} and G_{ie} in Eqs 2.6-8, -9 may appear to be inconsistent because they do not have the same poles. The exact G_{ib} and G_{ie} transfer functions must have the same poles, which are the closed-loop poles of the feedback loop. The reason for this apparent inconsistently is that G_{ib} has low-frequency zeros which approximately cancel its low-frequency closed-loop poles, and G_{ie} has high-frequency zeros which approximately cancel its high-frequency closed-loop poles.

To demonstrate this principle, convert all of the factors of Eqs 2.6-8, -9 to the form $(s + \omega_x)$ in the following manner:

$$1 + \frac{s}{\omega_x} = \frac{\omega_x + s}{\omega_x} = \frac{1}{\omega_x}(s + \omega_x) \qquad (2.6\text{-}10)$$

$$1 + \frac{\omega_x}{s} = \frac{s + \omega_x}{s} = \frac{1}{s}(s + \omega_x) \qquad (2.6\text{-}11)$$

Applying Eqs 2.6-10, -11 to Eqs 2.6-7 to -9 gives the following exact transfer function for G and approximate transfer functions for G_{ib} and G_{ie}:

$$G = \frac{\omega_c \omega_3 \omega_4 (s + \omega_2)}{s(s + \omega_1)(s + \omega_3)(s + \omega_4)} \qquad (2.6\text{-}12)$$

$$G_{ib} \cong \frac{\omega_c \omega_3 \omega_4}{(s + \omega_c)(s + \omega_3)(s + \omega_4)} \qquad (2.6\text{-}13)$$

$$G_{ie} \cong \frac{s(s + \omega_1)}{(s + \omega_2)(s + \omega_c)} \qquad (2.6\text{-}14)$$

Multiply the numerator and denominator of G_{ib} in Eq 2.6-13 by $(s + \omega_2)$ and multiply the numerator and denominator of G_{ie} in Eq 2.6-14 by $(s + \omega_3)(s + \omega_4)$. The approximations for G_{ib} and G_{ie} become

$$G_{ib} \cong \frac{\omega_c \omega_3 \omega_4 (s + \omega_2)}{(s + \omega_2)(s + \omega_c)(s + \omega_3)(s + \omega_4)} \qquad (2.6\text{-}15)$$

$$G_{ie} \cong \frac{s(s + \omega_1)(s + \omega_3)(s + \omega_4)}{(s + \omega_2)(s + \omega_c)(s + \omega_3)(s + \omega_4)} \qquad (2.6\text{-}16)$$

These approximations for G_{ib}, G_{ie} have poles at: $s = -\omega_2, -\omega_c, -\omega_3, -\omega_4$. The actual closed-loop poles are shifted from these approximate poles. Let us denote the exact closed-loop poles as: $s = -\omega_2', -\omega_c', -\omega_3', -\omega_4'$, where $\omega_2', \omega_c', \omega_3', \omega_4'$ may be complex. The exact equations for G_{ib}, G_{ie} can then be expressed as

$$G_{ib} = \frac{\omega_c \omega_3 \omega_4 (s + \omega_2)}{(s + \omega_2')(s + \omega_c')(s + \omega_3')(s + \omega_4')} \tag{2.6-17}$$

$$G_{ie} = \frac{s(s + \omega_1)(s + \omega_3)(s + \omega_4)}{(s + \omega_2')(s + \omega_c')(s + \omega_3')(s + \omega_4')} \tag{2.6-18}$$

By Eqs 2.6-5, -6 the ratio G_{ib}/G_{ie} is equal to G. Dividing Eq 2.6-17 by Eq 2.6-16 gives

$$G = \frac{G_{ie}}{G_{ib}} = \frac{\omega_c \omega_3 \omega_4 (s + \omega_2)}{s(s + \omega_1)(s + \omega_3)(s + \omega_4)} \tag{2.6-19}$$

This is the same as the expression for G given in Eq 2.6-49. Hence, the approximate G_{ib}, G_{ie} transfer functions are consistent.

In Eqs 2.6-17, -18 the closed-loop poles, which are represented as $s = -\omega_2'$, $-\omega_c'$, $-\omega_3'$, $-\omega_4'$, are approximately equal to $s = -\omega_2$, $-\omega_c$, $-\omega_3$, $-\omega_4$. This is the basis for the G_{ib} and G_{ie} approximations. It is justified by the principle that the closed-loop poles of a feedback control loop can be approximated by:

1.   The zeros of G that have break frequencies lower in frequency than the ω_c/s asymptote (the low-frequency zeros).
2.   The poles of G that have break frequencies higher in frequency than the ω_c/s asymptote (the high-frequency poles).
3.   The approximate pole: $s = -\omega_c$.

These approximations are developed in Ref [1.2] (Chapter 13).

The loop transfer function, which was given in Eq 2.6-19, has one low-frequency zero of break frequency ω_2 and two high-frequency poles of break frequency ω_3, ω_4. Hence the approximate closed-loop poles are

$$\text{Low-frequency zeros:} \quad s = -\omega_2 \tag{2.6-20}$$

$$\text{High-frequency poles:} \quad s = -\omega_3, -\omega_4 \tag{2.6-21}$$

$$\text{Pole at gain crossover:} \quad s = -\omega_c \tag{2.6-22}$$

These are the approximate closed-loop poles given in the approximations for G_{ib} and G_{ie} in Eqs 2.6-15, -16.

In Eqs 2.6-15, -16, G_{ib} has a zero at $s = -\omega_2$, which approximately cancels a closed-loop pole at $s = -\omega_2'$; and G_{ie} has zeros at $s = -\omega_3$, $-\omega_4$, which approximately cancel closed-loop poles at $s = -\omega_3'$, $-\omega_4'$. These closed-loop poles are only approximately canceled because the closed-loop poles are shifted relative to the corresponding low-frequency zeros and high-frequency poles of G. As is shown in Ref [1.2] (Chapter 13), the per-unit shift of a closed-loop pole from a high-frequency pole of G is approximately equal to the magnitude asymptote of G at the corresponding break frequency. For example, if the G asymptote value is 0.1, the shift is approximately 0.1 times the break frequency. Similarly, the per-unit shift of a closed-loop pole from a low-frequency zero of G is approximately equal to the reciprocal of the magnitude asymptote of G at the corresponding break frequency.

The equations for the approximate G_{ib} and G_{ie} transfer functions have been derived from magnitude asymptote plots, based on the assumption that each break frequency corresponds to a negative real pole of zero. When a pole or zero of G is complex or in the right-half plane, this does not work. In Ref [1.1] (Section 3.3.5) a more general approach is presented for developing the approximate equations for G_{ib} and G_{ie} that avoids this limitation.

2.7 APPROXIMATION OF FEEDBACK LOOP RESPONSE TO LOW-FREQUENCY INPUT

2.7.1 Step and Ramp Responses of Loop with Single Integration at Low Frequencies

A feedback control system operates accurately only when it is following a low-frequency input. Although feedback loop transfer functions may be very complicated in the high-frequency region, they are generally quite simple at low frequencies. Hence, the response of a feedback control loop to a low-frequency input can be readily approximated.

This section develops simple principles to explain the response of a feedback control system to a low-frequency input. A more detailed approach using error coefficients is presented in Chapter 3.

Consider again the loop transfer functions G_A and G_B studied in Section 2.4.3:

$$G_A = \frac{\omega_c}{s} \tag{2.7-1}$$

$$G_B = \frac{\omega_c}{s(1 + s/\omega_f)} \tag{2.7-2}$$

Magnitude plots of these loop transfer functions are shown by the solid curves in Fig 2.7-1. The magnitude plot for G_A is also the magnitude asymptote plot. The dashed curve is the magnitude asymptote plot of G_B. The break frequency ω_f for G_B is set equal to $0.7225\omega_c$, which is the condition at which $\text{Max}|G_{ib}| = 1.300$. For this setting, the asymptotic and exact gain crossover frequencies for G_B are related by

$$\omega_c = 1.40\omega_{gc} \tag{2.7-3}$$

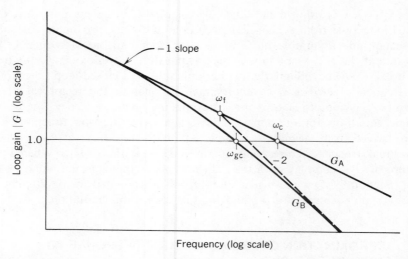

Figure 2.7-1   Magnitude asymptote plots of simple loop G_A and second-order loop G_B.

The error and feedback-signal transient responses of loops G_A and G_B to a unit step and a unit ramp are shown in Figs 2.7-2, -3. The ramp is typical of a low-frequency input signal, and so the ramp response helps to explain the response to a general low-frequency input.

In the steady state, the feedback signal responses for a ramp input, shown in Figs 2.7-2c, -3d, lag the input by a time delay equal to

$$\text{Steady-state ramp time delay} = \frac{1}{\omega_c} \qquad (2.7\text{-}4)$$

The maximum time delay between the input and feedback signals is approximately equal to

$$\text{Maximum ramp time delay} \cong \frac{1}{\omega_{gc}} \qquad (2.7\text{-}5)$$

For loop G_A, ω_{gc} is equal to ω_c, and so the maximum time delay is also equal to the final time delay $1/\omega_c$. For loop G_B, ω_{gc} is lower than ω_c by the factor 1.4, and so the maximum time delay is about 40% greater than the final time delay. The maximum ramp time delay for G_B is exactly equal to

$$\text{Maximum ramp time delay for } G_B = \frac{1.04}{\omega_{gc}} = \frac{1.46}{\omega_c} \qquad (2.7\text{-}6)$$

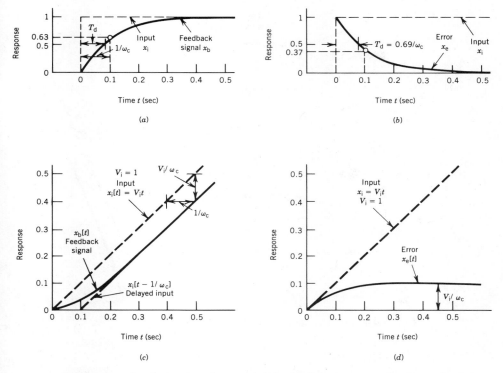

Figure 2.7-2 Feedback-signal and error-signal transient responses for loop G_A: (*a*) feedback response to unit step; (*b*) error response to unit step; (*c*) feedback response to unit ramp; (*d*) error response to unit ramp.

During a time delay, the ramp input changes by an amount equal to the slope (or velocity) V_i of the ramp multiplied by the time delay. Hence, during the steady-state time delay $1/\omega_c$, the input changes by

$$\Delta x_i = (\text{time delay}) \, V_i = \frac{1}{\omega_c} V_i = \frac{V_i}{\omega_c} \qquad (2.7\text{-}7)$$

As illustrated in Fig 2.7-2*c*, this is equal to the steady-state error for a ramp:

$$\text{Steady-state} \, [x_e] = \Delta x_i = \frac{V_i}{\omega_c} \qquad (2.7\text{-}8)$$

Since the maximum time delay between the input x_i and the feedback signal

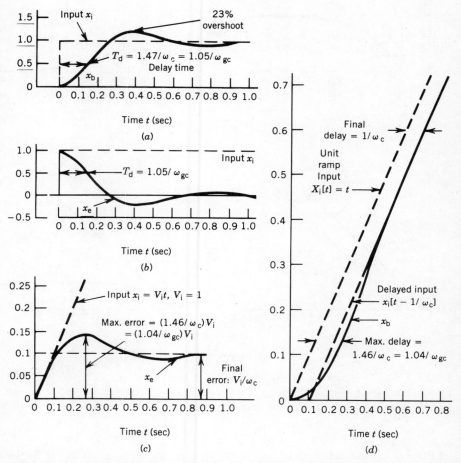

Figure 2.7-3 Feedback-signal and error-signal transient responses for second-order feedback loop G_B: (a) feedback response to unit step; (b) error response to unit step; (c) error response to unit ramp; (d) feedback response to unit ramp.

x_b is approximately $1/\omega_{gc}$, the maximum error for a ramp is approximately

$$\text{Max}[x_e] \cong \frac{V_i}{\omega_{gc}} \tag{2.7-9}$$

As shown in Fig 2.7-3c, the maximum ramp error for loop G_B is $1.04V_i/\omega_{gc}$. Figure 2.7-2d shows that the maximum ramp error for loop G_A is the final error, which is V_i/ω_c.

2.7.2   Approximating Feedback Loop by a Delay Line

Let us compare the transient responses of loops G_A and G_B with the responses of a delay line. Diagram a of Fig 2.7-4 shows the step response of a delay line of time delay T_d, and diagram b shows the response to a ramp input of velocity V. Let us define the difference between the input and output of the delay line as the delay line error. Diagram c shows the error response of the delay line to the ramp input.

The feedback and error responses for G_A and G_B in Figs 2.7-2, -3 to a ramp input are quite similar to the output and error responses in Fig 2.7-4 of the delay line to a ramp input, if the delay time T_d is set equal to $1/\omega_c$. In Fig

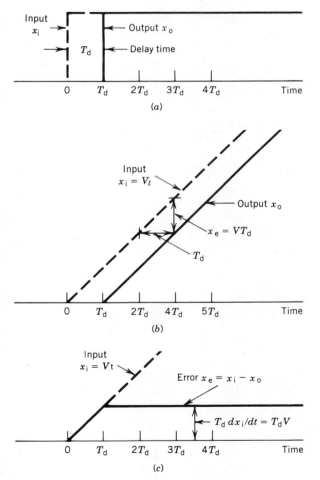

Figure 2.7-4   Step and ramp responses of ideal delay line: (a) step response; (b) ramp response; (c) error response to ramp.

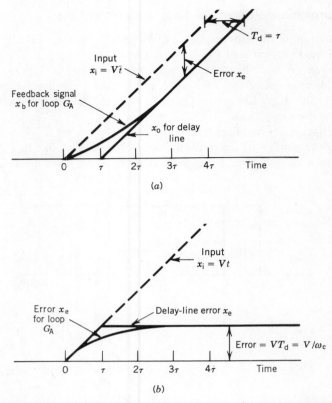

Figure 2.7-5   Comparison of output (feedback) and error responses to ramp input for simple feedback loop G_A and delay line: (a) feedback or output signal x_o; (b) error signal x_e.

2.7-5, diagram a compares the x_b response of G_A with the output responses of the delay line for the ramp input, and diagram b compares the error response of G_A with the error response of the delay line. The fact that the ramp response of the delay line is similar to that for loop G_A indicates that the frequency responses of the loop and delay line are nearly the same at low frequencies.

On the other hand, the frequency response of the delay line differs markedly from that of loop G_A at high frequencies. This high frequency difference becomes apparent by comparing the step responses. The step response of the delay line in Fig 2.7-4a is radically different from that of loop G_A in Fig 2.7-2a. Since a step is the derivative of a ramp, the step response is much more strongly affected by high frequency characteristics than the ramp response.

Let us now consider the responses of the feedback loops and the delay line to a low-frequency input of arbitrary shape, shown by the solid curve in Fig

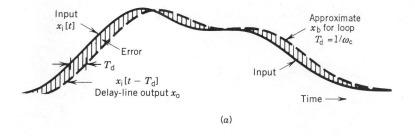

(a)

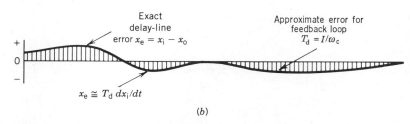

(b)

Figure 2.7-6 Exact response of delay line to an arbitrary low-frequency input, which is the approximate response for the feedback loop G_A: (a) arbitrary low-frequency input, and delay-line output; (b) error of delay line for arbitrary low-frequency input.

2.7-6a. The response to the delay line is the dashed curve, which is the input delayed by the time delay T_d of the delay line. This is also the approximate feedback response of loop G_A for $T_d = 1/\omega_c$.

For loop G_B, the maximum time delay is about 1.4 times the steady-state time delay. Consequently, the feedback signal response of loop G_B to the arbitrary input cannot be accurately approximated by simply delaying the input. Nevertheless, as a rough approximation this variation of time delay can be ignored, and so the dashed curve in Fig 2.7-6b also approximates the response for loop G_B.

As shown in Fig 2.7-5b, for both the delay line and feedback loop the error response to a ramp has a final value equal to $T_d V = (1/\omega_c)V$. Hence, the error for an arbitrary low-frequency input is approximately equal to

$$x_e \cong T_d \frac{dx_i}{dt} = \frac{1}{\omega_c} \frac{dx_i}{dt} \qquad (2.7\text{-}10)$$

where dx_i/dt is the derivative of the low-frequency input (the input velocity). This approximation is illustrated in Fig 2.7-6b.

This approximation can be derived in another manner by examining the signal-flow diagram of loop G_A in Fig 2.7-7a. The signal $\dot{X}_b$ is the Laplace transform of the derivative of the feedback signal dx_b/dt. The figure shows that $\dot{X}_b$ is proportional to the error signal X_e, as given by

$$\dot{X}_b = \omega_c X_e \qquad (2.7\text{-}11)$$

Taking the inverse transform of this gives

$$\frac{dx_b}{dt} = \omega_c x_e \qquad (2.7\text{-}12)$$

Solving for the error x_e gives

$$x_e = \frac{1}{\omega_c} \frac{dx_b}{dt} \qquad (2.7\text{-}13)$$

For a low-frequency input, the derivative of the feedback signal, dx_b/dt, is approximately equal to the derivative of the input, dx_i/dt. Hence, for a low-frequency input, Eq 2.7-13 can be approximated by

$$x_e = \frac{1}{\omega_c} \frac{dx_b}{dt} \cong \frac{1}{\omega_c} \frac{dx_i}{dt} \qquad (2.7\text{-}14)$$

This is the same as the approximation derived earlier in Eq 2.7-10.

2.7.3 Integral Network to Reduce Low-Frequency Error

At low frequencies, loops G_A and G_B approximate ω_c/s, which is a single integration. As has been shown, for such loops the low-frequency error is approximately equal to the input velocity divided by ω_c.

Much lower error for a low-frequency input can be achieved without increasing the gain crossover frequency by inserting an integral network into the loop, as shown in Fig 2.7-7. Diagram a shows the simple loop G_A, and diagram b shows loop G_C, which is formed by adding an integral network of transfer function $(1 + \omega_i/s)$ to loop G_A. Thus, the loop transfer function for loop G_C is

$$G_C = G_A\left(1 + \frac{\omega_i}{s}\right) = \frac{\omega_c(1 + \omega_i/s)}{s} \qquad (2.7\text{-}15)$$

Loop G_C was studied earlier in Section 2.4.4, which showed that the integral network produces a double integration at low frequencies and thereby increases the low-frequency loop gain. As indicated in diagram b, the integral network generates a correction signal X_{cor}, related to the error X_e by

$$X_{cor} = \frac{\omega_i}{s} X_e \qquad (2.7\text{-}16)$$

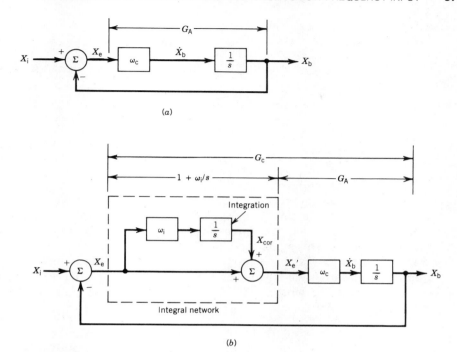

(a)

(b)

Figure 2.7-7 Signal-flow diagrams of feedback loops showing effect of integral network: (a) loop G_A; (b) loop G_C.

The inverse transform of this is

$$x_{cor} = \omega_i \int x_e \, dt \qquad (2.7\text{-}17)$$

Thus, the integral network correction signal x_{cor} is proportional to the integral of the error.

Figure 2.7-7b shows that $\dot{X}_b$ is proportional to the signal X_e', which is equal to the sum $(X_e + X_{cor})$. Thus,

$$\dot{X}_b = \omega_c X_e' = \omega_c(X_e + X_{cor}) \qquad (2.7\text{-}18)$$

Take the inverse transform of this and solve for x_e'. This gives

$$x_e' = x_e + x_{cor} = \frac{1}{\omega_c} \frac{dx_b}{dt} \qquad (2.7\text{-}19)$$

For a low-frequency input, the feedback-signal derivative dx_b/dt is approximately equal to the derivative of the input dx_i/dt. Hence, for a low-frequency input Eq 2.7-19 can be approximated by

$$x_e + x_{cor} = \frac{1}{\omega_c}\frac{dx_b}{dt} \cong \frac{1}{\omega_c}\frac{dx_i}{dt} \qquad (2.7\text{-}20)$$

Solving this for the error x_e gives

$$x_e \cong \frac{1}{\omega_c}\frac{dx_i}{dt} - x_{cor} = \frac{1}{\omega_c}\frac{dx_i}{dt} - \omega_i\int x_i\,dt \qquad (2.7\text{-}21)$$

The expression for the correction signal in Eq 2.7-17 was substituted for x_{cor}.

For a low-frequency input, Eq 2.7-21 shows that the correction signal x_{cor} subtracts from the error term $(1/\omega_c)(dx_i/dt)$ and thereby reduces the error. If the input velocity is constant, the correction signal x_{cor}, which is proportional to the integral of the error x_e, changes until the error x_e is reduced to zero. For a more general low-frequency input, the error x_e is not exactly zero, but nevertheless is much smaller than the correction signal x_{cor}. Hence, for a low-frequency input, the term x_e in Eq 2.7-20 can be neglected, and so the correction signal is approximated by

$$x_{cor} \cong \frac{1}{\omega_c}\frac{dx_b}{dt} \cong \frac{1}{\omega_c}\frac{dx_i}{dt} \qquad (2.7\text{-}22)$$

Comparing Eqs 2.7-14, -22 shows that for a low-frequency input the integral-network correction signal x_{cor} for loop G_C is approximately equal to the error signal x_e of loop G_A.

To approximate the error for loop G_C, differentiate Eq 2.7-17 and solve for x_e, to express x_e in terms of the correction signal x_{cor}. This gives

$$x_e = \frac{1}{\omega_i}\frac{dx_{cor}}{dt} \qquad (2.7\text{-}23)$$

Combining Eqs 2.7-22, -23 gives the following approximation of the error for a low-frequency input:

$$x_e = \frac{1}{\omega_c}\frac{dx_{cor}}{dt} \cong \frac{1}{\omega_i}\frac{d}{dt}\left[\frac{1}{\omega_c}\frac{dx_i}{dt}\right] = \frac{1}{\omega_c\omega_i}\frac{d^2x_i}{dt^2} \qquad (2.7\text{-}24)$$

Thus, for a low-frequency input, the error for the integral compensated loop G_C is approximately proportional to the second derivative of the input. This error is generally much smaller than that for loop G_A, which was approximated by Eq 2.7-10.

Equation 2.7-22 shows that for a low-frequency input the correction signal x_{cor} is approximately proportional to the input velocity. Hence, the integral-

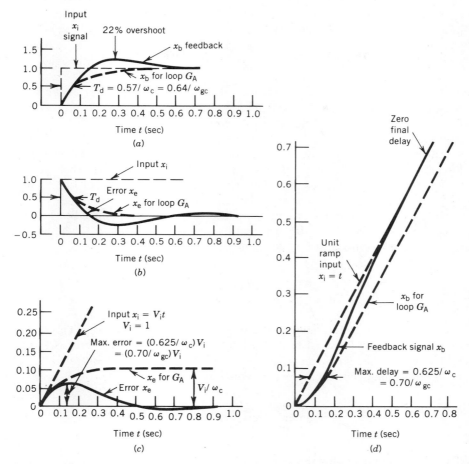

Figure 2.7-8 Feedback-signal and error-signal transient responses for integral-compensated loop G_C: (a) feedback response to unit step; (b) error response to unit step; (c) error response to unit ramp; (d) feedback response to unit ramp.

network correction signal provides a measure of input velocity, and so gives the feedback loop *velocity memory*. This concept is emphasized in Chapter 8, where feedback loops of this type are used to process and filter target tracking data. If a second integral network is added to the loop, to produce a triple integration at low frequencies, the loop has *acceleration memory* as well as velocity memory. The effect of a triple integration is discussed in Chapter 3, Section 3.2.6, and in Chapter 8, Section 8.2.2.

In Fig 2.7-8 the solid curves are the step and ramp responses of the integral-compensated loop G_C for the setting $\omega_i = 0.57\omega_c$. For comparison, the dashed curves give the responses for the simple loop G_A. As shown in diagram

c, the ramp error response for loop G_C initially rises much like the error for loop G_A, reaching a peak error that is reasonably close to the peak error for loop G_A, which is V_i/ω_c. Then the effect of the integral network takes over and forces the error x_e to zero. When the error x_e is zero, the feedback signal x_b follows the input x_i exactly.

As shown in Fig 2.7-8a, the integral network causes overshoot in the step response. For this parameter setting ($\omega_i = 0.57\omega_c$), the peak overshoot of the G_C step response is 22%. The exact gain crossover frequency ω_{gc} of loop G_C is equal to $1.123\omega_c$.

The peak overshoot of the step response can be decreased by reducing the integral-network gain parameter ω_i relative to ω_c. However, when ω_i is reduced, the ramp error response falls to zero more slowly than in Fig 2.7-8c, and the settling time of the step response is lengthened.

Since an integral network cannot reduce the ramp error immediately, the maximum error for a ramp input is approximately equal to V_i/ω_{gc}, whether or not the loop has integral compensation:

$$\text{Maximum ramp error} \cong \frac{V_i}{\omega_{gc}} \tag{2.7-25}$$

For loop G_C the exact value of the maximum error is $0.7V_i/\omega_{gc}$. This is an extreme case, and for most feedback loops the maximum ramp error is closer to V_i/ω_{gc}. As shown in diagram d, this maximum error is equivalent to a maximum time delay approximately equal to

$$\text{Maximum ramp time delay} \cong \frac{1}{\omega_{gc}} \tag{2.7-26}$$

The exact time delay for this case is $0.7/\omega_{gc}$.

The transfer function of the integral network is denoted F_i. For loop G_C, this is equal to

$$F_i = 1 + \frac{\omega_i}{s} \tag{2.7-27}$$

A magnitude asymptote plot of this transfer function is shown in Fig 2.7-9b. To avoid poor stability under saturated conditions, an integral network with the magnitude asymptote plot in Fig 2.7-9a may be used, which has a finite gain α at zero frequency. The transfer function is

$$F_i = \frac{1 + \omega_i/s}{1 + \omega_i/\alpha_s} \tag{2.7-28}$$

This reduces to Eq 2.7-27 if α is infinite. Circuits for implementing the

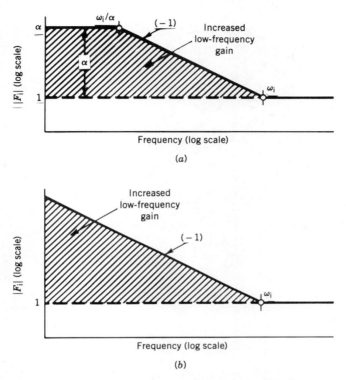

Figure 2.7-9 Magnitude asymptote plots of integral-network transfer functions F_i: (a) for finite zero-frequency gain α; (b) for infinite zero-frequency gain α.

integral-network transfer functions of Eqs 2.7-27, -28 were shown earlier in Fig 2.4-7.

Figure 2.7-10a shows magnitude asymptote plots for the basic loop G_A (plot 1) and for the integral compensated loops (plots 2, 3), which use the integral networks illustrated in diagrams a, b of Fig 2.7-9. Curve (3) is the plot for loop G_C given in Eq 2.7-11. These ideal loop transfer functions approximate ω_c/s at high frequencies. This is not a restriction because the following discussion applies regardless of the high-frequency response of the loop if the loop has good stability.

Curve (2) in Fig 2.7-10 is the loop frequency response G of a loop with an integral network having a finite zero-frequency gain α, and curve (3) is the response for infinite α. Diagram b of Fig 2.7-10 shows the magnitude asymptote plots of the approximate G_{ie} transfer functions that correspond to the magnitude asymptote plots of G in diagram a. These approximations of G_{ie} were derived using the techniques described in Section 2.6.

The cross-hatched area in diagram a of Fig 2.7-10 shows that the integral network increases the loop gain at frequencies below the integral-network

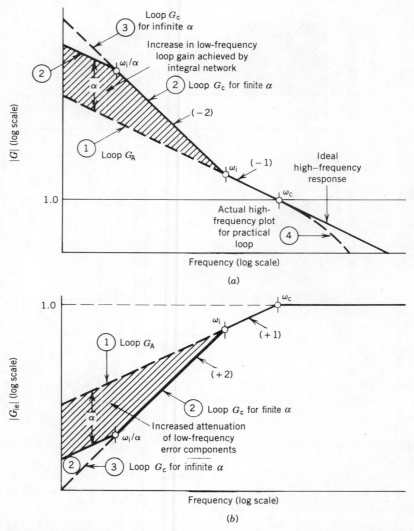

Figure 2.7-10   Magnitude asymptote plots for loops A and B of (a) loop transfer function G and (b) approximate error transfer function G_{ie}.

break frequency ω_i. The corresponding cross-hatched area in diagram b shows that this increase in the low-frequency loop gain results in a corresponding decrease of the magnitude of the G_{ie} error response at low frequencies.

Therefore, the effect of the integral network for a low-frequency input signal can be approximated as shown in Fig 2.7-11. Diagram a gives the magnitude asymptote plots of the transfer function $1/F_i$, which is the reciprocal of the integral-network transfer function F_i illustrated in Fig 2.7-9.

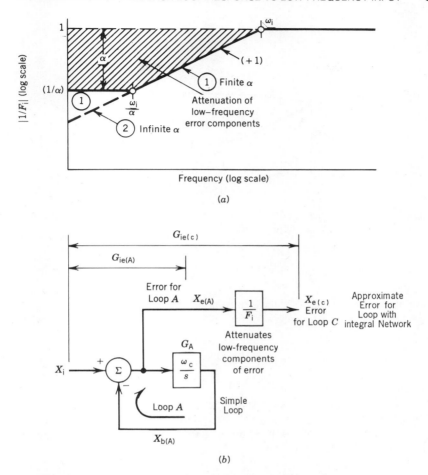

Figure 2.7-11 Approximate effect of an integral network on the error response to a low-frequency input: (*a*) magnitude asymptote plot of $1/F_i$; (*b*) approximate signal-flow diagram relating loop-*A* error to loop-*B* error.

Diagram *b* is a signal-flow diagram showing the approximate error response of a loop with integral compensation. The signal $X_{e(A)}$ is the error response of the single-integration loop G_A. This error signal is fed through the transfer function $1/F_i$, illustrated in diagram a, to obtain the approximate error for the integral-compensated loop. The transfer function $1/F_i$ acts like a highpass filter, which attenuates the low-frequency components of the error (at frequencies below ω_i).

Figure 2.7-12 shows the effect of integral compensation on the error response to the ramp input $X_i = Vt$ indicated by the dashed line. Curve (3) is the error response for the simple loop G_A. Curve (1) is the error response for

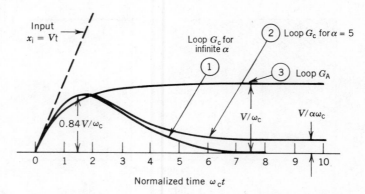

Figure 2.7-12   Comparison of ramp error responses for simple loop A and integral-compensated loop B, with infinite α and $\alpha = 5$.

the integral compensated loop with infinite α, and curve (2) is the error response for $\alpha = 5$.

As shown in Fig 2.7-12, the final value of error for the single-integration loop G_A is V/ω_c. With integral compensation, the final error is reduced to

$$\text{Final ramp error} = \frac{V}{\alpha\omega_c} \qquad (2.7\text{-}29)$$

When the integral-network zero-frequency gain α is infinite, as shown in curve (1), the final error is zero. When $\alpha = 5$, as shown in curve (2), the final error is $V/5\omega_c$, which is $1/5$ of the final error for the basic loop G_A. The maximum error is approximately the same for all three cases:

$$\text{Maximum ramp error} \cong \frac{V}{\omega_{gc}} \qquad (2.7\text{-}22)$$

The integral-network break frequency ω_i determines the rate at which the ramp error is decreased. If ω_i is reduced, the error responses (1) and (2) decay more slowly to the final values. (This effect is illustrated later by simulated responses in Chapter 5, Fig 5.3-7.)

Including integral compensation in a feedback loop adds overshoot to the step response. This principle is illustrated in Fig 2.7-13. Diagram a shows the error response of an integral compensated loop to a unit step, and diagram b shows the error response to a unit ramp. It is assumed that the zero-frequency gain α of the integral network is infinite, and so the final value of the ramp error is zero.

Since a unit ramp is the integral of a unit step, the error response of a feedback loop to a unit ramp is the integral of the error response to a unit

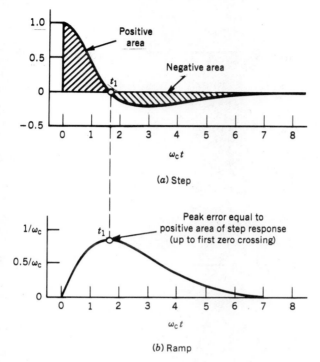

Figure 2.7-13 Comparison of transient error responses of integral-compensated loop for (a) unit step input, and (b) unit ramp input.

step. Therefore, the peak ramp error (at time t_1) occurs at the first zero crossing of the step error response. The peak ramp error is equal to the cross-hatched positive area of the step error response, up to the first zero crossing. If the final ramp error is to be zero, the net area under the step error response must be zero. Hence the negative area of the step error response must equal the positive area.

This shows that a feedback loop with integral compensation must have overshoot in the step response. If α is reasonably large, the area of overshoot (the negative area) must be approximately equal to the area of the step error response up to the first zero crossing (the positive area). The amount of overshoot can be reduced by decreasing the value of ω_i, but the area of overshoot cannot change. Hence, decreasing ω_i (while keeping ω_c constant) reduces the peak overshoot but adds a long tail to the overshoot transient.

Integral compensation can considerably reduce the error for a low-frequency input. However, this error reduction is achieved at the expense of waveform distortion, which may sometimes be undesirable.

This point is illustrated in Fig 2.7-14, which shows the responses of loops G_A and G_C to a truncated ramp input, starting at time t_1 and ending (0.6 sec

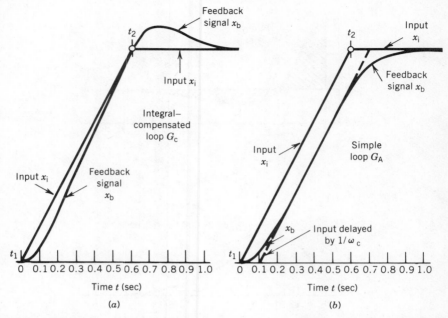

Figure 2.7-14 Comparison of responses to ramp segment: (a) for integral compensated loop G_C; (b) for single-integration loop G_A.

later) at time t_2. The response for loop G_A in diagram b is approximately equal to the input delayed in time by $1/\omega_c$, and so has nearly the same shape as the input. However, the shape of the response for loop G_C in diagram a is quite different. In diagram a the integral network has reduced the error nearly to zero by time t_2, when the ramp ends. After t_2 there is strong overshoot in the response as the correction signal x_{cor} stored in the integral network gradually decays to zero. Thus, in diagram a the action of the integral network appreciably distorts the waveform shape.

In some applications the integral-compensated loop G_C is preferred, because this loop minimizes the error. However, in other cases, duplication of waveform shape is important, and so a single-integration loop like G_A or G_B is better.

2.7.4 Summary of Response to a Low-Frequency Input

As was shown in Fig 2.7-10, for a feedback loop with a single integration at low frequencies (such as G_A or G_B) the error for a low-frequency input is approximately equal to the derivative of the input (the input velocity) divided by ω_c:

$$x_e \cong \frac{1}{\omega_c}\frac{dx_i}{dt} \qquad (2.7\text{-}31)$$

As was shown in Eq 2.7-24, when an integral network of infinite α is included in the loop, the error for a low-frequency input is approximately equal to the second derivative of the input (the input acceleration) divided by $\omega_c \omega_i$:

$$x_e \cong \frac{1}{\omega_c \omega_i} \frac{d^2 x_i}{dt^2} \qquad (2.7\text{-}32)$$

When the integral network has a finite low-frequency gain α, the error response has an additional term proportional to velocity, which is equal to the error for the single-integration loop given in Eq 2.7-31, divided by α. Thus, the approximate error response is

$$x_e \cong \frac{1}{\alpha \omega_c} \frac{d x_i}{dt} + \frac{1}{\omega_c \omega_i} \frac{d^2 x_i}{dt^2} \qquad (2.7\text{-}33)$$

When α is infinite, the first term is zero, and this reduces to Eq 2.7-32. These issues are discussed further in Chapter 3.

2.8 ANALYSIS OF A MULTI-LOOP FEEDBACK CONTROL SYSTEM

This section illustrates the use of signal-flow diagrams and magnitude asymptote plots in feedback control analysis, and applies these techniques to analyze a multi-loop feedback control system. The signal-flow diagrams for common opamp circuits and for a DC servo motor are developed. These are applied to a servo that positions an optical instrument stage through a rack-and-pinion drive. In Chapter 5, Section 5.3, this servo is used as a model for computer simulation, and its step and ramp responses are calculated. This discussion is a summary of a much more detailed analysis of this servo given in Ref [1.1] (Chapter 7).

2.8.1 Signal-Flow Diagrams of Ideal Opamp Circuits

As was shown in Section 2.4.4, modern operational amplifiers (opamps) generally have (1) very high DC gain; (2) very wide bandwidth; and (3) very high input impedance. Consequently, for most servo-type feedback control applications, opamps can be considered to be ideal elements such that (1) no current flows into or out of the input terminals and (2) the opamp continually varies its output voltage to keep the difference between the ($+$) and ($-$) input voltages equal to zero.

Figures 2.8-1 to -3 show three common types of opamp circuits. For ideal opamp performance these circuits can be represented by the signal-flow diagrams given in the figures.

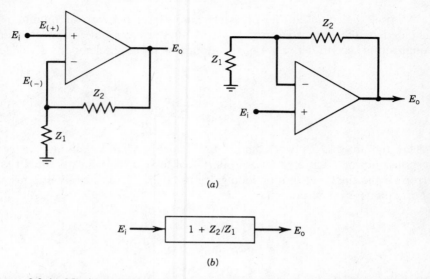

(a)

(b)

Figure 2.8-1 Noninverting opamp circuit: (a) two forms for representing the circuit; (b) signal-flow diagram.

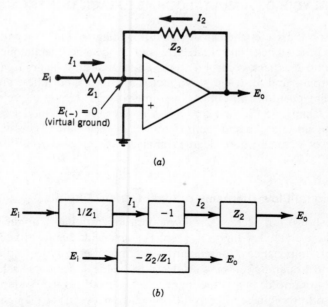

(a)

(b)

Figure 2.8-2 Inverting opamp circuit: (a) circuit; (b) signal-flow diagram.

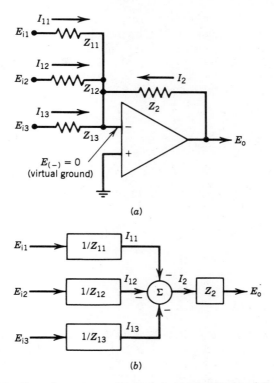

Figure 2.8-3 Summing opamp circuit: (a) circuit; (b) signal-flow diagram.

For the noninverting opamp circuit of Fig 2.8-1a, the voltage $E_{(-)}$ on the $(-)$ opamp input terminal is related as follows to the voltage E_o at the opamp output:

$$E_{(-)} = \frac{Z_1}{Z_1 + Z_2} E_o \qquad (2.8\text{-}1)$$

The opamp continually changes the output voltage E_o until the voltage $E_{(-)}$ is equal to the voltage $E_{(+)}$ on the $(+)$ input terminal, which is equal to the circuit input voltage E_i. Thus,

$$E_{(-)} = E_{(+)} = E_i \qquad (2.8\text{-}2)$$

Combining Eqs 2.8-1, -2 gives

$$E_o = \left(1 + \frac{Z_2}{Z_1}\right) E_i \qquad (2.8\text{-}3)$$

Equation 2.8-3 is expressed as a signal-flow diagram in Fig 2.8-1b.

Figure 2.8-2a shows an inverting opamp circuit. As was shown in Section 2.4.4, in an inverting opamp circuit, the opamp varies its output voltage E_o to keep the voltage on the $(-)$ input terminal essentially at ground potential. Hence, in diagram b, the currents I_1, I_2 flowing in impedances Z_1, Z_2 are equal to

$$I_1 = \frac{E_i}{Z_1} \tag{2.8-4}$$

$$I_2 = \frac{E_o}{Z_2} \tag{2.8-5}$$

Since no current flows into the $(-)$ input terminal of the opamp, the sum of the currents I_1, I_2 must be zero:

$$I_1 + I_2 = 0 \tag{2.8-6}$$

Combining Eqs 2.8-4 to -6 gives

$$-E_o = \left(\frac{Z_2}{Z_1}\right) E_i \tag{2.8-7}$$

This is expressed as a signal-flow diagram in Fig 2.8-2b.

The summing opamp circuit in Fig 2.8-3a is an extension of the inverting opamp circuit in Fig 2.8-2a. This circuit has three input voltages E_{i1}, E_{i2}, E_{i3}, and three input impedances Z_{11}, Z_{12}, Z_{13}. The voltage on the $(-)$ opamp terminal is controlled to be at ground potential. Hence, the output voltage E_o is related as follows to the three input voltages:

$$
\begin{aligned}
-E_o &= \frac{Z_2}{Z_{11}} E_{i1} + \frac{Z_2}{Z_{12}} E_{i2} + \frac{Z_2}{Z_{13}} E_{i3} \\
&= \left(\frac{E_{i1}}{Z_{11}} + \frac{E_{i2}}{Z_{12}} + \frac{E_{i2}}{Z_{12}} \right) Z_2
\end{aligned} \tag{2.8-8}
$$

This equation is expressed as a signal-flow diagram in Fig 2.8-3b.

2.8.2 Signal-Flow Diagram for DC Servo Motor with Inertia Load

The following symbols are used to denote angular motion:

$$\theta = \text{angle (rad)}$$

$$\Omega = \frac{d\theta}{dt} = \text{angular velocity (rad/sec)} \tag{2.8-9}$$

$$\alpha = \frac{d\Omega}{dt} = \text{angular acceleration } (\text{rad/sec}^2) \tag{2.8-10}$$

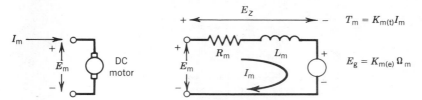

Figure 2.8-4   Circuit diagram of DC servo motor.

Generally, lowercase letters are used in this book for time variables, and uppercase letters for the Laplace transforms of time variables. However, this approach is cumbersome for variables denoted by Greek symbols. Hence the same symbols, θ, Ω, α, are used to represent both the time variables and their transforms.

The circuit for a DC servo motor is shown in Fig 2.8-4. Voltage E_m is the voltage applied across the motor armature circuit terminals, and I_m is the motor armature current. The resistance and inductance of the motor armature winding are denoted R_m and L_m. The voltage E_g is the motor-generated voltage (often called the back EMF), which is proportional to motor angular velocity Ω_m:

$$E_g = K_{m(e)}\Omega_m \qquad (2.8\text{-}11)$$

Parameter $K_{m(e)}$ is the motor voltage constant. The torque T_m generated in the motor is proportional to motor current I_m:

$$T_m = K_{m(t)}I_m \qquad (2.8\text{-}12)$$

where $K_{m(t)}$ is the motor torque constant. To satisfy conservation of energy, the voltage and torque constants must be equal, and so they are represented by a single constant K_m:

$$K_m = K_{m(e)} = K_{m(t)} \qquad (2.8\text{-}13)$$

It is usually not obvious that the voltage and torque constants of a motor are equal, because they are generally not expressed in consistent units. For example, $K_{m(e)}$ may be expressed in volts per 1000 rpm; while $K_{m(t)}$ may be expressed in inch-ounces per ampere. These constants are numerically equal when consistent MKS units are used, with $K_{m(e)}$ expressed in volt/(rad/sec) and $K_{m(t)}$ expressed in newton-meter/Amp.

In Fig 2.8-4, the voltage drop across the motor winding impedance, which is equal to $(E_m - E_g)$, is denoted E_z. The motor current I_m can be expressed in

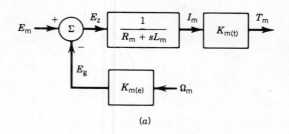

(a)

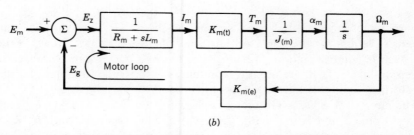

(b)

Figure 2.8-5 Signal-flow diagram of motor and load: (a) signal-flow diagram of motor equations; (b) motor with mechanical load.

terms of E_z as follows:

$$I_m = \frac{E_z}{R_m + sL_m} = \frac{E_m - E_g}{R_m + sL_m} \tag{2.8-14}$$

Equations 2.8-11 to -14 are represented as a signal-flow diagram in Fig 2.8-5a.

Assume that a motor of inertia J_m drives an output inertia J_o through a gear train of gear reduction N. The total effective inertia reflected to the motor shaft, denoted $J_{(m)}$, is equal to

$$J_{(m)} = J_m + \frac{J_o}{N^2} \tag{2.8-15}$$

If there are no other mechanical loads, the motor torque T_m is equal to

$$T_m = J_{(m)}\alpha_m \tag{2.8-16}$$

where α_m is the motor angular acceleration. The motor angular velocity Ω_m is the integral of the motor angular acceleration α_m. Hence, the transforms of these variables are related by

$$\Omega_m = \frac{1}{s}\alpha_m \tag{2.8-17}$$

Applying Eqs 2.8-16, -17 to diagram a of Fig 2.8-5 gives diagram b, which is the complete signal-flow diagram of the motor plus inertia load. The diagram shows that the motor generated voltage (or back EMF) closes a feedback loop, which is called the *motor back-EMF loop*, or simply the *motor loop*.

The loop transfer function of the motor loop, denoted $G_{(m)}$, is obtained by multiplying all of the transfer function blocks within the feedback loop of Fig 2.8-5c, which gives

$$G_{(m)} = \frac{K_m^2}{(R_m + sL_m)J_{(m)}s} \tag{2.8-18}$$

In accordance with Eq 2.8-13, $K_{m(e)}$ and $K_{m(t)}$ were set equal to K_m. The factor $(R_m + sL_m)$ is a high-frequency factor relative to the gain crossover frequency of this feedback loop, and so should be expressed in the form $(1 + s/\omega_x)$. Hence, the factor $(R_m + sL_m)$ becomes

$$(R_m + sL_m) = R_m\left(1 + \frac{sL_m}{R_m}\right) = R_m\left(1 + \frac{s}{\omega_e}\right) \tag{2.8-19}$$

The parameter ω_e is called the electrical break frequency of the motor, which is equal to

$$\omega_e = \frac{R_m}{L_m} = \text{Motor electrical break frequency} \tag{2.8-20}$$

The reciprocal of ω_e is denoted τ_e, which is the electrical time constant of the motor:

$$\tau_e = \frac{1}{\omega_e} = \text{Motor electrical time constant} \tag{2.8-21}$$

Combining Eqs 2.8-18, -19 gives

$$G_{(m)} = \frac{K_m^2}{R_m J_{(m)}s(1 + s/\omega_e)} = \frac{\omega_{cm}}{s(1 + s/\omega_e)} \tag{2.8-22}$$

The parameter ω_{cm} is the asymptotic gain crossover frequency of the motor back-EMF loop. Equation 2.8-22 shows that ω_{cm} is equal to

$$\omega_{cm} = \frac{K_m^2}{R_m J_{(m)}} \tag{2.8-23}$$

The asymptotic gain crossover frequency ω_{cm} of the motor loop is commonly denoted ω_m, and is called the *mechanical break frequency* of the motor. The reciprocal of ω_m is called the mechanical time constant of the motor:

$$\tau_m = \frac{1}{\omega_m} = \frac{1}{\omega_{cm}}$$

$$= \text{Mechanical time constant of motor} \qquad (2.8\text{-}24)$$

A magnitude asymptote plot for the loop transfer function $G_{(m)}$ of the motor loop is shown in diagram a of Fig 2.8-6. In accordance with Section 2.6, diagrams b and c are the approximate magnitude asymptote plots of the

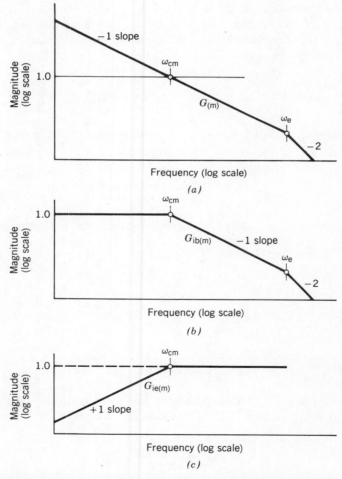

Figure 2.8-6 Magnitude asymptote plots for (a) loop transfer function $G_{(m)}$ of motor loop; (b) approximate feedback transfer function $G_{ib(m)}$; (c) approximate error transfer function $G_{ie(m)}$.

feedback and error transfer functions of the motor loop, denoted $G_{ib(m)}$ and $G_{ie(m)}$. The equations corresponding to these approximations are

$$G_{ib(m)} \cong \frac{1}{(1 + s/\omega_{cm})(1 + s/\omega_e)} \tag{2.8-25}$$

$$G_{ie(m)} \cong \frac{1}{1 + \omega_{cm}/s} \tag{2.8-26}$$

Usually ω_e is much larger than ω_{cm}, and so these approximations of $G_{ib(m)}$ and $G_{ie(m)}$ are very accurate.

2.8.3 Signal Flow Diagram of Stage-Positioning Servo

Figure 2.8-7 is the circuit diagram of a servo that moves an instrument stage in a single dimension through a rack-and-pinion drive. The controlled variable, denoted X_c, is the linear displacement of the stage. The variable X_c is measured by a position sensor to close a position feedback loop. A tachometer, mounted on the motor shaft, generates a signal proportional to motor velocity, which is used to close a velocity feedback loop. A current-sense resistor R_c is placed in series with the servo motor. The voltage across resistor R_c is proportional to motor current and is used to close a current feedback loop.

The servo motor drives a pinion through a gear train, and the pinion drives a geared rack fastened to the instrument stage. The displacement X_c of the stage is related as follows to the angular rotation θ_m of the motor shaft:

$$X_c = \left(\frac{r_p}{N}\right)\theta_m \tag{2.8-27}$$

where r_p is the radius of the pinion and N is the speed reduction ratio of the gear train. Neglecting the inertia of the gear train (which is usually small), the total inertia reflected to the motor shaft is

$$J_{(m)} = J_m + \left(\frac{X_c}{\theta_m}\right)^2 M_c = J_m + \left(\frac{r_p}{N}\right)^2 M_c \tag{2.8-28}$$

where M_c is the mass of the stage and J_m is the inertia of the servo motor.

The signal-flow diagram of the stage positioning servo of Fig 2.8-7 is shown in Fig 2.8-8. The signal-flow diagrams for opamps U1, U2, and U3 are obtained from Fig 2.8-3 and that for opamp U4 is obtained from Fig 2.8-1. Table 2.8-1 shows the impedances for the U1, U2, U3 summing opamp circuits, which are defined in accordance with Fig 2.8-3.

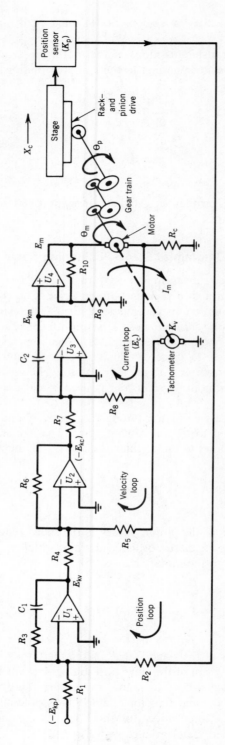

Figure 2.8-7 Circuit diagram of stage-positioning servo.

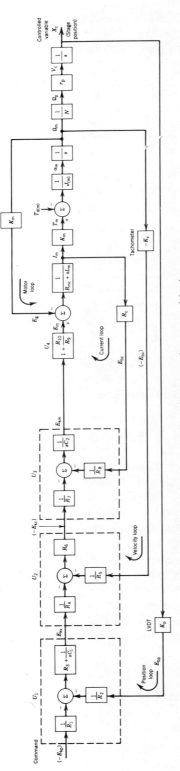

Figure 2.8-8 Signal-flow diagram of stage-positioning servo.

TABLE 2.8-1   Impedances for U1, U2, U3 Summing Opamp Circuits

Opamp	Z_{11}	Z_{12}	Z_2
U1	R_1	R_2	$R_3 + 1/sC_1$
U2	R_4	R_5	R_6
U3	R_7	R_8	$1/sC_1$

The signal-flow diagram for the servo motor is the same as that in Fig 2.8-5b except that the motor resistance R_m is replaced by R_{mc}, which is equal to

$$R_{mc} = R_m + R_c \qquad (2.8\text{-}29)$$

Resistor R_c is the current-sense resistor in series with the motor. In this servo, the voltage E_m is the total voltage applied to the series combination of motor plus current-sense resistor. The voltage E_{bc} is the current feedback voltage, which is proportional to motor current I_m:

$$E_{bc} = R_c I_m \qquad (2.8\text{-}30)$$

This voltage is applied to resistor R_8 of opamp circuit U3.

The voltage E_{bv} is the velocity feedback voltage obtained from a tachometer coupled to the motor shaft and is proportional to motor angular velocity Ω_m:

$$E_{bv} = \pm K_v \Omega_m \qquad (2.8\text{-}31)$$

The parameter K_v is the tachometer sensitivity in volt/(rad/sec). A tachometer has a floating winding, and the sign of Eq 2.8-31 depends on which terminal of the winding is grounded. In the signal-flow diagram, the negative sign is selected in order to achieve negative feedback in the velocity feedback loop. The voltage E_{bv} from the appropriate tachometer terminal is fed to resistor R_5 of opamp circuit U2.

The position sensor is assumed to be a linear variable differential transformer (LVDT). Although this is an AC transducer, modern LVDT devices usually contain integrated electronics for generating the AC reference signal and demodulating the resultant AC displacement signal. Thus, the complete device provides a DC voltage proportional to displacement. The voltage from the LVDT device is the position feedback voltage E_{bx}, which is equal to

$$E_{bx} = K_p X_c \qquad (2.8\text{-}32)$$

Parameter K_p is the sensitivity of the LVDT position transducer and X_c is the displacement of the stage (the controlled member). This position feedback voltage is applied to the resistor R_2 of opamp circuit U1.

A position command voltage $-E_{kp}$ is applied to resistor R_1 of opamp circuit U1. This drives the servo to produce a stage displacement X_c that is proportional to E_{kp}. This position command voltage is represented with a negative sign so that positive E_{kp} produces positive stage displacement X_c.

A motor torque disturbance signal $T_{d(m)}$ is shown in Fig 2.8-6, which subtracts from the motor torque T_m. This disturbance torque represents friction in the motor, gear train, and stage.

The diagram shows that the angular velocity of the pinion, Ω_p, is equal to the angular velocity of the motor, Ω_m, divided by the gear ratio N:

$$\Omega_p = \frac{\Omega_m}{N} \tag{2.8-33}$$

The velocity V_c of the controlled member (the stage) is equal to the angular velocity of the pinion Ω_p multiplied by the radius of the pinion r_p:

$$V_c = r_p \Omega_p \tag{2.8-34}$$

The displacement X_c of the controlled member (the stage) is the integral of the controlled member velocity V_c:

$$X_c = \frac{1}{s} V_c \tag{2.8-35}$$

Thus, the preceding transfer functions allow the complete signal-flow diagram of Fig 2.8-8 to be constructed.

2.8.4   Simplification of Signal-Flow Diagram

The signal-flow diagram of a feedback control system is simplified by expressing all feedback loops either in the G_{ib} or the G_{ie} form. For this case, the G_{ib} form is desired for all loops. To achieve this, the elements in the feedback path are shifted to the forward path in the manner shown in Fig 2.8-9.

Diagram *a* of Fig 2.8-9 represents the following equation:

$$C = A - KB \tag{2.8-36}$$

In diagram *b* the K block is shifted to the forward path. To compensate for this, the transfer function $1/K$ is placed in series with the A signal. The transfer function for diagram *b* is

$$C = K\{(1/K)A - B\} = A - KB \tag{2.8-37}$$

Since Eqs 2.8-36, -37 are the same, the signal-flow diagram in diagram *b* is equivalent to that in diagram *a*.

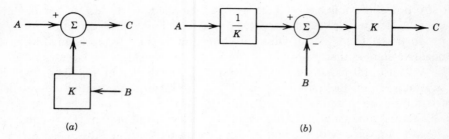

Figure 2.8-9 Signal-flow diagram manipulations to move element from feedback path to forward path: (*a*) original diagram; (*b*) modified diagram.

By applying Fig 2.8-9 to Fig 2.8-8, the signal-flow diagram of Fig 2.8-10 is formed. The feedback elements $1/R_2$, $1/R_5$, and $1/R_8$ are moved to the forward paths of the corresponding feedback loops, and their reciprocals are placed in front of the loops. To clarify the diagram manipulations, the feedback elements K_p, K_v, and R_c were not shifted to the forward path in this step. The minus sign of the $-K_v$ block was shifted to the forward path to convert the $(-)$ signs to the $(+)$ signs shown in parentheses.

The dynamic effect of the motor loop is small in this system and so is ignored. This is indicated by showing the feedback path of the motor loop as a dashed line in Fig 2.8-10. When the dashed motor loop is ignored, Fig 2.8-10 can be simplified to Fig 2.8-11 by applying the signal-flow diagram manipulations of Fig 2.8-9. The friction torque disturbance $T_{d(m)}$ is ignored in the following discussion. The effect of friction is studied in Ref [1.1] (Chapters 6 and 7).

Figure 2.8-11 shows that the loop transfer function of the current feedback loop is

$$G_{(c)} = \frac{R_c(R_9 + R_{10})}{sC_2 R_8 R_9 (R_{mc} + sL_m)} \tag{2.8-38}$$

The loop transfer function of the velocity loop is

$$G_{(v)} = \frac{K_v R_6 R_8}{R_5 R_7 R_c} G_{ib(c)} \frac{K_m}{sJ_{(m)}} \tag{2.8-39}$$

where $G_{ib(c)}$ is the feedback transfer function of the current loop, which is equal to

$$G_{ib(c)} = \frac{G_{(c)}}{1 + G_{(c)}} \tag{2.8-40}$$

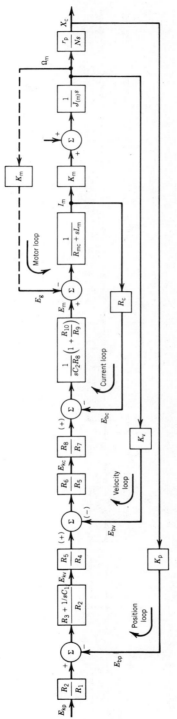

Figure 2.8-10 Simplified signal-flow diagram of stage-positioning servo in Fig 8-8.

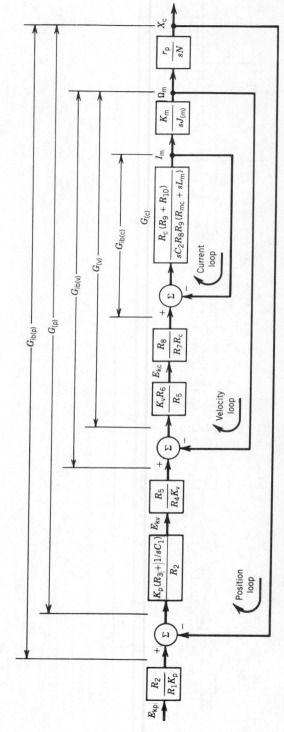

Figure 2.8-11 Final simplification of signal-flow diagram of stage-positioning servo, which ignores motor back-EMF loop and torque disturbance.

The loop transfer function of the position loop is

$$G_{(p)} = \frac{K_p(R_3 + 1/sC_1)R_5}{R_2 R_4 K_v} G_{ib(v)} \frac{r_p}{sN} \tag{2.8-41}$$

where $G_{ib(v)}$ is the feedback transfer function of the velocity loop, which is equal to

$$G_{ib(v)} = \frac{G_{(v)}}{1 + G_{(v)}} \tag{2.8-42}$$

The transfer function between the position command voltage E_{kp} and the stage position X_c (the controlled variable) is

$$\frac{X_c}{E_{kp}} = \frac{R_2}{R_1 K_p} G_{ib(p)} \tag{2.8-43}$$

where $G_{ib(p)}$ is the feedback transfer function of the position loop, which is equal to

$$G_{ib(p)} = \frac{G_{(p)}}{1 + G_{(p)}} \tag{2.8-44}$$

At low frequencies, $G_{ib(p)}$ is approximately unity. Hence, by Eq 2.8-43, X_c is approximately proportional to E_{kp} at low frequencies:

$$X_c \cong \frac{R_2}{R_1 K_p} E_{kp} \tag{2.8-45}$$

The factor $(R_{mc} + sL_m)$ in the current loop transfer function of Eq 2.8-38 is a high-frequency factor relative to the gain crossover of the current loop. Hence this factor is expressed in the form $(1 + s/\omega_x)$ as follows:

$$(R_{mc} + sK_m) = R_{mc}\left(1 + \frac{sL_m}{R_{mc}}\right) = R_{mc}\left(1 + \frac{s}{\omega_e}\right) \tag{2.8-46}$$

where ω_e is the electrical break frequency of the motor circuit, which is equal to

$$\omega_e = \frac{R_{mc}}{L_m} = \frac{R_m + R_c}{L_m} \tag{2.8-47}$$

Hence $G_{(c)}$ in Eq 2.8-38 is expressed as

$$G_{(c)} = \frac{R_c(R_9 + R_{10})}{sC_2 R_8 R_9 R_{mc}(1 + s/\omega_e)} = \frac{\omega_{cc}}{s(1 + s/\omega_e)} \qquad (2.8\text{-}48)$$

where ω_{cc} is the asymptotic gain crossover frequency of the current loop, which is equal to

$$\omega_{cc} = \frac{R_c(R_9 + R_{10})}{C_2 R_8 R_9 R_{mc}} \qquad (2.8\text{-}49)$$

A magnitude asymptote plot of $G_{(c)}$ of Eq 2.8-47 is shown as the solid curve in Fig 2.8-12a. This loop has good stability if the phase lag at the gain crossover frequency does not exceed 135°. It can be shown that this limit occurs approximately when $\omega_{cc} = \omega_e$. However, to simplify the discussion, the plots are drawn with ω_{cc} set below ω_e.

The dynamic effect of motor back EMF, which has been ignored in this analysis, is studied in Ref [1.1] (Section 7.4). The reference shows that this can be closely approximated by inserting into the current loop $G_{(c)}$ the error transfer function of the motor loop, $G_{ie(m)}$. The error transfer function of the motor loop was given in Eq 2.8-26 and plotted in Fig 2.8-6c. The effect of motor back-EMF loop is indicated in Fig 2.8-12a. The cross-hatched area shows that the motor loop decreases the low-frequency loop gain of the current loop.

In accordance with the approximation described in Section 2.6, if the current loop has good stability, the G_{ib} transfer function of the current loop can be approximated as shown by the dashed magnitude asymptote plot in Fig 2.8-12a. The corresponding approximate feedback transfer function of the current loop is

$$G_{ib(c)} \cong \frac{1}{(1 + s/\omega_{cc})(1 + s/\omega_e)} \qquad (2.8\text{-}50)$$

Substituting Eq 2.8-50 into Eq 2.8-39 gives the following approximate loop transfer function of the velocity loop:

$$G_{(v)} \cong \frac{K_v R_6 R_8 K_m}{R_5 R_7 R_c J_{(m)}s(1 + s/\omega_{cc})(1 + s/\omega_e)} \qquad (2.8\text{-}51)$$

This has the form

$$G_{(v)} \cong \frac{\omega_{cv}}{s(1 + s/\omega_{cc})(1 + s/\omega_e)} \qquad (2.8\text{-}52)$$

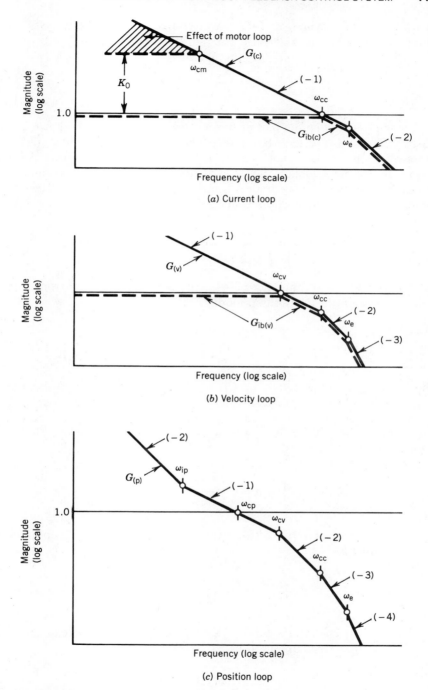

Figure 2.8-12 Magnitude asymptote plots of G and G_{ib} for current, velocity, and position loops.

where ω_{cv} is the asymptotic gain crossover frequency of the velocity loop. Comparing Eqs 2.8-52, -53 gives for ω_{cv}

$$\omega_{cv} = \frac{K_v R_6 R_8 K_m}{R_5 R_7 R_c J_{(m)}} \qquad (2.8\text{-}53)$$

A magnitude asymptote plot of this approximation of $G_{(v)}$ is shown in Fig 2.8-12b. This velocity loop has good stability if the phase lag at the gain crossover frequency does not exceed 135°. If the velocity feedback loop has good stability, its feedback transfer function $G_{ib(v)}$ can be approximated as shown by the dashed magnitude asymptote plot in Fig 2.8-12b. The corresponding approximate feedback transfer function for the velocity loop is

$$G_{ib(v)} \cong \frac{1}{(1 + s/\omega_{cv})(1 + s/\omega_{cc})(1 + s/\omega_e)} \qquad (2.8\text{-}54)$$

Substituting Eq 2.8-54 into Eq 2.8-41 gives the following approximate loop transfer function for the position loop:

$$G_{(p)} \cong \frac{K_p(R_3 + 1/sC_1)R_5 r_p}{R_2 R_4 K_v Ns(1 + s/\omega_{cv})(1 + s/\omega_{cc})(1 + s/\omega_e)} \qquad (2.8\text{-}55)$$

The numerator factor $(R_3 + 1/sC_1)$ is the response of the position-loop integral network, which increases the loop gain of the position loop at low frequencies. This is a low-frequency factor relative to the gain crossover frequency of the position loop, and so is expressed in the form $(1 + \omega_x/s)$. Hence, this factor becomes

$$\left(R_3 + \frac{1}{sC_1}\right) = R_3\left(1 + \frac{1}{sC_1 R_3}\right) = R_3\left(1 + \frac{\omega_{ip}}{s}\right) \qquad (2.8\text{-}56)$$

The parameter ω_{ip} is the break frequency of the position loop integral network, which is equal to

$$\omega_{ip} = \frac{1}{C_1 R_3} \qquad (2.8\text{-}57)$$

Substituting Eq 2.8-56 into Eq 2.8-55 gives the following approximate loop transfer function for the position loop:

$$G_{(p)} \cong \frac{K_p R_3 R_5 r_p(1 + \omega_{ip}/s)}{R_2 R_4 K_v Ns(1 + s/\omega_{cv})(1 + s/\omega_{cc})(1 + s/\omega_e)} \qquad (2.8\text{-}58)$$

This has the form

$$G_{(p)} \cong \frac{\omega_{cp}(1 + \omega_{ip}/s)}{s(1 + s/\omega_{cv})(1 + s/\omega_{cc})(1 + s/\omega_e)} \qquad (2.8\text{-}59)$$

The parameter ω_{cp} is the asymptotic gain crossover frequency of the position loop, which is equal to

$$\omega_{cp} = \frac{K_p R_3 R_5 r_p}{R_2 R_4 K_v N} \qquad (2.8\text{-}60)$$

A magnitude asymptote plot of the approximate loop transfer function of the position loop is shown in Fig 2.8-12c.

In Ref [1.1] (Section 3.4), approximation methods are developed for optimizing the parameters of a feedback loop. In Chapter 7 of that reference, these are applied to the stage-positioning servo. The gain crossover frequencies of the individual loops were set at the maximum values consistent with Max$|G_{ib}|$ = 1.30 for each loop. The resultant relative values of the loop parameters are

$$\omega_{cc} = \omega_e \qquad (2.8\text{-}61)$$

$$\omega_{cv} = \frac{\omega_{cc}}{2.43} = \frac{\omega_e}{2.43} \qquad (2.8\text{-}62)$$

$$\omega_{cp} = \frac{\omega_{cv}}{2.31} = \frac{\omega_e}{7.81} \qquad (2.8\text{-}63)$$

$$\omega_{ip} = \frac{\omega_{cp}}{4.34} = \frac{\omega_e}{33.9} \qquad (2.8\text{-}64)$$

In this example, the motor electrical break frequency ω_e is the primary dynamic constraint, which limits all of the other dynamic parameters of the control system. However, in most practical servo applications, dynamic performance is limited primarily by resonances in the mechanical structure, which were ignored in this analysis. Structural dynamics is studied in Ref [1.2] (Chapter 10).

2.8.5 Expressing Signal-Flow Diagram in Terms of the Asymptotic Gain Crossover Frequencies

The signal-flow diagram of Fig 2.8-11 is simulated later in Chapter 5, Section 5.2, to obtain its transient responses. As shown in Figs 2.8-13, -14, to simplify the simulation the parameters of the diagram are expressed in terms of the asymptotic gain crossover frequencies of the control loops.

Figure 2.8-13a shows the first step of the simplification. The loop transfer function of the current loop, $G_{(c)}$, is set equal to the right-hand expression of Eq 2.8-48, and $(R_3 + 1/sC_1)$ is set equal to the right-hand expression of Eq

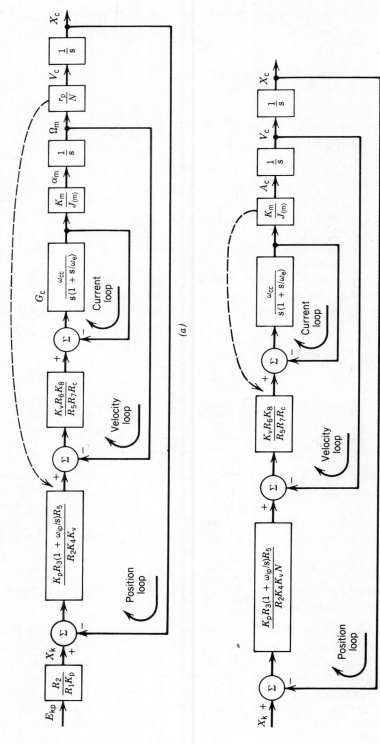

Figure 2.8-13 Signal-flow diagram manipulations to express feedback signals in terms of the controlled variable X_c and its derivatives:8. (*a*) modification of Fig 8-11; (*b*) modification of diagram *a*.

118

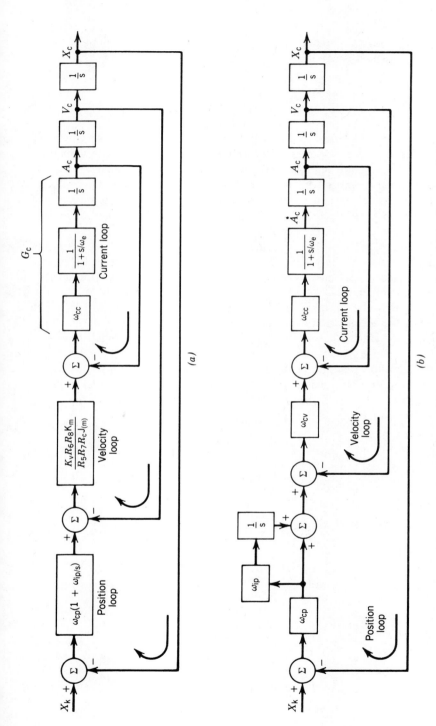

Figure 2.8-14 Manipulations to express gain parameters of signal-flow diagram of stage-positioning servo in terms of asymptotic gain crossover frequencies:8. (*a*) modification of Fig 8-13; (*b*) modification of diagram *a*.

2.8-56. The voltage command E_{kp} is replaced by the equivalent position command X_k, which is equal to

$$X_k = \frac{R_2}{R_1 K_p} E_{kp} \qquad (2.8\text{-}65)$$

The next step is to shift the block r_p/N to the input of the velocity loop, as shown by the dashed arrow in diagram a of Fig 2.8-13. This produces diagram b. With this change, the velocity feedback signal is now the linear velocity V_c of the controlled member (the stage) rather than the motor angular velocity Ω_m. As shown by the dashed arrow, the block $K_m J_{(m)}$ is shifted to the input of the current loop to produce Fig 2.8-14a. Also, Eq 2.8-60 is applied to replace the constant expression of the position loop by ω_{cp}.

Diagram a of Fig 2.8-14 is simplified to form diagram b by replacing the constant expression of the velocity loop by ω_{cv}, in accordance with Eq 2.8-53. With the change of signal-flow diagram, the feedback signal of the current feedback loop in diagram b now becomes A_c, the acceleration of the controlled variable. The variable $\dot{A}_c$ is the rate of acceleration, which is the derivative of the acceleration of the controlled variable. Also, in diagram b the expression $(1 + \omega_{ip}/s)$ has been replaced by an equivalent signal-flow diagram.

These signal-flow diagram manipulations illustrate the general principles for analyzing a multiloop feedback control system. These principles will be applied later in Chapter 8, Section 8.4, to study the dynamics of the Altair tracking filters.

Chapter 3

Tracking Error Due to Target Motion

This chapter develops techniques for calculating the errors of a tracking system that are caused by target motion. Target motion is of low frequency relative to the gain crossover frequency of the tracking loop. If it were not, the tracking control system could not follow it. Therefore, this chapter addresses the general problem of calculating the error of a feedback control loop to a low-frequency input.

3.1 SUMMARY OF CHAPTER

Error coefficients have long been used to characterize the response of a feedback control system to a low-frequency input. The low-frequency error is expressed as a sum of the derivatives of the input, multiplied by parameters called the error coefficients of the control system. The traditional error-coefficient theory is based on an infinite-series expansion of the system response. However, experience shows that only the first few derivative terms of the expansion are meaningful. This theory does not tell how many terms to use, or the accuracy of the resultant error-coefficient expansion.

This chapter extends the traditional error-coefficient concept by using a truncated expansion, which adds transients to the error-coefficient terms. As will be shown, this truncated expansion provides a much more effective method for characterizing the response of a feedback control system to a low-frequency input.

The general error-coefficient theory is developed in Section 3.3. As an introduction to this material, a much simpler approach is given in Section 3.2, which derives an equivalent truncated error-coefficient expansion by applying the approximation of the G_{ie} transfer function presented in Section 2.6. Although the method of Section 3.2 is not as rigorous as that of Section 3.3, it is quite adequate for engineering computations and is much easier to use.

In Section 3.2, the truncated error-coefficient expansion is used to calculate the angular tracking error for a radar that is tracking a target following a straight-line constant-velocity course. Various types of tracking-loop transfer functions are assumed, and the resultant tracking error responses are calculated.

As part of the error-coefficient theory, Section 3.3 presents a general theory for bounding the inaccuracy of the resultant calculated response. The error-coefficient theory is based on straight-line approximations of the input derivatives. The inaccuracy bounds are proportional to the deviations of the input derivatives from the straight-line approximations. These bounds are obtained by multiplying the deviations by parameters called the *remainder coefficients* for the particular derivatives. The remainder-coefficient concept is generalized to obtain the *remainder transient*, which describes the time interval over which the remainder-coefficient error-bound applies.

The remainder-coefficient concept is used later in Chapter 8 to produce a simple optimal tracking theory, which helps to explain the operation of Kalman filter theory in the Altair tracking system. In the Altair tracker, target motion is characterized by placing an upper bound on the uncertainty of target acceleration. The theory of Section 3.3 shows that the corresponding tracking-error bound is equal to the acceleration uncertainty multiplied by the acceleration remainder coefficient of the tracking loop.

3.2 ERROR COEFFICIENTS FOR APPROXIMATING THE LOW-FREQUENCY RESPONSE TO TARGET MOTION OF A RADAR ANGLE-TRACKING SERVO

In Chapter 2, Section 2.7, approximate equations were developed for the error response of different feedback loops to general low-frequency inputs, expressed in terms of the derivatives of the input. That approach will now be extended and applied to calculate the tracking errors for radar angle tracking servos with various loop transfer functions. As will be shown, the error response can be calculated by multiplying the derivatives of the input by constants called *error coefficients*. This derivation is based on the approximations of the G_{ie} error transfer function presented in Section 2.6. A rigorous error-coefficient approach that does not approximate the system transfer function is given in Section 3.3.

3.2.1 Error-Coefficient Expansions for Feedback Loops with Infinite Zero-Frequency Loop Gain

If a feedback-control loop has good stability, a good approximation of its response to a low-frequency input can be obtained by assuming that the loop has ideal high-frequency response, i.e., that G approximates ω_c/s for $\omega > \omega_c$. Hence the loop transfer function can be assumed to have the form $G =$

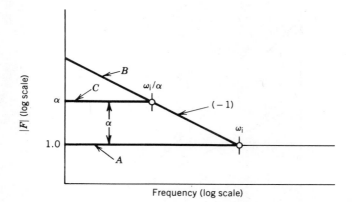

Figure 3.2-1 Magnitude asymptote plots of integral-network transfer function F for loops A, B, and C.

$F(\omega_c/s)$, where the transfer function F is defined as having unity-magnitude asymptote at frequencies equal to or greater than the gain crossover frequency ω_c. The function F is called the low-frequency compensation factor. Let us consider the following three transfer functions for F:

$$F_A = 1 \qquad (3.2\text{-}1)$$

$$F_B = 1 + \frac{\omega_i}{s} \qquad (3.2\text{-}2)$$

$$F_C = \frac{1 + \omega_i/s}{1 + \omega_i/\alpha s} \qquad (3.2\text{-}3)$$

Magnitude asymptote plots of these three transfer functions are shown in Fig 3.2-1. Since F_B and F_C increase the low-frequency loop gain over the basic ω_c/s feedback loop, the resultant feedback loop has less error in response to a low-frequency signal. The total loop transfer function G is equal to $F(\omega_c/s)$, and so the loop transfer functions for the three loops are

$$G_A = F_A \frac{\omega_c}{s} = \frac{\omega_c}{s} \qquad (3.2\text{-}4)$$

$$G_B = F_B \frac{\omega_c}{s} = \frac{\omega_c(1 + \omega_i/s)}{s} \qquad (3.2\text{-}5)$$

$$G_C = F_C \frac{\omega_c}{s} = \frac{\omega_c(1 + \omega_i/s)}{s(1 + \omega_i/\alpha s)} \qquad (3.2\text{-}6)$$

Magnitude asymptote plots of these are shown in Fig 3.2-2.

Magnitude asymptote plots of the approximate error transfer functions for

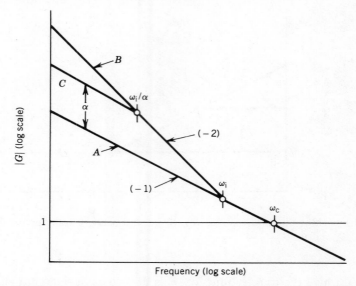

Figure 3.2-2 Magnitude asymptote plots of loop transfer function G for loops A, B, and C.

these three loops are shown in Fig 3.2-3. These were constructed using the techniques described in Section 2.6. The corresponding approximate equations for G_{ie} are

$$*G_{ie(A)} = \frac{1}{1 + \omega_c/s} \tag{3.2-7}$$

$$*G_{ie(B)} = \frac{1}{(1 + \omega_i/s)(1 + \omega_c/s)} \tag{3.2-8}$$

$$*G_{ie(C)} = \frac{1 + (\omega_i/\alpha s)}{(1 + \omega_i/s)(1 + \omega_c/s)} \tag{3.2-9}$$

The asterisk * preceding G represents "approximation of." Modify these transfer functions by changing all of the denominator factors from the form $(1 + \omega_x/s)$ to the form $(1 + s/\omega_x)$ as follows:

$$\left(1 + \frac{\omega_x}{s}\right) = \frac{\omega_x}{s}\left(\frac{s}{\omega_x} + 1\right) = \frac{\omega_x}{s}\left(1 + \frac{s}{\omega_x}\right) \tag{3.2-10}$$

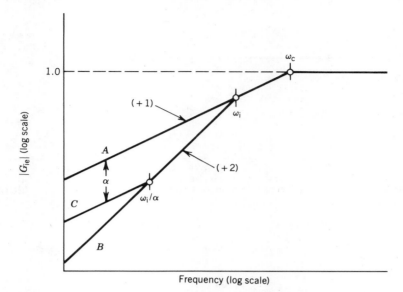

Figure 3.2-3   Magnitude asymptote plots of approximate error transfer function G_{ie} for loops A, B, and C.

The error transfer functions for loops A and B in Eqs 3.2-7, -8 become

$$^*G_{ie\,(A)} = \frac{s}{\omega_c} \frac{1}{(1 + s/\omega_c)} \tag{3.2-11}$$

$$^*G_{ie\,(B)} = \frac{s^2}{\omega_c \omega_i} \frac{1}{(1 + s/\omega_i)(1 + s/\omega_c)} \tag{3.2-12}$$

The transfer functions H_c and H_{ic} are defined as

$$H_c = \frac{1}{1 + s/\omega_c} \tag{3.2-13}$$

$$H_{ic} = \frac{1}{(1 + s/\omega_i)(1 + s/\omega_c)} \tag{3.2-14}$$

Figure 3.2-4 gives magnitude asymptote plots of these transfer functions, which show that they are lowpass filters. Expressing Eqs 3.2-11, -12 in terms of H_c, H_{ic} gives

$$^*G_{ie\,(A)} = \frac{s}{\omega_c} H_c \tag{3.2-15}$$

$$^*G_{ie\,(B)} = \frac{s^2}{\omega_i \omega_c} H_{ic} \tag{3.2-16}$$

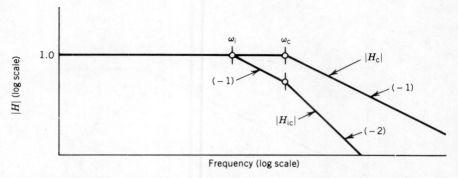

Figure 3.2-4 Magnitude asymptote plots of frequency responses for lowpass filters H_c and H_{ic}.

Since $*G_{ie(C)}$ has the same denominator as $*G_{ie(B)}$, Eq 3.2-9 becomes

$$*G_{ie(C)} = \left(1 + \frac{\omega_i}{\alpha s}\right) *G_{ie(B)}$$

$$= \left(1 + \frac{\omega_i}{\alpha s}\right) \frac{s^2}{\omega_i \omega_c} H_{ic} = \left(\frac{s^2}{\omega_c \omega_i} + \frac{s}{\alpha \omega_c}\right) H_{ic} \qquad (3.2\text{-}17)$$

The transform of the error X_e is $G_{ie} X_i$. Hence the approximate transforms of the error signals are

$$X_{e(A)} = *G_{ie(A)} X_i = \frac{s X_i}{\omega_c} H_c \qquad (3.2\text{-}18)$$

$$X_{e(B)} = *G_{ie(B)} X_i = \frac{s^2 X_i}{\omega_i \omega_c} H_{ic} \qquad (3.2\text{-}19)$$

$$X_{e(C)} = *G_{ie(C)} X_i = \left(\frac{s^2 X_i}{\omega_i \omega_c} + \frac{s X_i}{\alpha \omega_c}\right) H_{ic} \qquad (3.2\text{-}20)$$

Let us denote the nth derivative of $x[t]$ as $D^n x[t]$, and the Laplace transform of that derivative as $D^n X$. Hence

$$s X_i = D X_i = \mathscr{L}[D x_i] = \mathscr{L}\left[\frac{dx_i}{dt}\right] \qquad (3.2\text{-}21)$$

$$s^2 X_i = D^2 X_i = \mathscr{L}[D^2 x_i] = \mathscr{L}\left[\frac{d^2 x_i}{dt^2}\right] \qquad (3.2\text{-}22)$$

Equations 3.2-18 to -20 become

$$X_{e(A)} = \frac{DX_i}{\omega_c} H_c \tag{3.2-23}$$

$$X_{e(B)} = \frac{D^2 X_i}{\omega_i \omega_c} H_{ic} \tag{3.2-24}$$

$$X_{e(C)} = \left(\frac{D^2 X_i}{\omega_i \omega_c} + \frac{DX_i}{\alpha \omega_c} \right) H_{ic} \tag{3.2-25}$$

For input signals that are low in frequency relative to the break frequencies ω_i and ω_c of the lowpass transfer functions H_c and H_{ic}, the transfer functions H_c and H_{ic} can be approximated by unity. With this approximation, the resultant approximate error transforms are

$$X'_{e(A)} = \frac{DX_i}{\omega_c} \tag{3.2-26}$$

$$X'_{e(B)} = \frac{D^2 X_i}{\omega_i \omega_c} \tag{3.2-27}$$

$$X'_{e(C)} = \frac{D^2 X_i}{\omega_i \omega_c} + \frac{DX_i}{\alpha \omega_i} \tag{3.2-28}$$

The error transform is denoted X'_e to indicate that this represents an approximation that assumes that H_c, H_{ic} are unity. The inverse Laplace transforms of these are

$$x'_{e(A)} = \frac{1}{\omega_c} Dx_i = \frac{1}{\omega_c} \frac{dx_i}{dt} \tag{3.2-29}$$

$$x'_{e(B)} = \frac{1}{\omega_i \omega_c} D^2 x_i = \frac{1}{\omega_i \omega_c} \frac{d^2 x_i}{dt^2} \tag{3.2-30}$$

$$x'_{e(C)} = \frac{1}{\alpha \omega_c} Dx_i + \frac{1}{\omega_i \omega_c} D^2 x_i = \frac{1}{\alpha \omega_c} \frac{dx_i}{dt} + \frac{1}{\omega_i \omega_c} \frac{d^2 x_i}{dt^2} \tag{3.2-31}$$

The approximate error response to a low-frequency input can be written in the following general form:

$$x_e = c_0 x_i + c_1 Dx_i + c_2 D^2 x_i + c_3 D^3 x_i + \cdots \tag{3.2-32}$$

The parameters $c_0, c_1, c_2, c_3, \ldots$ are called the *error coefficients* of the feedback control loop. The coefficient c_0 is frequently called the position error

TABLE 3.2-1   Error Coefficients for Loops A, B, and C

	A	B	C
Position error coefficient c_0	0	0	0
Velocity error coefficient c_1	$1/\omega_c$	0	$1/\alpha\omega_c$
Acceleration error coefficient c_2	—	$1/\omega_i\omega_c$	$1/\omega_i\omega_c$

coefficient, c_1 the velocity error coefficient, c_2 the acceleration error coefficient, etc. For a loop with infinite gain at zero frequency (which applies to loops A, B, and C) the position error coefficient c_0 is zero. The number of terms that are included in this series depends on the order of the maximum low-frequency integration of the feedback loop, as will be explained later. The error coefficients for the three cases are shown in Table 3.2-1.

In Table 3.2-1, the velocity error coefficient c_1 for loop A is $1/\omega_c$. Since loop C has α times the loop gain of loop A at low frequencies, its velocity error coefficient is reduced from $1/\omega_c$ by the factor α, so that c_1 is equal to $1/\alpha\omega_c$. Loop B is equivalent to loop C with the factor α set equal to infinity. Hence, the velocity error coefficient c_1 for loop B is zero. The acceleration error coefficients c_2 for loops B and C are both equal to $1/\omega_i\omega_c$. For loop A the acceleration error coefficient c_2 is not included in the expansion, because the velocity error term adequately characterizes the low-frequency error response.

The preceding discussion assumed that the lowpass filter responses H_c and H_{ic} can be approximated by unity. Their effect can be included in the following manner. In Fig 3.2-5, the transform expressions for Eqs 3.2-23, -24, -25 are represented in signal-flow diagram form. The x'_e signals are the error responses that are calculated by the simplified analysis given above in Eqs 3.2-26, -27, -28. These signal-flow diagrams show that the complete x_e response can be obtained by passing the x'_e signal through the lowpass filter H_c or H_{ic}.

The lowpass filter H_c has the same ramp response shown in Chapter 2, Fig 2.7-5a. Hence for a low-frequency input, the output from H_c is approximately equal to its input delayed in time by the time constant τ_c, which is equal to $1/\omega_c$:

$$\tau_c = 1/\omega_c \tag{3.2-33}$$

Thus, for a low-frequency input, the error response for loop A is approximated by

$$x_e[t] = x'_e[t - \tau_c] \tag{3.2-34}$$

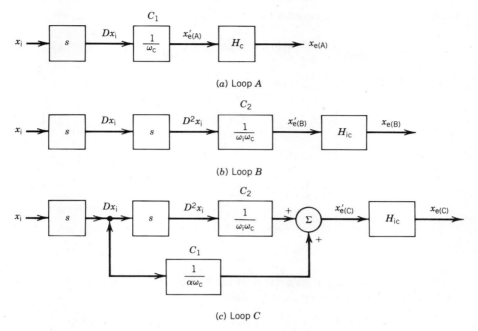

Figure 3.2-5 Approximate signal-flow diagrams for low-frequency input to loops A, B, and C.

Since $x'_e = c_1 Dx_i$, the error response for loop A is

$$x_{e(A)}[t] = c_1 Dx_i[t - \tau_c] = \frac{1}{\omega_c} Dx_i[t - \tau_c] \qquad (3.2\text{-}35)$$

This is a good approximation for the response of loop A to a low-frequency input. The error at time t is equal to $1/\omega_c$ multiplied by the input velocity Dx_i evaluated at time $(t - \tau_c)$, where $\tau_c = 1/\omega_c$.

The lowpass filter H_{ic} is equivalent to two single-order lowpass filters in cascade, one having a time constant $\tau_c = 1/\omega_c$ and the other having the time constant τ_i given by

$$\tau_i = 1/\omega_i \qquad (3.2\text{-}36)$$

For a low-frequency input, each filter produces a time delay equal to its time constant, and so the total time delay through the two filters, which is denoted τ_{ic}, is the sum of the time constants of the two filters. Thus,

$$\tau_{ic} = \tau_i + \tau_c = \frac{1}{\omega_i} + \frac{1}{\omega_c} \qquad (3.2\text{-}37)$$

Therefore, the error responses for loops B and C are approximated by

$$x_e[t] = x'_e[t - \tau_{ic}] \tag{3.2-38}$$

Combining this with Eqs 3.2-30, -31 gives

$$x_{e(B)}[t] = c_2 D^2 x_i[t - \tau_{ic}] = \frac{1}{\omega_i \omega_c} D^2 x_i[t - \tau_{ic}] \tag{3.2-39}$$

$$x_{e(C)}[t] = c_1 Dx_i[t - \tau_{ic}] + c_2 D^2 x_i[t - \tau_{ic}]$$

$$= \frac{1}{\alpha \omega_c} Dx_i[t - \tau_{ic}] + \frac{1}{\omega_i \omega_c} D^2 x_i[t - \tau_{ic}] \tag{3.2-40}$$

These provide good approximations to the responses of loops B and C to low-frequency inputs.

3.2.2 Error-Coefficient Expansions for Feedback Loops with Finite Zero-Frequency Loop Gain

The loop transfer functions for loops A, B, and C have infinite loop gain at zero frequency. Now let us consider two loops, D and E, with finite zero-frequency loop gain, which have the following loop transfer functions:

$$G_{(D)} = \frac{\omega_c}{s(1 + \omega_0/s)} \tag{3.2-41}$$

$$G_{(E)} = \frac{\omega_c(1 + \omega_i/s)}{s(1 + \omega_0/s)(1 + \omega_i/\alpha s)} \tag{3.2-42}$$

Magnitude asymptote plots of these are shown in diagram a of Fig 3.2-6. The corresponding approximate G_{ie} plots are shown in diagram b. The approximate equations for G_{ie} are:

$$*G_{ie(D)} = \frac{1 + \omega_0/s}{1 + \omega_c/s} \tag{3.2-43}$$

$$*G_{ie(E)} = \frac{(1 + \omega_0/s)(1 + \omega_i/\alpha s)}{(1 + \omega_i/s)(1 + \omega_c/s)} \tag{3.2-44}$$

The values of G at zero frequency are

$$G_{(D)}[0] = K_{0(D)} = \omega_c/\omega_0 \tag{3.2-45}$$

$$G_{(E)}[0] = K_{0(E)} = \alpha \omega_c/\omega_0 \tag{3.2-46}$$

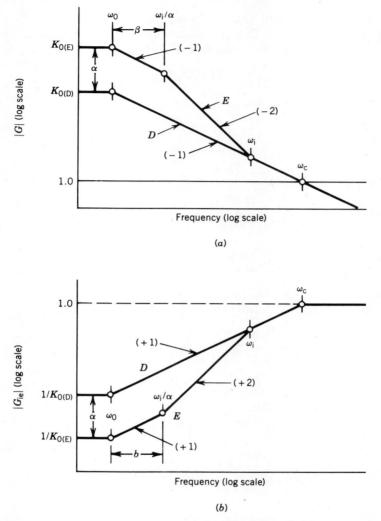

Figure 3.2-6 Magnitude asymptote plots of G and approximate G_{ie} for loops D and E, which have finite loop gain at zero frequency: (a) G; (b) G_{ie}.

As shown in Fig 3.2-6a, the ratio of the two lowest break frequencies of loop E is denoted by the symbol β:

$$\omega_i/\alpha = \beta\omega_0 \qquad (3.2\text{-}47)$$

Hence,

$$\alpha\beta = \omega_i/\omega_0 \qquad (3.2\text{-}48)$$

Convert the denominator factors of Eqs 3.2.43, -44 from the form $(1 + \omega_x/s)$ to the form $(1 + s/\omega_x)$. These equations can then be expressed as follows in terms of H_c and H_{ic}, which were defined in Eqs 3.2-13, -14:

$$*G_{ie\,(D)} = \frac{(s/\omega_c)(1 + \omega_0/s)}{1 + s/\omega_c} = \left(\frac{s}{\omega_c} + \frac{\omega_0}{\omega_c}\right)H_c \qquad (3.2\text{-}49)$$

$$*G_{ie\,(E)} = \frac{\left(\dfrac{s^2}{\omega_i\omega_c}\right)\left(1 + \dfrac{\omega_0}{s}\right)\left(1 + \dfrac{\omega_i}{\alpha s}\right)}{\left(1 + \dfrac{s}{\omega_i}\right)\left(1 + \dfrac{s}{\omega_c}\right)}$$

$$= \left(\frac{s^2}{\omega_i\omega_c}\right)\left(1 + \frac{\omega_0}{s} + \frac{\omega_i}{\alpha s} + \frac{\omega_0\omega_i}{\alpha s^2}\right)H_{ic}$$

$$= \left\{\frac{s^2}{\omega_i\omega_c} + s\left(\frac{\omega_0}{\omega_i\omega_c} + \frac{1}{\alpha\omega_c}\right) + \frac{\omega_0}{\alpha\omega_c}\right\}H_{ic} \qquad (3.2\text{-}50)$$

Substitute into Eqs 3.2-49, -50 the values for K_0 given in Eqs 5.3-45, -46 and the value for $\alpha\beta$ in Eq 3.2-48. This gives

$$*G_{ie\,(D)} = \left(\frac{s}{\omega_c} + \frac{1}{K_{0(D)}}\right)H_c \qquad (3.2\text{-}51)$$

$$*G_{ie\,(E)} = \left\{\frac{s^2}{\omega_c\omega_i} + s\left(\frac{1}{\alpha\beta\omega_c} + \frac{1}{\alpha\omega_c}\right) + \frac{1}{K_{0(E)}}\right\}H_{ic}$$

$$= \left\{\frac{s^2}{\omega_c\omega_i} + \frac{s}{\alpha\omega_c}\left(1 + \frac{1}{\beta}\right) + \frac{1}{K_{0(E)}}\right\}H_{ic} \qquad (3.2\text{-}52)$$

The approximate error transforms are therefore

$$X_{e(D)} = *G_{ie\,(D)}X_i = \left(\frac{sX_i}{\omega_c} + \frac{X_i}{K_{0(D)}}\right)H_c$$

$$= \left(\frac{DX_i}{\omega_c} + \frac{X_i}{K_{0(D)}}\right)H_c \qquad (3.2\text{-}53)$$

$$X_{e(E)} = *G_{ie\,(E)}X_i$$

$$= \left\{\frac{s^2X_i}{\omega_i\omega_c} + sX_i\frac{1}{\alpha\omega_c}\left(1 + \frac{1}{\beta}\right) + \frac{X_i}{K_{0(E)}}\right\}H_{ic}$$

$$= \left\{\frac{D^2X_i}{\omega_i\omega_c} + DX_i\frac{1}{\alpha\omega_c}\left(1 + \frac{1}{\beta}\right) + \frac{X_i}{K_{0(E)}}\right\}H_{ic} \qquad (3.2\text{-}54)$$

TABLE 3.2-2   Error Coefficients for Loops D and E

	D	E
Position error coefficient c_0	$1/K_0 = \omega_0/\omega_c$	$1/K_0 = \omega_0/\alpha\omega_c$
Velocity error coefficient c_1	$1/\omega_c$	$(1/\alpha\omega_c)(1 + 1/\beta)$
Acceleration error coefficient c_2	—	$1/\omega_i\omega_c$

If H_c and H_{ic} are approximated by unity, the inverse transforms of Eqs 3.2-53, -54 can be taken to obtain the following first approximations to the low-frequency error response:

$$x'_{e(D)} = \frac{1}{\omega_c} Dx_i + \frac{1}{K_{0(D)}} x_i \qquad (3.2\text{-}55)$$

$$x'_{e(E)} = \frac{1}{\omega_i\omega_c} D^2 x_i + \left\{ \frac{1}{\alpha\omega_c}\left(1 + \frac{1}{\beta}\right)\right\} Dx_i + \frac{1}{K_{0(E)}} x_i \qquad (3.2\text{-}56)$$

The error-coefficient values for these expansions are listed in Table 3.2-2. For each of these cases the position error coefficient c_0 is nonzero, and is equal to the reciprocal of the zero-frequency loop gain K_0.

Since these error-coefficient values are based on the expansion of the approximate error transfer function, they are approximations. The exact position error coefficient c_0 is equal to the value of G_{ie} at zero frequency, which is

$$c_0 = G_{ie}[0] = \frac{1}{1 + G[0]} = \frac{1}{1 + K_0} \qquad (3.2\text{-}57)$$

For a reasonably large value of the zero-frequency loop gain K_0, this is very nearly equal to the value $c_0 = 1/K_0$ given in Table 3.2-2.

Equations 3.2-53, -54 are expressed in signal-flow diagram form in Fig 3.2-7. When the effects of the time delays of H_c and H_{ic} are included, the error responses corresponding to Eqs 3.2-55, 56 are approximated quite accurately as follows for low-frequency input signals:

$$x_{e(D)}[t] = c_1 Dx_i[t - \tau_c] + c_0 x_i[t - \tau_c] \qquad (3.2\text{-}58)$$

$$x_{e(E)}[t] = c_2 D^2 x_i[t - \tau_{ic}] + c_1 Dx_i[t - \tau_{ic}] + c_0 x_i[t - \tau_{ic}] \qquad (3.2\text{-}59)$$

The error-coefficient expansions of the error responses include terms up to a specified derivative. The order of that specified derivative is equal to the maximum logarithmic slope of the magnitude asymptote of G_{ie}. Alternatively, it is equal to the negative of the maximum logarithmic slope of the magnitude asymptote of G at frequencies below ω_c. Since $|G|$ for loops A and D has a

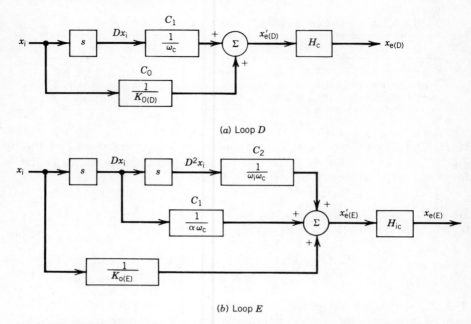

(a) Loop D

(b) Loop E

Figure 3.2-7 Approximate signal-flow diagram for low-frequency input to loops D and E.

maximum slope of -1 below ω_c (a single integration at low frequencies), terms up to the first derivative are included in the error-coefficient expansion for these loops. For loops B, C, and E, the maximum asymptote slope of $|G|$ below ω_c is -2 (representing a double integration at low frequencies), and so terms up to the second derivative are included in the error-coefficient expansion. When the asymptote of $|G|$ has a slope of -3 at frequencies below ω_c (a triple integration), terms up to the third derivative are included in the error-coefficient expansion. This issue is discussed in a more general fashion in Section 3.3.

3.2.3 Error Coefficients for Five Radar Tracking Loops

Let us consider a classic radar tracking problem, a radar tracking a target following a straight-line constant-velocity course. Five different radar tracking loops are assumed that have (at low frequencies) the loop transfer functions described by the five magnitude asymptote plots shown in Fig 3.2-8. (The loops are assumed to have good stability. Hence we can ignore the high-frequency characteristics and approximate all of the loop transfer functions by ω_c/s at frequencies above ω_c.)

Loops B and C are the only ones that are practical radar tracking loops. The others are included to illustrate the low-frequency error components of a feedback loop.

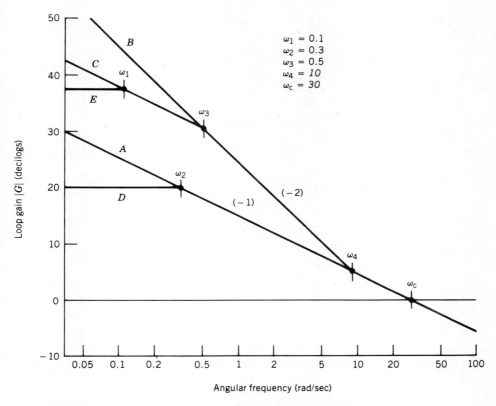

Figure 3.2-8   Magnitude asymptote plots of G for radar-tracking loops A to E.

Let us calculate the error coefficients for these five loop transfer functions. The loop transfer functions corresponding to the magnitude asymptote plots in Fig 3.2-8 are

$$G_A = \frac{\omega_c}{s} \tag{3.2-60}$$

$$G_B = \frac{\omega_c(1 + \omega_4/s)}{s} \tag{3.2-61}$$

$$G_C = \frac{\omega_c(1 + \omega_4/s)}{s(1 + \omega_3/s)} \tag{3.2-62}$$

$$G_D = \frac{\omega_c}{s(1 + \omega_2/s)} \tag{3.2-63}$$

$$G_E = \frac{\omega_c(1 + \omega_4/s)}{s(1 + \omega_1/s)(1 + \omega_3/s)} \tag{3.2-64}$$

The break frequencies and gain crossover frequency are

$$\omega_1 = 0.1 \; \text{sec}^{-1} \tag{3.2-65}$$

$$\omega_2 = 0.3 \; \text{sec}^{-1} \tag{3.2-66}$$

$$\omega_3 = 0.5 \; \text{sec}^{-1} \tag{3.2-67}$$

$$\omega_4 = 10 \; \text{sec}^{-1} \tag{3.2-68}$$

$$\omega_c = 30 \; \text{sec}^{-1} \tag{3.2-69}$$

The parameters ω_i, α, and β defined in Figs 3.2-2 and -6 are

$$\omega_i = \omega_4 = 10 \; \text{sec}^{-1} \tag{3.2-70}$$

$$\alpha = \frac{\omega_4}{\omega_3} = \frac{10}{0.5} = 20 \tag{3.2-71}$$

$$\beta = \frac{\omega_3}{\omega_1} = \frac{0.5}{0.1} = 5 \tag{3.2-72}$$

The values of zero-frequency loop gain K_0 for loops D and E are

Loop D:
$$K_0 = \frac{\omega_c}{\omega_2} = \frac{30}{0.3} = 100 \tag{3.2-73}$$

Loop E:
$$K_0 = \frac{\omega_c}{\omega_1} \frac{\omega_4}{\omega_3} = \frac{30}{0.1} \frac{10}{0.5} = 6000 \tag{3.2-74}$$

From Tables 3.2-1 and -2, the pertinent error coefficients for the five feedback control loops are

Loop A:

$$c_0 = 0 \tag{3.2-75}$$

$$c_1 = \frac{1}{\omega_c} = \frac{1}{30 \; \text{sec}^{-1}} = \tfrac{1}{30} \; \text{sec} \tag{3.2-76}$$

Loop B:

$$c_0 = 0 \tag{3.2-77}$$

$$c_1 = 0 \tag{3.2-78}$$

$$c_2 = \frac{1}{\omega_i \omega_c} = \frac{1}{(10 \; \text{sec}^{-1})(30 \; \text{sec}^{-1})} = \tfrac{1}{300} \; \text{sec}^2 \tag{3.2-79}$$

Loop C:

$$c_0 = 0 \tag{3.2-80}$$

$$c_1 = \frac{1}{\alpha \omega_c} = \frac{1}{(20)(30 \text{ sec}^{-1})} = \tfrac{1}{600} \text{ sec} \tag{3.2-81}$$

$$c_2 = \frac{1}{\omega_i \omega_c} = \frac{1}{(10 \text{ sec}^{-1})(30 \text{ sec}^{-1})} = \tfrac{1}{300} \text{ sec}^2 \tag{3.2-82}$$

Loop D:

$$c_0 = \frac{1}{1 + K_0} = \frac{1}{1 + 100} \cong \tfrac{1}{100} \tag{3.2-83}$$

$$c_1 = \frac{1}{\omega_c} = \frac{1}{30 \text{ sec}^{-1}} = \tfrac{1}{30} \text{ sec} \tag{3.2-84}$$

Loop E:

$$c_0 = \frac{1}{1 + K_0} = \frac{1}{1 + 6000} \cong \frac{1}{6000} \tag{3.2-85}$$

$$c_1 = \frac{1 + 1/\beta}{\alpha \omega_c} = \frac{1 + \tfrac{1}{5}}{(20)(30 \text{ sec}^{-1})} = \tfrac{1}{500} \text{ sec} \tag{3.2-86}$$

$$c_2 = \frac{1}{\omega_c \omega_i} = \frac{1}{(30 \text{ sec}^{-1})(10 \text{ sec}^{-1})} = \tfrac{1}{300} \text{ sec}^2 \tag{3.2-87}$$

Note that the values of the velocity error coefficient c_1 are the same for loops A and D, and the values of the acceleration error coefficient c_2 are the same for loops B, C, and E. The time-delay parameters for these five loops are

Loops A and D:

$$\tau_c = \frac{1}{\omega_c} = \frac{1}{30 \text{ sec}^{-1}} = 0.033 \text{ sec} \tag{3.2-88}$$

Loops B, C, and E:

$$\tau_{ic} = \frac{1}{\omega_i} + \frac{1}{\omega_c} = \frac{1}{10 \text{ sec}^{-1}} + \frac{1}{30 \text{ sec}^{-1}}$$
$$= 0.100 \text{ sec} + 0.033 \text{ sec} = 0.133 \text{ sec} \tag{3.2-89}$$

TABLE 3.2-3 Error-Coefficient Parameters for Loops of Fig 3.2-8 for Calculating Response to Arbitrary Low-Frequency Input

Loop	c_0	c_1 (sec)	c_2 (sec²)	Time Delay (sec)
A	0	$\frac{1}{30}$	—	$\tau_c = 0.033$
B	0	0	$\frac{1}{300}$	$\tau_{ic} = 0.133$
C	0	$\frac{1}{600}$	$\frac{1}{300}$	$\tau_{ic} = 0.133$
D	$\frac{1}{100}$	$\frac{1}{30}$	—	$\tau_c = 0.033$
E	$\frac{1}{6000}$	$\frac{1}{500}$	$\frac{1}{300}$	$\tau_{ic} = 0.133$

The error coefficients and time-delay parameters for the five loops are shown in Table 3.2-3.

As a simple example for applying error coefficients, let us assume that the command angular input to the radar tracking servo is a sinusoid of angular frequency $\omega = 2$ rad/sec with an amplitude Θ_m of 10°, given by

$$\Theta_i = \Theta_m \sin[\omega t] = (10 \text{ deg})\sin[2t] \tag{3.2-90}$$

The command angular velocity and acceleration are obtained by differentiating Eq 3.2-90:

$$\Omega_i = \frac{d\Theta_i}{dt} = \Theta_m \omega \cos[\omega t] = (20 \text{ deg/sec})\cos[2t] \tag{3.2-91}$$

$$\alpha_i = \frac{d\Omega_i}{dt} = -\Theta_m \omega^2 \sin[\omega t]$$

$$= -(40 \text{ deg/sec}^2)\sin[\omega t] \tag{3.2-92}$$

Consider the error response of loop E to this input. From the values for loop E in Table 3.2-3, this error is

$$\Theta_e[t] = c_0 \Theta_i[t - \tau_{ic}] + c_1 \Omega_i[t - \tau_{ic}] + c_2 \alpha_i[t - \tau_{ic}]$$

$$= \tfrac{1}{6000}(10 \text{ deg})\sin[2(t - 0.133)]$$

$$+ \left(\tfrac{1}{500} \text{ sec}\right)(20 \text{ deg/sec})\cos[2(t - 0.133)]$$

$$+ \left(\tfrac{1}{300} \text{ sec}^2\right)(-40 \text{ deg/sec}^2)\sin[2(t - 0.133)]$$

$$= \{0.00167\sin[2t - \phi]$$

$$+ 0.0400\cos[2t - \phi] - 0.1333\sin[2t - \phi]\} \text{ deg} \tag{3.2-93}$$

The phase ϕ is

$$\phi = \omega\tau_{ic} \text{ rad} = 2(0.133) \text{ rad} = 0.266 \text{ rad} = 15.2° \tag{3.2-94}$$

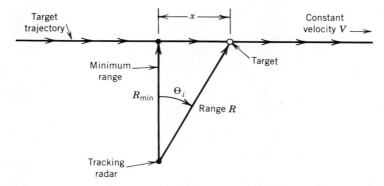

Figure 3.2-9   Geometry for target-tracking example.

3.2.4   Angular Input and Derivatives for a Constant-Velocity Target Course

Let us apply the error-coefficient expansion to a classic target-tracking problem. As shown in Fig 3.2-9, a target that is moving in a straight-line course, at a constant velocity V, is being tracked by a radar tracking servo. The instantaneous range from the tracking servo to the target is denoted R, and the minimum range between the tracking servo and the target trajectory is denoted R_{min}. The time axis is chosen so that time $t = 0$ occurs when the target is at minimum range. The position of the target along the trajectory is denoted x, and is measured from the point where the target is at minimum range. This distance x is thus equal to

$$x = Vt \qquad (3.2\text{-}95)$$

A zero reference vector is drawn from the tracking servo to the point on the trajectory corresponding to minimum range. The angle between this reference vector and the vector drawn from the tracking servo to the target is denoted Θ_i, and is characterized by

$$\tan[\Theta_i] = x/R_{min} \qquad (3.2\text{-}96)$$

The tracking servo operates in angular coordinates. To develop a plot of the tracking error by means of error coefficients, expressions for the angle Θ_i, the angular velocity $\Omega_i = d\Theta_i/dt$, and the angular acceleration $\alpha_i = d\Omega_i/dt$ are needed, which are derived in the following manner. Differentiating Eq 3.2-96 gives

$$\frac{d\tan[\Theta_i]}{dt} = \frac{d\Theta_i/dt}{\cos^2[\Theta]} = \frac{dx/dt}{R_{min}} = \frac{V}{R_{min}} \qquad (3.2\text{-}97)$$

Note that $d \tan[\Theta] = d\Theta/\cos^2[\Theta]$. Hence, $d\Theta_i/dt$ is

$$\Omega_i = \frac{d\Theta_i}{dt} = \frac{V}{R_{min}}\cos^2[\Theta_i] = \frac{V}{R_{min}}\left(\frac{R_{min}}{R}\right)^2 \qquad (3.2\text{-}98)$$

Figure 3.2-9 shows that $\cos[\Theta_i] = R/R_{min}$. Since R^2 equals $(x^2 + R_{min}^2)$, Eq 3.2-98 becomes

$$\Omega_i = \frac{V}{R_{min}}\frac{R_{min}^2}{\left(R_{min}^2 + x^2\right)} = \frac{V/R_{min}}{1 + \left(x/R_{min}\right)^2} \qquad (3.2\text{-}99)$$

The maximum angular velocity Ω_{max} of the target occurs when the target is at minimum range (at time $t = 0$), and is equal to the target velocity V divided by the minimum range:

$$\Omega_{max} = \frac{V}{R_{min}} \qquad (3.2\text{-}100)$$

Since $x = Vt$, the ratio x/R_{min} can be expressed as

$$\frac{x}{R_{min}} = \frac{Vt}{R_{min}} = \Omega_{max}t \qquad (3.2\text{-}101)$$

Substituting Eqs 3.2-100, -101 into Eq 3.2-99 gives the following expressions for the angular velocity of the target experienced by the tracking servo:

$$\Omega_i = \frac{d\Theta_i}{dt} = \frac{\Omega_{max}}{1 + \Omega_{max}^2 t^2} \qquad (3.2\text{-}102)$$

Differentiating this gives the target angular acceleration, which is

$$\alpha_i = \frac{d\Omega_i}{dt} = \frac{d\Theta_i^2}{dt^2} = \frac{-2\Omega_{max}^3 t}{\left(1 + \Omega_{max}^2 t^2\right)^2} \qquad (3.2\text{-}103)$$

The derivative of this is the rate of acceleration of the target, which is

$$\frac{d\alpha_i}{dt} = -\frac{2\Omega_{max}^3\left(1 - 3\Omega_{max}^2 t^2\right)}{\left(1 + \Omega_{max}^2 t^2\right)^3} \qquad (3.2\text{-}104)$$

This rate of acceleration is zero when $(1 - 3\Omega_{max}^2 t^2)$ is zero, which occurs at the following value of time t:

$$t = t_m = \pm \frac{1}{\sqrt{3}\,\Omega_{max}} \qquad (3.2\text{-}105)$$

This time t_m is the time of the maximum absolute value of acceleration, which occurs when the angle Θ_i (defined in Fig 3.2-9) is $\pm 30°$. Substituting Eq 3.2-105 into Eq 3.2-103 gives the following for the maximum absolute value of acceleration:

$$\text{Max}|\alpha_i| = \tfrac{3}{8}\sqrt{3}\,\Omega_{\max}^2 = 0.650\Omega_{\max}^2 \qquad (3.2\text{-}106)$$

From Eqs 3.2-61, -66 the input angle is

$$\Theta_i = \arctan[\Omega_{\max}t] \qquad (3.2\text{-}107)$$

Let us assume the following parameters for the tracking example:

$$R_{\min} = 1000 \text{ ft} \qquad (3.2\text{-}108)$$

$$V = 1000 \text{ ft/sec} \qquad (3.2\text{-}109)$$

This velocity is approximately Mach 1. By Eq 3.2-100, the maximum angular velocity is

$$\Omega_{\max} = \frac{V}{R_{\min}} = \frac{1000 \text{ ft/sec}}{1000 \text{ ft}} = 1.0 \text{ rad/sec} \qquad (3.2\text{-}110)$$

The expressions for angle, angular velocity, angular acceleration, and angular rate of acceleration given in Eqs 3.2-107, -102, -103, and -104 are

$$\Theta_i = \arctan[t] \qquad \text{rad} \qquad (3.2\text{-}111)$$

$$\Omega_i = \frac{1}{(1 + t^2)} \qquad \text{rad/sec} \qquad (3.2\text{-}112)$$

$$\alpha_i = -\frac{2t}{(1 + t^2)^2} \qquad \text{rad/sec}^2 \qquad (3.2\text{-}113)$$

$$\frac{d\alpha_i}{dt} = -\frac{2(1 - 3t^2)}{(1 + t^2)^3} \qquad \text{rad/sec}^3 \qquad (3.2\text{-}114)$$

By Eq 3.2-106, the maximum absolute value of angular acceleration is

$$\text{Max}|\alpha_i| = 0.650\Omega_{\max}^2 = 0.650 \text{ rad/sec}^2 \qquad (3.2\text{-}115)$$

By Eq 3.2-105, this occurs at a time

$$t_m = \pm 1/\sqrt{3}\,(1 \text{ rad/sec}) = \pm 0.577 \text{ sec} \qquad (3.2\text{-}116)$$

Figures 3.2-10 to -12 show plots of the input angle, angular velocity, and angular acceleration given in Eqs 3.2-111 to -113.

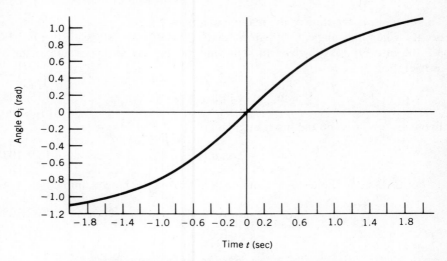

Figure 3.2-10   Plot of target angle Θ_i for example.

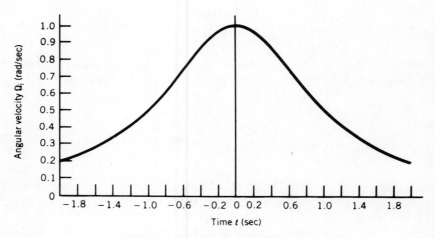

Figure 3.2-11   Angular velocity of target.

3.2.5   Calculation of Tracking Servo Error for a Constant-Velocity Target Course

Loop A. As shown in Table 3.2-3, the error response of loop A is characterized by the velocity error coefficient $c_1 = \frac{1}{30}$ sec, with a time delay $\tau_c = 0.033$ sec. Hence, the error response (ignoring time delay) is

$$\Theta_e' = c_1 \frac{d\Theta_i}{dt} = c_1\Omega_i = \left(\tfrac{1}{30}\ \text{sec}\right)\Omega_i \qquad (3.2\text{-}117)$$

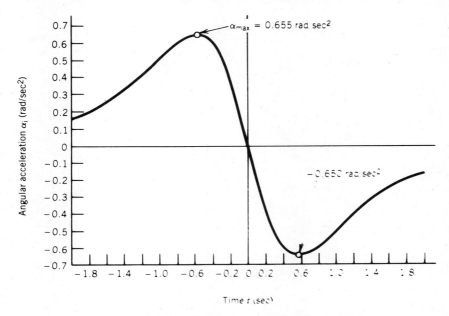

Figure 3.2-12 Angular acceleration of target.

The error response for loop A is constructed in Fig 3.2-13. The plot of input velocity Ω_i in Fig 3.2-11 is multiplied by $\frac{1}{30}$ sec to obtain the plot of Θ_e' shown as curve ① in Fig 3.2-13. To provide a more convenient error scale, the angular error in radians is multiplied by 1000 to express it in milliradians. The effect of time delay τ_c is included by delaying curve ① by $\tau_c = 0.033$ sec, to form curve ②, which is the final calculated error.

As shown in Fig 3.2-11, the maximum angular velocity occurs at time $t = 0$, and is 1.0 rad/sec, or 1000 mrad/sec. Hence, the maximum approximate error Θ_e' for loop A occurs at time $t = 0$, and is

$$\text{Max}[\Theta_e'] = \left(\tfrac{1}{30} \text{ sec}\right)\text{max}[\Omega_i]$$
$$= \left(\tfrac{1}{30} \text{ sec}\right)(1000 \text{ mrad/sec}) = 33.3 \text{ mrad} \qquad (3.2\text{-}118)$$

This is the error at point a in Fig 3.2-13. When the time delay τ_c is included, the maximum error is still 33.3 mrad, but occurs at time $t = 0.033$ sec, which is the time of the peak for curve ②.

Loop B. As shown in Table 3.2-3, the error response of loop B is character-ized by the acceleration error coefficient $c_2 = \frac{1}{300}$ sec^2, with a time delay $\tau_{ic} = 0.133$ sec. Hence, the error response (ignoring time delay) is

$$\Theta_e' = c_2 \frac{d^2\Theta_i}{dt^2} = c_2\alpha_i = \left(\tfrac{1}{300} \text{ sec}^2\right)\alpha_i \qquad (3.2\text{-}119)$$

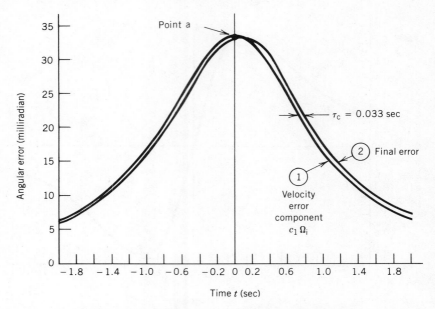

Figure 3.2-13   Calculation of tracking error for loop A.

The error response for loop B is constructed in Fig 3.2-14. Note that the error scale for Fig 3.2-14 is expanded relative to that for Fig 3.2-13, because the error for loop B is much smaller than that for loop A. The plot of input acceleration α_i in Fig 3.2-12 is multiplied by $\frac{1}{300}$ sec^2 to obtain the plot of Θ_e' shown as curve ① in Fig 3.2-14. The effect of time delay τ_{ic} is included by delaying curve ① by $\tau_{ic} = 0.133$ sec, to form curve ②, which is the final calculated error.

As shown in Fig 3.2-12, the maximum angular acceleration is 0.65 rad/sec^2 (or 650 mrad/sec^2), and occurs at time $t = -0.577$ sec. Hence, the maximum approximate error Θ_e' for loop B occurs at time $t = -0.577$ sec, and is

$$\text{Max}[\Theta_e'] = \left(\tfrac{1}{300} \sec^2\right)\text{Max}[\alpha_i]$$
$$= \left(\tfrac{1}{300} \sec^2\right)(6500 \text{ mrad/sec}^2) = 2.17 \text{ mrad} \quad (3.2\text{-}120)$$

This is the error for point a in Fig 3.2-14. When the time delay τ_{ic} is included to form curve ②, the maximum error is still 2.17 mrad, but occurs at the following time:

$$t = -0.577 \sec + \tau_{ic}$$
$$= -0.577 \sec + 0.133 \sec = -0.444 \sec \quad (3.2\text{-}121)$$

A comparison of Figs 3.2-13, -14 shows that the integral network employed in

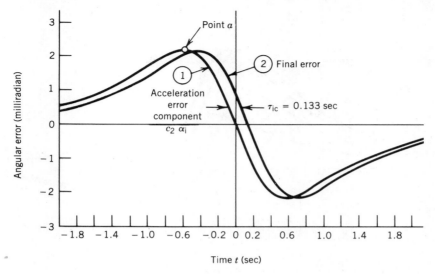

Figure 3.2-14 Calculation of tracking error for loop B.

loop B reduces the peak error relative to loop A by a factor of $15 : 1$, from 33.3 to 2.17 mrad.

Loop C. As shown in Table 3.2-3, the error response of loop C is characterized by the velocity error coefficient $c_1 = \frac{1}{600}$ sec and the acceleration error coefficient $c_2 = \frac{1}{300}$ sec^2, with a time delay $\tau_{ic} = 0.133$ sec. Hence, the error response (ignoring time delay) is

$$
\Theta_e' = c_1 \frac{d\Theta_i}{dt} + c_2 \frac{d^2\Theta_i}{dt^2}
$$

$$
= c_1 \Omega_i + c_2 \alpha_i = \left(\tfrac{1}{600} \text{ sec} \right) \Omega_i + \left(\tfrac{1}{300} \text{ sec}^2 \right) \alpha_i \qquad (3.2\text{-}122)
$$

The error response for loop C is constructed in Fig 3.2-15. The input velocity curve of Fig 3.2-11 is multiplied by $\frac{1}{600}$ sec to form curve ① of Fig 3.2-15, and the input acceleration curve α_i in Fig 3.2-12 is multiplied by $\frac{1}{300}$ sec^2 to form curve ②. Note that curve ②, which is the acceleration component of error, is the same as the undelayed error response for loop B, shown as curve ① in Fig 3.2-14. Curves ① and ② of Fig 3.2-15 are summed to form curve ③. Curve ③ is delayed by $\tau_{ic} = 0.133$ sec, to form the final error, curve ④.

Since the velocity error coefficient c_1 for loop C is lower than that for loop A by the factor $\alpha = 20$, the velocity error component for loop C (curve ① in Fig 3.2-15) is $\frac{1}{20}$ of the approximate error Θ_e' for loop A (curve ① in Fig 3.2-13). Hence, the maximum value for the velocity component of error for

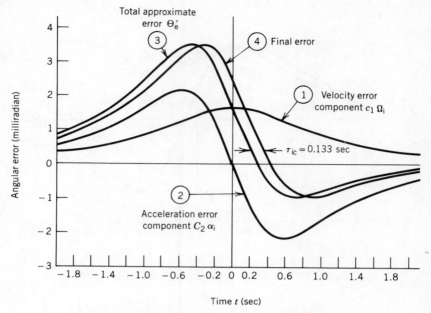

Figure 3.2-15 Calculation of tracking error for loop C.

loop C is

$$\text{Max[Velocity error]} = \tfrac{1}{20}(33.3 \text{ mrad})$$

$$= 1.67 \text{ mrad} \qquad (3.2\text{-}123)$$

Loop D. As shown in Table 3.2-3, the error response of loop D is characterized by the position error coefficient $c_0 = \tfrac{1}{100}$ and the velocity error coefficient $c_1 = \tfrac{1}{30}$ sec, with a time delay $\tau_c = 0.033$ sec. Hence, the error response (ignoring time delay) is

$$\Theta_e' = c_0\Theta_i + c_1\frac{d\Theta_i}{dt} = c_0\Theta_i + c_1\Omega_i$$

$$= \tfrac{1}{100}\Theta_i + \left(\tfrac{1}{30} \text{ sec}\right)\Omega_i \qquad (3.2\text{-}124)$$

The error response for loop D is constructed in Fig 3.2-16. The plot of input velocity Ω_i in Fig 3.2-11 is multiplied by $\tfrac{1}{30}$ sec to form curve ① of Fig 3.2-16, and the input angle curve Θ_i of Fig 3.2-10 is multiplied by $\tfrac{1}{100}$ to form curve ② of Fig 3.2-16. Note that curve ①, which is the velocity component of error, is the same as curve ① in Fig 3.2-13, which is the total undelayed error for loop A. Curves ① and ② of Fig 3.2-16 are summed to form curve

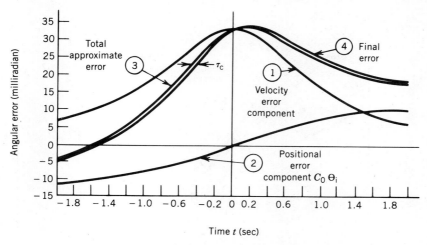

Figure 3.2-16   Calculation of tracking error for loop D.

③, which represents Θ_e'. Curve ③ is delayed by $\tau_c = 0.033$ sec to form the final error, curve ④.

This shows that the error for loop D is approximately equal to the error for loop A plus the position error term $c_0\Theta_i$, which is the input angle Θ_i multiplied by the position error coefficient.

Loop E. The error response of loop E is not calculated here. It is approximately equal to the error for loop C, shown in Fig 3.2-15, but is somewhat greater because of the following two effects:

1. Loop E has a position component of error equal to the input angle θ_i multiplied by the position error coefficient $c_0 = \frac{1}{6000}$.
2. The velocity error coefficient c_1 of loop E exceeds that for loop C by the factor $(1 + 1/\beta)$, which is equal to 1.2. Hence, the velocity component of error for loop E is equal to 1.2 times the velocity component of error for loop C.

3.2.6   Effect of Low-Frequency Triple Integration

The radar tracking example showed that adding the integral network to the single-integration loop A to produce the double-integration loop B or C greatly reduced the angular tracking error. An obvious question to ask is: Why not add another integral network to produce a triple integration at low frequencies, and thereby decrease the error further? This question is examined in Ref [1.1] (Section 5.4.6). That reference shows that a triple integration rarely achieves a significant error reduction. The following summarizes the reasons for this.

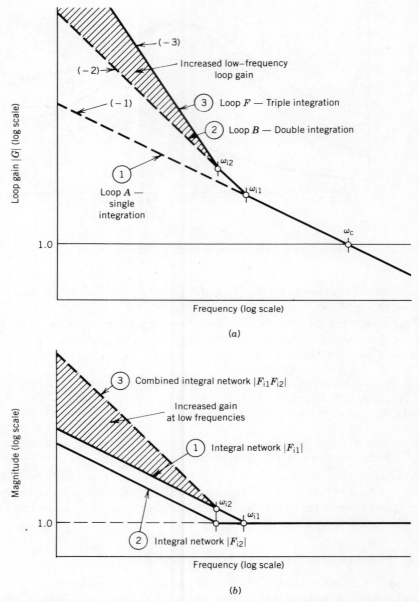

Figure 3.2-17 Asymptote plots of loop gain $|G|$ and integral-network gains for triple-integration loop F, compared with those for loops A and B: (a) plots of $|G|$; plots of integral-network gains $|F_{i1}|$, $|F_{i2}|$.

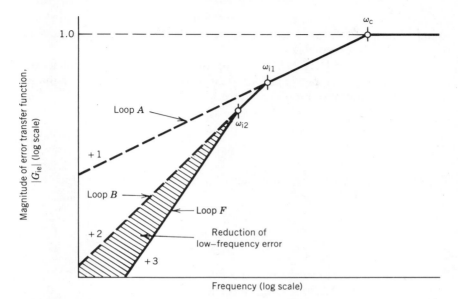

Figure 3.2-18   Magnitude asymptote plot of error transfer function G_{ie} for triple-integration loop F, compared with those for loops A and B.

In Fig 3.2-17a, curves (1) and (2) are the asymptote plots of loop gain for the single-integration loop A and the double-integration loop B. Curve (3) is the plot for the following loop transfer function, called loop F, which has a triple integration at low frequencies:

$$G_F = \frac{\omega_c(1 + \omega_{i1}/s)(1 + \omega_{i2}/s)}{s} \qquad (3.2\text{-}125)$$

The following discussion applies to any feedback loop having this response at low frequencies, regardless of the high-frequency response, provided that the loop has good stability.

In Fig 3.2-17, diagram b shows magnitude asymptote plots of the transfer functions of the integral networks for loops B and F. Curve (1) is the plot for the integral network of loop B; curve (2) is the plot for the second integral network added to loop F; and curve (3) is the total response for the two integral networks of loop F. The cross-hatched area shows the additional gain that is achieved by the extra loop-F integration. The corresponding increase in loop gain is shown by the cross-hatched area in diagram a.

Figure 3.2-18 shows the magnitude asymptote plots of the approximate G_{ie} transfer functions for loops A, B, and F. The cross-hatched area between the G_{ie} plots for loop B and loop F shows the reduction of error achieved by the triple integration relative to that for the double integration. This shows that the additional integral network in loop F reduces the components of the error signal of loop B that are lower in frequency than the break frequency ω_{i2} of the second integral network.

As shown in Ref [1.1] (Section 5.4.6), the approximate error for loop (F) for a low-frequency input is proportional to the third derivative of the input (the rate of acceleration), in accordance with the relation

$$x'_{e(F)} = c_3 \frac{d^3 x_i}{dt^3} \tag{3.2-126}$$

The parameter c_3 is the error coefficient for the third derivative, which is equal to

$$c_3 = \frac{1}{\omega_c \omega_{i1} \omega_{i2}} \tag{3.2-127}$$

For the radar tracking example, the peak magnitude of error for loop F, calculated by Eq 3.2-126, is roughly equal to the peak magnitude of the error that was calculated for loop B, which is proportional to the acceleration of the input. This result is typical of most practical cases.

A triple integration in usually ineffective in achieving error reduction relative to a double integration, because the error components at very low frequencies are generally quite small for the double-integration loop B. Consequently, further reduction of these error components does not appreciably change the peak value of the error.

An important consideration in including a second integral network in a feedback loop is that the network adds phase lag to the loop near gain crossover, and so the break frequency of the first integral network must be lowered to maintain good stability. Define ω_i as the break frequency of a loop with a double integration at low frequencies. The corresponding break frequencies ω_{i1}, ω_{i2} of a loop with a triple integration at low frequencies should be related as follows to ω_i, if the two loops are to have comparable stability:

$$\omega_{i1} + \omega_{i2} = \omega_i \tag{3.2-128}$$

It is reasonable to set $\omega_{i1} = 2\omega_{i2}$. For this condition, the values for ω_{i1}, ω_{i2} are

$$\omega_{i1} = \tfrac{2}{3}\omega_i \qquad \omega_{i2} = \tfrac{1}{3}\omega_i \tag{3.2-129}$$

Thus, for this condition, the break frequency ω_{i1} of the first integral network must be reduced by a factor of $\tfrac{2}{3}$ when the second integral network is added. The increased error caused by the reduced break frequency of the first integral network usually offsets any error reduction achieved by the second integral network.

As was shown in Section 2.7, a loop with a double integration at low frequency has *velocity memory*. After the initial transient has settled, the correction signal generated by the integral network is proportional to the input velocity. When a second integral network is added to produce a triple integration at low frequency, the feedback loop has *acceleration memory* as well as velocity memory.

The discussion in Chapter 8, Section 8.2.2, will show that the feedback loops of the Altair tracker use double-integration feedback loops in the angle tracker (which tracks in three dimensions) and a triple-integration feedback loop in the range tracker. Thus, the three angle-track feedback loops have velocity memory, while the range-track loop has both acceleration memory and velocity memory.

The reason for using a triple integration in the Altair range tracker is to improve its operation in the "coast" mode. When the warhead being tracked reenters the atmosphere, the resultant ionization wave may obscure the radar echo pulse. If this occurs, the range tracker "coasts" until the target echo pulse reappears. The triple integration gives the range tracker acceleration memory, which allows the tracker to coast for a longer period while maintaining acceptable tracking error. Thus, a triple integration is desirable in this feedback control loop even though it probably does not reduce tracking error under normal operating conditions.

3.3 GENERAL ANALYSIS OF THE RESPONSE TO AN ARBITRARY INPUT

3.3.1 Summary of Approach

Section 3.2 presented an error-coefficient method for calculating the response to an arbitrary input that was based on approximations of the G_{ie} frequency response. This section gives a more general approach, which does not approximate the system transfer function, and addresses high-frequency inputs as well as low-frequency inputs. It provides a basic philosophy for the approximations of system response given in Chapter 2. This section is a condensation of Chapter 14 in Ref [1.2].

If one has an input signal that is prescribed explicitly either by a set of equations or by tabulated data, the simplest way to calculate the response of a feedback system to that input is to feed the data into a simulation of the system. However, this situation rarely occurs in practice. One is usually faced with vague descriptions of tasks the system must perform, and disturbances it will experience. The response of the system must be characterized in a general fashion so that the effects of these vaguely defined commands and disturbances can be evaluated. Therefore, one needs insight and general quantitative means of characterizing the response to various inputs. The approach of this chapter is directed to that end.

Most methods for calculating the response of a system to an arbitrary input require that the input be approximated by a series of impulses, or steps, or straight lines, or parabolas. These may be considered to represent steps of derivatives or integrals of the input: A straight line is a step of the first derivative, a parabola is a step of the second derivative, and an impulse is a step of the first integral. By using a sufficient number of any one of these elements, any of these methods can approximate the response to the accuracy desired.

This section shows how to break an input down optimally into a sum of not just one type of element, but a number of different types, using for each portion of the input curve the order of derivative step that best approximates that portion. For example, an impulse is used to approximate a pulselike portion of the input, while a parabola is used to approximate a slowly varying portion. Approximating the input in this manner requires much fewer approximating elements than when one type is used. Even more important, it gives insight into why the system responds the way it does, and how the system transfer function can be changed to improve the response.

Thus, any curve can be represented as a sum of steps of various derivatives and integrals of the input. The steps of derivatives approximate the low-frequency portions of the input, and the steps of integrals approximate the high-frequency portions. To calculate the response to an arbitrary input, general expressions are needed for the responses to steps of derivatives and integrals of the input. In feedback-control applications, low-frequency inputs are usually of more interest, and so the response to a step of a derivative is considered first.

3.3.2 Response to a Step of a Derivative

The following analysis yields an expression for the response to a step of a derivative, which contains a series of steady-state coefficients and a transient term proportional to the size of the step. To simplify the symbolism, the nth time derivative is denoted D^n:

$$D^n x_i[t] = \frac{d^n x_i[t]}{dt^n} \qquad (3.3\text{-}1)$$

Similarly, the mth time integral of $x_i[t]$ is denoted $D^{-m}x_i[t]$.

As shown in Ref [1.2] (Chapter 14), the error response of a feedback loop to a step of the nth derivative of amplitude $D^n x_i$ can be expressed as

$$x_e = c_0 x_i + c_1 D x_i + c_2 D^2 x_i + \cdots + (c_n + T_n[t]) D^n x_i \qquad (3.3\text{-}2)$$

where c_0, c_1, etc. are the steady-state error coefficients, and $T_n[t]$ is the

nth-derivative transient term for the error response. The variables x_i, Dx_i, ... $D^{n-1}x_i$ are the lower order derivatives that correspond to this nth-derivative step $D^n x_i$.

The error coefficients are equal to

$$c_n = \frac{1}{n!} \frac{d^n G_{ie}}{ds^n}\bigg|_{s=0} \tag{3.3-3}$$

If the feedback loop has only single-order closed-loop poles, the nth-derivative transient term can be expressed as

$$T_n[t] = \sum_{k=1}^{K} K_{kn} e^{s_k t} \tag{3.3-4}$$

The closed loop has K closed-loop poles, and s_k is the value of a particular pole (which may be complex). The transient coefficient K_{kn} for the zeroth derivative ($n = 0$) for a particular pole s_k is

$$K_{k0} = (s - s_k) \frac{G_{ie}}{s}\bigg|_{s=s_k} \tag{3.3-5}$$

The transient coefficients for derivatives are related to this by

$$K_{kn} = \frac{K_{k0}}{s_k^n} \tag{3.3-6}$$

It can be shown from Eq 3.3-6 that the transient terms for successive derivatives are related by

$$T_{n-1}[t] = \frac{dT_n[t]}{dt} \tag{3.3-7}$$

Although Eq 3.3-6 does not apply to transfer functions with multiple-order poles, it can be shown that Eq 3.3-7 applies to all transfer functions.

Equation 3.3-6 shows that as n is increased, the coefficients for high-frequency poles (i.e., the poles with large values of $|s_k|$) are decreased at a much faster rate than the coefficients for low-frequency poles. Consequently, if n is sufficiently large, the transient coefficient for the pole of lowest frequency is much greater than all the other coefficients. Thus, for high-order derivatives, the transient terms are determined almost entirely by the lowest frequency closed-loop pole (or poles). If this pole is real, which is often the case, the transient terms for the higher derivatives are simply real exponentials.

The quantity $(c_n + T_n[t])$ in Eq 3.3-2 is called the nth-derivative composite transient term. As will be explained, by successively differentiating an input signal, and applying straight-line approximations to the result, an input signal can be broken down into a sum of steps of derivatives. Equation 3.3-2 shows that the resultant error response can be calculated by (1) including a composite transient term for each step of derivative, proportional to the size of the step; and (2) adding to the sum of these transients a steady-state portion, consisting of the lower-order derivatives, multiplied by the corresponding steady-state coefficients.

The expansion for the error response in Eq 3.3-2 can also be expressed in terms of the feedback response x_b by applying the relation

$$x_b = x_i - x_e \tag{3.3-8}$$

The expansion of Eq 3.3-2 becomes

$$x_b = c_0' x_i + c_1' D x_i + c_2' D^2 x_i + \cdots + c_n' D^n x_i + T_n'[t] D^n x_i \tag{3.3-9}$$

The parameters $c_0', c_1', \ldots$ are the steady-state coefficients for the feedback response. These are related as follows to the steady-state error coefficients:

$$c_n' = -c_n \qquad \text{for } n > 0 \tag{3.3-10}$$

$$c_0' = 1 - c_0 \tag{3.3-11}$$

The term $T_n'[t]$ is the nth-derivative transient term for the feedback response. This is the negative of the error-response transient term $T_n[t]$:

$$T_n'[t] = -T_n[t] \tag{3.3-12}$$

Practical feedback loops have zero gain at infinite frequency. For such loops, the feedback signal x_b cannot change instantaneously for a step of the input, or for a step of a derivative of the input. Hence for $n \geq 0$, the expansion of Eq 3.3-9 can be set equal to zero at $t = 0+$. This requires that

$$c_n' + T_n'[0] = 0 \qquad \text{for } n \geq 0 \tag{3.3-13}$$

For the error response, the same relation holds for $n > 0$:

$$c_n + T_n[0] = 0 \qquad \text{for } n > 0 \tag{3.3-14}$$

Thus, for the derivatives, the composite transient terms for both the error and feedback responses start at zero at time $t = 0$, and for the zero derivative ($n = 0$) the composite transient term for the feedback signal starts at zero. Integrating Eq 3.3-12 gives

$$T_n[t] - T_n[0] = \int_0^t T_{n-1}[t]\, dt \tag{3.3-15}$$

Applying Eqs 3.3-13, -14 to this gives

$$c_n' + T_n'[t] = \int_0^t T_{n-1}'[t]\, dt \qquad \text{for } n \geq 0 \qquad (3.3\text{-}16)$$

$$c_n + T_n[t] = \int_0^t T_{n-1}[t]\, dt \qquad \text{for } n > 0 \qquad (3.3\text{-}17)$$

3.3.3 Feedback Loop Example

The following loop transfer function is used as an example:

$$G = \frac{\omega_c(1 + \omega_2/s)}{s(1 + \omega_1/s)(1 + s/\omega_3)} = \frac{\omega_c\omega_3(s + \omega_2)}{s(s + \omega_1)(s + \omega_3)} = \frac{N}{D} \qquad (3.3\text{-}18)$$

A magnitude asymptote plot of this is shown in Fig 3.3-1. The assumed

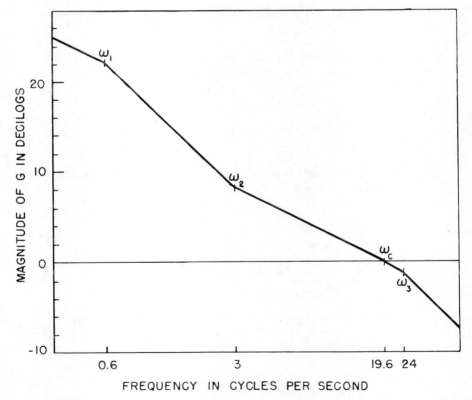

Figure 3.3-1 Magnitude asymptotes for illustrative feedback loop. Reprinted with permission from Ref [3.1] © 1955 AIEE (now IEEE).

TABLE 3.3-1   **Parameters of Feedback-Loop Transfer Function**

Open-Loop Parameters	Closed-Loop Parameters
$\omega_c = 123.0 \text{ sec}^{-1}$ (19.6 Hz)	$\omega_a = 21.53 \text{ sec}^{-1}$
$\omega_1 = 3.77 \text{ sec}^{-1}$ (0.30 Hz)	$\omega_n = 127.0 \text{ sec}^{-1}$
$\omega_2 = 18.55 \text{ sec}^{-1}$ (6.00 Hz)	$\zeta = 0.522$
$\omega_3 = 150.8 \text{ sec}^{-1}$ (24.0 Hz)	$\omega_o = 108.3 \text{ sec}^{-1}$
	$\zeta\omega_n = 66.3 \text{ sec}^{-1}$

values of the system parameters are shown in the left-hand part of Table 3.3-1. The G_{ib} transfer function is

$$G_{ib} = \frac{G}{1+G} = \frac{N}{N+D}$$

$$= \frac{\omega_c\omega_3(s+\omega_2)}{s^3 + s^2(\omega_1+\omega_3) + s\omega_3(\omega_1+\omega_c) + \omega_c\omega_2\omega_3} \quad (3.3\text{-}19)$$

The closed-loop poles can be calculated using the techniques of Ref [1.2] (Section 13.2). This allows the denominator to be factored, to give

$$G_{ib} = \frac{\omega_c\omega_3(s+\omega_2)}{(s+\omega_a)(s^2 + 2\zeta\omega_n s + \omega_n^2)} \quad (3.3\text{-}20)$$

The values for ω_a, ω_n, and ζ are given in the right-hand part of Table 3.3-1. The equation for G_{ie} is

$$G_{ie} = \frac{1}{1+G} = \frac{D}{N+D}$$

$$= \frac{s(s+\omega_1)(s+\omega_3)}{s^3 + s^2(\omega_1+\omega_3) + s\omega_3(\omega_1+\omega_c) + \omega_c\omega_2\omega_3}$$

$$= \frac{s(s+\omega_1)(s+\omega_3)}{(s+\omega_a)(s^2 + 2\zeta\omega_n s + \omega_n^2)} \quad (3.3\text{-}21)$$

The error-response nth-derivative transient term has the form

$$T_n[t] = K_{an}e^{-\omega_a t} + |K_{bn}|e^{-\zeta\omega_n t}\cos[\omega_o t + \text{Ang}[K_{bn}]] \quad (3.3\text{-}22)$$

The coefficients for a unit step of the input are

$$K_{a0} = (s + \omega_a)G_{ie}|_{s=-\omega_a} = \frac{(s + \omega_1)(s + \omega_3)}{(s^2 + 2\zeta\omega_n s + \omega_n^2)}\bigg|_{s=-\omega_a} \tag{3.3-23}$$

$$K_{b0} = \frac{1}{j\omega_o}(s^2 + 2\zeta\omega_n s + \omega_n^2)\frac{G_{ie}}{s}\bigg|_{s=-\zeta\omega_n+j\omega_o}$$

$$= \frac{(s + \omega_1)(s + \omega_3)}{(j\omega_o)(s + \omega_a)}\bigg|_{s=-\zeta\omega_n+j\omega_o} \tag{3.3-24}$$

The coefficients for derivatives of the input are related to these by

$$K_{an} = \frac{K_{a0}}{(-\omega_a)^n} \tag{3.3-25}$$

$$K_{bn} = \frac{K_{b0}}{(-\zeta\omega_n + j\omega_o)^n} \tag{3.3-26}$$

The magnitude and phase of the expression $(-\zeta\omega_n + j\omega_o)$ are

$$|-\zeta\omega_n + j\omega_o| = \omega_n \tag{3.3-27}$$

$$\text{Ang}[-\zeta\omega_n + j\omega_0] = \pi - \arccos[\zeta] \text{ radian} \tag{3.3-28}$$

Hence, by Eq 3.3-6 the magnitude and phase of K_{bn} are related as follows to the magnitude and phase of K_{b0}:

$$|K_{bn}| = |K_{b0}|/\omega_n^n \tag{3.3-29}$$

$$\text{Ang}[K_{bn}] = \text{Ang}[K_{b0}] - n(\pi - \arccos[\zeta]) \text{ radian.} \tag{3.3-30}$$

Table 3.3-2 shows the values of K_{an}, $|K_{bn}|$, and Ang[K_{bn}] for values of n from -2 to $+3$. The values for $n = 0$ are calculated from Eqs 3.3-23, -24, and the values for other values of n are obtained by applying Eqs 3.3-25, -29, and -30.

The simplest way to calculate the error coefficients is to expand the transfer function G_{ie} in long division as described in Appendix F. By Eq 3.3-21 the expression for G_{ie} using the values of Table 3.3-1 is

$$G_{ie} = \frac{s^3 + (\omega_1 + \omega_3)s^2 + \omega_1\omega_3 s}{s^3 + (\omega_1 + \omega_3)s^2 + \omega_3(\omega_1 + \omega_c)s + \omega_c\omega_2\omega_3}$$

$$= \frac{s^3 + 154.57s^2 + 568.52s}{s^3 + 154.57s^2 + 19{,}117s + 349{,}640} \tag{3.3-31}$$

TABLE 3.3-2   Coefficients for Error-Response Transient Terms $T_n[t]$ for Steps of Derivatives and Integrals of the Input

n	K_{an}	$|K_{bn}|$	Ang[K_{bn}] (rad)
-2	-76.9 sec^{-2}	$2.17 \times 10^4 \text{ sec}^{-2}$	-2.571
-1	3.57 sec^{-1}	171.5 sec^{-1}	1.593
0	-0.166	1.35	-0.529
1	$7.71 \times 10^{-3} \text{ sec}$	$1.06 \times 10^{-2} \text{ sec}$	-2.649
2	$-3.58 \times 10^{-4} \text{ sec}^2$	$8.37 \times 10^{-5} \text{ sec}^2$	1.513
3	$1.66 \times 10^{-5} \text{ sec}^3$	$6.59 \times 10^{-7} \text{ sec}^3$	-0.607

The expressions in the numerator and the denominator are arranged in ascending powers of s, and long division is used to divide the numerator by the denominator:

$$a_0 + a_1 s + a_2 s^2 + s^3 \enspace \overline{\smash{\big)}\, \begin{array}{c} 1.6260 \times 10^{-3}s + 3.5319 \times 10^{-4}s^2 - 1.717 \times 10^{-5}s^3 \\ \hline 568.52s + 154.57s^2 + s^3 \\ 568.52s + 31.08s^2 + 0.2513s^3 \\ \hline 123.49s^2 + 0.7487s^3 \\ 123.49s^2 + 6.7519s^3 \\ \hline - 6.0032s^3 \end{array}}$$

$$(3.3\text{-}32)$$

where $a_0 = 349{,}640$, $a_1 = 19{,}117$, and $a_2 = 154.57$. This yields the following error-coefficient expansion:

$$G_{ie} = c_1 s + c_2 s^2 + c_3 s^3 + \cdots$$

$$= 1.6260 \times 10^{-3}s + 3.5319 \times 10^{-4}s^2 - 1.717 \times 10^{-5}s^3 + \cdots \quad (3.3\text{-}33)$$

These error-coefficient values c_1, c_2, and c_3 are listed in the first column of Table 3.3-3. The second column gives the approximate error-coefficient values based on the equations derived in Section 3.2. The equations for the approxi-

TABLE 3.3-3   Accurate Values of Error Coefficients for Example Compared with Approximate Values

	Accurate Value	Approximate Value
c_1	$1.626 \times 10^{-3} \text{ sec}$	$1/\alpha\omega_c = 1.626 \times 10^{-3} \text{ sec}$
c_2	$3.53 \times 10^{-4} \text{ sec}^2$	$1/\omega_c\omega_i = 4.31 \times 10^{-4} \text{ sec}^2$
c_3	$-1.717 \times 10^{-5} \text{ sec}^3$	$-\tau_{ic}c_2 = -2.64 \times 10^{-5} \text{ sec}^3$

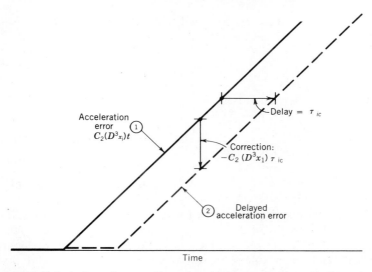

Figure 3.3-2 Calculation of a rate-of-acceleration error that is equivalent to the time delay applied to the acceleration error.

mate velocity and acceleration error coefficients, as given in Section 3.2, are

$$c_1 = \frac{1}{\alpha\omega_c} = \frac{1}{(\omega_2/\omega_1)\omega_c} = \frac{\omega_1}{\omega_2\omega_c} \tag{3.3-34}$$

$$c_2 = \frac{1}{\omega_c\omega_i} = \frac{1}{\omega_c\omega_2} \tag{3.3-35}$$

Section 3.2 did not include an error term for the rate-of-acceleration error coefficient c_3 for this class of loop transfer function. However, the time delay applied to the error-coefficient terms has an effect that is equivalent to a rate-of-acceleration error component. If the third derivative D^3x_i is constant, the second derivative (acceleration) $D^2X_i[t]$ is a ramp equal to $(D^3x_i)t$, and the acceleration error is

$$\text{Acceleration error} = c_2\big(D^2x_i[t]\big) = c_2\big(D^3x_i\big)t \tag{3.3-36}$$

This error is shown in Fig 3.3-2 as curve ①. Curve ② is the corrected acceleration error, which is obtained by delaying curve ① by the time delay τ_{ic}. As shown, curve ② could be formed by adding a correction to curve ① equal to

$$\text{Correction} = -\tau_{ic}\big(c_2D^3x_i\big) = c_3D^3x_i \tag{3.3-37}$$

This is an error component proportional to the third derivative, and so can be described by a rate-of-acceleration error coefficient equal to

$$c_3 = -\tau_{ic}c_2 \tag{3.3-38}$$

For this example, the value for τ_{ic} is

$$\tau_{ic} = \frac{1}{\omega_i} + \frac{1}{\omega_c} = \frac{1}{\omega_2} + \frac{1}{\omega_c} = 0.0612 \text{ sec} \tag{3.3-39}$$

The approximate value for c_3 shown in Table 3.3-3 is obtained by multiplying the approximate value of c_2 by the negative of Eq 3.3-39.

The approximate and exact values for c_1 in Table 3.3-3 are the same, but those for c_2 differ by about 20%, and those for c_3 differ by about 50%. This loop has a value for α of 5, which is quite low. If α were higher, the approximate and exact values for these error coefficients would be much closer.

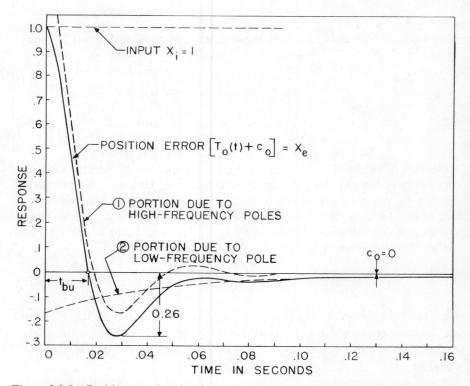

Figure 3.3-3 Position transient for the error response. Reprinted with permission from Ref [3.1] © 1955 AIEE (now IEEE).

The composite error-response transient terms for the example are shown in Figs 3.3-3 to -6 for the input (position), the first derivative (velocity), the second derivative (acceleration), and the third derivative (rate-of-acceleration).

Position Transient. The solid curve of Fig 3.3-3 is the error response to a unit step and is the position transient term $T_0[t]$. Note that the position error coefficient c_0 is zero, and so the final value of the response is zero. The dashed curve ① shows the component due to the underdamped pole pair, and the dashed curve ② shows the component due to the low-frequency real pole.

Velocity Transient. The solid curve of Fig 3.3-4 is the error response to a unit ramp, and is the composite velocity transient term $(c_1 + T_1[t])$. This has a final value equal to the velocity error coefficient c_1. The dashed curves ① and ② show the components due to the underdamped poles and the low-frequency real pole. For this transient, the amplitudes of the transient components for the two sets of poles are approximately equal, whereas for the position transient the component for the high-frequency pole pair is much larger than that for the low-frequency pole.

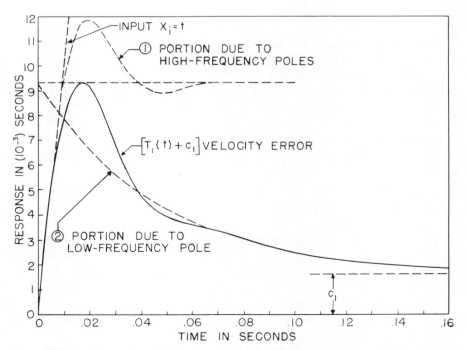

Figure 3.3-4 Composite velocity transient for the error response. Reprinted with permission from Ref [3.1] © 1955 AIEE (now IEEE).

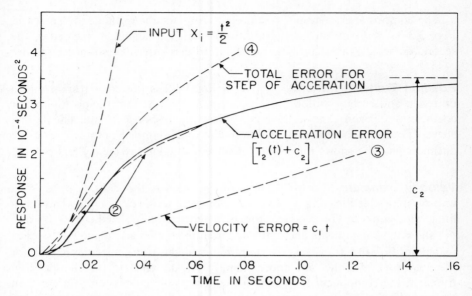

Figure 3.3-5 Composite acceleration transient for the error response. Reprinted with permission from Ref [3.1] © 1955 AIEE (now IEEE).

Acceleration Transient. The solid curve of Fig 3.3-5 is the composite acceleration transient term $(c_2 + T_2[t])$. The final value of this is the acceleration error coefficient c_2. A unit step of acceleration produces a unit ramp of velocity, and so there is also a velocity error term $c_1 t$, shown by the dashed curve ③, which is equal to the velocity t multiplied by the velocity error coefficient c_1. Curve ③ is added to the solid composite transient term to obtain the total error for the step of acceleration, shown as curve ④. The dashed curve ② is the component due to the low-frequency real pole. The small difference between curve ② and the solid curve is the component for the high-frequency pole pair. This shows that the acceleration transient is almost entirely characterized by the low-frequency real pole.

Rate-of-Acceleration Transient. The solid curve of Fig 3.3-6 is the composite transient term for a step of rate-of-acceleration, denoted $(c_3 + T_3[t])$. This composite transient has a final value equal to the rate-of-acceleration error coefficient c_3. The dashed curve ② is the component for the low-frequency real pole, and the very small difference between curve ② and the solid curve is the component for the high-frequency complex pole pair. Note that the values of this plot are negative.

The total error for a step of rate-of-acceleration is shown in Fig 3.3-7. Curve ① is the rate-of-acceleration composite transient obtained from Fig 3.3-6. The unit step of rate-of-acceleration produces a ramp of acceleration

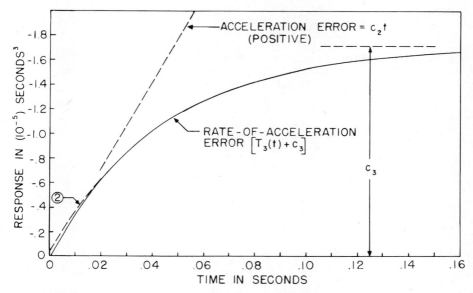

Figure 3.3-6 Composite rate-of-acceleration transient for the error response. Reprinted with permission from Ref [3.1] © 1955 AIEE (now IEEE).

equal to t, and a parabolic velocity equal to $t^2/2$. Curve ② is the acceleration error c_2t, and curve ③ is the velocity error term $c_1t^2/2$. Curve ④ is the sum of the acceleration error ② and the rate-of-acceleration error ①. This shows that the rate-of-acceleration error term has approximately the effect of a time delay applied to the acceleration error ②. The velocity error ③ is added to curve ④ to obtain the total error response, shown as curve ⑤. The input is shown as curve ⑥.

3.3.4 Calculation of Response to Arbitrary Input

Curve ① in Fig 3.3-8 shows the arbitrary input position signal considered for the example. The system is initially at rest at an angle of $0°$. At time $t = 0$, the angular position jumps discontinuously to $-1.5°$, and then follows the smooth curve ①. This input is broken down as follows.

First the step at $t = 0$ is subtracted from the rest of the curve and considered separately. Then the smooth portion of this position curve is differentiated to yield the velocity curve ②. At time $t = 0.32$ sec, the velocity curve has a steep drop. Since the fall time of this drop is about 0.008 sec, which is much less than the rise time of the velocity transient, the drop has approximately the effect of a step of velocity. Hence, the drop is approximated as a step.

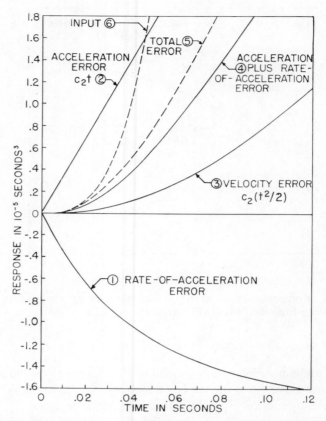

Figure 3.3-7 Components of response for a step of rate-of-acceleration. Reprinted with permission from Ref [3.1] © 1955 AIEE (now IEEE).

The smooth portion of the approximated velocity curve is differentiated to yield curve ③, the acceleration. At times $t = 0$ and $t = 0.45$ sec, the acceleration curve rises significantly faster than the acceleration transient, and so is approximated by steps at these points. The rest of the acceleration curve is approximated by straight-line segments, as shown by the long broken lines. When the broken-line approximation is differentiated, it yields the rate-of-acceleration plot, curve ④, which is a series of steps. Thus, except for the portions neglected in the various approximations, the input is broken down into a series of steps of various derivatives of the input.

The composite transient plots in Figs 1.3-3 to -6 are used to construct the response of the loop to the arbitrary input of Fig 3.3-8. In Fig 3.3-9, the approximated curves of velocity, acceleration, and rate of acceleration are multiplied by the corresponding error coefficients to yield curves ①', ②', and ③'. The velocity curve includes a single step approximation at $t = 0.32$ sec;

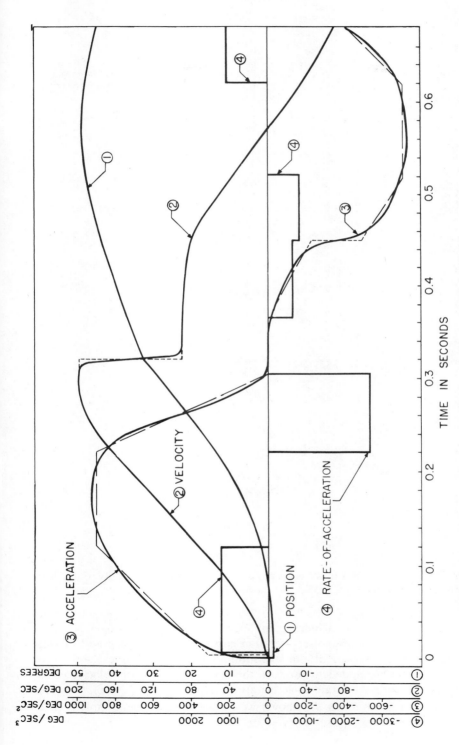

Figure 3.3-8 Illustrative arbitrary input and its derivatives. Reprinted with permission from Ref [3.1] © 1955 AIEE (now IEEE).

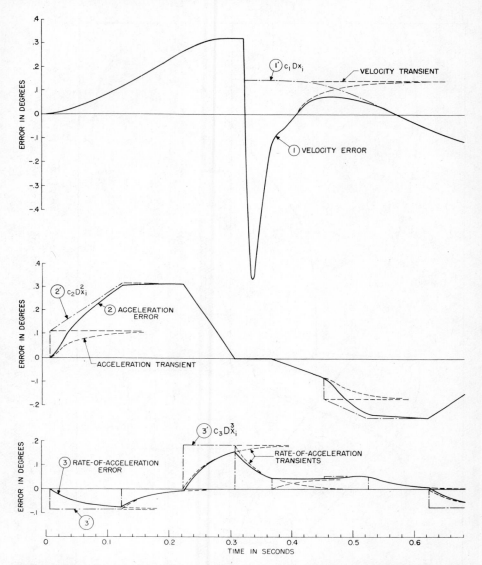

Figure 3.3-9 Components of the error response to the arbitrary input. Reprinted with permission from Ref [3.1] © 1955 AIEE (now IEEE).

the acceleration curve includes step approximations at $t = 0.03$ sec and $t = 0.45$ sec; and the rate-of-acceleration curve is entirely composed of step approximations. At each of the discontinuities of these curves, composite transient terms are added to produce the resultant curves of velocity error, acceleration error, and rate-of-acceleration error labeled ①, ②, ③. The individual transient terms are shown as dashed curves; the complete error

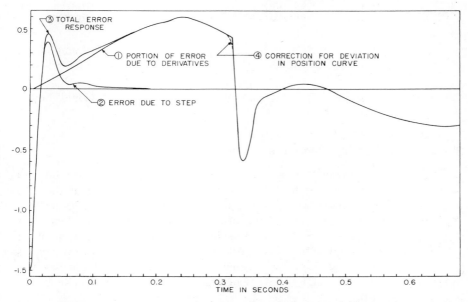

Figure 3.3-10 Total error response to the arbitrary input. Reprinted with permission from Ref [3.1] © 1955 AIEE (now IEEE).

components are shown as solid curves; and curves (1'), (2'), and (3') are shown as dot-dashed curves in the regions where they do not correspond to the actual error curves (1), (2), (3).

The largest transient in Fig 3.3-9 occurs at 0.32 sec and is caused by the approximate step of velocity at that point. The transient is obtained by multiplying the curve of $(T_1[t] + c_1)$ in Fig 3.3-4 by the step of velocity, which is -108 deg/sec. When the transient is added to curve (1') (which is $c_1 D x_i[t]$), the resultant curve (1) does not have a discontinuity. The reason is that the discontinuity in curve (1') is c_1 times the step of velocity, which is the final value of the transient.

The acceleration error curve has two transients, at 0.03 and 0.45 sec. Adding the two transients to the discontinuous curve (2') gives the continuous acceleration error curve (2). The rate-of-acceleration error is entirely composed of steps, there being eight in all. For each of these steps there is a transient which adds to curve (3') to form the continuous rate-of-acceleration curve (3).

Curves (1), (2), and (3), the three error components in Fig 3.3-9, are added together to form curve (1) in Fig 3.3-10, which is the total error caused by the derivatives of the input. Curve (2) in Fig 3.3-10 is the error component caused by the $-1.5°$ step of position that occurs at $t = 0$. Curve (2) is obtained by multiplying the position transient term in Fig 3.3-3 by the value of the step,

$-0.15°$. The total error response, curve ③, is obtained by adding curve ②
to curve ①.

Although discontinuities did not occur in the error-response components
for the derivatives, the step of position produces a discontinuity in the position
component, curve ②, and in the total error response, curve ③. If the
feedback response were plotted instead of the error, there would be no
discontinuity.

3.3.5   Response to a Step of an Integral of the Input

If the input signal contains a high-frequency component, which rises and falls
like a pulse in a period significantly shorter than the rise time of the step
response, this component should be approximated by an impulse, a doublet,
etc., which can be regarded as a step of an integral of the input. Hence, for
high-frequency signals, one should consider the expansion for a step of an
integral of the input. The following discussion gives only the basic principles
for approximating the response to high-frequency input components. The
application of these principles to particular examples is presented in Reference
[1.2] (Chapter 14).

For a high-frequency input signal, one is generally interested in the feedback response, not the error response. The feedback signal response to a step
of the mth integral $D^{-m}x_i$ is

$$x_b = a_0 x_i + a_1 D^{-1}x_i + \cdots + a_{m-1}D^{-m+1}x_i + (D^{-m}x_i)T'_{-m}[t] \quad (3.3\text{-}40)$$

The transient term $T'_{-m}[t]$ is obtained by replacing n by $-m$ in the expression
for $T'_n[t]$. The coefficients $a_0, a_1, \ldots, a_m$ are called the initial-value coefficients.

The simplest general way to calculate the initial-value coefficients is to
expand the G_{ib} transfer function by long division, as shown in Appendix F,
with the terms arranged in the reverse order used for calculating the steady-state
coefficients. The first coefficient a_0 is zero if G_{ib} has zero gain at infinite
frequency. The initial-value coefficients with zero values, and the first nonzero
coefficient, can be calculated from

$$a_m = s^m G_{ib}|_{s=\infty} = s^m G|_{s=\infty} \quad (3.3\text{-}41)$$

The a_m coefficients are called initial-value coefficients because they are the
initial values of the corresponding transient terms:

$$a_m = T'_{-m}[0] \quad (3.3\text{-}42)$$

In accordance with Eq 3.3-41, the initial-value coefficients can be calculated
by examining the loop transfer function G as s approaches infinity, which by

Eq 3.3-18 is

$$G \cong \frac{\omega_c \omega_3}{s^2} \text{ as } s \to \infty \tag{3.3-43}$$

Hence, by Eq 3.3-41, the first three initial-value coefficients are

$$a_0 = G[s]\Big|_{s=\infty} = \frac{\omega_c \omega_3}{s^2}\Big|_{s=\infty} = 0 \tag{3.3-44}$$

$$a_1 = sG[s]\Big|_{s=\infty} = \frac{\omega_c \omega_3}{s}\Big|_{s=\infty} = 0 \tag{3.3-45}$$

$$a_2 = s^2 G[s]\Big|_{s=\infty} = \omega_c \omega_3 \big|_{s=\infty} = \omega_c \omega_3 = 1.85 \times 10^4 \text{ sec}^{-2} \tag{3.3-46}$$

If the higher-order initial-value coefficients are desired, they can be calculated by long-division expansion of G_{ib}, as described in Appendix F.

Equation 3.3-40 shows that the feedback response to high-frequency portions of the input can be calculated as follows. Integrate each high-frequency portion successively, and at an appropriate integral, approximate the curve by steps. Multiply each input curve, except the last, by the corresponding initial-value coefficient a_m, and add the resultant curves to the calculated response. For each step, add to the response a transient proportional to the magnitude

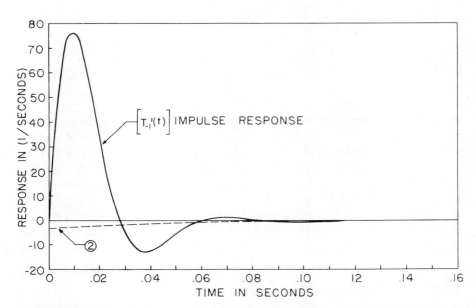

Figure 3.3-11 Impulse transient. Reprinted with permission from Ref [3.1] © 1955 AIEE (now IEEE).

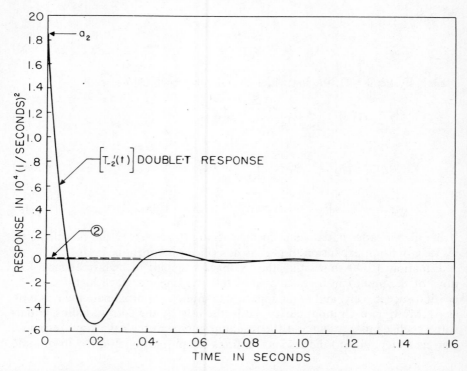

Figure 3.3-12 Doublet transient. Reprinted with permission from Ref [3.1] © 1955 AIEE (now IEEE).

of the step. These are the transient responses to a unit impulse, doublet, triplet, etc.

Impulse and Doublet Response. For the example, the feedback signal responses to a unit impulse and unit doublet are shown in Figs 3.3-11 and -12. These are the transient terms $T'_{-1}[t]$ and $T'_{-2}[t]$ for unit steps of the first and second integrals of the input. The initial values of these are the initial-value coefficients. As shown in Eqs 3.3-44 to -46, the initial value a_1 of the impulse response is zero, and the initial value a_2 of the doublet response is 1.86×10^4 sec^{-2}. The very small dashed curves labeled ② are the components of these responses due to the low-frequency real pole. Thus, the impulse and doublet responses are characterized primarily by the high-frequency complex pole pair.

3.3.6 The Remainder Coefficient

The calculation of the response to an arbitrary input requires that approximations be made on the input. The effect of these approximations on the accuracy of the calculated response can be determined by calculating an upper bound to the inaccuracy caused by the approximation. If the approximation is performed properly, this upper bound can be kept small with respect to the total error response, and hence the per-unit accuracy of the calculated response is guaranteed to be high.

The feedback-signal response to a deviation neglected in the nth derivative can be calculated from

$$x_b = c_0' x_i + c_1' D x_i + \cdots + c_{n-1} D^{n-1} x_i + \mathscr{R}_n[t] \qquad (3.3\text{-}47)$$

The last term, $\mathscr{R}_n[t]$, is the real convolution of $T_{n-1}'[t]$ and $D^n x_i[t]$, and can be expressed as

$$\mathscr{R}_n[t] = \int_0^\infty d\tau \, T_{n-1}[\tau] D^n x_i[t - \tau] \qquad (3.3\text{-}48)$$

Equation 3.3-48 represents the exact value of the remainder of an error-coefficient expansion. Equation 3.3-47 shows that the effect on the calculated response of neglecting a deviation of the nth derivative is given by the remainder $\mathscr{R}_n[t]$ for that deviation. The steady-state coefficient terms in this expansion are automatically included in the calculated response when the procedure described previously is followed. Hence, it is only the term $\mathscr{R}_n[t]$ that is disregarded when the nth-derivative deviation is neglected.

An exact calculation of $\mathscr{R}_n[t]$ for a given deviation would be very difficult, and is not necessary because $\mathscr{R}_n[t]$ should be small with respect to the total response $x_e[t]$ when the approximations are performed properly. An upper bound on the magnitude of $\mathscr{R}_n[t]$ is a measure of the accuracy of the calculated response. As shown in Ref [1.2] (Chapter 14), this upper bound is

$$|\mathscr{R}_n[t]| \le r_n \operatorname{Max}|D^n x_i| \qquad (3.3\text{-}49)$$

where r_n is defined as the nth-derivative remainder coefficient, which is equal to

$$r_n = \int_0^\infty d\tau \, |T_{n-1}[\tau]| \qquad (3.3\text{-}50)$$

The quantity $\operatorname{Max}|D^n x_i|$ is the maximum magnitude of the nth-derivative deviation that is being neglected. This shows that the upper bound for the magnitude of the error caused by neglecting a deviation at the nth derivative is equal to the nth-derivative remainder coefficient r_n multiplied by the maximum value of the magnitude of the deviation.

The remainder coefficient r_n for the nth derivative can be readily derived from a plot of the transient term of the nth derivative. Equation 3.3-15 gives

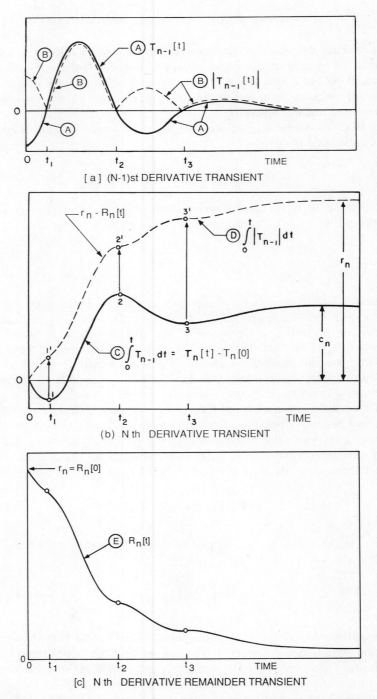

Figure 3.3-13 Method for calculating the remainder coefficient and the remainder transient: (*a*) transient term for $(n-1)$st derivative; (*b*) transient term for nth derivative; (*c*) remainder transient for nth derivative. Reprinted with permission from Ref [3.1] © 1955 AIEE (now IEEE).

the following expression for the nth-derivative transient term:

$$T_n[t] - T_n[0] = \int_0^t T_{n-1}[t] \, dt \qquad (3.3\text{-}51)$$

Comparing this with Eq 3.3-50 shows that $(T_n[t] - T_n[0])$ is the integral of $T_{n-1}[t]$, while r_n is the integral of the magnitude of $T_{n-1}[t]$, evaluated at time $t = \infty$. Hence the remainder coefficient can be calculated as shown in Fig 3.3-13. In diagram a, the solid curve Ⓐ is a plot of the $(n-1)$th derivative transient term $T_{n-1}[t]$, while the dashed curve Ⓑ is a plot of the magnitude of this, which is $|T_{n-1}[t]|$. In diagram b, the solid curve Ⓒ is the integral of the solid curve Ⓐ of diagram a, and so by Eq 3.3-15 is a plot of $(T_n[t] - T_n[0])$. Curve Ⓒ can be formed by plotting the nth-derivative transient term $T_n[t]$, and shifting the curve vertically until it is zero at time $t = 0$.

The dashed curve Ⓓ in diagram b is a plot of the integral of the dashed curve Ⓑ in diagram a, which is $|T_{n-1}[t]|$. Hence, by Eq 3.3-50, the final value of curve Ⓓ is the nth-derivative remainder coefficient r_n, as shown. Curve Ⓓ can be readily constructed from the solid curve Ⓒ as follows.

In the regions where curve Ⓒ has a positive slope, curve Ⓓ has the same shape as curve Ⓒ, and in the regions where curve Ⓒ has a negative slope, curve Ⓓ has the shape of the mirror image of curve Ⓒ reflected about the horizontal axis. Thus to form curve Ⓓ, the segments of curve Ⓒ with positive slope, and the mirror images of the segments with negative slope, are shifted vertically to form a curve that starts at zero and progresses as a monotonically increasing curve.

An equivalent bound can be calculated as follows for the inaccuracy caused by neglecting a deviation of the mth integral. The transform of the feedback signal can be expressed as

$$x_b = a_0 x_i + a_1 D^{-1} x_i + \cdots + a_m D^{-m} x_i + \mathscr{R}_{-m}[t] \qquad (3.3\text{-}52)$$

where $\mathscr{R}_{-m}[t]$ is equal to the expression for $\mathscr{R}_n[t]$ given in Eq 3.3-48, if n is replaced by $-m$. The magnitude of this is bounded by the mth integral remainder coefficient r_{-m} multiplied by $\text{Max}|D^{-m} x_i|$:

$$|\mathscr{R}_{-m}[t]| \le r_{-m} \text{Max}|D^{-m} x_i| \qquad (3.3\text{-}53)$$

3.3.7   Calculation of Accuracy for Example

Figure 3.3-14 shows the constructions for calculating the remainder coefficients for the feedback loop example. These are calculated from the transient terms given previously for steps of position, velocity, acceleration, and rate of

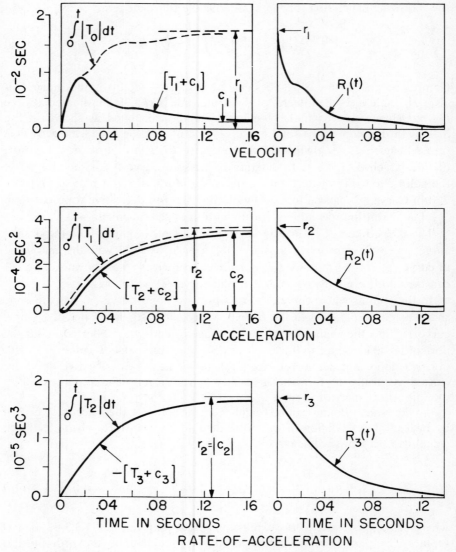

Figure 3.3-14 Calculation of remainder coefficients for illustrative system. Reprinted with permission from Ref [3.1] © 1955 AIEE (now IEEE).

acceleration, and for steps of the first and second integral (impulse response and doublet response). These yield the remainder coefficients r_n given in Table 3.3-4 for values of n from -2 to $+3$.

Note that the remainder coefficient r_n is very much greater than the magnitude of the error coefficient c_n for steps of position and velocity.

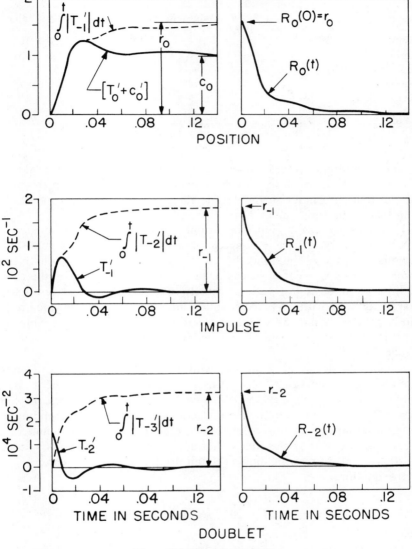

Figure 3.3-14   (*Continued*)

However, for a step of acceleration and all higher derivatives, r_n is approximately equal to $|c_n|$. This illustrates the general principle that in an adequately damped feedback loop, the magnitude of the error coefficient becomes roughly equal to the remainder coefficient at a certain derivative, and often approaches even closer at a higher derivative, but can never become greater.

TABLE 3.3-4   Remainder Coefficients for Illustrative Feedback Loop

Response	Coefficient	Value
Doublet	r_{-2}	$3.10 \times 10^4 \text{ sec}^{-2}$
Impulse	r_{-1}	182 sec^{-1}
Position	r_0	1.55
Velocity	r_1	$1.707 \times 10^{-2} \text{ sec}$
Acceleration	r_2	$3.57 \times 10^{-4} \text{ sec}^2$
Rate of acceleration	r_3	$1.715 \times 10^{-5} \text{ sec}^3$

The final straight-line approximation of the signal should be made on the first derivative at which $|c_n|$ is approximately equal to r_n. When this is done, the per-unit inaccuracy in the calculated response is no greater (essentially) than the per-unit inaccuracy of the approximation.

In the example, the final straight-line approximations were made on the acceleration curve. There is an acceleration component in the computed error response equal to c_2 multiplied by the approximate acceleration curve, and an acceleration component of inaccuracy equal to r_2 multiplied by the acceleration deviation. Since $|c_2|$ is approximately equal to r_2, if the acceleration curve is approximated within 5%, the resultant inaccuracy in the calculated response can be no greater than 5% of the acceleration component of error. Since r_n can never be less than $|c_n|$, nothing is gained by performing the approximation at a higher derivative.

Plots of the deviations associated with the error-coefficient expansion of the input are shown in Fig 3.3-15. Let us use the remainder-coefficient values of Table 3.3-4 to calculate the error bounds corresponding to these deviations. Figure 3.3-15c gives a plot of the acceleration deviation corresponding to the straight-line approximation of the input acceleration. The maximum value of the magnitude of the deviation is 55 deg/sec². Using the value of r_2 in Table 3.3-4 gives the following upper bound to the inaccuracy caused by this approximation:

$$r_2 \text{ Max}|D^2x_i| = (3.57 \times 10^{-4} \text{ sec}^2)(55 \text{ deg} / \text{sec}^2) = 0.0196° \quad (3.3-54)$$

The acceleration-error contribution of curve ② in Fig 3.3-9 was obtained by multiplying c_2 by the approximated acceleration curve, not the exact curve. The reason for this is explained by the remainder expansion of Eq 3.3-47. The expansion shows that the remainder coefficient bounds the inaccuracy from neglecting a deviation only when that deviation is not included in the response. Thus, the inaccuracy bound of Eq 3.3-49 holds only if the approximated acceleration curve is used to calculate the acceleration component of error.

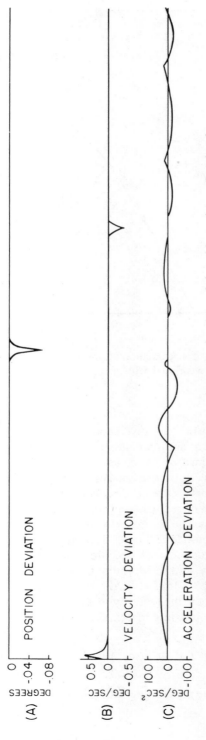

Figure 3.3-15 Deviation components for approximations of the arbitrary input. Reprinted with permission from Ref [3.1] © 1955 AIEE (now IEEE).

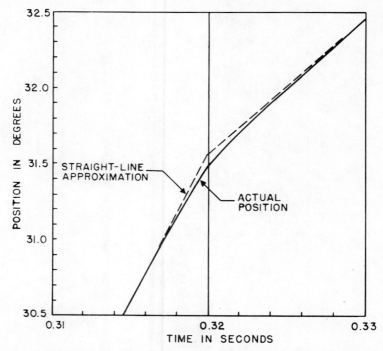

Figure 3.3-16   Line-segment approximation of position, which corresponds to the step approximation of velocity. Reprinted with permission from Ref [3.1] © 1955 AIEE (now IEEE).

Step approximations of a derivative should be considered to be the integrated effect of line-segment approximations of the lower-order derivative. For example, the step approximation of the velocity curve in Fig 3.3-8 at 0.32 sec should be considered to be a straight-line approximation of the position curve. In Fig 3.3-16, the solid curve shows the actual input position, and the dashed straight lines show the line-segment approximation, which yields the step approximation of velocity. The deviation of the position curve corresponding to this approximation is plotted in Fig 3.3-15a. Similarly, the step approximations of acceleration at 0 and 0.45 sec are considered to be line-segment approximations of velocity, and the deviations corresponding to these approximations are shown in Fig 3.3-15b.

The peak magnitude of the velocity deviation of Fig 3.3-15b is 0.57 deg/sec, and so the corresponding inaccuracy bound for this deviation is

$$r_1 \, \text{Max}|Dx_i| = (1.707 \times 10^{-2} \, \text{sec}^2)(0.57 \, \text{deg/sec}) = 0.0097° \quad (3.3-55)$$

The peak magnitude of the position deviation in Fig 3.3-15a is 0.072°, and so

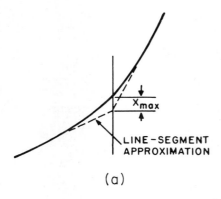

(a)

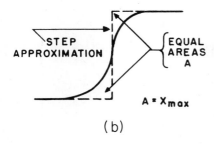

(b)

Figure 3.3-17 Comparison of a step approximation of a derivative with the equivalent line-segment approximation of the lower-order derivative: (*a*) line-segment approximation of $(n-1)$st derivative; (*b*) step approximation of the nth derivative. Reprinted with permission from Ref [3.1] © 1955 AIEE (now IEEE).

the corresponding inaccuracy bound is

$$r_0 \, \text{Max}|x_i| = 1.55(0.072°) = 0.122° \tag{3.3-56}$$

Although it is convenient to consider a step approximation to be a line segment approximation of the lower order derivative, it is not necessary to perform the line segment approximation directly. Consider for example Fig 3.3-17, which shows a portion of the $(n-1)$th derivative (diagram *a*), which is approximated by straight lines, and the nth derivative (diagram *b*), with the corresponding step approximation. The step approximation can be performed directly on the nth derivative by making the areas between the approximate curve and the exact curve equal, as shown in diagram *b*. This area A is equal to the maximum deviation of the $(n-1)$th derivative from the line-segment approximation. Hence the inaccuracy bound is equal to the area A, obtained from the nth derivative, multiplied by the remainder coefficient for the $(n-1)$th derivative.

3.3.8 The Remainder Transient

A basic limitation on the use of the remainder coefficient is that the upper bound to the inaccuracy calculated from it applies for all values of time. This can be remedied by generalizing the remainder-coefficient concept to obtain the remainder transient, which provides a quantitative measure of the period of time over which the inaccuracy applies.

As shown in Fig 3.3-13b, the dashed curve ⓓ is denoted $(r_n - R_n[t])$, where $R_n[t]$ is defined as the remainder transient. The plot of the remainder transient $R_n[t]$ is shown as curve ⓔ of diagram c, and is found by inverting curve ⓓ and shifting it vertically until its final value is zero. Since curve ⓓ is equal to the integral of $|T_{n-1}[t]|$, this gives

$$r_n - R_n[t] = \int_0^t |T_{n-1}[t]| \, dt \qquad (3.3\text{-}57)$$

Solve for $R_n[t]$, and set r_n equal to the expression of Eq 3.3-50:

$$R_n[t] = r_n - \int_0^t |T_{n-1}[t]| \, dt$$

$$= \int_0^\infty |T_{n-1}[t]| \, dt - \int_0^t |T_{n-1}[t]| \, dt = \int_t^\infty |T_{n-1}[t]| \, dt \quad (3.3\text{-}58)$$

It can be seen from this equation, as well as from the plot of curve ⓔ in Fig 3.3-13c, that the initial value of the remainder transient (at $t = 0$) is equal to the remainder coefficient.

The remainder transient can be applied by assuming that the nth derivative is approximated as shown in Fig 3.3-18a. The magnitude of the deviation is assumed to be no greater than $\text{Max}|D^n x_i|$ for values of time less than t_1, and zero thereafter:

$$|D^n x_i[t]| \le \text{Max}|D^n x_i| \quad \text{for } t \le t_1 \qquad (3.3\text{-}59)$$

$$D_n x_i[t] = 0 \qquad \qquad \text{for } t > t_1 \qquad (3.3\text{-}60)$$

(Note that $\text{Max}|D^n x_i|$ is represented by N in Fig 3.3-18.) The general expression for the remainder associated with any approximation is given by Eq 3.3-47 as

$$\mathscr{R}_n[t] = \int_0^\infty d\tau \, T_{n-1}[\tau] D^n x_i[t - \tau] \qquad (3.3\text{-}61)$$

As shown in Fig 3.3-18a, τ represents a time variable measured in the direction of negative time from the present time t. At values of τ less than $\tau_1 = (t - t_1)$, the nth-derivative deviation $D^n x_i$ is zero. Hence the integral of Eq 3.3-61 should be split into two regions as follows:

$$\mathscr{R}_n[t] = \int_0^{\tau_1} d\tau \, T_{n-1}[\tau] D^n x_i[t - \tau] + \int_{\tau_1}^\infty d\tau \, T_{n-1}[\tau] D^n x_i[t - \tau] \quad (3.3\text{-}62)$$

$$N = \text{Max} \mid D^n x_i \mid$$

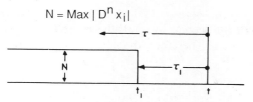

(A) A DEVIATION BOUND FOR nth DERIVATIVE

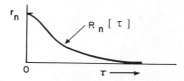

(B) A TYPICAL REMAINDER TRANSIENT

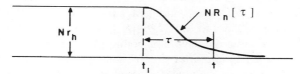

(C) UPPER BOUND TO INACCURACY FOR (A)

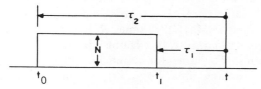

(D) A DEVIATION BOUND FOR nth DERIVATIVE

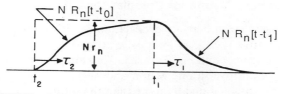

(E) UPPER BOUND TO INACCURACY FOR (D)

Figure 3.3-18 Application of remainder transient. Reprinted with permission from Ref [3.1] © 1955 AIEE (now IEEE).

The first integral is zero, because the nth-derivative deviation $D^n x_i$ is zero for that range of τ. In the second integral, the magnitude of the deviation is no greater than $\text{Max}|D^n x_i|$. Therefore, the upper bound on the magnitude of the remainder at time t is

$$|\mathcal{R}_n[t]| \le \text{Max}|D^n x_i| \int_{\tau_1}^{\infty} d\tau |T_{n-1}[\tau]| \qquad (3.3\text{-}63)$$

Since $\tau_1 = t - t_1$, the integral is equal to $R_n[t - t_1]$. Hence

$$|\mathcal{R}_n[t]| \le \text{Max}|D^n x_i| R_n[t - t_1] \qquad (3.3\text{-}64)$$

In Fig 3.3-18, diagram b is a plot of a typical remainder transient $R_n[t]$. Diagram c is the upper bound to the inaccuracy, obtained by applying this remainder transient to the deviation bound of diagram a. Prior to time t_1, the remainder bound is equal to $\text{Max}|D^n x_i| r_n$. After the deviation ends at time t_1, the remainder bound decays along a curve proportional to the remainder transient, which is equal to $\text{Max}|D^n x_i| R_n[t - t_1]$.

Let us now assume the deviation bound of diagram d of Fig 3.3-18, which is zero before time t_0 and after time t_1, and is limited by $\text{Max}|D^n x_i|$ between t_0 and t_1. By extending the preceding analysis, it can be shown that the bound on the remainder is as given in diagram e. This remainder bound rises after time t_0 along a curve equal to $\text{Max}|D^n x_i|$ times ($r_n - R_n[t - t_0]$), and reaches a final value of $\text{Max}|D^n x_i| r_n$. After time t_1, the bound decays from this value along the curve equal to $\text{Max}|D^n x_i|$ times $R_n[t - t_1]$.

The remainder transients for the example are plotted in Fig 3.3-14 for all of the transient terms. The time for a remainder transient to settle can be regarded as the "settling time" for a step of that particular derivative or integral of the input.

3.3.9 Lower Error Bound for Unipolar Deviation

In many cases the nth-derivative deviation is either completely positive or completely negative (i.e., it is unipolar), at least in a local region. For such cases, a lower error bound can be used. This lower bound also applies when the deviation is positive in one region and negative in another provided that the two regions are separated by more than the settling time for the remainder transient of that derivative.

The remainder bound for a unipolar deviation of a derivative of the input is

$$|\mathcal{R}_n[t]| \le \tfrac{1}{2}(r_n + |c_n|)\text{Max}|D^n x_i| \qquad (3.3\text{-}65)$$

The remainder bound for a unipolar deviation of an integral of the input is

$$|\mathcal{R}_{-m}[t]| \le \tfrac{1}{2}(r_{-m} + |a_m|)\text{Max}|D^{-m} x_i| \qquad (3.3\text{-}66)$$

When the deviation of the mth integral is unipolar, the error bound stays the same whether or not the mth-integral initial-value term is included in the calculated response.

Thus, if the deviation for a derivative or integral is unipolar (either positive or negative), a reduced bound for the remainder can be obtained by replacing the remainder coefficient r_n (or r_{-m}) by $\frac{1}{2}(r_n + |c_n|)$ for an nth-derivative deviation, or by $\frac{1}{2}(r_{-m} + |a_m|)$ for an mth-integral deviation. The resultant bound for an integral holds regardless of whether the mth-integral initial-value term is included in the calculated response. In applying this principle, two deviations of opposite sign can be regarded as unipolar, if they are separated by a time interval greater than the settling time of the remainder transient for that derivative or integral.

Chapter 4

Modulation and Demodulation of Signals

A number of signals in a radar tracking system are conveyed as amplitude modulation of an AC carrier. This chapter examines the processes of amplitude modulation and demodulation, and the effect that a filter operating on the AC-modulated waveform produces on the signal information.

4.1 SUMMARY OF CHAPTER

The following are some examples of AC-modulated signals in radar tracking systems:

1. In tracking servos, AC-modulated signals often occur in transducers, such as synchros and resolvers, that measure the antenna angles.
2. To derive angular tracking signals, the antenna beam is sometimes scanned in a conical pattern. This conical scan applies a modulation to the received signal, which conveys angular tracking error information.
3. The radar pulse train transmitted to the target is a pulse-modulated AC carrier signal at the radar frequency. The resultant RF echo pulses received from the target are heterodyned to an intermediate frequency. This produces a pulse-modulated AC signal at the IF frequency, which is amplified and filtered in the receiver.

This chapter examines AC modulation and demodulation. It analyzes the effect that a filter, operating at the AC (or carrier) frequency, has on the modulation. It presents a transformation to determine the signal-frequency filter that is equivalent to a given carrier-frequency filter, and vice-versa. A carrier-frequency filter, operating on a modulated AC signal, has the same effect on the modulation that the equivalent signal-frequency filter would have if it processed the modulating waveform directly.

This carrier-frequency to signal-frequency transformation is applied in Chapter 7 to show how the IF bandpass filter of a radar receiver affects the envelope of the received pulse train.

The chapter describes the phase-sensitive detector, which is commonly used to demodulate the control signals of AC angle transducers. As will be shown in Chapter 7, the same circuit is used in the receiver of a conical-scan tracking radar, to derive tracking error signals from the conical scan modulation. The circuit is also used (at a much higher frequency) in a monopulse tracking radar, to derive tracking error signals from the IF error-channel signals. These phase-sensitive detector circuits use the sum-channel IF signal as a reference.

4.2   AMPLITUDE MODULATION AND DEMODULATION

4.2.1   Analysis of Ideal Amplitude Modulation and Demodulation Processes

Let us examine the mathematical processes associated with ideal amplitude modulation and demodulation. An amplitude-modulated signal has the form

$$e_{ac}[t] = x[t]\cos[\omega_r t] = x[t]\cos[2\pi f_r t] \tag{4.2-1}$$

where $x[t]$ is the modulating signal, ω_r is the frequency of the modulating carrier (or reference) in radian/sec, and f_r is that carrier frequency in hertz.

Figure 4.2-1 illustrates the amplitude modulation process. Diagram a shows the signal $x[t]$, and diagram b shows an AC carrier (or reference) signal $R[t]$. The signals of diagrams a and b are multiplied together to form the AC-modulated signal $e_{ac}[t]$ shown in diagram c. Note that at point (A), where the input $x[t]$ is positive, the AC-modulated signal $e_{ac}[t]$ of diagram c is in phase with the reference $R[t]$ of diagram b. At point (B), where $x[t]$ is negative, the signal $e_{ac}[t]$ of diagram c is 180° out of phase with respect to the reference of diagram b.

Let us assume that the modulating signal $x[t]$ is a sinusoid of frequency ω_m and amplitude X_m:

$$x[t] = X_m \cos[\omega_m t] \tag{4.2-2}$$

Combining Eqs 4.2-1, -2 gives for the modulated signal $e_{ac}[t]$:

$$e_{ac}[t] = X_m \cos[\omega_m t]\cos[\omega_r t] \tag{4.2-3}$$

This can be expanded by applying the formula

$$\cos[a]\cos[b] = \tfrac{1}{2}(\cos[a + b] + \cos[a - b]) \tag{4.2-4}$$

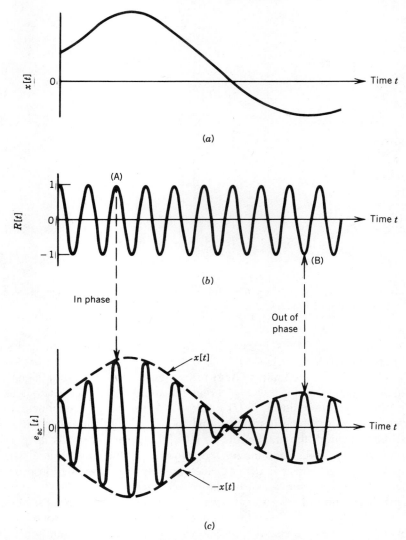

Figure 4.2-1   Waveforms illustrating AC modulation process: (*a*) modulating signal $x[t]$; (*b*) sinusoidal reference (or carrier) $R[t]$; (*c*) modulated signal $e_{ac}[t]$.

Equation 4.2-3 becomes

$$e_{ac}[t] = \tfrac{1}{2}X_m\{\cos[(\omega_r + \omega_m)t] + \cos[(\omega_r - \omega_m)t]\} \qquad (4.2\text{-}5)$$

This shows that the amplitude-modulated signal $e_{ac}[t]$ consists of two sidebands, at the frequencies $(\omega_r + \omega_m)$ and $(\omega_r - \omega_m)$. These sidebands are displaced above and below the carrier reference frequency ω_r by the modula-

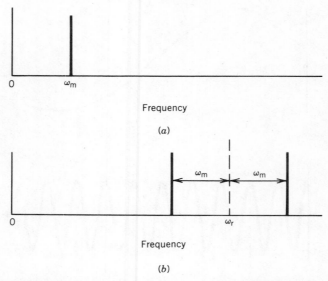

Figure 4.2-2   Effect of AC modulation on signal spectrum: (*a*) spectrum of signal $x[t]$; (*b*) spectrum of modulated signal $e_{ac}[t]$.

tion frequency ω_m. The spectra of the signal $x[t]$ and the modulated signal $e_{ac}[t]$ are shown in Fig 4.2-2.

In an amplitude-modulated (AM) radio signal, the modulating signal $x[t]$ in Eq 4.2-2 is the sum of a DC level plus the audio signal. Because of this DC level, the modulated signal $e_{ac}[t]$ of Eq 4.2-5 also has a component at the carrier frequency, which is proportional to $\cos[\omega_r]$. On the other hand, in a control application, the modulating signal $x[t]$ does not normally have a fixed DC component, and so the modulated signal $e_x[t]$ does not normally have a component at the carrier frequency. Such an amplitude-modulation process is sometimes called *suppressed-carrier amplitude modulation*.

The modulation of an AM radio wave is extracted by an envelope detector, which follows the envelope of the received signal. An envelope detector ignores the phase of the signal, and so cannot derive the modulation from a suppressed-carrier modulated waveform. A phase-sensitive detector is needed, which responds to phase as well as amplitude.

As will be shown in Chapter 7, a phase-sensitive detector provides coherent detection, whereas an envelope detector provides noncoherent detection. If the signal/noise ratio at the input to a detector is less than unity, a coherent detector yields a much better signal than a noncoherent detector.

Thus, to extract the modulation $x[t]$, the signal $e_{ac}[t]$ can be demodulated by feeding it through a phase-sensitive detector, which uses the carrier as a reference signal. This reference can be expressed as

$$R[t] = \cos[\omega_r t] \qquad (4.2\text{-}6)$$

A mathematically ideal phase-sensitive detector is one that multiplies the modulated signal by the reference and filters the result. Multiplying the signal $e_{ac}[t]$ in Eq 4.2-5 by twice the reference $R[t]$ in Eq 4.2-6 gives

$$
\begin{aligned}
e[t] &= 2e_{ac}[t]R[t] \\
&= X_m\{\cos[(\omega_r + \omega_m)t] + \cos[(\omega_r - \omega_m)t]\}\cos[\omega_r t] \\
&= X_m\cos[(\omega_r + \omega_m)t]\cos[\omega_r t] + X_m\cos[(\omega_r - \omega_m)t]\cos[\omega_r t] \quad (4.2\text{-}7)
\end{aligned}
$$

As will be seen, the factor of 2 is included to achieve unity gain in the process. Applying the trigonometric identity of Eq 4.2-4 to this gives

$$
\begin{aligned}
e[t] &= \tfrac{1}{2}X_m\{\cos[(2\omega_r + \omega_m)t] + \cos[\omega_m t]\} \\
&\quad + \tfrac{1}{2}X_m\{\cos[(2\omega_r - \omega_m)t] + \cos[\omega_m t]\} \\
&= X_m\cos[\omega_m t] + \tfrac{1}{2}X_m\{\cos[(2\omega_r + \omega_m)t] + \cos[(2\omega_r - \omega_m)t]\} \quad (4.2\text{-}8)
\end{aligned}
$$

This signal $e[t]$ is fed through a lowpass filter, which passes the signal at the frequency ω_m essentially unchanged, and blocks the double harmonic signals at the frequencies $(2\omega_r + \omega_m)$ and $(2\omega_r - \omega_m)$. The filter output, which is denoted $\tilde{x}[t]$, is approximately

$$
\tilde{x}[t] \cong X_m\cos[\omega_m t] \quad (4.2\text{-}9)
$$

If the modulation frequency ω_m is high relative to the reference frequency ω_r, the lowpass filter degrades the modulation information somewhat in order to provide adequate attenuation of harmonics.

Thus the modulation and demodulation processes can be represented as shown in Fig 4.2-3. The signal is modulated by multiplying the signal by the reference $\cos[\omega_r t]$. It is demodulated by multiplying the modulated signal by the same reference, multiplied by 2, and feeding the result through a lowpass

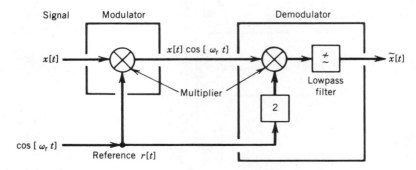

Figure 4.2-3   Mathematically ideal processes for modulation and demodulation.

filter. The output is represented by the signal $\tilde{x}[t]$, where the ($\sim$) indicates that the $x[t]$ information has been filtered by the lowpass filter.

Note that multiplication is represented by the symbol $\otimes$. Some writers use this same symbol to represent summation in feedback-control signal-flow diagrams, but this is poor practice. This book uses the symbol $\textstyle\sum$ to represent summation and reserves $\otimes$ for multiplication.

4.2.2   Practical AC Modulators

AC Transducers. The linear variable differential transformer (LVDT) is a convenient example for illustrating the amplitude modulation processes implemented in an AC transducer, because it is easy to explain. The basic LVDT transducer consists of three coils wound about a hollow tube, which contains a movable iron slug. As shown in Fig 4.2-4, the reference winding A is excited by a constant AC reference voltage E_r. This signal couples magnetically through the iron slug to the two output windings B_1 and B_2. The windings B_1, B_2 are connected in series opposition, so that the AC voltages generated in these windings tend to cancel.

The iron slug is allowed to move axially. When the slug is in the central (null) position, there is equal magnetic coupling from winding A to windings B_1, B_2. Hence voltages E_1, E_2 are equal, and the winding output voltage E_x is zero. When the slug is displaced upward from the null position, the magnetic coupling to winding B_1 increases and that to winding B_2 decreases. Voltage E_1 is now greater than E_2, and the difference between them, which is E_x, is in phase with the reference voltage E_r applied to winding A. The amplitude of E_x is proportional to the displacement of the slug from null. When the slug is

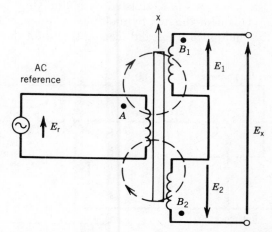

Figure 4.2-4   Circuit using linear variable differential transformer (LVDT).

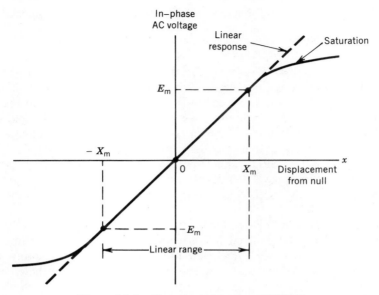

Figure 4.2-5   Characteristic curve of LVDT.

displaced from null in the opposite direction, voltage E_2 becomes greater than E_1, and so E_x is 180° out of phase with the reference E_r.

Figure 4.2-5 shows a plot of the in-phase AC voltage E_x versus the displacement x of the slug from the null position. A positive value indicates that the AC voltage is in phase with the reference, whereas a negative value indicates that it is 180° out of phase with the reference. There is a large central region where the device is very linear, which is labeled the linear range. In this range, the response of the LVDT can be described by

$$e_{ac} = E_m \cos[\omega_r t] \frac{x}{X_m} \tag{4.2-10}$$

The variable x is the displacement from null, $\pm X_m$ is the extent of the linear range, and E_m is the peak value of the AC signal which occurs when the LVDT is at either limit of this range.

Phase shift in the circuit generates a small quadrature output voltage that is proportional to $\sin[\omega_r t]$. The demodulation process, which converts the AC signal e_x to a DC signal, discriminates against this quadrature signal, and ideally rejects it.

Modern LVDT devices often contain electronic circuitry, which generates the AC reference signal and demodulates the signal e_x, to provide a DC signal proportional to displacement. An excitation of several kilohertz is generally used. The high carrier frequency improves the magnetic efficiency of the device, and minimizes the dynamic effects associated with modulation and demodulation.

AC Modulation of DC Electrical Signal.

A chopping device is generally employed to convert a DC signal to an AC-modulated signal. Vibrating mechanical switches were once commonly used for this purpose, but the most common modulating device used today is the integrated-circuit field effect transistor (FET) switch. Figure 4.2-6 shows an AC modulator employing FET switches. This could be implemented with a Siliconix DG191 integrated circuit, which has two SPDT (single-pole double-throw) FET switches in a 16-pin dual-inline package (DIP). Each switch is controlled by a standard TTL digital signal, and has one switch state for a TTL HIGH (2–5 V) and the other state for a TTL LOW (0–0.8 V).

The TTL command for controlling the FET switch can be derived by feeding the AC reference signal into a comparator with a TTL output (for example, a Harris HA4900, which has four comparators in a 16-pin DIP). Diagram a of Fig 4.2-7 shows the AC reference signal input to the comparator. The output, shown in diagram b, is a TTL-compatible square wave in synchronism with the reference. In diagram c, the dashed curve ① is the

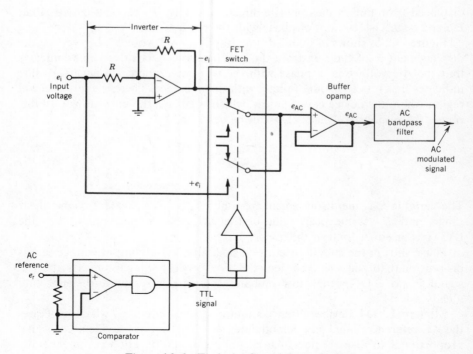

Figure 4.2-6 Typical AC modulator circuit.

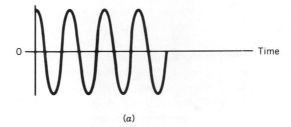

(a)

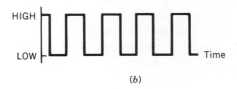

(b)

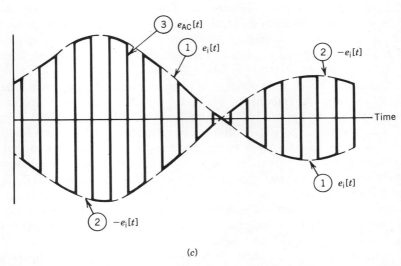

(c)

Figure 4.2-7 Modulated DC signal generated by circuit of Fig 4.2-6: (a) AC reference; (b) TTL signal; (c) modulated signal.

input DC signal $e_i[t]$ that is being modulated, which is the same as the signal shown in Fig 4.2-1a. The dashed curve ② is the output $-e_i[t]$ from the inverter opamp, which is the negative of the input $e_i[t]$. The solid curve ③ is the square-wave modulated voltage $e_{AC}[t]$ at the outputs of the FET switches. This voltage $e_{AC}[t]$ is fed through a buffer opamp to an AC bandpass filter tuned to the reference frequency. The bandpass filter attenuates the harmonics of the square-wave modulated signal (curve ③) to produce a smooth wave-form similar to that shown in Fig 4.2-1c.

Modulators in Communication Systems. In communication and radar signal-processing equipment, signal modulation is usually provided by a mixer, which adds the modulating signal to the reference, and feeds the result into a nonlinear device, typically a diode. A balanced circuit is commonly used, called a balanced mixer, which rejects unwanted components. The output from a balanced mixer is a close approximation to the product of the signal multiplied by the reference.

A mixer is generally not used for modulation in control systems, because it operates at low signal levels and does not allow accurate control of DC offset. In contrast, a switching modulator can operate at high signal levels while providing a very accurate zero. However, its output has much higher harmonic content.

4.2.3   Practical AC Demodulators

The same circuit used to modulate the AC signal can be used to demodulate it, except that the filtering is different. Figure 4.2-8 shows a typical demodulating circuit, called a phase-sensitive detector. Two demodulator approachers are included in the figure: one uses a lowpass filter to attenuate the AC harmonics, and the other uses a sample-and-hold circuit.

Figure 4.2-9 shows the waveforms of the demodulation process. Diagram *a* shows the input AC modulated waveform, which is the same as the waveform shown earlier in Fig 4.2-1c. Diagram *b* shows the AC reference signal, and diagram *c* shows the TTL switch control waveform that is derived from the AC reference. Diagram *d* is the demodulated waveform at the output of the FET switches. This is equal to the waveform *a* when the TTL command *c* is HIGH, and it is equal to the negative of the waveform *a* when the TTL switch command *c* is LOW.

The signal of diagram *d* is fed through the lowpass filter, which attenuates the harmonics of the waveform. The resultant filtered signal is approximately the same as the original DC signal, which was shown in Fig 4.2-1a.

Another way to eliminate harmonics is to employ a sample-and-hold circuit. The sampling pulse train is shown in diagram *e* of Fig 4.2-9. These are short pulses that are timed to occur at the peaks of the AC waveform. The sampling pulses can be derived using digital circuitry from the TTL switch-control pulses of diagram *c*. These sampling pulses (*e*) sample the peak values of the demodulated waveform (*d*), which are indicated by the points ○. Each sampled value is stored on the capacitor *C* in Fig 4.2-8, which holds the output voltage fixed between sampling instants. Diagram *f* in Fig 4.2-9 shows the waveform at the output of the sample-and-hold circuit. A small amount of additional lowpass filtering is generally needed after the sample-and-hold circuit to attenuate the remaining harmonics.

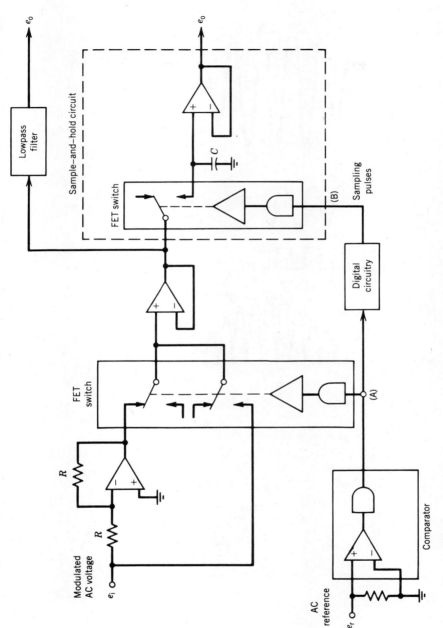

Figure 4.2-8 Phase-sensitive detector, showing two methods of attenuating harmonics.

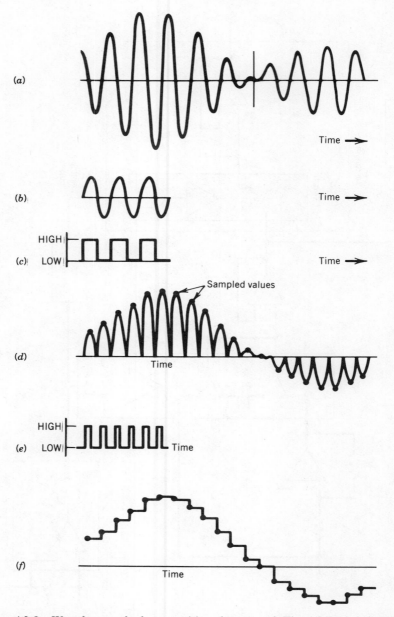

Figure 4.2-9 Waveforms of phase-sensitive detector of Fig 4.2-8: (*a*) input AC modulated waveform; (*b*) AC reference; (*c*) TTL pulses controlling FET switches; (*d*) demodulated signal before lowpass filtering; (*e*) TTL sampling pulses; (*f*) output from sample-and-hold circuit.

A sample-and-hold circuit is a very effective means of reducing AC harmonics in a demodulator. To achieve equivalent harmonic reduction with a lowpass filter results in large phase lag in the demodulated signal, which may severely degrade feedback-control performance.

4.3 EFFECT OF CARRIER-FREQUENCY (AC) NETWORKS ON THE SIGNAL-FREQUENCY RESPONSE

4.3.1 Equivalent Signal-Frequency Transfer Function for Carrier-Frequency (AC) Network

Let us now consider the dynamic effect on the signal information of a network placed within the AC path. The transfer function of an element within the AC path is denoted $F[\omega]$, while the transfer function of an element within the DC signal path is denoted $H[\omega]$. To simplify the analysis, the input is represented as the real part of a rotating vector. The input is assumed to have unity magnitude and zero phase. Hence the vector input signal is

$$X_i = e^{j\omega_m t} \tag{4.3-1}$$

The time-domain input signal is the real part of this:

$$x_i[t] = \text{Re}[X_i] = \text{Re}[e^{j\omega_m t}] = \cos[\omega_m t] \tag{4.3-2}$$

This is multiplied by the following reference signal:

$$r[t] = 2\cos[\omega_r t] = (e^{j\omega_r t} + e^{-j\omega_r t}) \tag{4.3-3}$$

The factor of 2 is included to achieve a net gain of unity in the AC section. The modulated vector signal is

$$X_{i(AC)} = X_i r[t] = (e^{j\omega_m t})(e^{j\omega_r t} + e^{-j\omega_r t})$$
$$= e^{j(\omega_m + \omega_r)t} + e^{j(\omega_m - \omega_r)t} \tag{4.3-4}$$

This has two frequency components at the frequencies $(\omega_m + \omega_r)$ and $(\omega_m - \omega_r)$. This AC signal is fed through a filter of transfer function $F[\omega]$. Each component is modified by the complex value of the transfer function $F[\omega]$ at its frequency $(\omega_m + \omega_r)$ or $(\omega_m - \omega_r)$. Hence, the output from the filter is

$$X_{o(AC)} = e^{j(\omega_m + \omega_r)t}F[\omega_m + \omega_r] + e^{j(\omega_m - \omega_r)t}F[\omega_m - \omega_r] \tag{4.3-5}$$

This signal is demodulated by multiplying it by the reference $\cos[\omega_r t]$. This gives the DC output signal X_o, which is

$$
\begin{aligned}
X_o &= \cos[\omega_r t] X_{o(AC)} = \tfrac{1}{2}(e^{j\omega_r t} + e^{-j\omega_r t}) X_{o(AC)} \\
&= \tfrac{1}{2}(e^{j\omega_r t} + e^{-j\omega_r t}) \\
&\quad \times \{ e^{j(\omega_m + \omega_r)t} F[\omega_m + \omega_r] + e^{j(\omega_m - \omega_r)t} F[\omega_m - \omega_r] \} \\
&= \tfrac{1}{2}(F[\omega_m + \omega_r] + F[\omega_m - \omega_r]) e^{j\omega_m t} \\
&\quad + \tfrac{1}{2} F[\omega_m + \omega_r] e^{j(\omega_m + 2\omega_r)t} + \tfrac{1}{2} F[\omega_m - \omega_r] e^{j(\omega_m - 2\omega_r)t}
\end{aligned} \quad (4.3\text{-}6)
$$

For the moment, we shall consider only the frequency components of X_o at the modulation frequency ω_m, and assume that the components at the frequencies $(\omega_m + 2\omega_r)$ and $(\omega_m - 2\omega_r)$ are eliminated by filtering. By Eq 4.3-1, the vector input signal X_i is $e^{j\omega_m t}$. Hence the transfer function X_o/X_i for the $e^{j\omega_m t}$ component is

$$
H[\omega] = \frac{X_o}{X_i} = \tfrac{1}{2}(F[\omega_m - \omega_r] + F[\omega_m - \omega_r]) \quad (4.3\text{-}7)
$$

This transfer function X_o/X_i, which is denoted $H[\omega]$, is the signal-frequency equivalent of the AC transfer function $F[\omega]$.

When $\omega_m < \omega_r$, the frequency $(\omega_m - \omega_r)$ is negative. Let us consider what is meant by a negative frequency. If the frequency parameter ω in any transfer function is replaced by $-\omega$, each factor $j\omega$ is replaced by $-j\omega$. The resultant transfer function has the same magnitude, but the sign of the phase is reversed. Thus, for a transfer function $F[\omega]$, the magnitude and phase of $F[-\omega]$ are

$$
|F[-\omega]| = |F[\omega]| \quad (4.3\text{-}8)
$$

$$
\text{Ang}[F[-\omega]] = -\text{Ang}[f[\omega]] \quad (4.3\text{-}9)
$$

Hence $F[-\omega]$ is the conjugate of $F[\omega]$:

$$
F[-\omega] = F[\omega]^* \quad (4.3\text{-}10)
$$

where the asterisk (*) denotes the conjugate. If this principle is applied to Eq 4.3-7, the transfer function $F[\omega_m - \omega_r]$ can be replaced by $F[\omega_r - \omega_m]^*$ when the quantity $(\omega_m - \omega_r)$ is negative. Hence, Eq 4.3-7 can be expressed in the following forms, depending on whether ω_m is greater or less than ω_r:

$$
H[\omega_m] = \tfrac{1}{2}(F[\omega_m + \omega_r] + F[\omega_r - \omega_m]^*) \quad \text{for } \omega_m < \omega_r \quad (4.3\text{-}11)
$$

$$
H[\omega_m] = \tfrac{1}{2}(F[\omega_m + \omega_r] + F[\omega_m - \omega_r]) \quad \text{for } \omega_m > \omega_r \quad (4.3\text{-}12)
$$

This representation eliminates negative frequencies. On the other hand, during analysis it is convenient to ignore the issue of whether the frequency is positive or negative. Hence the form $F[\omega_m - \omega_r]$ is used even when ω_m is less than ω_r. Thus for analysis purposes it is convenient to deal with a fictitious negative frequency.

4.3.2 Signal-Frequency to Carrier-Frequency Transformation

Let us find the ideal filter response $F[\omega]$ that operates on the AC modulated signal to produce the same effect as a specified signal-frequency response $H[\omega]$. This could theoretically be achieved by replacing every frequency ω in $H[\omega]$ by $(\omega - \omega_r)$. In Fig 4.3-1, curves ① and ② are the magnitude and phase plots of the desired signal-frequency transfer function $H[\omega]$. The dashed portions show the frequency response at negative frequencies. If ω is replaced by $(\omega - \omega_r)$ in the transfer function $H[\omega]$, the frequency-response plots ① and ② are shifted upward in frequency to form curves ③ and ④, which are the plots of the ideal AC frequency response $F[\omega]$. Hence

$$F[\omega_r + \omega_m] = H[\omega_m] \tag{4.3-13}$$

$$F[\omega_r - \omega_m]^* = H[-\omega_m]^* = H[\omega_m] \tag{4.3-14}$$

Comparing Eqs 4.3-13, -14 with Eq 4.3-7 shows that $F[\omega]$ satisfies the requirements of Eq 4.3-7.

This ideal frequency response is not physically realizable, because $j\omega$ would be replaced by $j(\omega - \omega_r)$. This is equivalent to $(s - j\omega_r)$, which is not real for

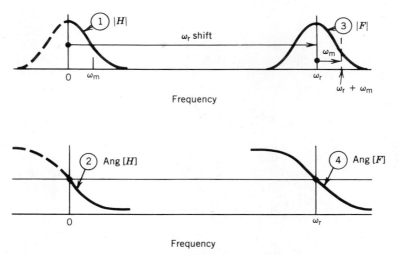

Figure 4.3-1 Shift of signal-frequency transfer function (lowpass filter) to reference (or carrier) frequency, to produce equivalent carrier-frequency transfer function (bandpass filter).

real values of s. On the other hand, the ideal frequency response can be closely approximated at frequencies close to ω_r by making the following substitution in the signal-frequency transfer function $H[\omega]$:

$$\omega \to \frac{(\omega - \omega_r)(\omega + \omega_r)}{2\omega} = \frac{\omega^2 - \omega_r^2}{2\omega} = \frac{\omega}{2} - \frac{\omega_r^2}{2\omega} \tag{4.3-15}$$

For frequencies near the reference frequency ω_r, the factor $(\omega + \omega_r)/2\omega$ is approximately unity, and so this transformation approximates the ideal: $\omega \to (\omega - \omega_r)$. In terms of the complex frequency $j\omega$, the transformation of Eq 4.3-15 is

$$j\omega \to j\frac{\omega}{2} - j\frac{\omega_r^2}{2\omega} = \frac{j\omega}{2} + \frac{\omega_r^2}{2j\omega} \tag{4.3-16}$$

Expressing this in terms of s gives

$$s \to \frac{s}{2} + \frac{\omega_r^2}{2s} = \frac{s^2 + \omega_r^2}{2s} \tag{4.3-17}$$

This substitution converts a signal-frequency (or DC) transfer function into an equivalent carrier-frequency (or AC) transfer function. The transfer function can be implemented directly in terms of circuit components in the following manner. Let us define the impedance of an inductor L_1 in the signal-frequency (DC) transfer function $H[\omega]$ as Z_{DC}:

$$Z_{DC} = sL_1 \tag{4.3-18}$$

To form the equivalent AC filter, s is replaced by the expression of Eq 4.3-17, and so the signal-frequency impedance Z_{DC} is replaced by the following AC impedance Z_{AC}:

$$Z_{AC} = \left(\frac{s}{2} + \frac{\omega_r^2}{2s}\right)L_1 = s\frac{L_1}{2} + \frac{\omega_r^2 L_1}{2s} \tag{4.3-19}$$

As shown in Fig 4.3-2a, this equivalent impedance Z_{AC} in the AC filter is an inductor of inductance $L_{AC} = L_1/2$ in series with a capacitor of capacitance $C_{AC} = 2/\omega_r^2 L_1$. The impedance Z_{AC} experiences series resonance at the frequency $1/\sqrt{L_{AC}C_{AC}}$, which is equal to

$$\frac{1}{\sqrt{L_{AC}C_{AC}}} = \frac{1}{\sqrt{(L_1/2)(2/\omega_r^2 L_1)}} = \omega_r \tag{4.3-20}$$

Thus, the series resonant frequency of the AC impedance Z_{AC} is equal to the

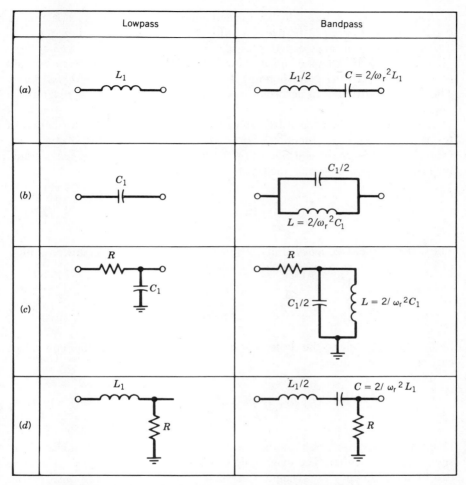

Figure 4.3-2 Equivalent circuit elements to achieve signal-frequency to carrier-frequency transformation.

reference (or carrier) frequency ω_r. In similar fashion, let us define the admittance of a capacitor C_1 of the signal-frequency (DC) filter $H[\omega]$ as Y_{DC}:

$$Y_{DC} = sC_1 \qquad (4.3\text{-}21)$$

In the equivalent AC filter, the capacitor C_1 is replaced by the following admittance Y_{AC}:

$$Y_{AC} = \left(\frac{s}{2} + \frac{\omega_r^2}{2s} \right) C_1 = s\frac{C_1}{2} + \frac{\omega_r^2 C_1}{2s} \qquad (4.3\text{-}22)$$

As shown in Fig 4.3-2b, this equivalent admittance Y_{AC} is a capacitor of capacitance $C_{AC} = C_1/2$ in parallel with an inductor of inductance $L_{AC} = 2/\omega_r^2 C_1$. This admittance experiences parallel resonance at the frequency ω_r.

In Fig 4.3-2, diagrams c and d show RL and RC signal-frequency lowpass filters, along with the corresponding AC (carrier-frequency) bandpass filters. These are constructed in accordance with the transformations of diagrams a and b.

Let us consider this signal-frequency to carrier-frequency transformation in terms of the transfer function. For every single-order factor of the form $(s + \omega_x)$ in the signal-frequency transfer function, the carrier-frequency transfer function has a quadratic factor of the form

$$(s + \omega_x) \rightarrow \frac{s^2 + 2\omega_x s + \omega_r^2}{2s} \tag{4.3-23}$$

Thus, the carrier-frequency transfer function for a single-order lowpass filter of break frequency ω_x is obtained as follows:

$$\frac{\omega_x}{s + \omega_x} \rightarrow \frac{2\omega_x s}{s^2 + 2\omega_x s + \omega_r^2} \tag{4.3-24}$$

Setting $s = j\omega$ gives the following frequency response of the equivalent carrier-frequency filter:

$$F[j\omega] = \frac{j2\omega\omega_x}{(\omega_r^2 - \omega^2) + j2\omega\omega_x} \tag{4.3-25}$$

Diagram a of Fig 4.3-3 shows the magnitude of the frequency response of the signal-frequency filter. The magnitude of this lowpass response is -1.5 dg at the break frequency ω_x. Diagram b shows the magnitude response of the equivalent carrier-frequency filter. The bandwidth of this bandpass response between the two -1.5 dg points is 2ω.

It can be shown from Eq 4.3-25 that the carrier-frequency filter has a low-frequency magnitude asymptote equal to $2\omega\omega_x/\omega_r^2$ and a high-frequency magnitude asymptote equal to $2\omega_x/\omega$. These asymptotes are indicated as dashed lines in Fig 4.3-3b. At the reference frequency ω_r, the magnitude asymptote value is $2\omega_x/\omega_r$, while the actual magnitude value is unity. The Q of the circuit is equal to the ratio of the peak magnitude value divided by the asymptote value at the resonant frequency ω_r, which is equal to

$$Q = \frac{\omega_r}{2\omega_x} \tag{4.3-26}$$

Hence the width of the bandpass frequency response, between the half-power

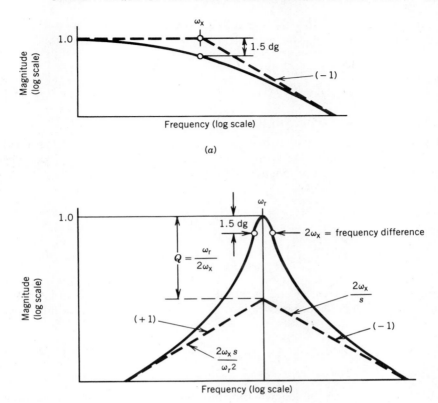

Figure 4.3-3 Frequency-response magnitude plots for (a) a single-order lowpass filter and (b) the equivalent bandpass filter.

points (the points of -1.5-dg magnitude), is equal to the resonant frequency ω_r divided by Q:

$$2\omega_x = \frac{\omega_r}{Q} \tag{4.3-27}$$

When the signal-frequency filter has an underdamped quadratic factor, the corresponding carrier-frequency filter has a pair of quadratic factors stagger-tuned about the reference frequency. Consider for example the following lowpass signal-frequency transfer function:

$$H[s] = \frac{\omega_n^2}{s^2 + 2\zeta\omega_n s + \omega_n^2} \tag{4.3-28}$$

It can be shown, by using the signal-frequency to carrier-frequency transformation of Eq 4.3-17, that the equivalent AC transfer function is as follows:

$$F[s] = \frac{2(\omega_n/K)s}{\left[s^2 + (1/Q)(\omega_r/K)s + (\omega_r/K)^2\right]} \frac{2(K\omega_n)s}{\left[s^2 + (1/Q)(K\omega_r)s + (K\omega_r)^2\right]}$$

$$(4.3\text{-}29)$$

This transfer function consists of two tuned factors of the same Q, one tuned to the frequency ω_r/K and the other tuned to the frequency $K\omega_r$. The parameters of the transfer functions of Eqs 4.3-28, -29 are related by

$$4\zeta\omega_n = \frac{\omega_r(K + 1/K)}{Q} \qquad (4.3\text{-}30)$$

$$(2\omega_n)^2 = \left(\frac{\omega_r}{Q}\right)^2 + \left(K\omega_r - \frac{\omega_r}{K}\right)^2 \qquad (4.3\text{-}31)$$

Solve Eq 4.3-30 for ω_r/Q and substitute this into Eq 4.3-31. This gives the following for the difference between the two tuned frequencies ω_r/K and $K\omega_r$:

$$K\omega_r - \frac{\omega_r}{K} = 2\omega_n\sqrt{1 - \frac{4\zeta^2}{(K + 1/K)^2}} \qquad (4.3\text{-}32)$$

Equation 4.3-30 gives the following for the amplification Q, which is the same for the two factors:

$$Q = \frac{\omega_r(K + 1/K)}{4\zeta\omega_n} \qquad (4.3\text{-}33)$$

If the bandwidth of the stagger-tuned filter is reasonable narrow relative to the reference frequency ω_r, the factor $(K + 1/K)$ can be closely approximated by

$$K + \frac{1}{K} \cong 2 \qquad (4.3\text{-}34)$$

Substituting this approximation into Eqs 4.3-32, -33 gives the approximations

$$K\omega_r - \frac{\omega_r}{K} \cong 2\omega_n\sqrt{1 - \zeta^2} \qquad (4.3\text{-}35)$$

$$Q \cong \frac{\omega_r}{2\zeta\omega_n} \qquad (4.3\text{-}36)$$

In Fig 4.3-4, diagram a shows the magnitude of a quadratic lowpass signal-frequency filter of damping ratio $\zeta = 0.5$, and diagram b shows the

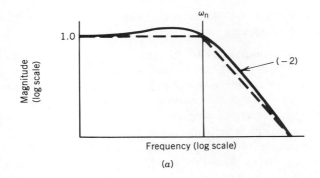

(a)

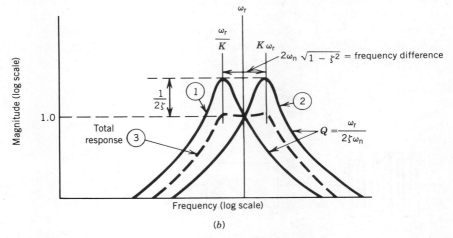

(b)

Figure 4.3-4 Frequency-response magnitude plots for (a) a second-order lowpass filter and (b) the equivalent bandpass filter.

dashed curve ③, the magnitude response of the resultant bandpass carrier-frequency filter. This bandpass response is the cascade effect of two responses ① and ②, tuned below and above the reference frequency ω_r at the frequencies ω_r/K and $K\omega_r$. These responses have the same Q, which is approximately equal to $\omega_r/2\zeta\omega_n$. The difference in frequency between $K\omega_r$ and ω_r/K is approximately equal to $2\omega_n\sqrt{1-\zeta^2}$. (Remember that on a logarithmic scale the spacing between frequencies ω_r/K and ω_r is equal to the ratio of the frequencies ω_r and $K\omega_r$, not their difference.) The gains of the two factors at the reference frequency ω_r are the same and are equal to

$$\text{Gain} = \frac{\omega_r}{2\omega_n Q} = \frac{2\zeta}{K + 1/K} \cong \zeta \qquad (4.3\text{-}37)$$

4.3.3 Application of AC Modulation Principles to Signal Processing in Communication Systems

The principles of modulation, AC filtering, and demodulation that have been described apply both to AC servos and to signal processing at much higher frequencies in communication and radar systems. The signal-frequency to carrier-frequency (lowpass to bandpass) filter transformation is a powerful concept for explaining signal processing. In Chapter 7, it will be applied in the discussions of acquisition and tracking in a radar system.

The application of these concepts is illustrated in Fig 4.3-5. In diagram *a*, a pulsed waveform is fed through a bandpass filter. The waveform is a pulse-modulated carrier, and the bandpass filter is tuned to the frequency of the carrier. One would like to know the envelope of the waveform at the output of the bandpass filter.

This can be found as shown in diagram *b*. Using the approach of Section 4.3.2, one calculates the signal-frequency lowpass filter that is equivalent to the carrier-frequency bandpass filter. The envelope of the input pulse waveform is

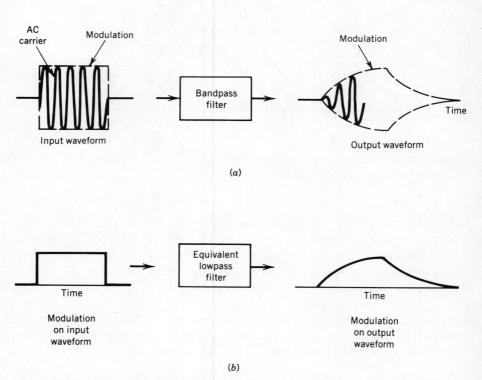

Figure 4.3-5 Application of signal-frequency to carrier-frequency transformation to radar and communication systems: (*a*) pulse-modulated carrier signal fed through bandpass filter; (*b*) equivalent effect at signal frequency.

fed into this equivalent signal-frequency filter, and the output is determined. This output from the signal-frequency lowpass filter is the same as the envelope of the output from the carrier-frequency bandpass filter.

4.3.4 Exact Effect of Amplitude Modulation and Demodulation

The signal-frequency to carrier-frequency transformation described in Section 4.3.2 applies exactly only when the modulation frequency is much less than the carrier frequency. However, as shown in Ref [1.2] (Chapter 8), the deviations of the actual upper and lower sidebands from the ideal response tend to compensate for one another. Consequently, the response of the realizable carrier-frequency filter approximates the ideal response for modulation frequencies that are rather large relative to the carrier frequency.

The primary practical effect of modulation frequencies that are large relative to the carrier frequency is that a system that can pass such high modulation frequencies must also transmit an appreciable amount of harmonics from the demodulation process. (These harmonics were ignored in Eq 4.3-7.) In a feedback system, such harmonics may severely degrade the stability of the feedback loop.

An exact analysis of the effect of harmonics in a feedback loop that contains an AC modulation process is very complicated. However, an approximate analysis can be achieved by assuming that the demodulation process employs sampling in the manner described in Section 4.2.3. The sampling demodulator samples the AC waveform twice per cycle, and so converts the carrier-frequency signal to a sampled-data signal, with a sampling frequency equal to twice the modulation frequency.

Sampled-data systems are analyzed in Chapter 5. As shown in Section 5.5, the maximum gain crossover frequency that can be achieved in a sampled-data feedback loop with acceptable stability is 1/8 of the sampling frequency. Hence, for a feedback loop having an AC-modulated signal, with a sampled-data demodulator, the maximum gain crossover frequency achievable with acceptable stability is 1/4 of the reference (or carrier) frequency. This also appears to be a good practical limit even when the AC demodulator does not use sampling.

Chapter 5

Sampled Data and
Computer Simulation

When a signal is carried in the form of digital data, each digital quantity represents the value of the signal at a particular instant of time. Between those instants, the signal is not specified. Such a signal, which is defined only at discrete points in time, is called a sampled signal.

When signal processing is performed using digital computation, the signal is handled in terms of sampled data. Hence, sampled-data theory is needed to understand digital signal processing. This chapter develops the principles of sampled-data theory and applies them to digital signal processing and to digital computer simulation.

Much of the signal processing in modern radar tracking systems is implemented in terms of digital data. Hence, sampled-data theory is essential for understanding radar tracking.

5.1 SUMMARY OF CHAPTER

Section 5.2 explains the basic principles associated with sampling and the conversion of sampled data to a continuous signal. It shows that a sampled signal can only convey frequency information up to half the sampling frequency. If the signal being sampled has components of higher frequency, *spectral foldover* or *aliasing* occurs, which transforms any high-frequency signal component into a distortion component of lower frequency.

Digital sampled data is usually converted to a continuous signal by means of a simple-hold circuit, which holds the signal fixed between sampling instants. This adds a time delay of half the sampling period, and still transmits a significant amount of sampling harmonics. Subsequent filtering is usually required to attenuate these sampling harmonics, and this adds more time delay. Much lower harmonics can be achieved by a detector that provides linear interpolation between sampling points. However, a linear-interpolator detector adds a time delay of one sample period.

Section 5.3 explains the principles of dynamic signal processing using digital sampled data. Dynamics are introduced into a digital computer program by employing algorithms in which the output variable y is a weighted sum of the input x and past values of the input and output. Such an algorithm is called a difference equation and has the form

$$y = a_0 x + a_1 x_p + a_2 x_{pp} + \cdots + b_1 y_p + b_2 y_{pp} + \cdots \qquad (5.1\text{-}1)$$

where x_p, y_p are the past values of x, y computed in the previous cycle, x_{pp}, y_{pp} are the past values of x, y computed in the cycle before that, etc. The parameters $a_0, a_1, a_2, \ldots, b_1, b_2, \ldots$ are constants. The Laplace transform of this difference equation of Eq 5.1-1 is

$$Y = a_0 X + a_1 \bar{z} X + a_2 \bar{z}^2 X + \cdots + b_1 \bar{z} Y + b_2 \bar{z}^2 Y + \cdots \qquad (5.1\text{-}2)$$

where X, Y are the Laplace transforms of x, y. The variable $\bar{z}$ is the Laplace transform corresponding to a time delay of one sampling period (denoted T) and is equal to

$$\bar{z} = e^{-sT} \qquad (5.1\text{-}3)$$

The Laplace transform of Eq 5.1-2 can be converted to a more useful transfer function by making the substitution

$$\bar{z} = \frac{1 - pT/2}{1 + pT/2} \qquad (5.1\text{-}4)$$

The variable p is a pseudo complex-frequency, which is approximately equal to the Laplace transform complex-frequency s for frequencies below $1/4$ of the sampling frequency.

By means of Eq 5.1-4, one can readily determine the frequency response that corresponds to a particular difference equation, and one can synthesize a difference-equation computer algorithm to satisfy a required frequency response. The relations between the pseudo complex-frequency p and the true complex frequency s (throughout the s-plane) are explained in Section 5.4.

In Chapter 8, these principles are applied to determine the equivalent frequency responses of the tracking algorithms in the Altair radar system. This allows the dynamic characteristics of the Kalman-filter tracking system to be evaluated in a simple manner.

The preceding principles are applied to develop a technique called *serial simulation*, which allows one to simulate the time response of a very complex dynamic system using no more than BASIC computer language on a personal computer. The computations are performed in an iterative, or feedback, manner. There is a time delay of one sample period around each feedback loop, which introduces the transfer function $\bar{z} = e^{-sT}$ into the loop. With serial simulation, this time delay is compensated for by including digital computation that inserts the transfer function $(1 + pT)$ in series with each time delay $\bar{z}$. The $(1 + pT)$ factor adds phase lead, which cancels the phase lag of the $\bar{z}$ factor at low frequencies. This results in accurate simulation at frequencies below $1/16$ of the sampling frequency, and reasonable simulation up to $1/8$ of the sampling frequency.

The serial-simulation technique is developed in Section 5.3 and extended in Section 5.5. Theory to explain the proper use of serial simulation is presented in Section 5.4.

Another approach to digital computer simulation is based on the *state variable* and *state equation*. These concepts are explained in Section 5.5.

To apply the state variable approach, the dynamic equations of a system are formulated in a signal-flow diagram so that all dynamic effects are expressed in terms of integrators. The outputs from the integrators are defined as the *state variables* of the system. The corresponding inputs to the integrators are the derivatives of the state variables. The state equations express each state-variable derivative as a function of the state variables, and the input forcing functions applied to the system. The state equations do not themselves involve dynamics; they express the instantaneous values of the state-variable derivatives in terms of the instantaneous values of the state variables and input forcing functions. On the other hand, the state equations can have time-varying coefficients and can be nonlinear.

Section 5.5 provides simple explanations of the state variable and the state equation. It explains Runge–Kutta integration, which is used to integrate state variables in a computer-simulation program. The section also explains the transition matrix, which is a key element in Kalman filter theory.

Section 5.6 describes the effects of sampling in a feedback loop that has both continuous and sampled data. It summarizes conclusions developed in Ref [1.2] (Section 11.6), which were derived from studies using the z-transform in frequency response and transient response analyses.

5.2 SAMPLED-DATA MODULATION AND DEMODULATION

5.2.1 Spectrum of Sampled Signal

A fundamental limitation of a sampled signal is that it cannot convey information at frequencies greater than half the sampling frequency. This can be demonstrated by examining the spectrum of the sampled signal. Chapter 4 described the spectrum of an AC modulated signal, and showed that the modulation process generates two sidebands displaced above and below the carrier frequency by the modulation frequency ω_m. The spectrum of a sampled-data signal has an infinite series of sidebands. These sidebands are displaced above and below the sampling frequency, and all harmonics of the sampling frequency, by the modulation frequency ω_m.

Sampling a signal is equivalent to modulating (or multiplying) the signal by a train of very narrow pulses having unit amplitude, as shown in Fig 5.2-1. Diagram *a* shows the input signal, and diagram *b* shows the modulating train of unit-amplitude pulses, which is called the reference signal. The sampling pulse period is denoted T, and the pulse width is denoted τ. The pulses are made so narrow there is negligible variation of the continuous signal within the duration of the pulse, and so the signal is effectively sampled at discrete instants of time.

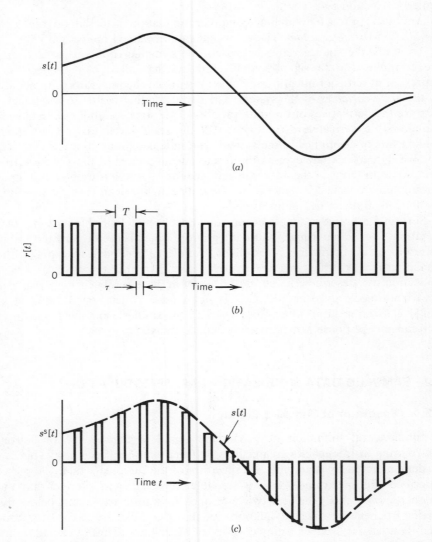

Figure 5.2-1 Waveforms illustrating the sampling process: (*a*) continuous signal $s[t]$; (*b*) sampling reference signal $r[t]$; (*c*) sampled signal $s^s[t]$.

The continuous signal $s[t]$ of diagram a is multiplied by the reference-signal pulse train $r[t]$ of diagram b, to obtain the sampled signal $s^s[t]$, shown in diagram c:

$$s^s[t] = s[t]r[t] \qquad (5.2\text{-}1)$$

The symbol $s*[t]$ is commonly used to represent a sampled signal. This book uses the superscript s instead of the asterisk, so that the latter can be reserved to represent the conjugate of a complex quantity.

The reference pulse train can be represented as follows by a Fourier series. The general expression for a Fourier series is

$$r[t] = \frac{a_0}{2} + \sum_{n=1}^{\infty} \left\{ a_n \cos\frac{2\pi nt}{T} + b_n \sin\left[\frac{2\pi nt}{T}\right] \right\} \qquad (5.2\text{-}2)$$

The Fourier coefficients of Eq 5.2-2 are

$$a_n = \frac{2}{T} \int_{-T/2}^{+T/2} r[t]\cos\left[\frac{2\pi nt}{T}\right] dt \qquad (5.2\text{-}3)$$

$$b_n = \frac{2}{T} \int_{-T/2}^{+T/2} r[t]\sin\left[\frac{2\pi nt}{T}\right] dt \qquad (5.2\text{-}4)$$

The $t = 0$ point is set at the center of a pulse, which makes the sine coefficients b_n zero. The reference $r[t]$ is unity for $-\tau/2 < t < +\tau/2$, and is zero elsewhere within the $\pm T/2$ range of the integration. Hence Eq 5.2-3 gives for the coefficients a_n

$$a_n = \frac{2}{T} \int_{-\tau/2}^{+\tau/2} \cos\left[\frac{2\pi nt}{T}\right] dt \qquad (5.2\text{-}5)$$

Solving this integral gives

$$a_0 = \frac{2\tau}{T} \qquad (5.2\text{-}6)$$

$$a_n = \frac{2}{\pi n} \sin\left[\frac{\pi n\tau}{T}\right] \qquad (\text{for} \quad n > 0) \qquad (5.2\text{-}7)$$

If the pulse width τ is made vanishingly small, $\sin[\pi n\tau/T]$ can be replaced by $\pi n\tau/T$. For $n > 0$, this gives for a_n

$$a_n = \frac{2\tau}{T} \qquad (\text{for} \quad n > 0) \qquad (5.2\text{-}8)$$

This is the same as the expression for a_0 in Eq 5.2-6. Hence, all of the coefficients a_n are equal to a_0, and so the Fourier series of Eq 5.2-2 becomes

$$r[t] = \frac{\tau}{T}\left\{1 + 2\sum_{n=1}^{\infty}\cos\left[\frac{2\pi nt}{T}\right]\right\} \tag{5.2-9}$$

In this infinite series, each component is proportional to τ and so is vanishingly small.

Assume that the continuous signal $s[t]$ is a sinusoid of unit amplitude, given by

$$s[t] = \cos[\omega_m t] \tag{5.2-10}$$

By Eq 5.2-1, the sampled signal is the product of Eqs 5.2-9, -10, which is

$$s^s[t] = s[t]r[t] = \frac{\tau}{T}\left\{\cos[\omega_m t] + \sum_{n=1}^{\infty}2\cos[n\omega_s t]\cos[\omega_m t]\right\} \tag{5.2-11}$$

where ω_s is the sampling frequency in rad/sec. The following sampling parameters are defined:

T = sampling period
$F = 1/T$ = sampling frequency in hertz
$\omega_s = 2\pi F$ = sampling frequency in rad/sec

Apply the following trigonometric identity to Eq 5.2-11:

$$\cos[a]\cos[b] = \tfrac{1}{2}(\cos[a+b] + \cos[a-b]) \tag{5.2-12}$$

Equation 5.2-11 becomes

$$s^s[t] = \frac{\tau}{T}\cos[\omega_m t]$$

$$= +\frac{\tau}{T}\sum_{n=1}^{\infty}\{\cos[(n\omega_s - \omega_m)t] + \cos[(n\omega_s + \omega_m)t]\} \tag{5.2-13}$$

This has frequency components at the modulation frequency ω_m and at the harmonic sideband frequencies $(n\omega_s \pm \omega_m)$, where n is an integer varying from 1 to infinity. Figure 5.2-2 shows the spectra of the continuous and sampled signals expressed in hertz. Diagram a shows the spectrum of the continuous signal $s[t]$, which is a unit-amplitude line at the frequency $f_m = \omega_m/2\pi$. Diagram b shows the spectrum of the sampled signal $s^s[t]$. This has an infinite series of lines of amplitude τ/T, at the frequencies f_m, $(F - f_m)$, $(F + f_m)$, $(2F - f_m)$, $(2F + f_m)$, etc.

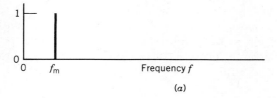

(a)

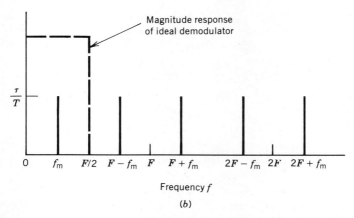

(b)

Figure 5.2-2   Spectra of continuous and sampled signals: (a) continuous signal $s[t]$; (b) sampled signal $s^s[t]$.

5.2.2   Ideal Demodulation of Sampled Signal

To recreate the continuous signal $s[t]$, the sampled signal $s^s[t]$ can be demodulated by filtering it to pass the signal at the frequency f_m, while eliminating all the other lines. As the signal frequency f_m is increased relative to the sampling frequency F, the two sidebands at the frequencies f_m and $(F - f_m)$ move toward one another. When the signal frequency f_m is equal to $F/2$, these two sidebands coincide and so cannot be separated. Hence information in the continuous signal can only be recovered if the signal frequency is less than half the sampling frequency.

If all of the frequency components of the continuous signal are at less than half the sampling frequency, the signal can theoretically be recovered exactly by passing the sampled data through a filter with no phase lag, having a flat magnitude response out to half the sample frequency and zero response at higher frequencies. The required filter response $H[f]$ is shown by the dashed plot in Fig 5.2-2b, and is defined by

$$H(f) = T/\tau \qquad \text{for} \quad 0 < f < F/2 \qquad (5.2\text{-}14)$$
$$H(f) = 0 \qquad \text{for} \quad f > F/2 \qquad (5.2\text{-}15)$$

Since this filter has magnitude attenuation but no phase shift, it cannot be realized in real time. However, it can be implemented by non-real-time processing if the complete set of sampled-data values is available.

The inverse Fourier transform of this filter frequency-response is the unit-impulse response, which is

$$h[t] = \mathscr{F}^{-1}[H[f]] = \frac{1}{\tau} \frac{\sin[\pi t/T]}{\pi t/T} \qquad (5.2\text{-}16)$$

Each sampling pulse has unit amplitude and a pulse width τ, and so has an area equal to τ. As the pulse width τ is made vanishingly small, the pulse becomes an impulse of area τ. Hence the response of the filter $H[f]$ to a sampling pulse is equal to the unit-impulse response of Eq 5.2-16 multiplied by the impulse area τ. This response, which is denoted $p[t]$, is

$$p[t] = \tau h[t] = \frac{\sin[\tau t/T]}{\pi t/T} \qquad (5.2\text{-}17)$$

A plot of $p[t]$ is shown in Fig 5.2-3a. This is the ideal demodulated response for a single sampled value equal to unity.

The ideal continuous output signal corresponding to a series of sampled values can be calculated as shown in Fig 5.2-3b. Each of the sampled values is multiplied by the response $p[t]$ of diagram a, with the time axis of each $p[t]$ shifted to coincide with the sampling instant. The contributions for all of the sampled values are added to obtain the total plot. The points (a), (b), (c) in diagram b are the sampled values. The curves Ⓐ, Ⓑ, Ⓒ are the plots of $p[t]$ multiplied by these sampled values. These curves, along with the curves for all of the other points, are added to produce the ideally demodulated signal shown by the dashed curve.

The figure shows only the contributions from three sampled values, over a limited time interval. The contributions from a great many more sampled values must be included in the summation before an accurate representation of the continuous signal over this time interval can be obtained. Such a summation would be very difficult to perform with hand calculation, but is reasonable with a digital computer.

This ideal demodulation process exactly duplicates the original continuous signal $s[t]$ if the signal has no components at frequencies greater than half the sampling frequency. When this condition is not satisfied, an effect called "foldover" or "aliasing" occurs in the sampling process. This distorts the information, and prohibits the original continuous data from being recovered exactly.

Foldover (or aliasing) is illustrated in Fig 5.2-4. The solid curve shows the spectrum of a continuous signal. When that signal is sampled, the portion of this spectrum at frequencies greater than $F/2$ is folded over into lower

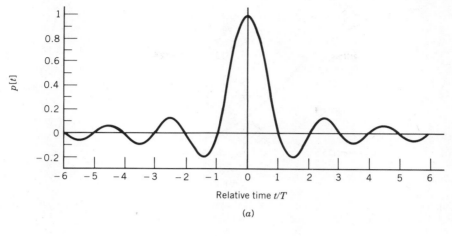

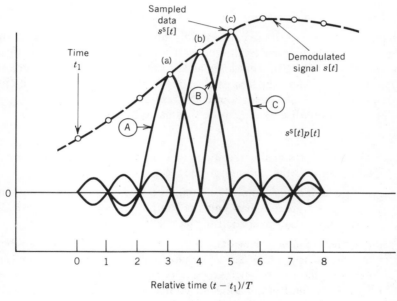

Figure 5.2-3   Ideal demodulation of sampled data: (a) response $p[t]$ of ideal sampled-data demodulator to a sample value equal to unity; (b) calculation of the ideal continuous signal that corresponds to a set of sampled values.

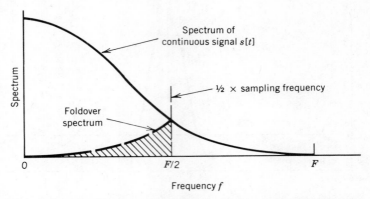

Figure 5.2-4 Spectrum foldover (or aliasing) caused by sampling, when signal has components at frequencies greater than half the sampling frequency.

frequencies, to produce the contribution shown by the cross-hatched area. The components represented by the cross-hatched foldover spectrum distort the signal. The sampling process loses all information at frequencies greater than $F/2$. Any components at frequencies greater than $F/2$ that are present in the continuous signal add distortion to the resultant sampled data.

The preceding discussion can be summarized as follows:

1. The sampled signal can only convey information at frequencies that are less than half the sampling frequency.
2. Before a signal is sampled, it should be filtered to remove components at frequencies greater than half the sampling frequency, because any such components add distortion to the resultant data.
3. An ideal sampled-data demodulation process (which cannot be implemented in real time) can be achieved by transmitting, without attenuation or phase shift, all frequency components out to half the sampling frequency, and eliminating all components at higher frequencies. This ideal sampled-data demodulator would exactly recover the continuous signal that is sampled, provided that the signal has no components at frequencies greater than half the sampling frequency. When that condition is not satisfied, this demodulation process still gives the best possible *a priori* approximation of the continuous signal.

The ideal demodulated signal is a smooth curve that passes through the sampled points. One can obtain a reasonable approximation of this ideal response by using a French-curve template to draw a smooth plot through the data points.

5.2.3   Simple-Hold Circuit for Demodulation of Sampled Data

A digital sampled signal is usually converted to analog form by a digital-to-analog (D/A) converter, which has a circuit that holds the output fixed between sampling instants. We call this a *simple-hold* circuit, but it is commonly called a "zero-order hold" circuit in the sampled-data literature. Figure 5.2-5 shows the response $p[t]$ of a simple-hold circuit to a sampled value of unity occurring at time $t = 0$. This response $p[t]$ can be represented as a positive unit step occurring at time $t = 0$, followed by a negative unit step occurring at $t = T$. Hence $p[t]$ can be expressed as

$$p[t] = u[t] - u[t - T]   \tag{5.2-18}$$

where $u[t]$ is a unit step occurring at time $t = 0$. The Laplace transform of this is

$$P[s] = \mathscr{L}[p[t]] = \frac{1}{s} - \frac{1}{s}e^{-sT} = \frac{1}{s}(1 - e^{-sT})   \tag{5.2-19}$$

The sampled value $s^s[t]$ is considered to be a pulse of unit amplitude and infinitesimal pulse width τ occurring at time $t = 0$. This can be represented as an impulse of amplitude τ occurring at time $t = 0$:

$$s^s[t] = \tau\delta[t]   \tag{5.2-20}$$

where $\delta[t]$ is a unit impulse occurring at $t = 0$. The Laplace transform of this is

$$S^s[s] = \mathscr{L}[s^s[t]] = \tau   \tag{5.2-21}$$

Dividing Eq 5.2-19 by Eq 5.2-21 gives the transfer function $H[s]$ of the

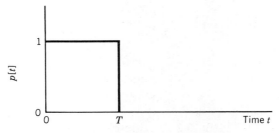

Figure 5.2-5   Response of simple-hold circuit to a sampled value of unity at time $t = 0$.

simple-hold circuit:

$$H[s] = \frac{P[s]}{S^s[s]} = \frac{1 - e^{-sT}}{s\tau} \tag{5.2-22}$$

Setting $s = j\omega$ gives the frequency response of the simple-hold circuit:

$$H[j\omega] = \frac{1 - e^{-j\omega T}}{j\omega\tau} \tag{5.2-23}$$

Factoring $e^{-j\omega T/2}$ from the expression gives

$$H[j\omega] = e^{-j\omega T/2} \frac{\left(e^{j\omega T/2} - e^{-j\omega T/2}\right)}{j\omega\tau} \tag{5.2-24}$$

Now $\sin[\omega T/2]$ is equal to

$$\sin\left[\frac{\omega T}{2}\right] = \frac{e^{j\omega T/2} - e^{-j\omega T/2}}{2j} \tag{5.2-25}$$

Combining Eqs 5.2-24, -25 gives

$$H[j\omega] = \frac{T}{\tau} e^{-j\omega T/2} \frac{\sin[\omega T/2]}{\omega T/2} \tag{5.2-26}$$

This can be expressed in the form

$$H[j\omega] = \frac{T}{\tau} M[\omega] e^{-j\omega T/2} \tag{5.2-27}$$

where $M[\omega]$ represents the normalized frequency response magnitude of the demodulation process, which is

$$M[\omega] = \frac{\sin[\omega T/2]}{\omega T/2} = \frac{\sin[\pi f/F]}{\pi f/F} \tag{5.2-28}$$

A plot of $M[\omega]$ is shown as curve ② in Fig 5.2-6. For comparison the dashed curve ① shows the response of an ideal sampled-data demodulator, which passes all data out to $\frac{1}{2}$ of the sampling frequency unchanged, and eliminates everything at higher frequencies.

The factor $e^{-j\omega T/2}$ in Eq 5.2-27 represents a time delay equal to $\frac{1}{2}$ of the sampling period T. This has unity magnitude response and a phase lag

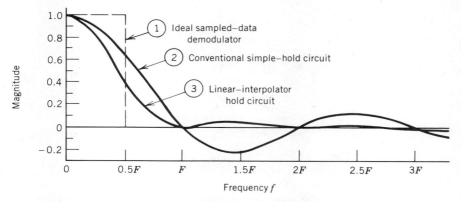

Figure 5.2-6 Frequency-response magnitude plots for different techniques of demodulating sampled data.

proportional to frequency given by

$$\text{Ang}[e^{-j\omega T/2}] = -\frac{\omega T}{2} \text{ rad} = -\frac{f}{F} 180° \qquad (5.2\text{-}29)$$

This time delay of half a sampling period is equivalent to a phase lag of half a cycle (or 180°) at the sampling frequency $f = F$.

Figure 5.2-7 illustrates this digital-to-analog conversion process in the time domain. Curve ① is the original continuous signal $s[t]$. Points a, b, c, d are

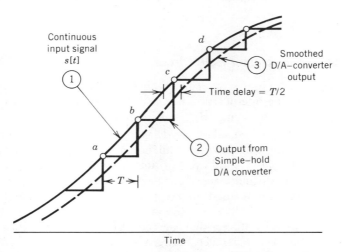

Figure 5.2-7 Time response of conventional simple-hold D/A converter.

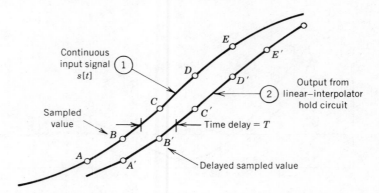

Figure 5.2-8 Time response of linear-interpolator hold circuit.

the sampled-data values defining $s^s[t]$. Curve ② is the output from the simple-hold D/A converter that demodulates $s^s[t]$. When the demodulated signal is filtered to attenuate the high-frequency components, the smoothed plot shown by the dashed curve ③ remains. (The time delay of the filter is not included in the plot ③.) The smoothed signal lags the continuous signal $s^s[t]$ by $\frac{1}{2}$ of a sampling period. This illustrates the principle that the simple-hold D/A conversion causes a time delay equal to $\frac{1}{2}$ of the sampling period. The filter that attenuates the harmonics of the stepped waveform ② to form the smoothed curve ③ adds additional time delay to the signal. The dashed curve ③ is a theoretical abstraction, that shows only the time delay of the simple-hold demodulation process, and omits subsequent time delays due to filtering.

5.2.4   Linear-Interpolator Hold Circuit

A much smoother output signal can be obtained by using a hold circuit that provides straight-line interpolation between the sampled points. In Fig 5.2-8, curve ① is the input signal $s[t]$, and points A, B, C, D are the sampled values $s^s[t]$. These sampled values are delayed by one period T to produce the values A', B', C', D'; and the demodulation process connects straight lines between adjacent points. A full cycle of time delay is required in the demodulation in order to calculate the difference between successive values. This difference information is used to compute the slopes of the straight-line interpolations.

Figure 5.2-9 shows a process that theoretically could act as a linear-interpolator hold circuit. The difference between successive samples $(y_n - y_{n-1})$ is calculated and converted to analog form in the D/A converter. The output

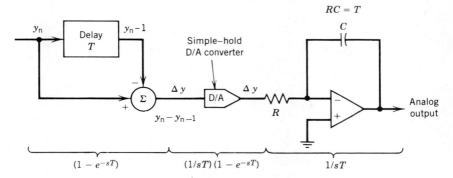

Figure 5.2-9   Theoretical linear-interpolator hold circuit.

from the D/A converter is fed to an integrator having the transfer function $1/Ts$. The integrator output is the demodulated analog signal. This circuit can theoretically provide linear interpolation, but is not practical because inaccuracy in the integrator would cause the output to drift. This drift can be corrected by resetting the integrator voltage to the correct output at each sampling instant, as is done in the circuit of Fig 5.2-10.

Figure 5.2-10 shows a practical linear-interpolator hold circuit. Two D/A converters are used: DAC1, which provides conventional simple-hold demodulation of the sampled signal, and DAC2, which converts to analog form the differential values between successive samples of the signal. The differential signal from DAC2 is fed to an opamp integrator circuit, having an RC time constant equal to the sample period T. By integrating this differential information, the integrator provides linear interpolation of the sample data at its output. At each sampling instant, the integrator output is compared with the output from DAC1. The FET switch is closed at each sampling instant, and the feedback circuit adjusts the capacitor voltage to correct for drift and make the integrator output agree with the output from DAC1.

There is one sample period of time delay in the formation of the demodulated signal by the integrator. To compensate for this delay, the output from DAC1 is appropriately delayed relative to the input data. The operation of DAC1 is actually delayed by only half a sample period. This eliminates the effect of transients in DAC1 because the output from DAC1 is allowed to settle by half a sample period before it is used to reset the voltage on the integrator capacitor.

To illustrate the timing of this demodulation process, it is assumed in Fig 5.2-10 that a continuous signal $y[t]$ is sampled and the sampled data are converted by the linear-interpolator hold circuit back to a continuous signal equivalent to $y[t]$. As is shown, there is a time delay of one sample period between the input continuous signal $y[t]$ and the reconstruction of $y[t]$ provided at the integrator output.

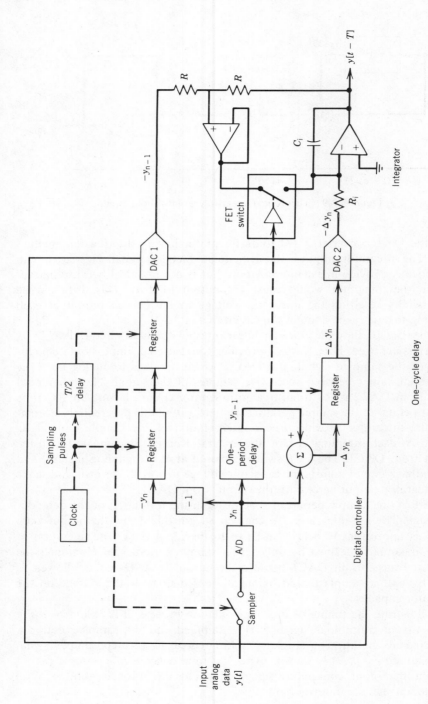

Figure 5.2-10 Practical linear-interpolator hold circuit.

The transfer function of the linear-interpolator hold circuit can be derived from the ideal process shown in Fig 5.2-9. This transfer function is

$$H[s] = (1 - e^{-sT})\left\{ \frac{1}{s\tau}(1 - e^{-sT}) \right\} \frac{1}{sT}$$

$$= \frac{T}{\tau}\left(\frac{1 - e^{-sT}}{sT} \right)^2 \tag{5.2-30}$$

The factor $(1 - e^{-sT})$ is the transfer function of the subtraction process $(y_n - y_{n-1})$. The factor $1/sT$ is the transfer function of the analog integrator. The factor within the braces $\{\ \}$ is the transfer function of the D/A converter, which was given in Eq 5.2-22. Setting $s = j\omega$ in Eq 5.2-30 gives the frequency response. This can be simplified to the following form, which is similar to Eq 5.2-26:

$$H[j\omega] = \frac{T}{\tau}\left(\frac{1 - e^{-j\omega T}}{j\omega T} \right)^2$$

$$= \frac{T}{\tau}e^{-j\omega T}\left(\frac{\sin[\omega T/2]}{\omega T/2} \right)^2 \tag{5.2-31}$$

This can be expressed as

$$H[j\omega] = \frac{T}{\tau}M[\omega]e^{-j\omega T} \tag{5.2-32}$$

The function $M[\omega]$ is the normalized magnitude response, which is

$$M[\omega] = \left(\frac{\sin[\omega T/2]}{\omega T/2} \right)^2 = \left(\frac{\sin[\pi f/F]}{\pi f/F} \right)^2 \tag{5.2-33}$$

The exponential $e^{-j\omega T}$ in Eq 5.2-32 represents the delay of one sample period in the process. This produces twice the phase lag of the simple-hold circuit. The magnitude response $M[\omega]$ of Eq 5.2-33 is the square of the magnitude response of the simple-hold circuit, which was given in Eq 5.2-28. A plot of this is given as curve ③ in Fig 5.2-6.

A simple-hold circuit, provided by a D/A converter, may require appreciable subsequent filtering to achieve adequate reduction of the sampling harmonics. A linear-interpolator hold circuit, with its much lower harmonic output, may not need subsequent filtering. Consequently, the overall phase lag of the simple-hold circuit and its filtering may actually be greater than the phase lag of a linear-interpolator hold circuit.

5.2.5 Oversampling

The effect of a linear-interpolator hold circuit can be approximated with a simpler circuit by using oversampling, a technique that is commonly used in compact-disk audio players. In Fig 5.2-11, diagram a shows a circuit that can provide times-4 oversampling. The corresponding waveforms for a single digital value are shown in diagram b.

The digital input is fed to a simple-hold D/A converter, which forms the analog signal A. The signal A is sampled by three sample-and-hold circuits to form signals B, C, and D. The sampling instants of the sample-and-hold circuits are delayed relative to those of the D/A converter by $\frac{1}{4}$, $\frac{1}{2}$, and $\frac{3}{4}$ sample period. The four analog signals A, B, C, D for a single digital value are

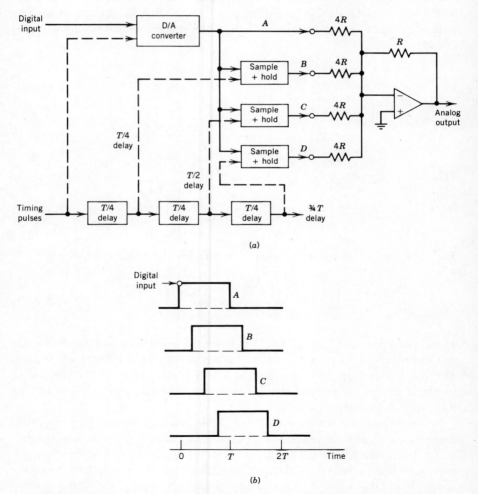

(a)

(b)

Figure 5.2-11 Times-4 oversampled hold circuit: (a) circuit; (b) waveforms.

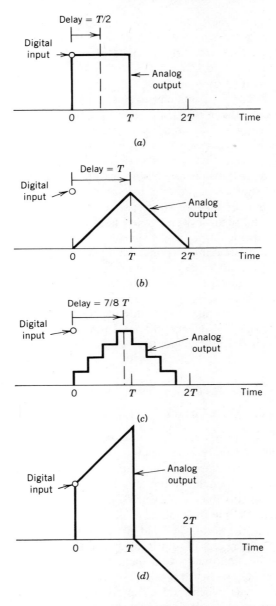

Figure 5.2-12 Analog responses of various hold circuits to a single digital value: (a) simple-hold circuit; (b) linear-interpolator hold circuit; (c) times-4 oversampled hold circuit; (d) first-order hold circuit.

shown in diagram b. These four signals are summed in an opamp, with a gain of $\frac{1}{4}$, to form the analog output voltage. The resultant analog output voltage (ignoring the phase reversal of the opamp) is shown in Fig 5.2-12c. This response (relative to the digital input value) is the sum of the four waveforms of Fig 5.2-11b, divided by 4.

Figure 5.2-12 shows the analog responses, for a single digital value, of the simple-hold circuit (diagram a), the linear-interpolator hold circuit (diagram b), and the times-4 oversampled hold circuit of Fig 5.2-11 (diagram c). (Diagram d shows the response for the first-order hold circuit, to be discussed in Section 5.2.6.) Note that the analog output with times-4 oversampling in diagram c approximates that for the linear-interpolator hold in diagram b.

Figure 5.2-12 shows that the response for the simple-hold circuit has a time delay of $T/2$; that for the linear-interpolator hold circuit has a time delay of T; and that for the times-4 oversampled hold circuit has a time delay of $\frac{7}{8}T$.

If the output from the times-4 oversampled hold circuit is fed through a lowpass filter, the resultant response can closely approximate that for the linear-interpolator hold circuit. Assume, for example, a second-order filter with the transfer function $1/(1 + \tau s)^2$, where $\tau = T/8$. This filter would provide $11 : 1$ attenuation of the ripple-frequency fundamental of Fig 5.2-12c, which is the fourth harmonic of the sampling frequency. The time delay of this filter is 2τ, which is $T/4$. Hence, the total time delay of the hold circuit plus filter is $\frac{9}{8}T$. This is only 12.5% greater than that of the linear-interpolator hold circuit.

5.2.6 First-Order Hold Circuit

The sampled-data literature makes little reference to the linear-interpolator hold circuit, but frequently discusses the "first-order hold" circuit. These two circuits should not be confused. The first-order hold circuit provides linear *extrapolation* of the sampled data, not linear *interpolation*. The difference between two sampled values is measured, to set the slope of the output signal in the subsequent sample interval.

The response of the first-order hold circuit to a single digital value is shown in Fig 5.2-12d. Figure 5.2-13 shows the response of this hold circuit to the sampled values of a continuous sinusoidal input, with a frequency of $\frac{1}{12}$ of the sampling frequency. The figure shows that the first-order hold circuit generates large output harmonics when the frequency content of the input signal is at all high. Consequently, this circuit has little practical value. This conclusion is consistent with the findings of Ragazzini and Franklin [5.1] (p. 39), who reported that the first-order hold circuit is "not commonly employed in feedback control systems".

In contrast, the linear-interpolator hold circuit has many potential applications. For example, Oppenheim, Willsky, and Young [5.2] discuss its use in the signal processing of digitized imagery data.

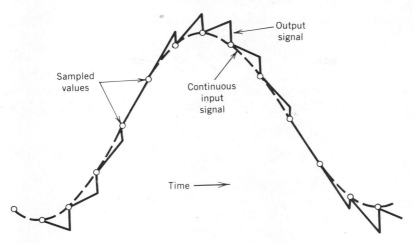

Figure 5.2-13   Response of first-order hold circuit to a sampled sinusoid at 1.12 of the sampling frequency.

5.2.7   Polynomial-Fit Demodulation of Sampled Data

The linear-interpolator hold circuit draws a straight line between successive sampled values. As shown in Fig 5.2-14, this approach can be extended by constructing a third-order polynomial that passes through four successive sampled values a, b, c, d. The section of that polynomial shown by the solid curve is the output signal between sampled points b, c. This polynomial has

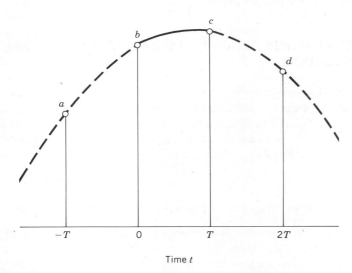

Figure 5.2-14   Four-point polynomial fit for demodulating sampled data.

the following form, where time t is zero at point b:

$$x = K_0 + K_1 t + K_2 t^2 + K_3 t^3 \qquad (5.2\text{-}34)$$

To solve for the coefficients, set $x = a$ at $t = -T$, $x = b$ at $t = 0$, $x = c$ at $t = +T$, and $x = d$ at $t = +2T$. Applying these relations to Eq 5.2-34 yields the coefficients:

$$K_0 = b \qquad (5.2\text{-}35)$$

$$K_1 = \frac{-2a + 5b + 6c - d}{6T} \qquad (5.2\text{-}36)$$

$$K_2 = \frac{a - 2b + c}{2T^2} \qquad (5.2\text{-}37)$$

$$K_3 = \frac{-a - 5b - 3c + d}{6T^3} \qquad (5.2\text{-}38)$$

These coefficients are substituted into Eq 5.2-34 to provide third-order interpolation between the sample points b, c. This approach is generally not practical in the actual detection stage. However, it can be implemented within the computer to increase the data rate by generating sample values between the original points. A linear-interpolator hold circuit can be used to provide straight-line interpolation between the resultant data points. This demodulation process can achieve very accurate reconstruction of the signal, but causes a time delay of two sample periods.

5.3   DIGITAL COMPUTATION FOR SAMPLED-DATA FILTERING AND SIMULATION

This section develops algorithms for processing digital data to achieve specified frequency-response characteristics. These are used to simulate dynamic systems on a digital computer. The simulation procedure developed in this section is called *serial simulation* to distinguish it from the parallel simulation procedure using the Runge–Kutta integration routine to be described in Section 5.5.

5.3.1   Signal-Flow Diagram of Servo for Computer Simulation

As an example for computer simulation, consider the stage-positioning servo studied in Section 2.8. Figure 5.3-1 is a simplified version of the signal-flow diagram of that servo shown earlier in Fig 2.8-12b. The motor electrical break frequency ω_e has been replaced by its reciprocal, the motor electrical time

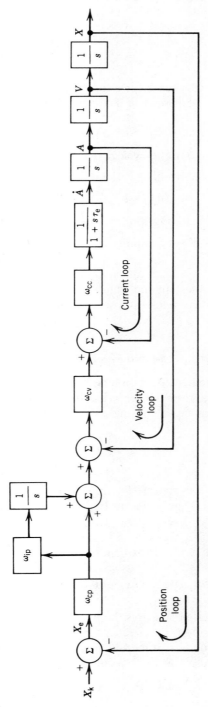

Figure 5.3-1 Simplified signal-flow diagram of stage-positioning servo to be used in simulation.

constant τ_e, defined by

$$\tau_e = 1/\omega_e \tag{5.3-1}$$

To simplify the diagram, the subscript c for the controlled variable is omitted. The following transform variables are defined:

$X = X_c$ = controlled position

$V = sX$ = controlled velocity $\tag{5.3-2}$

$A = sV$ = controlled acceleration $\tag{5.3-3}$

$\dot{A} = sA$ = controlled rate of change of acceleration $\tag{5.3-4}$

X_k = command position

$X_e = X_k - X$ = position error $\tag{5.3-5}$

To simulate this signal-flow diagram on a digital computer, computer algorithms are needed to implement the transfer functions $1/s$ for an integrator and $1/(1 + s\tau)$ for a lowpass filter. A method for deriving such algorithms will now be presented.

5.3.2 Algorithm for Sampled-Data Filter

To develop a general method for designing linear sampled-data filters, let us examine a sampled-data formula for achieving integration. Assume that a variable $y[t]$ is calculated by integrating a variable $x[t]$:

$$y = \int x \, dt \tag{5.3-6}$$

A well-known sampled-data routine for implementing this equation is trapezoidal integration, which is illustrated in Fig 5.3-2. The nth samples of x and y are defined as x_n and y_n, and the $(n - 1)$th samples are defined as x_{n-1} and y_{n-1}. As shown in Fig 5.3-2, the area under the x-curve between the $(n - 1)$th sample and the nth sample is approximately

$$\Delta y_n \cong \tfrac{1}{2}(x_n + x_{n-1})T \tag{5.3-7}$$

where T is the sample period, which is the time between the $(n - 1)$th and the nth sample. In this equation, the area under the x-curve between these two samples is approximated by a trapezoid formed by drawing a straight line between points x_{n-1} and x_n. Since y is the integral of x, its value at the nth sample is

$$y_n = y_{n-1} + \Delta y_n \tag{5.3-8}$$

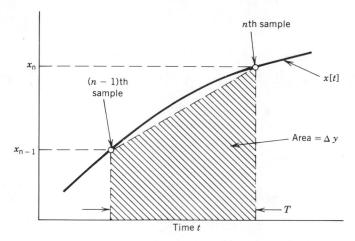

Figure 5.3-2 Illustration of trapezoidal integration.

where y_{n-1} is the integral of x up to the $(n - 1)$th sample. Substituting Eq 5.3-7 into Eq 5.3-8 gives the trapezoidal integration formula:

$$y_n = y_{n-1} + \tfrac{1}{2}T(x_n + x_{n-1}) \qquad (5.3\text{-}9)$$

To find the frequency response of this algorithm, the Laplace transform of the equation is taken, and s is replaced by $j\omega$. In terms of a continuous signal, x_n can be considered to represent $x[t]$, and x_{n-1} to represent $x[t - T]$. Thus

$$x_n = x[t], \qquad x_{n-1} = x[t - T] \qquad (5.3\text{-}10)$$

Hence, the Laplace transforms of x_{n-1} and x_n are related by

$$\mathscr{L}[x_{n-1}] = \mathscr{L}[x[t - T]] = e^{-sT}\mathscr{L}[x[t]] = e^{-sT}\mathscr{L}[x_n] \quad (5.3\text{-}11)$$

(Remember that e^{-sT} in the Laplace transform corresponding to a time delay T.) Using this principle, the Laplace transform of Eq 5.3-11 can be taken to give

$$\mathscr{L}[y_n] = \mathscr{L}[y_{n-1}] + \tfrac{1}{2}T\{\mathscr{L}[x_n] + \mathscr{L}[x_{n-1}]\}$$
$$= e^{-sT}\mathscr{L}[y_n] + \tfrac{1}{2}T\{\mathscr{L}[x_n] + e^{-sT}\mathscr{L}[x_n]\} \qquad (5.3\text{-}12)$$

The Laplace transforms of the x and y sample values are denoted

$$\mathscr{L}[y_n] = Y \qquad (5.3\text{-}13)$$
$$\mathscr{L}[x_n] = X \qquad (5.3\text{-}14)$$

Hence Eq 5.3-12 becomes

$$Y = e^{-sT}Y + \tfrac{1}{2}T(X + e^{-sT}X) \tag{5.3-15}$$

By algebraic manipulation of Eq 5.3-15, the transfer function Y/X is calculated as follows:

$$Y(1 - e^{-sT}) = \tfrac{1}{2}TX(1 + e^{-sT}) \tag{5.3-16}$$

$$\frac{Y}{X} = \frac{(T/2)(1 + e^{-sT})}{1 - e^{-sT}} \tag{5.3-17}$$

To obtain the frequency response, set $s = j\omega$. The exponentials become

$$e^{-sT} = e^{-j\omega T} = \cos[\omega T] - j\sin[\omega T] \tag{5.3-18}$$

Substituting this into Eq 5.3-17 gives the frequency response between X and Y:

$$\frac{Y}{X} = \frac{(T/2)(1 + \cos[\omega T] - j\sin[\omega T])}{1 - \cos[\omega T] + j\sin[\omega T]} \tag{5.3-19}$$

Multiply the numerator and denominator by $(1 - \cos[\omega T] - j\sin[\omega T])$ to rationalize the expression. The frequency response reduces to

$$\frac{Y}{X} = \frac{(T/2)(-2j\sin[\omega T])}{2(1 - \cos[\omega T])} = \frac{T\sin[\omega T]}{2j(1 - \cos[\omega T])} \tag{5.3-20}$$

For small values of ωT, the sine and cosine expressions can be approximated by

$$\sin[\omega T] \cong \omega T \tag{5.3-21}$$

$$\cos[\omega T] \cong 1 - \tfrac{1}{2}(\omega T)^2 \tag{5.3-22}$$

Substituting these into Eq 5.3-20 gives the approximate frequency response

$$\frac{Y}{X} \cong \frac{1}{j\omega} \tag{5.3-23}$$

Thus, at low frequencies, trapezoidal integration exhibits the frequency response $1/j\omega$, which is the transfer function of an ideal integrator.

This result can be derived much more simply by employing the "W-transform" method developed by Johnson, Lindorff, and Nordling [5.3], [5.4]. However, we replace their symbol W with U, so that W can be reserved to represent a pseudo angular frequency, which approximates ω at low frequencies. The first step is to make the substitution

$$\bar{z} = e^{-sT} \tag{5.3-24}$$

This is the transfer function for a time delay of one sample period. The sampled-data literature generally deals with the reciprocal of this (e^{sT}), which is denoted z. The author uses the variable $\bar{z}$ (which is equal to $1/z$) because time shifts only occur in terms of delays. Thus, when the variable z is used, it occurs naturally in the form z^{-1}, z^{-2}, etc. In terms of $\bar{z}$, the transform of the basic trapezoidal integration algorithm of Eq 5.3-9 is

$$Y = \bar{z}Y + \tfrac{1}{2}T(X + \bar{z}X) \tag{5.3-25}$$

Note that $\bar{z}$ is substituted for each time delay. Algebraic manipulation of Eq 5.3-25 yields the following transfer function:

$$\frac{Y}{X} = \frac{(T/2)(1 + \bar{z})}{1 - \bar{z}} \tag{5.3-26}$$

This could also be derived from Eq 5.3-17 by setting e^{-Ts} equal to $\bar{z}$.

In the Johnson–Lindorff–Nordling transform, U is defined as

$$U = \frac{1 - \bar{z}}{1 + \bar{z}} \tag{5.3-27}$$

It can be seen by substitution that this transformation applies both ways:

$$\bar{z} = \frac{1 - U}{1 + U} \tag{5.3-28}$$

In Eq 5.3-27, replace $\bar{z}$ by e^{-sT} and set $s = j\omega$ to obtain the frequency response. This yields

$$U = \frac{1 - e^{-sT}}{1 + e^{-sT}} = \frac{1 - e^{-j\omega T}}{1 + e^{-j\omega T}} \tag{5.3-29}$$

Multiply numerator and denominator by $e^{j\omega T/2}$, to obtain

$$U = \frac{e^{j\omega T/2} - e^{-j\omega T/2}}{e^{j\omega T/2} + e^{-j\omega T/2}} \tag{5.3-30}$$

Remember that $\cos[\theta]$ and $\sin[\theta]$ are expressed as follows in terms of complex exponentials:

$$\cos[\theta] = \frac{e^{j\theta} + e^{-j\theta}}{2} \tag{5.3-31}$$

$$\sin[\theta] = \frac{e^{j\theta} - e^{-j\theta}}{2j} \tag{5.3-32}$$

Hence, $j\tan[\theta]$ is equal to

$$j\tan[\theta] = j\frac{\sin[\theta]}{\cos[\theta]} = \frac{e^{j\theta} - e^{-j\theta}}{e^{j\theta} + e^{-j\theta}} \tag{5.3-33}$$

Comparing Eqs 5.3-30, -33 shows that

$$U = j\tan[\omega T/2] \tag{5.3-34}$$

For angles less than $45°$ (or $\pi/4$ rad) the function $\tan[\theta]$ can be approximated by

$$\tan[\theta] \cong \theta \qquad \text{for} \quad \theta < \pi/4 \text{ rad} = 45° \tag{5.3-35}$$

The maximum error of this approximation, which occurs at $45°$, is 21%. Applying this approximation to Eq 5.3-34 gives

$$U = j\tan\left[\frac{\omega T}{2}\right] \cong \frac{j\omega T}{2} = \frac{sT}{2} \qquad \text{for} \quad \frac{\omega T}{2} < \frac{\pi}{4} \tag{5.3-36}$$

In terms of frequency in hertz, the limit is

$$f = \frac{\omega}{2\pi} < \frac{1}{4T} = \frac{F}{4} \tag{5.3-37}$$

where F is the sampling frequency, which is equal to $1/T$. This shows that the approximation of Eq 5.3-36 holds with reasonable accuracy up to $\frac{1}{4}$ of the sample frequency. Since sampled data can only convey frequency information up to $\frac{1}{2}$ of the sample frequency, this approximation is good up to $\frac{1}{2}$ of the maximum frequency that can be conveyed by the sampled data.

Let us apply the U-transform to trapezoidal integration. Substituting Eq 5.3-28 into Eq 5.3-26 gives the following for the transfer function of trapezoidal integration, in terms of U:

$$\frac{Y}{X} = \frac{(T/2)(1 + \bar{z})}{1 - \bar{z}} = \frac{(T/2)[1 + (1 + U)/(1 - U)]}{1 - (1 + U)/(1 - U)}$$

$$= \frac{(T/2)[(1 + U) + (1 - U)]}{(1 + U) - (1 - U)} = \frac{(T/2)(2)}{2U} = \frac{T}{2U} \quad (5.3\text{-}38)$$

Apply the approximation of Eq 5.3-36 by replacing U with $sT/2$. Equation 5.3-38 becomes

$$\frac{Y}{X} = \frac{T}{2U} \cong \frac{T}{2(sT/2)} = \frac{1}{s} = \frac{1}{j\omega} \quad (5.3\text{-}39)$$

Thus, the frequency response for trapezoidal integration approximates $1/j\omega$ quite accurately up to $\frac{1}{4}$ of the sampling frequency.

The preceding discussion can be summarized by the following equations:

$$\bar{z} = e^{-sT} = \text{transform of one-cycle time delay} \quad (5.3\text{-}40)$$

$$U = \frac{1 - \bar{z}}{1 + \bar{z}} \quad (5.3\text{-}41)$$

$$\bar{z} = \frac{1 - U}{1 + U} \quad (5.3\text{-}42)$$

$$U \cong \frac{sT}{2}, \quad s \cong \frac{2U}{T} \quad \text{for} \quad f < \frac{F}{4} \quad (5.3\text{-}43)$$

To apply these equations to another example, let us calculate the sampled-data routine having the transfer function of a single-order lowpass filter of time constant τ:

$$\frac{Y}{X} = \frac{1}{1 + s\tau} \quad (5.3\text{-}44)$$

To simplify the calculations, the time constant τ is defined to be a factor B times the sample period T:

$$\tau = BT \quad (5.3\text{-}45)$$

Hence the lowpass-filter transfer function is

$$\frac{Y}{X} = \frac{1}{1 + sBT} \quad (5.3\text{-}46)$$

The relation of Eq 5.3-43 is used to obtain the approximate sampled-data transfer function in terms of U:

$$\frac{Y}{X} = \frac{1}{1 + (2U/T)BT} = \frac{1}{1 + 2BU} \qquad (5.3\text{-}47)$$

Substituting Eq 5.3-41 into this gives the sampled-data transfer function in terms of $\bar{z}$:

$$\frac{Y}{X} = \frac{1}{1 + [2B(1 - \bar{z})/(1 + \bar{z})]} = \frac{1 + \bar{z}}{(1 + \bar{z}) + 2B(1 - \bar{z})}$$

$$= \frac{1 + \bar{z}}{(1 + 2B) + \bar{z}(1 - 2B)} \qquad (5.3\text{-}48)$$

Multiply both sides by the denominators to obtain

$$Y(1 + 2B) + \bar{z}Y(1 - 2B) = X + \bar{z}X \qquad (5.3\text{-}49)$$

Solve this for Y:

$$Y = \frac{\bar{z}Y(2B - 1) + X + \bar{z}X}{2B + 1} \qquad (5.3\text{-}50)$$

The inverse transform of this is the sampled-data computer routine:

$$y_n = \frac{y_{n-1}(2B - 1) + x_n + x_{n-1}}{2B + 1} \qquad (5.3\text{-}51)$$

Assume for example that $B = 9.5$, so that $\tau = 9.5T$. Hence, $(2B + 1) = 20$, and $(2B - 1) = 18$. Equation 5.3-51 becomes

$$y_n = \frac{18y_{n-1} + x_n + x_{n-1}}{20} \qquad (5.3\text{-}52)$$

When implementing these computer routines in software, it is convenient to use the subscript p to represent "the past value of." Thus, x_p is the past value of x, and y_p is the past value of y. The trapezoidal integration routine of Eq 5.3-26 is expressed as

$$y = y_p + \tfrac{1}{2}T(x + x_p) \qquad (5.3\text{-}53)$$

Before the next cycle, the following equations should be implemented, to set the past values equal to the present values:

$$x_p = x \qquad (5.3\text{-}54)$$

$$y_p = y \qquad (5.3\text{-}55)$$

In like fashion, the computer routine for the lowpass transfer function the expression $1/(1 + s\tau)$ is, from Eq 5.3-51,

$$y = \frac{y_{\mathrm{p}}(2(\tau/T) - 1) + x + x_{\mathrm{p}}}{2(\tau/T) + 1} \qquad (5.3\text{-}56)$$

5.3.3 Digital Serial Simulation of Servo

The preceding section has developed computer routines for implementing the transfer functions $1/s$ and $1/(1 + s\tau)$. These can be applied to simulate the elements of the signal-flow diagram of Fig 5.3-1. However, the computations are performed in a feedback manner, and the feedback signals are derived from computations performed in the previous cycle. Hence, there is a one-cycle time delay around each feedback loop, which adds phase lag to the loop. To achieve accurate simulation, each loop must include lead compensation which offsets the phase lag caused by this one-cycle time delay.

Figure 5.3-3a shows the signal-flow diagram of the stage-positioning servo given in Fig 5.3-1, which is modified to include a one-cycle time delay in each of the three feedback paths. The transfer function of a one-cycle delay is e^{-sT}, which is represented as $\bar{z}$. The feedback loops provide feedback of position X, velocity V, and acceleration A. The delayed signals are the past values of X, V, and A, and so are represented as X_{p}, V_{p}, and A_{p}.

The frequency-response transfer function for a one-cycle time delay is $e^{-j\omega T}$. This has unity magnitude and a phase given by

$$\mathrm{Ang}[\bar{z}] = \mathrm{Ang}[e^{-j\omega T}] = -\omega T \qquad (5.3\text{-}57)$$

To compensate for this phase lag at low frequencies, the compensation transfer function $(1 + sT)$ is inserted in cascade with the time delay. For real frequencies ($s = j\omega$), the magnitude and phase of this compensation transfer function is

$$|1 + sT| = |1 + j\omega T| = \sqrt{1 + (\omega T)^2} \qquad (5.3\text{-}58)$$

$$\mathrm{Ang}[1 + sT] = \mathrm{Ang}[1 + j\omega T] = \arctan[\omega T] \qquad (5.3\text{-}59)$$

For $\omega T \ll 1$, these can be approximated by

$$|1 + j\omega T| \cong 1 \qquad (5.3\text{-}60)$$

$$\mathrm{Ang}[1 + j\omega T] \cong +\omega T \qquad (5.3\text{-}61)$$

Thus, at low frequencies, the compensation factor has approximately unity magnitude response and a phase lead of $+\omega T$ radian. This phase lead cancels the phase lag $-\omega T$ (given in Eq 5.3-57) produced by the one-cycle time delay.

The effect of this compensation factor at high frequencies is discussed in Section 5.4. As will be shown, the high-frequency response of the compensation factor can cause instability if the sampling rate is too low. This can be

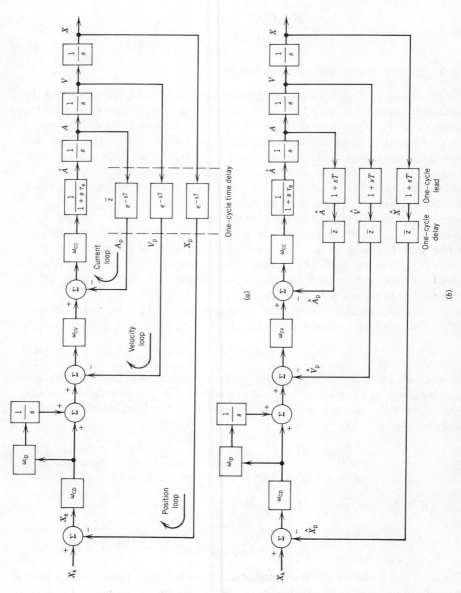

Figure 5.3-3 Method of compensating for phase lag due to one-cycle time delay around a digital feedback loop: (*a*) signal-flow diagram showing effect of one-cycle time delay; (*b*) system with time delay compensation.

avoided by using a sampling frequency that is 16 times greater than the maximum gain crossover frequency of any feedback loop.

In Fig 5.3-3b, a compensation factor of the form $(1 + sT)$ is placed in series with each one-cycle time-delay factor in the feedback loops (represented by the transfer function $\bar{z}$). The effect of a $(1 + sT)$ compensation factor is to provide at low frequencies a phase lead equivalent to a time shift in the future of one sample period T. Passing a signal X through a $(1 + sT)$ transfer function produces a signal labeled $\hat{X}$, which (for a low-frequency signal) is equal to the future value of X, one cycle in the future. Hence, the signal $\hat{X}$ is called the "predicted value of X" or the "future value of X." The signal $\hat{X}$ is passed through the one-cycle time delay $\bar{z}$ to form the signal $\hat{X}_p$, which is called the "past value of the predicted X" or the "past value of the future X." Thus, at low frequencies, $\hat{X}_p$ is equivalent to X.

To simplify the computation, it is convenient to normalize all of the $1/s$ integrations by dividing them by T, so as to provide transfer functions of the form $1/sT$. To compensate for this change, a gain constant at a prior point of each loop is multiplied by T. The resultant signal-flow diagram is shown in Fig 5.3-4a.

This normalization of the integration transfer functions changes the variables of the signal-flow diagram. The velocity V changes to TV, the acceleration A changes to T^2A, and the rate of acceleration $\dot{A}$ changes to $T^3\dot{A}$. All of these variables now have the same units as the output position X, and so are represented as follows by the symbol X with appropriate subscripts:

$$X_v = TV \qquad \text{(normalized velocity)} \qquad (5.3\text{-}62)$$

$$X_a = T^2A \qquad \text{(normalized acceleration)} \qquad (5.3\text{-}63)$$

$$X_{ra} = T^3\dot{A} \qquad \text{(normalized rate of acceleration)} \qquad (5.3\text{-}64)$$

The simulation is implemented in terms of the normalized variables X_v, X_a, X_{ra}. To obtain the actual values of velocity, acceleration, and rate of acceleration, these normalized values are divided by the sample period T raised to the appropriate power.

To simplify the calculations further, each $(1 + sT)$ compensation factor is shifted to the forward part of the loop, as indicated by the dashed lines in Fig 5.3-4a. Figure 5.3-4b shows the resultant signal-flow diagram after the factor $(1 + sT)$ of the current loop has been shifted to the forward part of that loop. The normalized integration $1/sT$ is replaced by $(1 + sT)/sT$. The factor $1/(1 + sT)$ is inserted in the velocity loop after this point to keep the signal-flow diagrams equivalent. When the $(1 + sT)$ factor in the feedback path of the velocity loop is placed in the forward path of the velocity loop, it cancels the factor $1/(1 + sT)$ and places a factor $1/(1 + sT)$ in the position loop. Moving the $(1 + sT)$ factor from the feedback path of the position loop to the forward path cancels the factor $1/(1 + sT)$ in the position loop.

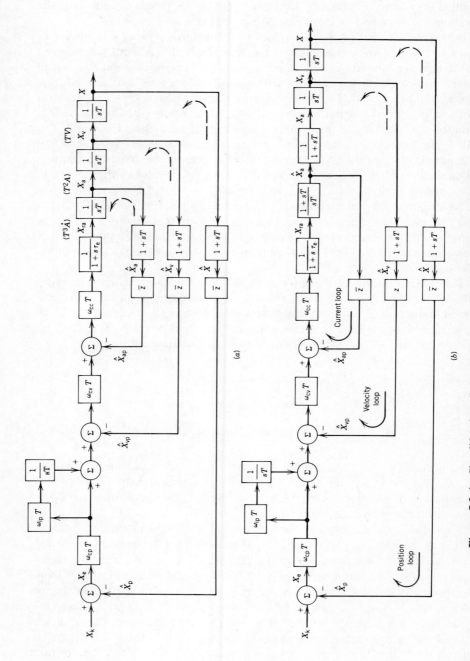

Figure 5.3-4 Simplification of simulation signal-flow diagram: (*a*) normalization of integrations; (*b*) effect of shifting time-delay compensation factor to forward path of current loop.

The resultant signal-flow diagram used for computation is shown in Fig 5.3-5. The $(1 + sT)$ compensation factors in the three feedback paths have been replaced by a single $(1 + sT)$ factor in the forward path of the current loop. This is combined with the integration of that loop to produce the net transfer function $(1 + sT)/sT$. Also the variables X, X_v, and X_a have been replaced by their predicted values $\hat{X}$, $\hat{X}_v$, $\hat{X}_a$, which correspond to values one sample period in the future. To obtain the actual signal X, the signal $\hat{X}$ can be passed through the transfer function $1/(1 + Ts)$ as shown. However, if X is not used subsequently in a feedback manner, an adequate representation of X can be derived much more simply by delaying $\hat{X}$ by one sample period. Either approach gives a good representation of X for low-frequency information.

Table 5.3-1 shows the algorithms for simulating the various transfer functions in the signal-flow diagram of Fig 5.3-5, along with others that are particularly useful. The following simple integration formula is often used to provide sampled-data integration:

$$y = y_p + Tx \qquad \text{(Simple Integration Formula)} \qquad (5.3\text{-}65)$$

Normalizing this gives $y = y_p + x$. As shown in the table, the transfer function for this is $(1/sT)(1 + s\bar{T}/2)$.

It was not absolutely necessary to convert the signal-flow diagram of Fig 5.3-4a to the form of Fig 5.3-5. This conversion was made to simplify the calculations performed in the simulation. Sometimes, it may not be desirable to shift the compensation factor $(1 + sT)$ from the feedback path to the forward path. In such a case, the transfer function $(1 + sT)$ could be simulated directly in the calculations. Therefore, Table 5.3-1 includes the algorithm for this transfer function.

One can write by inspection from the signal-flow diagram of Fig 5.3-5 the steps in a digital-computer program for simulating the system response. The variables labeled in the general form Y_1, Y_2, Y_3, etc. on the signal-flow diagram represent signals for which the physical significance is not specifically defined. All of the variables in the signal-flow diagram represent the transforms of the signals, whereas the variables used in the computer routine represent the actual time-domain signals. To be consistent with the standard symbolism of this book, the signals in the computer program should be represented by lowercase letters. However, to do this would confuse the process of translating the signal-flow diagram to a computer program. Accordingly, all of the variables in the computer program are represented by the same uppercase letters used in the signal-flow diagram, even though they represent time-domain signals in the computer program.

The following is an outline of a computer program expressed in BASIC computer-language format for simulating the response of the signal-flow

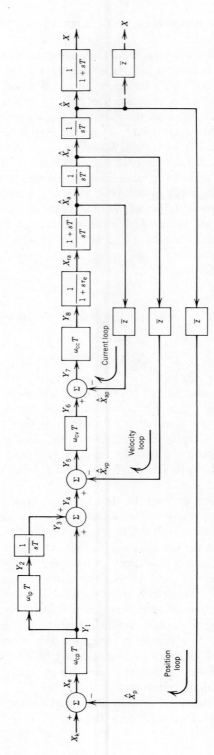

Figure 5.3-5 Final signal-flow diagram of stage-positioning servo for serial simulation.

TABLE 5.3-1   Algorithms for Simulating Common Normalized Transfer Functions

Transfer Function	Algorithm
(1) $\dfrac{Y}{X} = \dfrac{1}{sT}$	$y = y_p + \frac{1}{2}(x + x_p)$
(2) $\dfrac{Y}{X} = \dfrac{1 + sT}{sT}$	$y = y_p + \frac{3}{2}x - \frac{1}{2}x_p$
(3) $\dfrac{Y}{X} = \dfrac{1 + sT/2}{sT}$	$y = y_p + x$
(4) $\dfrac{Y}{X} = \dfrac{1}{1 + s\tau}$	$y = \dfrac{y_p(2\tau/T - 1) + x + x_p}{2\tau/T + 1}$
(5) $\dfrac{Y}{X} = \dfrac{1}{1 + \omega_x/s}$	$y = \dfrac{y_p(1 - \omega_x T/2) + x - x_p}{1 + \omega_x T/2}$
(6) $\dfrac{Y}{X} = \dfrac{1}{1 + sT}$	$y = \frac{1}{3}(y_p + x + x_p)$
(7) $\dfrac{Y}{X} = 1 + sT$	$y = -y_p + 3x - x_p$

diagram of Fig 5.3-5. The symbol T, which represents the sample period, is not to be confused with the variable TME, which represents the elapsed time. (The variable TIME should not be used, because it represents absolute time in BASIC.) The program steps are numbered in sequence to simplify the presentation. In the actual program, space should be left in the numbering sequence to allow for subsequent program changes.

SPECIFY PARAMETERS:

System parameters:

$$\omega_{cp} = ?, \ \omega_{ip} = ?, \ \omega_{cv} = ?, \ \omega_{cc} = ?, \ \tau_e = 1/\omega_e = ?$$

Program parameters:

$T = ?$ (sample period)
$\text{TME}_{\text{Max}} = ?$ (maximum elapsed time for computation)
$L_{\text{Max}} = ?$ (number of computation cycles per print cycle)

SPECIFY INITIAL CONDITIONS:

Set to Zero the Initial Values Required in Computation:

$$\hat{X}_\mathrm{p} = \hat{X}_\mathrm{vp} = \hat{X}_\mathrm{ap} = X_\mathrm{rap} = Y_\mathrm{2p} = Y_\mathrm{3p} = Y_\mathrm{8p} = 0$$

Specify Initial Values of the Program Parameters:

TME $= 0$ (initial time)

$L = 0$ (initiate print counter)

SPECIFY INPUT SIGNAL(S):

100 TME $=$ TME $+ T$

101 $X_\mathrm{k} =$ Function[TME] (given function of time)

SYSTEM EQUATIONS (Read from Fig 5.3-5 and Table 5.3-1):

200 $X_\mathrm{e} = X_\mathrm{k} - \hat{X}_\mathrm{p}$

201 $Y_1 = (\omega_\mathrm{cp} * T) * X_\mathrm{e}$

202 $Y_2 = (\omega_\mathrm{ip} * T) * Y_1$

203 $Y_3 = Y_\mathrm{3p} + (1/2)*(Y_2 + Y_\mathrm{2p})$

204 $Y_4 = Y_1 + Y_3$

205 $Y_5 = Y_4 - \hat{X}_\mathrm{vp}$

206 $Y_6 = (\omega_\mathrm{cv} * T) * Y_5$

207 $Y_7 = Y_6 - \hat{X}_\mathrm{ap}$

208 $Y_8 = (\omega_\mathrm{cc} * T) * Y_7$

209 $X_\mathrm{ra} = (X_\mathrm{rap} * (2 * (\tau_\mathrm{e}/T) - 1) + Y_8 + Y_\mathrm{8p})/(2 * (\tau_\mathrm{e}/T) + 1)$

210 $\hat{X}_\mathrm{a} = \hat{X}_\mathrm{ap} + (3/2) * X_\mathrm{ra} - (1/2) * X_\mathrm{rap}$

211 $\hat{X}_\mathrm{v} = \hat{X}_\mathrm{vp} + (1/2)*(\hat{X}_\mathrm{a} + \hat{X}_\mathrm{ap})$

212 $\hat{X} = \hat{X}_\mathrm{p} + (1/2)*(\hat{X}_\mathrm{v} + \hat{X}_\mathrm{vp})$

213 $X = \hat{X}_\mathrm{p}$

CHECK PRINT COUNT, PRINT DATA:

300 $L = L + 1$

301 IF $(L < L_\mathrm{Max})$ GOTO 400

302 $L = 0$

303 PRINT data (TME, X, X_e, $\hat{X}_\mathrm{v}$, etc.)

SET PAST VALUES EQUAL TO PRESENT VALUES:

400 $\hat{X}_p = \hat{X}$
401 $Y_{2p} = Y_2$
402 $Y_{3p} = Y_3$
403 $\hat{X}_{vp} = \hat{X}_v$
404 $\hat{X}_{ap} = \hat{X}_a$
405 $Y_{8p} = Y_8$
406 $X_{rap} = X_{ra}$

CHECK TIME LIMIT, REPEAT

500 IF (TME $<$ TME$_{Max}$) GOTO 100
501 END

This program prints data every L_{Max} cycles. Data for time TME $= 0$ are obtained by printing the initial (past) values of the variables.

To achieve accurate simulation, the sampling frequency $1/T$ must be much greater than the maximum gain crossover frequency of any feedback loop or the maximum break frequency of any transfer function. As will be shown in Section 5.4, accurate simulation is generally assured if the sampling frequency is at least a factor of 16 times greater than any gain crossover frequency or any break frequency. Thus, for any gain crossover frequency $f_c = \omega_c/2\pi$, the minimum allowable sampling frequency F is

$$\frac{1}{T} = F \geq 16 f_c = 16 \frac{\omega_c}{2\pi} = 2.5 \omega_c \qquad (5.3\text{-}66)$$

This can be simplified to the requirement

$$\omega_c T \leq 1/2.5 = 0.4 \qquad (5.3\text{-}67)$$

For any transfer function being simulated, this becomes:

$$\omega_x T \leq 0.4 \qquad (5.3\text{-}68)$$

where ω_x is the highest break frequency of the transfer function. Setting $\omega_x = 1/\tau$ gives

$$T \leq 0.4\tau \qquad (5.3\text{-}69)$$

Thus the sampling period T should be no greater than 40% of the smallest time constant τ being simulated.

On the other hand, when a factor having low damping is simulated, an even shorter sampling period T may be needed. As will be shown in Section 5.4 (Eq

5.4-55), the simulated damping ratio ζ of an underdamped quadratic factor of natural frequency ω_n is reduced from the desired value by the following damping ratio error:

$$\Delta\zeta = 0.01(\omega_n T/0.4)^3 \qquad (5.3\text{-}70)$$

For example, if $\omega_n T = 0.4$, the simulated damping ratio is 0.01 lower than the desired value.

Two convenient command input signals for studying the system response are the unit step and the unit ramp. For a unit step input, the function of time to be specified in step 100 for the command input X_k is

100   $X_k = 1$ (for unit step input)

For a unit ramp input, the function of time is

100   $X_k =$ TME   (for unit ramp input)

In designating the variables, the following symbolism has been used to represent one-cycle time shifts in the past and future:

$X_{ap} =$ *past value of* X_a, which is the value of X_a calculated in the previous cycle.

$\hat{X}_a =$ *future* (or *predicted*) *value of* X_a, which for a low-frequency signal is equal to the value of X_a one cycle in the future; obtained by passing X_a through the transfer function $(1 + sT)$.

$\tilde{X}_a =$ *delayed value of* X_a, which for a low-frequency signal is equal to the value of X_a one cycle in the past; obtained by passing X_a through the transfer function $1/(1 + sT)$.

The delayed value of X_a, represented by $\tilde{X}_a$, was not illustrated in Fig 5.3-5, but will be used in subsequent examples.

TABLE 5.3-2   Parameters Used in Simulating Signal-Flow Diagram of Fig 5.3-5

Parameter	Value (sec^{-1})
(1) ω_c	1109
(2) ω_{cc}	1109
(3) ω_{cv}	457
(4) ω_{cp}	142
(5) ω_{ip}	32.7

To simplify the computer program, the following are suggested for the names of the variables actually used in the program:

$$\text{XA} = X_a \qquad \text{(general variable } X_a)$$

$$\text{XAP} = X_{ap} \qquad \text{(past value of } X_a)$$

$$\text{XAPP} = X_{app} \qquad \text{(past value of } X_{ap})$$

$$\text{XAF} = \hat{X}_a \qquad \text{(future, or predicted, } X_a)$$

$$\text{XAFP} = \hat{X}_{ap} \qquad \text{(past value of future } X_a)$$

$$\text{XAD} = \tilde{X}_a \qquad \text{(delayed } X_a)$$

$$\text{XADP} = \hat{X}_{ap} \qquad \text{(past value of delayed } X_a)$$

$$\text{DXA} = \dot{X}_a \qquad \text{(derivative of } X_a)$$

$$\text{DDXA} = \ddot{X}_a \qquad \text{(second derivative of } X_a)$$

$$\text{TME} = \text{Time} \qquad \text{(elapsed time)}$$

$$\text{T} = T \qquad \text{(sample period)}$$

The BASIC programs used in small personal computers often limit the names of variables to two alphanumeric characters. For such computers, self-explanatory variable names are generally not possible, and so a cross-reference table may be needed to relate the variable names to meaningful symbols. This problem can be minimized by using one letter to designate the present value of X, and another to designate the past value of X. For example, B2 could be the past value of A2, and BA the past value of AA.

5.3.4 Results of Simulation

Table 5.3-2 gives illustrative values that were used to implement the serial-simulation program given in Section 5.3.3. These values were obtained from the analysis of the stage-positioning servo given in Ref [1.1] (Chapter 7, Table 7.2-1).

The maximum break frequency is ω_e or ω_{cc}, both of which are equal to 1109 sec^{-1}. Substituting this for ω_x in Eq 5.3-68 gives the following upper limit for the sampling period T:

$$T \leqq \frac{0.4}{\omega_x} = \frac{0.4}{\omega_{cc}} = \frac{0.4}{1109 \text{ sec}^{-1}}$$

$$\leqq 3.6 \times 10^{-4} \text{ sec} \qquad (5.3\text{-}71)$$

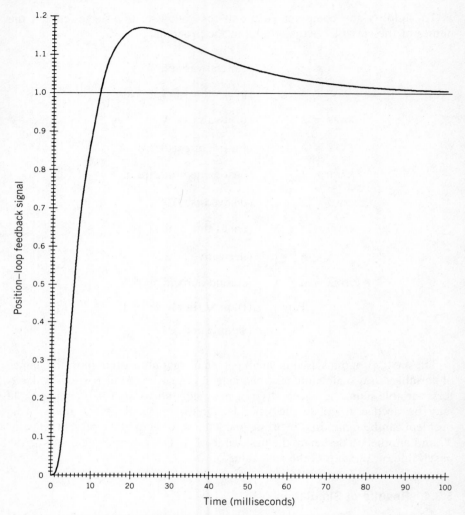

Figure 5.3-6   Simulated feedback-signal response to a unit step.

Accordingly, the following value is selected for the sample period:

$$T = 2.5 \times 10^{-4} \text{ sec} = 0.25 \text{ msec} \qquad (5.3\text{-}72)$$

This corresponds to a sampling frequency ($F = 1/T$) of 4000 Hz.

Figure 5.3-6 shows the simulated feedback response to a unit step input to the servo of Fig 5.3-5, using the parameters of Table 5.3-2. Figure 5.3-7 shows the simulated error responses to a ramp input for three specified values of ω_{ip}, including the value 32.7 sec^{-1} given in Table 5.3-2. The ramp input is

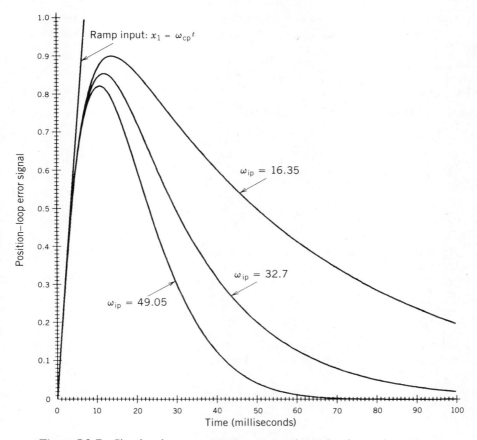

Figure 5.3-7   Simulated error response to a ramp input for three values of ω_{ip}.

normalized in terms of ω_{cp} and is given by

$$x_i = \omega_{cp}t = (142 \text{ sec}^{-1})t \qquad (5.3\text{-}73)$$

These responses were obtained from the serial-simulation program given in section 5.3.3 using a sample period T of 0.25 msec.

5.4 PSEUDO-FREQUENCY CONCEPT FOR EVALUATING THE DYNAMIC LIMITATIONS OF SAMPLED-DATA ALGORITHMS

5.4.1 Definition of Pseudo Frequency W

In Section 5.3, sampled-data routines were developed to achieve specified frequency-response characteristics by noting that the variable U, which is defined as $(1 - \bar{z})/(1 + \bar{z})$, is approximately equal to $sT/2$ at low frequencies.

In order to investigate the limitations of this approximation, it is convenient to define the complex pseudo frequency p, which is exactly related to U by

$$U = p\frac{T}{2} \qquad (5.4\text{-}1)$$

The real and imaginary parts of p are defined as

$$p = \alpha + jW \qquad (5.4\text{-}2)$$

Hence U is related to α and W by

$$U = \alpha\frac{T}{2} + jW\frac{T}{2} \qquad (5.4\text{-}3)$$

As was shown in Eq 5.3-34, for real frequencies (i.e., for $s = j\omega$), U is imaginary and is equal to

$$U = j\tan\left[\frac{\omega T}{2}\right] \qquad (5.4\text{-}4)$$

Equating the imaginary parts of U in Eqs 5.4-3, -4 gives the following expression, which holds for real frequencies ($s = j\omega$):

$$\frac{WT}{2} = \tan\left[\frac{\omega T}{2}\right] \qquad (5.4\text{-}5)$$

For $\omega T/2 \ll 1$, $\tan[\omega T/2]$ is approximately equal to $\omega T/2$, and so W is approximately equal to ω. At higher frequencies, Eq 5.4-5 can be solved as follows to calculate one of these variables from the other:

$$WT = 2\tan[\omega T/2] \qquad (5.4\text{-}6)$$

$$\omega T = 2\arctan[WT/2] \qquad (5.4\text{-}7)$$

Figure 5.4-1 shows a plot of the normalized true frequency ωT versus the normalized pseudo frequency WT. The vertical axis also shows f/F, which is the ratio of real frequency f in hertz, divided by the sampling frequency F. Note that ωT is related as follows to f/F:

$$\omega T = 2\pi f T = 2\pi\left(\frac{f}{F}\right) \qquad (5.4\text{-}8)$$

At low frequencies, ω is approximately equal to W. This approximation holds reasonably well up to the limit $WT = 2$. At this point, ωT and f/F are equal

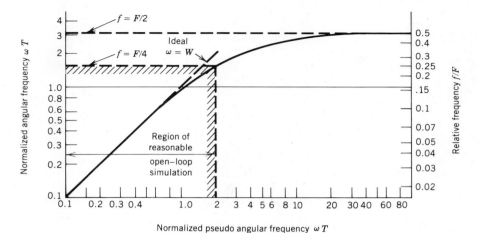

Figure 5.4-1 Plot of normalized true angular frequency (ωT) versus normalized pseudo angular frequency (WT).

to

$$\omega T = 2 \arctan\left[\frac{WT}{2}\right] = 2 \arctan[1] = \frac{\pi}{2} = 1.57 \qquad (5.4\text{-}9)$$

$$\frac{f}{F} = \frac{\omega T}{2\pi} = \frac{1}{4} \qquad (5.4\text{-}10)$$

Thus, W can be approximated by ω with reasonable accuracy up to $\frac{1}{4}$ of the sampling frequency, which is $\frac{1}{2}$ of the maximum frequency of the sampled information. Above $\frac{1}{4}$ of the sampling frequency, the variables W and ω depart strongly. The pseudo frequency W increases to infinity as the true frequency ω increases from $\frac{1}{4}$ to $\frac{1}{2}$ of the sampling frequency. When W is infinite, ωT is equal to π, and f/F is equal to 0.5.

To apply this plot of ω versus W, consider the exact frequency response of the normalized trapezoidal integration algorithm, which is

$$y = y_\mathrm{p} + \tfrac{1}{2}(x + x_\mathrm{p}) \qquad (5.4\text{-}11)$$

The Laplace transform of this is

$$Y = \bar{z}Y + \tfrac{1}{2}(X + \bar{z}X) \qquad (5.4\text{-}12)$$

Solving for the ratio Y/X gives

$$\frac{Y}{X} = \frac{1 + \bar{z}}{2(1 - \bar{z})} = \frac{1}{2U} \qquad (5.4\text{-}13)$$

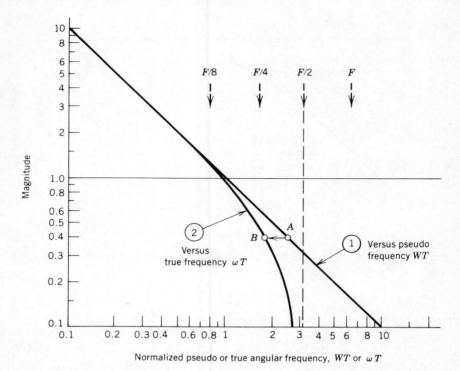

Magnitude

Normalized pseudo or true angular frequency, WT or ωT

Figure 5.4-2   Magnitude of frequency response of normalized trapezoidal integration.

This is expressed as follows in terms of the complex pseudo frequency p by setting $U = pT/2$:

$$\frac{Y}{X} = \frac{1}{pT} \qquad (5.4\text{-}14)$$

To obtain the pseudo-frequency response, p is replaced by jW:

$$\frac{Y}{X} = \frac{1}{jWT} \qquad (5.4\text{-}15)$$

In Fig 5.4-2, curve ① is a plot of the magnitude of this frequency response versus the normalized pseudo frequency WT.

The horizontal scale of Fig 5.4-2 represents both the normalized pseudo frequency WT and the normalized true frequency ωT. Curve ② shows the frequency response of the trapezoidal integration formula in terms of the true frequency ω; whereas curve ① is the corresponding response in terms of the pseudo frequency W. Curve ② is derived from curve ① by calculating for each value of W the corresponding value of ω. Consider, for example, point A on curve ①, which occurs at $WT = 2.5$. At this value of WT, the magnitude of the transfer function, which is $1/WT$, is equal to $1/2.5$ or 0.4, as shown on the vertical scale. The value of ωT at the corresponding point B on curve ② is derived from Eq 5.4-7 as follows:

$$\omega T = 2 \arctan\left[\frac{WT}{2}\right] = 2 \arctan\left[\frac{2.5}{2}\right] = 1.79 \qquad (5.4\text{-}16)$$

Thus point (A) lies at $\omega T = 2.5$, while point (B) lies at $\omega T = 1.79$.

The frequency response of an ideal integrator would have the same shape as curve ① of Fig 5.4-2, relative to the true frequency ω. Therefore, the departure of curve ② from the ideal curve ① represents the departure of the actual frequency response from the ideal. The figure shows that the actual response ② of the integrator follows the ideal response ① very accurately up to $\frac{1}{8}$ of the sampling frequency (up to $F/8$), and there is reasonable agreement up to $\frac{1}{4}$ of the sampling frequency (up to $F/4$). The frequency response extends only to $\frac{1}{2}$ of the sampling frequency ($F/2$), which is the limit of the information band. This limit corresponds to $\omega T = \pi$.

In a sampled-data signal, the base band extends from zero to half the sampling frequency. Signal components at frequencies above the base band are merely duplicates of the base-band information, and so are usually of no concern. However, when a sampled-data signal is demodulated, any signal in a high-frequency band that passes through the filtering of the demodulation process adds distortion to the demodulated signal. To determine the effect of this distortion, it is sometimes desirable to plot the frequency response of a sampled-data signal over much more than the base frequency band. This can be done in the following manner.

Figure 5.4-3 illustrates the relationships between the magnitude and phase values of a sampled-data transfer function (or a transform) for negative frequencies and for frequencies greater than half the sampling frequency, relative to the values in the base band. The fundamental relations are

$$H[-f] = H^*[f] \qquad (5.4\text{-}17)$$
$$H[NF + f] = H[f] \qquad (5.4\text{-}18)$$

The function H is a transfer function (or a transform), the variable f is the frequency in hertz, N is an integer, and F is the sampling frequency in hertz. Equation 5.4-17 states that the value of the transfer function (or transform) H at the negative frequency $-f$ is equal to the conjugate of H at the frequency f. This principle is expressed as follows in terms of magnitude and phase:

$$|H[-f]| = |H[f]| \qquad (5.4\text{-}19)$$
$$\text{Ang}[H[-f]] = -\text{Ang}[H[f]] \qquad (5.4\text{-}20)$$

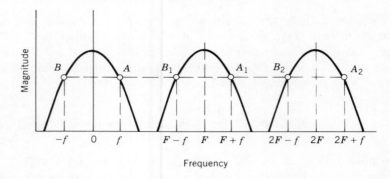

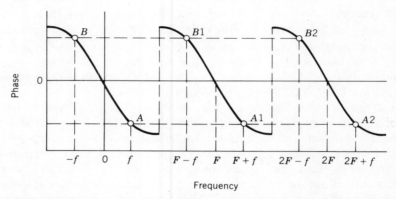

Figure 5.4-3 Magnitude and phase values of a simplified transfer function in the main band, related to the values at negative frequencies and at higher frequencies.

From Eqs 5.4-18, -19, the magnitude values in the different bands are related by

$$|H[NF + f]| = |H[f]| \qquad (5.4\text{-}21)$$

$$|H[NF - f]| = |H[-f]| = |H[f]| \qquad (5.4\text{-}22)$$

By Eqs 5.4-18, -20, the phase values in the different bands are related by

$$\text{Ang}[H[NF + f]] = \text{Ang}[H[f]] \qquad (5.4\text{-}23)$$

$$\text{Ang}[H[NF - f]] = \text{Ang}[H[-f]] = -\text{Ang}[H[f]] \qquad (5.4\text{-}24)$$

These equations are illustrated in Fig 5.4-3.

Consider the magnitude plots of Fig 5.4-3. The magnitude at the frequency f is indicated by point A. This is the same value as the magnitude at the frequency $-f$, indicated by point B. The magnitude also has this same value at the frequencies $(F + f), (F - f), (2F + f), (2F - f)$, which are indicated by points A_1, B_1, A_2, B_2, respectively.

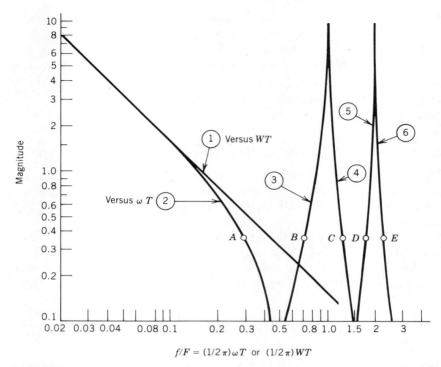

Figure 5.4-4 Magnitude of sampled frequency response for normalized trapezoidal integration, for frequencies up to 2.5 times the sampling frequency.

For the phase plots of Fig 5.4-3, the phase at frequency f is indicated by point A. The phase at the frequency $-f$, which is indicated by point B, is the negative of this phase at point A. The phase values at the frequencies $(F + f)$ and $(2F + f)$, indicated by points A_1 and A_2, are the same as the phase at point A at the frequency f. The phase values at the frequencies $(F - f)$ and $(2F - f)$, indicated by points B_1 and B_2, are the same as the phase at the frequency $-f$ indicated by point B, which is the negative of the phase at the frequency f, indicated by point A.

Using the principles illustrated in Fig 5.4-3, the magnitude of the frequency response of the normalized trapezoidal integration formula is plotted in Fig 5.4-4, out to 2.5 times the sampling frequency. To facilitate the construction of this plot, the frequency scale of Fig 5.4-2 is divided by 2π for both the variables ωT and WT. As shown in the following equation, this changes the ωT scale to the frequency ratio f/F:

$$\frac{1}{2\pi}\omega T = fT = \frac{f}{F} \tag{5.4-25}$$

Curves ① and ② of Fig 5.4-4 are derived from curves ① and ② of Fig

5.4-2 by sliding the horizontal frequency scale so that the normalized frequency $\omega T = \pi$ is shifted to $f/F = 0.5$.

Curve segments ③, ④, ⑤, and ⑥ are derived from curve ②. For example, at point A on curve ②, the magnitude is 0.36 and the frequency is $f = 0.3F$. This same magnitude (0.36) occurs at point B at the frequency $0.7F$, at point C at the frequency $1.3F$, at point D at the frequency $1.7F$, and at point E at the frequency $2.3F$.

5.4.2 Accuracy of Sampled-Data Simulation of a Feedback Loop

The preceding has shown that a sampled-data filter can provide, in an open-loop process, a very close approximation of a desired frequency response up to $\frac{1}{8}$ of the sampling frequency, and a reasonable approximation up to $\frac{1}{4}$ of the sampling frequency. On the other hand, in a feedback system, the frequency limits for good simulation are much lower, because of the one-cycle time delay around a feedback loop. This case is illustrated in Fig 5.4-5.

The transfer function for a one-cycle time delay can be expressed as follows in terms of the pseudo frequency W:

$$\bar{z} = \frac{1 - U}{1 + U} = \frac{1 - jW(T/2)}{1 + jW(T/2)} \tag{5.4-26}$$

The magnitude of this function is unity. The phase is plotted as curve ① in Fig 5.4-5 versus the normalized pseudo frequency WT, expressed on a logarithmic scale. Scales are also shown to give the corresponding values of the normalized angular frequency ωT and the relative frequency f/F.

As was shown in Section 5.3, to compensated for the phase lag caused by the one-cycle time delay around a loop, an algorithm is used that exhibits a transfer function approximating $(1 + sT)$. The exact compensation transfer function is

$$(1 + pT) = (1 + jWT) \tag{5.4-27}$$

Plots of the magnitude and phase of this compensation transfer function are shown as curves ② and ③ in Fig 5.4-5. Curve ④, which is the sum of the phase curves ① and ②, is the total phase. Thus, the magnitude plot ③ and phase plot ④ are the combined effect of the time delay and the time-delay compensation.

From Eqs 5.4-6, -7, the normalized pseudo-frequency WT can be related as follows to the relative frequency f/F:

$$WT = 2\tan[\pi f/F] \tag{5.4-28}$$

This shows that for frequency f equal to $F/4$, $F/8$, and $F/16$, the values of WT are 2.0, 0.828, and 0.398, respectively. These frequencies are indicated by

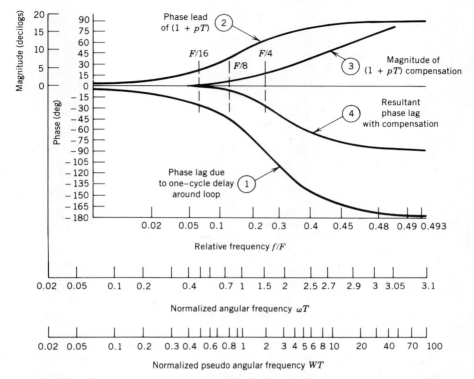

Figure 5.4-5 Frequency response of magnitude and phase versus normalized pseudo frequency WT, showing effect of $(1 + pT)$ compensation to offset phase lag of one-cycle time delay.

dashed vertical lines in Fig 5.4-5. At $f = F/8$ (or $WT = 0.828$), the magnitude deviation (curve ③) is $+1.1$ dg, and the phase deviation is $-5.4°$. This frequency is an upper limit to the range of acceptable simulation. At $f = F/16$ (or $WT = 0.398$), the magnitude deviation of curve ③ is $+0.3$ dg, and the phase deviation of curve ④ is $-0.8°$. These deviations are negligible for most applications.

Thus, when the $(1 + pT)$ digital compensation is added to a feedback loop to compensate for the unavoidable one-cycle time delay around the loop, the resultant loop provides very good simulation up to $\frac{1}{16}$ of the sampling frequency, and reasonable simulation up to $\frac{1}{8}$ of the sampling frequency.

5.4.3 Map of Complex Pseudo-Frequency *p*-Plane onto *s*-Plane

The relationship $U = j \tan[\omega T/2]$ given in Eq 5.4-4 holds only for real frequencies (for $s = j\omega$). An analysis of the general relationship between the variable U and the full s-plane is presented in Appendix A. The following

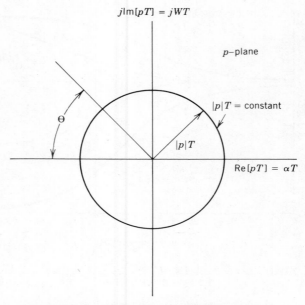

Figure 5.4-6 Complex pseudo-frequency p-plane.

equations are derived in that appendix, which express the real and imaginary parts of the complex frequency s (denoted σ and ω) as functions of the real and imaginary parts of the complex pseudo frequency p (denoted α and W):

$$\tan[\omega T] = \frac{WT}{1 - (|p|T/2)^2} \tag{5.4-29}$$

$$\tanh[\sigma T] = \frac{\alpha T}{1 + (|p|T/2)^2} \tag{5.4-30}$$

The variable $|p|$ is the magnitude of the complex pseudo frequency p, which is related to α and W by

$$|p|^2 = \alpha^2 + W^2 \tag{5.4-31}$$

The complex pseudo-frequency p-plane is illustrated in Fig 5.4-6. The real part of the complex pseudo frequency is defined as α and the imaginary part as W. Two sets of contours are drawn on this p-plane. One set is for constant values of $|p|$, which consists of circles about the origin. The second set is for constant values of the ratio W/α, which consists of radial lines emanating from the origin. As shown in the figure, the contours for constant values of W/α are defined in terms of the angle Θ between the radial line and the negative real p-axis.

In Fig 5.4-7, the p-plane contours of Fig 5.4-6 are mapped onto the s-plane over the following ranges: $0 < \omega T < \pi$, and $-6 < \alpha T < +6$. The frequency range in hertz corresponding to this range of ωT is $0 < f < F/2$. Thus, the plot covers the complete frequency range of the information band of the sampled data, which extends to half the sampling frequency.

The contour map of Fig 5.4-7 shows the following:

1. The p-plane maps onto the s-plane with high uniformity up to $|p|T = 1$, which corresponds to $\frac{1}{6}$ of the sampling frequency.

2. The right half of the p-plane maps onto the right half of the s-plane, and the jW axis of the p-plane maps onto the $j\omega$ axis of the s-plane.

3. As $|p|$ approaches infinity, the map of the p-plane converges to a point on the $j\omega$ axis at $\omega T = \pi$, which corresponds to $\frac{1}{2}$ of the sampling frequency ($f/F = 0.5$).

4. The contour for $|p|T = 2$ is a horizontal line of constant ωT, along which $\omega T = \pi/2$ and $f/F = 0.25$. For values of $|p|T$ greater than 2, a negative real root of p (for which $W = 0$) results in a complex root of s, for which $\omega T = \pi$, or $f = 0.5F$.

The following conclusions can be drawn from this contour map of Fig 5.4-7:

1. A stable sampled transfer function cannot have poles of p at infinity, because this results in oscillatory poles of s on the $j\omega$ axis at $\frac{1}{2}$ of the sampling frequency.

2. A pseudo Nyquist plot of the digital loop transfer function $G^s[jW]$ can be constructed relative to the pseudo frequency W, in the same manner that $G[j\omega]$ is constructed relative to the true frequency ω. If $G^s[p]$ has no poles at $p = \infty$, the loop will be stable if the $G^s[jW]$ Nyquist plot satisfies the requirements of a stable $G[j\omega]$ Nyquist plot. (The reason for this is that the right half of the p-plane maps onto the right half of the s-plane.)

3. To achieve good stability, the poles of p should not exceed the limit $|p|T = 2$. If this limit is exceeded, the loop transfer function $G[j\omega]$ has underdamped high-frequency poles, even when the poles of p are real. Underdamped high-frequency poles of G generate underdamped closed-loop poles that are close to them. These closed-loop poles would tend to make the loop oscillate at half the sampling frequency.

Because of conclusion (3), the break frequency of any pole relative to the pseudo frequency W should not exceed the normalized pseudo frequency $WT = 2$, which corresponds to $\omega T = \pi/2$, or $f = 0.25F$. It is possible to reach this limit, but the limit should not be exceeded. Remember that every feedback loop has the transfer function $\bar{z}$, caused by the one-cycle time delay around

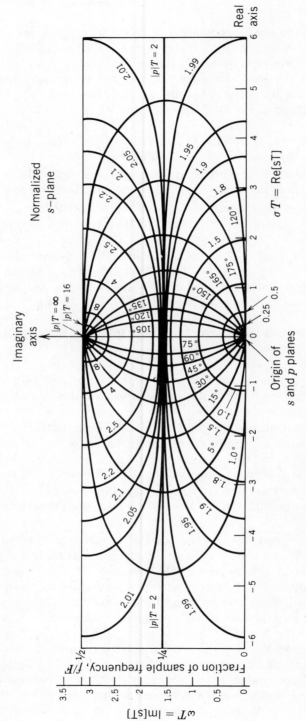

Figure 5.4-7 Map of complex pseudo-frequency p-plane onto s-plane.

the loop, which is expressed as follows in terms of p:

$$\bar{z} = \frac{1 - p(T/2)}{1 + p(T/2)} \tag{5.4-32}$$

This has a left-half-plane pole and a right-half-plane zero with the same pseudo break frequency, $WT = 2$.

5.4.4 Application of Pseudo Frequency in Design of a Sampled-Data Feedback Loop

The map of the complex pseudo-frequency p-plane onto the s-plane shows that a stable system can have no poles at infinite p. An important implication of this requirement is that a digital feedback loop should not be closed around a differentiator; it should only be closed around an integrator. The digital transfer function of a differentiator is p, which has a pole at infinite p; while the digital transfer function of an integrator is $1/p$, which has a zero at infinite p. This principle is applied in Section 5.5, which develops general approaches to the simulation of dynamic equations.

To develop quantitative requirements for achieving good stability with digital feedback, consider the simulation of the following simple loop transfer function:

$$G = \frac{\omega_c}{s} \tag{5.4-33}$$

The sampled-data transfer function for simulating this is

$$G^s = \frac{\omega_c}{p} \bar{z}(1 + pT) = \frac{\omega_c(1 - pT/2)(1 + pT)}{p(1 + pT/2)} \tag{5.4-34}$$

The factor ω_c/p is the ideal transfer function; the factor $\bar{z}$ is the transfer function for the one-cycle time delay around the loop; and $(1 + pT)$ is the compensation for this one-cycle delay. The factor $\bar{z}$ is expressed in terms of the complex pseudo frequency p in accordance with Eq 5.4-32. Setting $p = jW$ gives the pseudo-frequency response:

$$G^s = \frac{\omega_c(1 - jWT/2)(1 + jWT)}{jWT(1 + jWT/2)} \tag{5.4-35}$$

The magnitude and phase of this loop transfer function are plotted in Fig 5.4-8 versus the normalized pseudo frequency WT for the particular case $\omega_c T = 0.4$. The phase plot is the same as curve ④ of Fig 5.4-5, shifted by $-90°$. The G-locus obtained from this data is plotted on a simplified Nichols

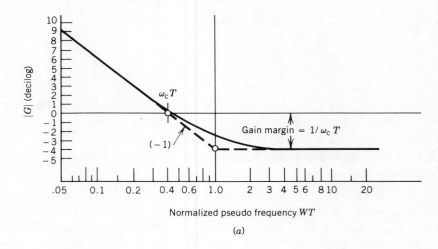

(a)

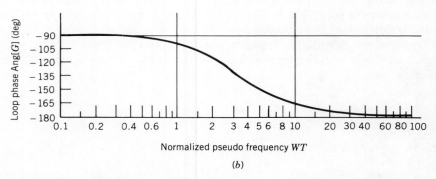

(b)

Figure 5.4-8 Pseudo-frequency response plots of sampled loop transfer function $G^s[jW]$, for simulating loop transfer function $G = \omega_c/s$: (a) magnitude plot; (b) phase plot.

chart in Fig 5.4-9. (The Nichols chart was explained in Section 2.3, Fig 2.3-11.) This simplified Nichols chart shows only two contours: $|G_{ib}| = 0$ dg and $|G_{ib}| = 1.0$ dg $= 2$ dB.

The G-locus (curve ③) on the simplified Nichols chart of Fig 5.4-9 shows the values of the normalized pseudo frequency WT along the lower and right side of the locus. The values of the true relative frequency f/F are shown along the upper and left side of this locus. The phase of G is $-180°$ at the point $WT = \infty$, or $f/F = 0.5$. The reciprocal of $|G|$ at this point is therefore the gain margin of the loop. The plot of $|G|$ in Fig 5.4-9 shows that the magnitude of G at infinite WT is equal to the magnitude asymptote of G at

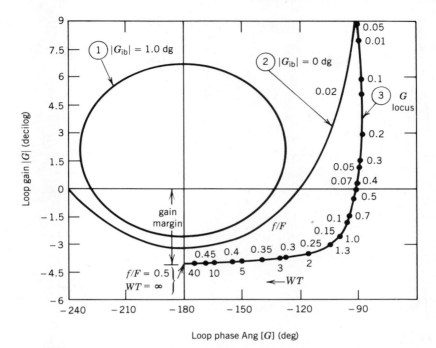

Figure 5.4-9 Locus of sampled loop transfer function of Fig 5.4-8 on abbreviated Nichols chart.

$WT = 1$, which is equal to $\omega_c T$. Thus, the gain margin is

$$\text{Gain margin} = \frac{1}{\omega_c T} \qquad (5.4\text{-}36)$$

As was stated in Chapter 2, Eq 2.3-26, the gain margin of a feedback loop should be at least a factor of 2.5 (or 4 dg) to assure good stability. Applying this requirement gives the following upper limit on $\omega_c T$:

$$\omega_c T \le 1/2.5 = 0.4 \qquad (5.4\text{-}37)$$

Set $\omega_c = 2\pi f_c$, and $T = 1/F$. This gives the corresponding upper limit on the gain crossover frequency f_c expressed in hertz:

$$f_c = \frac{\omega_c}{2\pi} \le \frac{0.4}{2\pi T} = \frac{F}{15.7} \qquad (5.4\text{-}38)$$

This is approximately $F/16$. Thus to assure a gain margin of at least 2.5, the

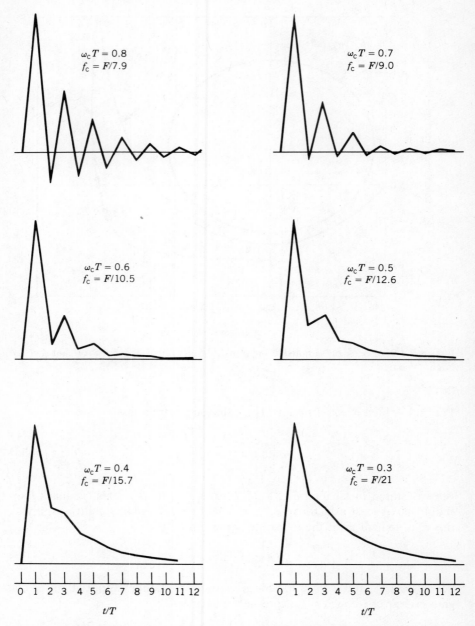

Figure 5.4-10 Unit step transient responses of sampled-data feedback loop having ideal transfer function ω_c/s, for different values of the parameter $\omega_c T$.

gain crossover frequency should not exceed $\frac{1}{16}$ of the sample frequency F, and the value of ωT should not exceed 0.4.

Let us compare this stability requirement based on gain margin with a direct measurement of stability derived from the transient response of the digital loop. Figure 5.4-10 shows the transient error responses of this digital loop for a unit-step input for values of $\omega_c T$ from 0.3 to 0.8. These responses were obtained by simulation. Since these are sampled-data responses, the values are specified only at the sampling instants. However for clarity, straight lines are drawn between the sampled points. The response for $\omega_c T = 1.0$, which is not shown, is a constant-amplitude oscillation, with a frequency equal to half the sampling frequency, and a period equal to twice the sample period T.

As shown in Fig 5.4-10, the response is highly oscillatory for $\omega_c T = 0.8$, and is poor for $\omega_c T$ greater than 0.5. The curve for $\omega_c T = 0.4$ represents a good engineering design. This result substantiates the requirement, based on gain margin, that $\omega_c T$ should not exceed 0.4.

The preceding discussion has shown that the maximum gain crossover frequency of any feedback loop in a digital simulation program should not exceed $\frac{1}{16}$ of the sampling frequency. This is equivalent to requiring that $\omega_c T$ should not exceed 0.4. A simple way to check the stability of a digital simulation is to run the simulation a second time with the sampling period T reduced by a factor of 2. If the response does not change significantly, the sampling frequency is adequate.

5.4.5 Required Sampling Rate for Simulating Factors with Low Damping

As explained in Ref [1.2] (Chapter 10), mechanical structural resonance in a servomechanism often exhibits poles and zeros having very low damping. When one is simulating a factor with low damping, a better criterion for the sampling rate is required. Consider the signal-flow diagram of Fig 5.4-11a. The exact loop transfer function G and feedback transfer function G_{ib} for the outer loop are

$$G = \frac{\omega_{c1}\omega_{c2}}{s(s + \omega_{c2})} \tag{5.4-39}$$

$$G_{ib} = \frac{1}{1 + G} = \frac{\omega_{c1}\omega_{c2}}{s^2 + s\omega_{c2} + \omega_{c1}\omega_{c2}} \tag{5.4-40}$$

This feedback transfer function has the form

$$G_{ib} = \frac{\omega_n^2}{s^2 + 2\zeta\omega_n s + \omega_n^2} \tag{5.4-41}$$

$$\omega_{c1} = \omega_n / 2\zeta \tag{5.4-42}$$

$$\omega_{c2} = 2\zeta\omega_n \tag{5.4-43}$$

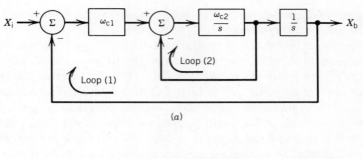

(a)

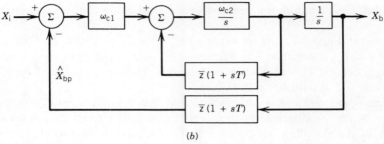

(b)

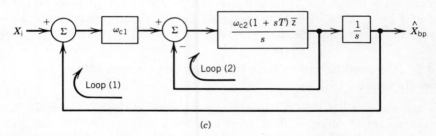

(c)

Figure 5.4-11 Serial simulation of feedback loop exhibiting underdamped response: (a) ideal signal-flow diagram; (b) serial simulation implementation of a; (c) modification of serial-simulation diagram b to express it in the form of ideal diagram a.

Diagram b of Fig 5.4-11 is the signal-flow diagram for serial simulation of diagram a. Diagram b can be manipulated to form diagram c, which differs from the ideal diagram a in that it has the following additional transfer function in the inner loop:

$$H = (1 + sT)\bar{z} = (1 + sT)e^{-sT} = (1 + j\omega T)e^{-j\omega T} \qquad (5.4\text{-}44)$$

The corresponding G_{ib} transfer function can be obtained from Eq 5.4-40 by multiplying each ω_{c2} factor by H. This yields

$$G_{ib} = \frac{\omega_{c1}\omega_{c2}H}{s^2 + s\omega_{c2}H + \omega_{c1}\omega_{c2}H} = \frac{\omega_n^2 H}{s^2 + s2\zeta\omega_n H + \omega_n^2 H} \qquad (5.4\text{-}45)$$

We are concerned with the characteristics of this simulated transfer function at angular frequencies ω that are much less than $1/T$. At such low frequencies, the magnitude of H is close to unity, and so only the phase of H is of concern. At any particular low frequency, H can be approximated by

$$H \cong 1 - j\phi \qquad (5.4\text{-}46)$$

where ϕ is the phase lag of H at that frequency. By Eq 5.4-44, ϕ is equal to

$$\phi = -\text{Ang}[H] = \omega T - \arctan[\omega T] \qquad (5.4\text{-}47)$$

Substitute the approximation for H of Eq 5.4-46 into Eq 5.4-45. This gives the following approximation of G_{ib} (for $s = j\omega$):

$$\begin{aligned}
G_{ib} &\cong \frac{\omega_n^2(1 - j\phi)}{-\omega^2 + j\omega 2\zeta\omega_n(1 - j\phi) + \omega_n^2(1 - j\phi)} \\
&\cong \frac{\omega_n^2(1 - j\phi)}{\left(\omega_n^2 - \omega^2 + 2\zeta\omega_n\omega\phi\right) + j\left(2\zeta\omega_n\omega - \omega_n^2\phi\right)} \qquad (5.4\text{-}48)
\end{aligned}$$

The factor $(1 - j\phi)$ in the numerator merely adds a very small phase shift to the transfer function, and so can be ignored. Let us denote the resonant frequency of G_{ib} as ω_n'. This is obtained by setting the real part of the denominator of Eq 5.4-48 equal to zero, which gives

$$\omega^2 = \omega_n'^2 = \omega_n^2 + 2\zeta\omega_n\omega_n'\phi \qquad (5.4\text{-}49)$$

The last term is quite small relative to ω_n^2. Hence the quantity ω_n' in this term can be approximated by ω_n without appreciably affecting the value of the equation. Equation 5.4-49 can then be solved to give

$$\omega_n' = \omega_n\sqrt{1 + 2\zeta\phi} \cong \omega_n(1 + \zeta\phi) \qquad (5.4\text{-}50)$$

The square root was approximated by noting that $\sqrt{1 + x}$ is approximately equal to $(1 + x/2)$ for $|x| \ll 1$. This simulated natural frequency ω_n' differs only slightly from the desired natural frequency ω_n. At the simulated natural frequency ω_n', G_{ib} of Eq 5.4-48 is equal to

$$G_{ib}[\omega_n'] = \frac{\omega_n^2}{j\left(2\zeta\omega_n\omega_n' - \omega_n^2\phi\right)} \cong \frac{1}{j(2\zeta - \phi)} \qquad (5.4\text{-}51)$$

where ω_n' is approximated by ω_n. The damping ratio of this simulated resonant factor is denoted ζ' and is related as follows to G_{ib} at the resonant frequency ω_n':

$$G_{ib}[\omega_n'] = \frac{1}{j2\zeta'} \qquad (5.4\text{-}52)$$

Setting Eqs 5.4-51, -52 equal gives for ζ'

$$\zeta' = \zeta - \frac{\phi}{2} = \zeta - \Delta\zeta \qquad (5.4\text{-}53)$$

The parameter $\Delta\zeta$ is the shift of the simulated damping ratio ζ' from the desired damping ratio ζ. Combining Eqs 5.4-47, -53 gives for $\Delta\zeta$

$$\Delta\zeta = \tfrac{1}{2}\phi = \tfrac{1}{2}\left(\omega_n'T - \arctan[\omega_n'T]\right)$$
$$\cong \tfrac{1}{2}\left(\omega_nT - \arctan[\omega_nT]\right) \qquad (5.4\text{-}54)$$

where ω_n' is approximated by ω_n.

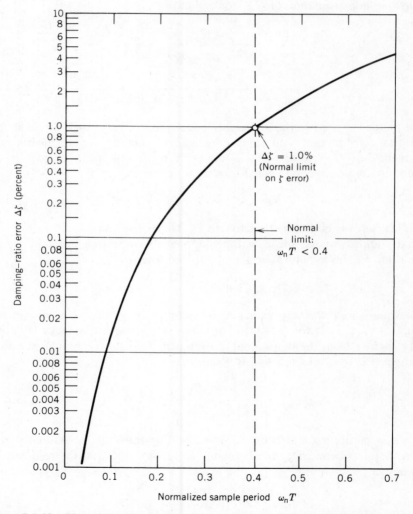

Figure 5.4-12   Plot of damping-ratio error (in percent) for serial simulation of underdamped feedback loop, versus normalized sampling period ω_nT.

The damping ratio ζ' of the simulation is less than the desired damping ratio ζ, and the parameter $\Delta\zeta$ is the error in damping ratio of the simulation. The damping ratio error $\Delta\zeta$ obtained from Eq 5.4-54 is plotted in Fig 5.4-12 versus the normalized sampling period $\omega_n T$.

According to Section 5.4.4, the parameter $\omega_{gc} T$ should not exceed 0.4, where ω_{gc} is the highest gain crossover frequency of the system. For a loop with low damping, ω_{gc} is approximately equal to ω_n. As shown in Fig 5.4-12, when $\omega_n T = 0.4$, the damping ratio of the simulated quadratic factor is in error by 1%. This damping-ratio error is reasonable if the damping ratio is high, but is quite unacceptable when the damping ratio is low. Thus, for simulating a factor with low damping, the normalized sampling period $\omega_n T$ must be appreciable less than 0.4.

For $\omega_n < 0.4$, the plot of Fig 5.4-12 can be approximated quite accurately by

$$\Delta\zeta \cong 0.01\left(\omega_n T/0.4\right)^3 \qquad (5.4\text{-}55)$$

Thus, the damping-ratio error $\Delta\zeta$ is proportional to the cube of the sampling period. For example, if the normalized sampling period $\omega_n T$ is reduced by a factor of 2 from 0.4 to 0.2, the damping-ratio error $\Delta\zeta$ is reduced by a factor of 2^3, or 8, from 1% to 0.125%.

5.5 GENERALIZED APPROACHES TO SIMULATION

5.5.1 ELF Transmitter Circuit

To develop more general approaches for dynamic simulation, this section considers the simulation of an electrical circuit. Two techniques are applied: (1) the serial simulation method developed in Section 5.3, and (2) the Runge–Kutta integration method, which is formulated in terms of state variables. The discussion explains the concepts of state variables and state equations.

To give this discussion physical significance, it is applied to a circuit investigated by the author in the study of a high-power communication transmitter that would operate in the extremely-low-frequency (ELF) band, from 30 to 300 Hz. The transmitter would be driven by a device called an inverter, which uses thyristors (silicon-controlled rectifiers), which act as switches to convert DC power to AC power with high efficiency. A *current-source inverter* would be used, powered by a constant-current source. Using appropriate electrical switching commands to the thyristors, the inverter controls the current delivered to the load, by switching the path of this constant current. The current-source inverter is used extensively in the control of large AC motors, and is described by Phillips [5.5].

To excite a sinusoidal current in the antenna, the current-source inverter would generate the current waveform shown in Fig 5.5-1. Figure 5.5-2 shows a

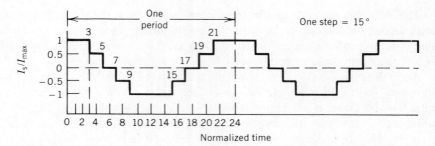

Figure 5.5-1 Waveform of source current I_s generated by current-source inverter.

circuit that was investigated during the transmitter study, for filtering the inverter current waveform and coupling it efficiently to the antenna. The current source I_s in Fig 5.5-2 has the inverter current waveform of Fig 5.5-1. The purpose of the circuit is to provide strong filtering of harmonics, while generating a sinusoidal antenna current I_a. The communications signal is modulated by shifting (or "keying") the frequency of the sinusoid. Communication efficiency is maximized if the frequency shift occurs, without phase change, when the antenna current passes through zero.

The antenna impedance is represented by an antenna resistance R_a in series with an antenna inductance L_a. The capacitance C_a is a tuning capacitor, which cancels the inductive antenna impedance at the mean transmitted frequency. Hence C_a is equal to

$$C_a = \frac{1}{\omega_o^2 L_a} \tag{5.5-1}$$

where ω_o is the mean transmitted frequency, expressed in rad/sec. The

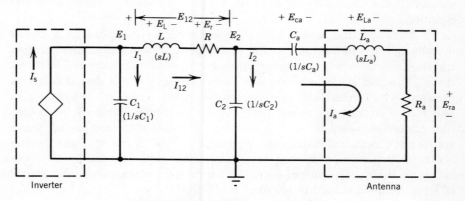

Figure 5.5-2 ELF transmitter circuit.

reactances of the inductor L and capacitors C_1 and C_2 at the frequency ω_o are set equal to the antenna resistance R_a. Hence

$$\frac{1}{\omega_o C_1} = \frac{1}{\omega_o C_2} = \omega_o L = R_a \qquad (5.5\text{-}2)$$

With these settings, the input impedance of the coupling network is resistive at the mean frequency ω_o, and is equal to the antenna resistance R_a. This input impedance is the load experienced by the inverter. The Q of the antenna at the mean frequency ω_o is defined as

$$Q_a = \frac{\omega_o L_a}{R_a} \qquad (5.5\text{-}3)$$

A typical value for Q_a is 5. The resistance R is the effective series resistance of the inductor L, which can be expressed as

$$R = \frac{\omega_o L}{Q} \qquad (5.5\text{-}4)$$

The parameter Q is the Q of the inductor, which is assumed to be 20.

To develop a computer program for simulating this ELF transmitter network, a signal-flow diagram of the network is constructed as shown in Fig 5.5-3. This figure presents in signal-flow form the Kirchhoff circuit equations, expressed with all dynamic calculations represented as integrals. Hence, the voltage E_c across a capacitor is derived from the current I_c flowing through the capacitor, while the current I_L flowing through an inductor is derived from the voltage E_L across the inductor. Thus the calculations for a capacitor

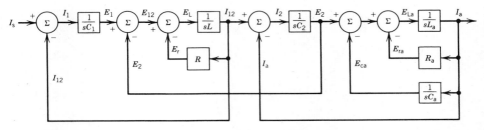

Figure 5.5-3 Signal-flow diagram of ELF transmitter circuit, showing Kirchhoff equations with all dynamics expressed as integrations.

and an inductor take the following form:

Calculation for capacitor:

$$E_c = \frac{1}{sC} I_c \tag{5.5-5}$$

Calculation for inductor:

$$I_L = \frac{1}{sL} E_L \tag{5.5-6}$$

For a resistor, either the voltage or the current can be used as the independent variable. With a little ingenuity, the student can readily express the Kirchhoff equations for any circuit in a signal-flow diagram of this form.

5.5.2 State Variables and State Equations

In Fig 5.5-4a, each transfer function of Fig 5.5-3 between voltage and current variables is split, so as to separate the $1/s$ integration from the factor $1/L$ or $1/C$, with the factor $1/s$ placed last. The outputs of the integrators are called the *state variables* of the system, and the inputs to the integrators are the derivatives of the state variables. Since the signal-flow diagram has five integrators, the system has five state variables. The transfer functions of the five integrators are

$$E_1 = \frac{1}{s}\dot{E}_1 \tag{5.5-7}$$

$$I_{12} = \frac{1}{s}\dot{I}_{12} \tag{5.5-8}$$

$$E_2 = \frac{1}{s}\dot{E}_2 \tag{5.5-9}$$

$$I_a = \frac{1}{s}\dot{I}_a \tag{5.5-10}$$

$$E_{ca} = \frac{1}{s}\dot{E}_{ca} \tag{5.5-11}$$

The variables E_1, I_{12}, E_2, I_a, and E_{ca} are the transforms of the state variables, and the variables $\dot{E}_1$, $\dot{I}_{12}$, $\dot{E}_2$, $\dot{I}_a$, and $\dot{E}_{ca}$ are the transforms of the derivatives

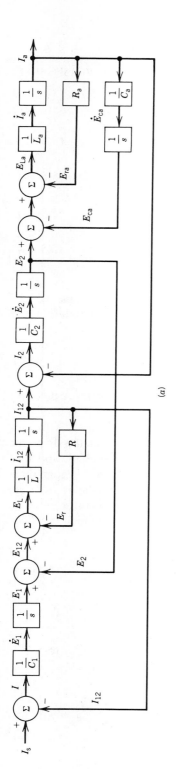

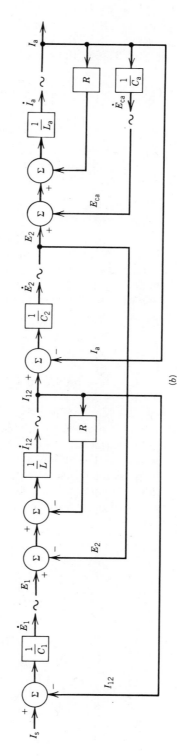

Figure 5.5-4 Modification of signal-flow diagram, with integrations expressed as separate elements to define the state variables and state equations: (*a*) modification of Fig 5.5-3, with integrations expressed separately; (*b*) modification of *a*, with integrations and secondary signals deleted to simplify specification of state equations.

of the state variables. Taking the inverse transforms of these equations gives

$$e_1 = \int \dot{e}_1 \, dt \qquad (5.5\text{-}12)$$

$$i_{12} = \int \dot{i}_{12} \, dt \qquad (5.5\text{-}13)$$

$$e_2 = \int \dot{e}_2 \, dt \qquad (5.5\text{-}14)$$

$$i_a = \int \dot{i}_a \, dt \qquad (5.5\text{-}15)$$

$$e_{ca} = \int \dot{e}_{ca} \, dt \qquad (5.5\text{-}16)$$

where e_1, i_{12}, e_2, i_a, and e_{ca} are the state variables, and $\dot{e}_1$, $\dot{i}_{12}$, $\dot{e}_2$, $\dot{i}_a$, and $\dot{e}_{ca}$ are the derivatives of the state variables. Hence, these equations, which calculate the state variables by integrating the derivatives of the state variables, have the general form

$$y_n = \int \dot{y}_n \, dt \qquad (5.5\text{-}17)$$

where y_n is the nth state variable, and $\dot{y}_n$ is the derivative of the nth state variable.

The signal-flow diagram of Fig 5.5-4a is redrawn in Fig 5.5-4b with the integrators omitted. All secondary variables are omitted, so that the diagram shows only the state variables, the derivatives of the state variables, and the input source current I_s. The following are the equations derived from that diagram, which express the transforms of the integrator inputs (the derivatives of the state variables) as functions of the transforms of the integrator outputs (the state variables) and the transform of the source current I_s (which is called a forcing function):

$$\dot{E}_i = \frac{1}{C_1}(I_s - I_{12}) \qquad (5.5\text{-}18)$$

$$\dot{I}_{12} = \frac{1}{L}(-RI_{12} + E_1 - E_2) \qquad (5.5\text{-}19)$$

$$\dot{E}_2 = \frac{1}{C_2}(I_{12} - I_a) \qquad (5.5\text{-}20)$$

$$\dot{I}_a = \frac{1}{L_a}(E_2 - E_{ca} - I_a R_a) \qquad (5.5\text{-}21)$$

$$\dot{E}_{ca} = \frac{1}{C_a}I_a \qquad (5.5\text{-}22)$$

Since these equations do not contain dynamic elements, their inverse transforms can be obtained by replacing the transforms of the signals by the signals themselves. The inverse transforms of Eqs 5.5-18 to 5.5-22 are

$$\dot{e}_i = \frac{1}{C_1}(i_s - i_{12}) \tag{5.5-23}$$

$$\dot{i}_{12} = \frac{1}{L}(-Ri_{12} + e_1 - e_2) \tag{5.5-24}$$

$$\dot{e}_2 = \frac{1}{C_2}(i_{12} - i_a) \tag{5.5-25}$$

$$\dot{i}_a = \frac{1}{L_a}(e_2 - e_{ca} - i_a R_a) \tag{5.5-26}$$

$$\dot{e}_{ca} = \frac{1}{C_a}i_a \tag{5.5-27}$$

These are called the *state equations*. They express the derivatives of the state variables as functions of the state variables and the input forcing functions. (For this case, there is only one input forcing function, the source current i_s.) An important reason for expressing the system equations in terms of state variables and state equations is that this allows the system time response to be calculated by means of a Runge–Kutta computer integration routine.

5.5.3 Runge–Kutta Integration Routine for Simulating Dynamic Systems

Most computer programs for simulating dynamic systems employ the Runge–Kutta integration routine. As was shown in Section 5.3, a digital computation loop experiences a time delay of one sample cycle, which produces error in the dynamic simulation unless compensation is included to correct for it. The Runge–Kutta integration routine compensates for this time delay by extrapolating the data in such a manner as to bring the signals in time synchronism with one another. The following is a description of the Runge–Kutta routine [5.6].

To perform a Runge–Kutta integration, the system equations are expressed in the state-equation form, which has the following general format:

1. State equations:

$$\dot{y}_n[t] = \text{function}\,[\,y_1[t],\, y_2[t],\ldots,\, x_1[t],\, x_2[t],\ldots\,] \tag{5.5-28}$$

2. Integrations:

$$y_1[t] = \int \dot{y}_1[t]\, dt, \qquad y_2[t] = \int \dot{y}_2[t]\, dt,\ldots \tag{5.5-29}$$

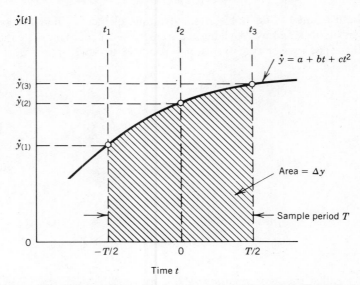

Figure 5.5-5 Calculation of integral of $\dot{y}[t]$ over sample period T for Runge–Kutta integration.

The functions $y_1[t]$, $y_2[t]$, etc. are the values of the state variables at a particular time t; and $x_1[t]$, $x_2[t]$, etc. are the values of the input forcing functions at that same time t. Often there is only a single input forcing function $x[t]$. The variables $\dot{y}_n[t]$ are the derivatives of the state variables $y_n[t]$. The state equations, which can be nonlinear, express the values of the derivatives $\dot{y}_n[t]$, at a particular instant of time, as functions of the values of the state variables $y_n[t]$ and the forcing functions $x_m[t]$ at that same instant of time.

The values of $y_n[t]$ at the beginning, middle, and end of a particular sample interval are denoted $y_{n(1)}$, $y_{n(2)}$, $y_{n(3)}$; and the values of the input forcing functions at these same instants are denoted $x_{m(1)}$, $x_{m(2)}$, $x_{m(3)}$. The Runge–Kutta algorithm operates by calculating the derivatives of the state variables at the beginning, middle, and end of each sample period. From these derivatives, denoted $\dot{y}_{n(1)}$, $\dot{y}_{n(2)}$, $\dot{y}_{n(3)}$, the integral over the sample interval, denoted Δy, is calculated as shown in Fig 5.5-5. For simplicity, the figure omits the n in the subscripts. The figure is a plot of the derivative $\dot{y}[t]$ over the sample interval T, which is characterized by the three values $\dot{y}_{(1)}$, $\dot{y}_{(2)}$, $\dot{y}_{(3)}$. Since only these three values are known, the best assumption one can make for the actual time function $\dot{y}[t]$ is a curve passing through the three points, described by a second-order equation of the form

$$\dot{y}[t] = a + bt + ct^2 \tag{5.5-30}$$

TABLE 5.5-1 Outline of Calculations of Runge–Kutta Integration

$$y_{n(1)}, x_{n(1)} \rightarrow \boxed{\text{State equations}} \rightarrow \dot{y}_{n(1)}$$

First estimate of $y_{n(2)}$: $y_{n(2)e1} = y_{n(1)} + (T/2)\dot{y}_{n(1)}$

$$y_{n(2)e1}, x_{n(2)} \rightarrow \boxed{\text{State equations}} \rightarrow \dot{y}_{n(2)e1}$$

Second estimate of $y_{n(2)}$: $y_{n(2)e2} = y_{n(1)} + (T/2)\dot{y}_{n(2)e1}$

$$y_{n(2)e2}, x_{n(2)} \rightarrow \boxed{\text{State equations}} \rightarrow \dot{y}_{n(2)e2}$$

Estimate of $y_{n(3)}$: $y_{n(3)e} = y_{n(1)} + (T)\dot{y}_{n(2)e2}.$

$$y_{n(3)e}, x_{n(3)} \rightarrow \boxed{\text{State equations}} \rightarrow \dot{y}_{n(3)e}$$

Average of estimates of $\dot{y}_{n(2)}$ is designated $\dot{y}_{n(2)e}$:

$$\dot{y}_{n(2)e} = \tfrac{1}{2}(\dot{y}_{n(2)e1} + \dot{y}_{n(2)e2})$$

Integral of $\dot{y}_n$ over sampling interval is

$$\Delta y_n = \tfrac{1}{6}T(\dot{y}_{n(1)} + 4\dot{y}_{n(2)e} + \dot{y}_{n(3)e})$$

Exact value for $y_{n(3)}$ at end of sample interval is

$$y_{n(3)} = y_{n(1)} + \Delta y_n$$

Variable $y_{n(3)}$ is the value of $y_{n(1)}$ for next sample interval.

The values of the three coefficients a, b, c are selected so that the curve passes through the three specified points. Equation 5.5-30 is integrated over the sample interval to obtain Δy, which is indicated by the cross-hatched area in Fig 5.5-5. Appendix B shows that the resultant expression for Δy is

$$\Delta y = \tfrac{1}{6}T(\dot{y}_{(1)} + 4\dot{y}_{(2)} + \dot{y}_{(3)}) \tag{5.5-31}$$

The calculations of the Runge–Kutta routine for one sample interval are outlined in Table 5.5-1. The following is a discussion of those calculations. The calculation for each sample interval starts with: (1) the values of the state variables $y_n[t]$ at time $t = t_i$ (at the start of that interval), and (2) the values of the input forcing functions at time t_i, $(t_i + T/2)$, and $(t_i + T)$ (at the beginning, middle, and end of that interval). Thus, at each sample interval, the calculation is started with known values of $y_{n(1)}$ for each state variable, and known values of $x_{m(1)}$, $x_{m(2)}$, $x_{m(3)}$ for each input forcing function.

The values of $y_{n(1)}$, $x_{m(1)}$, which are the values of the state variables and input forcing functions at time $t = t_i$ (at the start of the sample interval), are substituted into the state equations to obtain the values of the derivatives of the state variables at time $t = t_i$, which are denoted $\dot{y}_{n(1)}$. From these derivatives, the following equation is used to calculate the first estimates of the state variables at time $(t_i + T/2)$, which are denoted $y_{n(2)e1}$:

$$y_{n(2)e1} = y_{n(1)} + (T/2)\dot{y}_{n(1)} \qquad (5.5\text{-}32)$$

This calculation of the first estimate of $y_{n(2)}$ is illustrated in Fig 5.5-6a. These first estimates of $y_{n(2)}$, along with the values $x_{m(2)}$ of the forcing function at time $(t_i + T/2)$, are substituted into the state equations to obtain the first estimates of the derivatives at time $(t_i + T/2)$, which are denoted $\dot{y}_{n(2)e1}$. From these, the following equation is used to calculate the second estimates of the state variables at time $(t_i + T/2)$, which are denoted $y_{n(2)e2}$:

$$y_{n(2)e2} = y_{n(1)} + (T/2)\dot{y}_{n(2)e1} \qquad (5.5\text{-}33)$$

This calculation of the second estimate of $y_{n(2)}$ is illustrated in Fig 5.5-6b. These second estimates of $y_{n(2)}$, along with the values $x_{m(2)}$ of the forcing function at time $(t_i + T/2)$, are substituted into the state equations to obtain the second estimates of the derivatives at time $(t_i + T/2)$, which are denoted $\dot{y}_{n(2)e2}$. From these, the following equation is used to calculate the estimates of the state variables at time $(t_i + T)$, which are denoted $y_{n(3)e}$:

$$y_{n(3)e} = y_{n(1)} + (T)\dot{y}_{n(2)e2} \qquad (5.5\text{-}34)$$

This calculation of the estimate of $y_{n(3)}$ is illustrated in Fig 5.5-6c. These estimates of $y_{n(3)}$, along with the values $x_{m(3)}$ of the forcing function at time $(t_i + T)$, are substituted into the state equations to obtain the estimates of the derivatives at time $(t_i + T)$, which are denoted $\dot{y}_{n(3)e}$.

The estimates of the derivatives at point 2, at the middle of the interval, are averaged as follows to obtain the average estimated derivative, which is denoted $\dot{y}_{n(2)e}$:

$$\dot{y}_{n(2)e} = \tfrac{1}{2}\big(\dot{y}_{n(2)e1} + \dot{y}_{n(2)e2}\big) \qquad (5.5\text{-}35)$$

Thus, the calculations yield three derivatives for each state variable, which are the derivatives of y_n at the beginning, middle, and end of the sample interval. These three derivatives are denoted $\dot{y}_{n(1)}$, $\dot{y}_{n(2)e}$, and $\dot{y}_{n(3)e}$. In accordance with Eq 5.5-31, the integral of $\dot{y}_n[t]$ over the sample interval, which is denoted Δy, is calculated as follows from these three derivatives:

$$\Delta y_n = \tfrac{1}{6}T\big(\dot{y}_{n(1)} + 4\dot{y}_{n(2)e} + \dot{y}_{n(3)e}\big) \qquad (5.5\text{-}36)$$

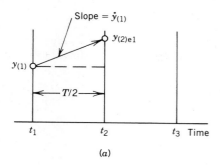

(a)

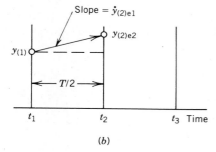

(b)

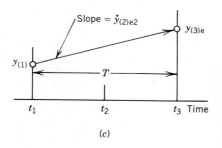

(c)

Figure 5.5-6 Runge–Kutta calculations of estimated state-variable values, at middle and end of sample interval: calculation of (a) first estimate of $y_{(2)}$; (b) second estimate of $y_{(2)}$; (c) estimate of $y_{(3)}$.

The true value for y_n at the end of the sample interval, which is denoted $y_{n(3)}$, is obtained by adding Δy_n to $y_{n(1)}$, the value at the beginning of the sample interval:

$$y_{n(3)} = y_{n(1)} + \Delta y_n \tag{5.5-37}$$

This value $y_{n(3)}$ now becomes the value of $y_{n(1)}$ for the next sample interval, and the process is repeated.

5.5.4 Matrix Representation of State Equations

To apply the Runge–Kutta integration procedure efficiently, the state equations must be expressed in a form that can be interpreted conveniently by a computer. This is achieved by writing the state equations in matrix format. The required matrix format can be simply explained by considering an example. Rules for matrix manipulation are summarized in Appendix C.

The variables of the state equations are designated by:

1. N state variables, labeled y_n.
2. N derivatives of state variables, labeled $\dot{y}_n$.
3. M forcing functions, labeled x_m.

Assume, for example, that $N = 4$ and $M = 2$, so that there are four state variables (y_1, y_2, y_3, y_4), four derivatives of state variables ($\dot{y}_1$, $\dot{y}_2$, $\dot{y}_3$, $\dot{y}_4$), and two forcing functions (x_1, x_2). The general form of the state equations is

$$\dot{y}_1 = a_{11}y_1 + a_{12}y_2 + a_{13}y_3 + a_{14}y_4 + b_{11}x_1 + b_{12}x_2 \qquad (5.5\text{-}38)$$

$$\dot{y}_2 = a_{21}y_1 + a_{22}y_2 + a_{23}y_3 + a_{24}y_4 + b_{21}x_1 + b_{22}x_2 \qquad (5.5\text{-}39)$$

$$\dot{y}_3 = a_{31}y_1 + a_{32}y_2 + a_{33}y_3 + a_{34}y_4 + b_{31}x_1 + b_{32}x_2 \qquad (5.5\text{-}40)$$

$$\dot{y}_4 = a_{41}y_1 + a_{42}y_2 + a_{43}y_3 + a_{44}y_4 + b_{41}x_1 + b_{42}x_2 \qquad (5.5\text{-}41)$$

These equations are expressed in matrix format as follows:

$$\begin{bmatrix} \dot{y}_1 \\ \dot{y}_2 \\ \dot{y}_3 \\ \dot{y}_4 \end{bmatrix} = \begin{bmatrix} a_{11} & a_{12} & a_{13} & a_{14} \\ a_{21} & a_{22} & a_{23} & a_{24} \\ a_{31} & a_{32} & a_{33} & a_{34} \\ a_{41} & a_{42} & a_{43} & a_{44} \end{bmatrix} \begin{bmatrix} y_1 \\ y_2 \\ y_3 \\ y_4 \end{bmatrix} + \begin{bmatrix} b_{11} & b_{12} \\ b_{21} & b_{22} \\ b_{31} & b_{32} \\ b_{41} & b_{42} \end{bmatrix} \begin{bmatrix} x_1 \\ x_2 \end{bmatrix} \qquad (5.5\text{-}42)$$

The matrix expression of Eq 5.5-42 is equivalent to Eqs 5.5-38 to -41. The coefficients a_{pq} for the state variables are arranged in an N-by-N matrix (N rows by N columns), while the coefficients b_{pq} for the forcing functions are arranged in an N-by-M matrix (N rows by M columns). The state variables y_n, and the derivatives of the state variables $\dot{y}_n$, are arranged in N-by-1 arrays, which are called N-component (column) vectors. The forcing functions x_m are arranged in an M-by-1 array, which is an M-component vector. In order for matrix multiplication to have meaning, the number of columns of the first matrix (or vector) must be equal to the number of rows of the second matrix (or vector).

The state equations for the ELF transmitter network were given in Eqs 5.5-23 to -27. These can be expressed as

$$\dot{e}_1 = -\frac{1}{C_1}i_{12} + \frac{1}{C_1}i_s \tag{5.5-43}$$

$$\dot{i}_{12} = \frac{1}{L}e_1 - \frac{R}{L}i_{12} - \frac{1}{L}e_2 \tag{5.5-44}$$

$$\dot{e}_2 = \frac{1}{C_2}i_{12} - \frac{1}{C_2}i_a \tag{5.5-45}$$

$$\dot{i}_a = \frac{1}{L_a}e_2 - \frac{R_a}{L_a}i_a - \frac{1}{L_a}e_{ca} \tag{5.5-46}$$

$$\dot{e}_{ca} = \frac{1}{C_a}i_a \tag{5.5-47}$$

These equations are represented in matrix format as follows:

$$
\begin{bmatrix} \dot{e}_1 \\ \dot{i}_{12} \\ \dot{e}_2 \\ \dot{i}_a \\ \dot{e}_{ca} \end{bmatrix} =
\begin{bmatrix}
0 & -1/C_1 & 0 & 0 & 0 \\
1/L & -R/L & -1/L & 0 & 0 \\
0 & 1/C_2 & 0 & -1/C_2 & 0 \\
0 & 0 & 1/L_a & -R_a/L_a & -1/L_a \\
0 & 0 & 0 & 1/C_a & 0
\end{bmatrix}
\begin{bmatrix} e_1 \\ i_{12} \\ e_2 \\ i_a \\ e_{ca} \end{bmatrix}
$$
$$
+ \begin{bmatrix} 1/C_1 \\ 0 \\ 0 \\ 0 \\ 0 \end{bmatrix} i_s \tag{5.5-48}
$$

The general matrix relation of Eq 5.5-42 can be condensed to the form

$$\underline{\dot{y}} = \underline{A}\underline{y} + \underline{B}\underline{x} \tag{5.5-49}$$

The underlines indicate that the variables $\underline{y}$, $\underline{\dot{y}}$, and $\underline{x}$ are vectors and the transfer blocks $\underline{A}$ and $\underline{B}$ are matrices. The variable $\underline{y}$ is a four-component vector that represents the four state variables y_1, y_2, y_3, y_4. The variable $\underline{\dot{y}}$ is a four-component vector that represents the four derivatives of state variables, $\dot{y}_1, \dot{y}_2, \dot{y}_3, \dot{y}_4$. The variable $\underline{x}$ is a two-component vector that represents the two forcing functions x_1, x_2. The transfer block $\underline{A}$ represents the matrix of the sixteen coefficients a_{pq}, and the transfer block $\underline{B}$ represents the matrix of the eight coefficients b_{pq}.

In Fig 5.5-7, the equations of the system are expressed in signal-flow diagram form, in accordance with the condensed matrix relation of Eq 5.5-49.

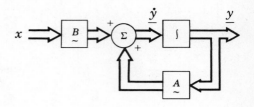

Figure 5.5-7 Signal-flow diagram representation of matrix form of state equations.

The signals y, $\dot{y}$, and $\underline{x}$ in the signal-flow diagram represent vector signals, with four components for y and $\dot{y}$, and two components for $\underline{x}$. The transfer blocks $\underline{A}$ and $\underline{B}$ represent matrices. The block between $\dot{y}$ and y indicates that each component of the state vector y is the integral of the corresponding component of $\dot{y}$.

Instead of using Runge–Kutta integration to implement the integrations between $\dot{y}$ and y, an approximate digital solution can be obtained by using the simple integration formula that was stated in Eq 5.3-65. This gives for any one of the components y_n

$$y_n = y_{np} + T(\dot{y}_n) \qquad (5.5\text{-}50)$$

where y_{np} is the past value of the component y_n, and $\dot{y}_n$ is the derivative, which is based on the y_n-values calculated in the previous cycle. Using this relation, the system equations can be solved in a sampled-data computation according to the matrix signal-flow diagram of Fig 5.5-8a. The y_p-values represent the past values of the state variables y computed in the previous cycle. The matrices $\underline{A}$ and $\underline{B}$ are multiplied by the sample period T to form

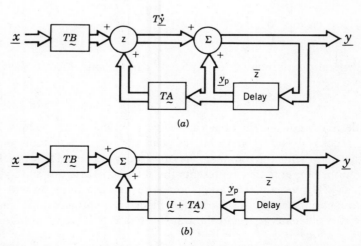

Figure 5.5-8 Matrix signal-flow diagram that approximates Fig 5.5-7 with sampled-data computations: (a) basic diagram; (b) simplified form of a.

the matrices $T\underset{\sim}{A}$ and $T\underset{\sim}{B}$. Since the sample period T is a scalar, all of the coefficients a_{pq}, b_{pq} of these matrices are multiplied by T.

Diagram a of Fig 5.5-8 can be simplified to form diagram b. The two feedback paths for y_p are combined by changing the matrix $T\underset{\sim}{A}$ to $(\underset{\sim}{I} + T\underset{\sim}{A})$. The matrix $\underset{\sim}{I}$ is an identity matrix, and is defined as follows for this example, which has four state variables:

$$\underset{\sim}{I} = \begin{bmatrix} 1 & 0 & 0 & 0 \\ 0 & 1 & 0 & 0 \\ 0 & 0 & 1 & 0 \\ 0 & 0 & 0 & 1 \end{bmatrix} \tag{5.5-51}$$

The identity matrix has unit values on the diagonal, and zero values for all other elements. The computations described by Fig 5.5-8b represent the following equations:

$$y_1 = (1 + Ta_{11})y_{1p} + Ta_{12}y_{2p} + Ta_{13}y_{3p} + Ta_{14}y_{4p} + Tb_{11}x_1 + Tb_{12}x_2 \tag{5.5-52}$$

$$y_2 = Ta_{21}y_{1p} + (1 + Ta_{22})y_{2p} + Ta_{23}y_{3p} + Ta_{24}y_{4p} + Tb_{21}x_1 + Tb_{22}x_2 \tag{5.5-53}$$

$$y_3 = Ta_{31}y_{1p} + Ta_{32}y_{2p} + (1 + Ta_{33})y_{3p} + Ta_{34}y_{4p} + Tb_{31}x_1 + Tb_{32}x_2 \tag{5.5-54}$$

$$y_4 = Ta_{41}y_{1p} + Ta_{42}y_{2p} + Ta_{43}y_{3p} + (1 + Ta_{44})y_{4p} + Tb_{41}x_1 + Tb_{42}x_2 \tag{5.5-55}$$

The sampled-data calculations indicated in Fig 5.5-8 provide only a first-order approximation of the differential equations described by Fig 5.5-7. The errors in the sampled-data calculations that are associated with the feedback loops have much more effect on the response than the errors associated with the forcing functions. Therefore, considerably better accuracy can be realized by correcting only the feedback sampled-data calculations. This can be achieved by replacing the matrix block $(\underset{\sim}{I} + T\underset{\sim}{A})$ with a matrix denoted $\underset{\sim}{\Phi}$, called the *transition matrix*, which is given by

$$\underset{\sim}{\Phi} = \underset{\sim}{I} + T\underset{\sim}{A} + \frac{(T\underset{\sim}{A})^2}{2!} + \frac{(T\underset{\sim}{A})^3}{3!} + \frac{(T\underset{\sim}{A})^4}{4!} + \cdots \tag{5.5-56}$$

A matrix signal-flow diagram of the resultant sampled-data calculations is shown in Fig 5.5-9. Equation 5.5-56 has the form of the expansion of an exponential. Hence the transition matrix can be expressed as

$$\underset{\sim}{\Phi} = \exp[T\underset{\sim}{A}] \tag{5.5-57}$$

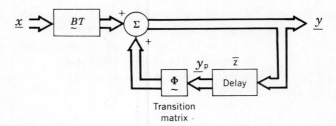

Figure 5.5-9 Matrix signal-flow diagram showing accurate sampled-data calculations for solving matrix equations of Fig 5.5-7.

The transition matrix can be derived as follows. The input forcing function vector $\underline{x}$ is assumed to be zero, and each state variable derivative $\dot{y}$ is expanded in a Taylor series, over the time interval of one sample period. In order to simplify the notation, the nth derivative of $y[t]$ is denoted $D^n y[t]$. The Taylor series expansion for $\dot{y}[t]$ is

$$\dot{y}[t] = Dy[t] = Dy[t_1] + D^2 y[t_1](t - t_1) + \frac{D^3 y[t_1](t - t_1)^2}{2!}$$

$$+ \frac{D^4 y[t_1](t - t_1)^3}{3!} + \cdots \tag{5.5-58}$$

Time t_1 designates the start of the sample interval. Integrating Eq 5.5-58 gives the integral of $\dot{y}[t]$ over the sample interval:

$$\int_{t_1}^{t_1 + T} \dot{y}[t] \, dt = Dy[t_1]T + \frac{D^2 y[t_1]T}{2!} + \frac{D^3 y[t_1]T^3}{3!} + \frac{D^4 y[t_1]T^4}{4!} + \cdots \tag{5.5-59}$$

This is added to $y[t_1]$ to obtain the value of $y[t]$ at the end of the sample interval:

$$y[t_1 + T] = y[t_1] + \int_{t_1}^{t_1 + T} \dot{y}[t] \, dt$$

$$= y[t_1] + Dy[t_1]T + \frac{D^2 y[t_1]T}{2!} + \frac{D^3 y[t_1]T^3}{3!}$$

$$+ \frac{D^4 y[t_1]T^4}{4!} + \cdots \tag{5.5-60}$$

The value $y[t_1 + T]$ is considered to be the present value of y. Hence $y[t_1]$ is the past value of y, denoted y_p; and $D^n y[t_1]$ is the past value of the nth derivative of y, denoted $D^n y_p$. Thus Eq 5.5-60 becomes

$$y = y_p + (T)Dy_p + \frac{T^2}{2!}D^2 y_p + \frac{T^3}{3!}D^3 y_p + \frac{T^4}{4!}D^4 y_p + \cdots \quad (5.5\text{-}61)$$

Since the input forcing function vector $\underset{\sim}{x}$ is set to zero, the past value of the derivative vector $\underset{\sim}{\dot{y}}$ is (by Eq 5.5-49)

$$\underset{\sim}{\dot{y}}_p = D\underset{\sim}{y}_p = \underset{\sim}{A}\underset{\sim}{y}_p \quad (5.5\text{-}62)$$

Differentiating this gives

$$D^2 \underset{\sim}{y}_p = \underset{\sim}{A}D\underset{\sim}{y}_p = \underset{\sim}{A}^2 \underset{\sim}{y}_p \quad (5.5\text{-}63)$$

$$D^n \underset{\sim}{y}_p = \underset{\sim}{A}D^{n-1}\underset{\sim}{y}_p = \underset{\sim}{A}^n \underset{\sim}{y}_p \quad (5.5\text{-}64)$$

Applying Eq 5.5-64 to Eq 5.5-61 gives the following matrix equation:

$$y = y_p + (T\underset{\sim}{A})\underset{\sim}{y}_p + \frac{(T\underset{\sim}{A})^2}{2!}\underset{\sim}{y}_p + \frac{(T\underset{\sim}{A})^3}{3!}\underset{\sim}{y}_p + \frac{(T\underset{\sim}{A})^4}{4!}\underset{\sim}{y}_p + \cdots \quad (5.5\text{-}65)$$

Factoring the vector $\underset{\sim}{y}_p$ from the right-hand side gives

$$y = \left[\underset{\sim}{I} + T\underset{\sim}{A} + \frac{(T\underset{\sim}{A})^2}{2!} + \frac{(T\underset{\sim}{A})^3}{3!} + \frac{(T\underset{\sim}{A})^4}{4!} + \cdots \right]\underset{\sim}{y}_p$$

$$= \underset{\sim}{\Phi}\underset{\sim}{y}_p \quad (5.5\text{-}66)$$

The expression in the brackets is the same as the equation for the transition matrix $\underset{\sim}{\Phi}$ given previously in Eq 5.5-56. The transition matrix is a key element in Kalman optimal-estimator theory, which is discussed in Chapter 8.

5.5.5 Expressing Transfer Functions in Terms of Integrations

To describe the signal-flow diagram of a control system in state-variable form, all of the transfer functions must be expressed in terms of simple integrators. This can be achieved by representing the transfer functions by feedback loops, as shown in Fig 5.5-10.

Diagrams a to c of Fig 5.5-10 employ a single feedback loop with the loop transfer function $G = \omega_c/s$. The transfer functions Y/X are

Diagram a:

$$\frac{Y}{X} = G_{ib} = \frac{G}{1 + G} = \frac{\omega_c/s}{1 + \omega_c/s} = \frac{\omega_c}{s + \omega_c} \quad (5.5\text{-}67)$$

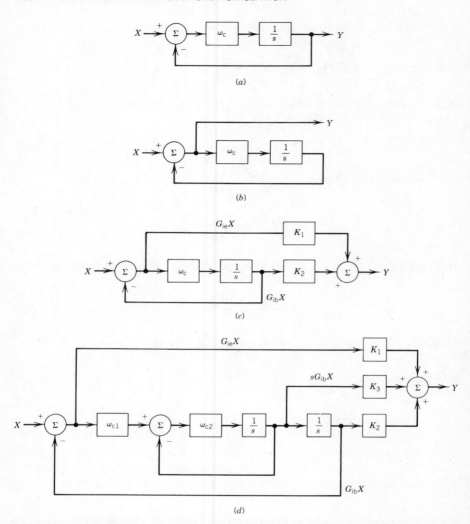

Figure 5.5-10 Signal-flow diagram employing integrations to implement the following transfer functions: (a) $\omega_c/(s + \omega_c)$; (b) $s/(s + \omega_c)$; (c) $(K_1 s + K_2\omega_c)/(s + \omega_c)$; (d) general transfer functions with underdamped poles.

Diagram b:

$$\frac{Y}{X} = G_{ie} = \frac{1}{1 + G} = \frac{1}{1 + \omega_c/s} = \frac{s}{s + \omega_c} \qquad (5.5\text{-}68)$$

Diagram c:

$$\frac{Y}{X} = K_1 G_{ie} + K_2 G_{ib} = \frac{K_1 s}{s + \omega_c} + \frac{K_2\omega_c}{s + \omega_c} = \frac{K_1 s + K_2\omega_c}{s + \omega_c} \qquad (5.5\text{-}69)$$

To achieve an underdamped transfer function, the signal-flow diagram must have two feedback loops. A general form of such a configuration is shown in Fig 5.5-10d. The loop transfer function of the outer loop is

$$G = \frac{\omega_{c1}}{s} \frac{\omega_{c2}}{(s + \omega_{c2})} = \frac{N}{D} \qquad (5.5\text{-}70)$$

The feedback and error transfer functions of the outer loop are therefore

$$G_{ib} = \frac{G}{1 + G} = \frac{N}{N + D} = \frac{\omega_{c1}\omega_{c2}}{s^2 + \omega_{c2}s + \omega_{c1}\omega_{c2}} \qquad (5.5\text{-}71)$$

$$G_{ie} = \frac{1}{1 + G} = \frac{D}{N + D} = \frac{s(s + \omega_{c2})}{s^2 + \omega_{c2}s + \omega_{c1}\omega_{c2}} \qquad (5.5\text{-}72)$$

The response Y/X is

$$\frac{Y}{X} = K_1 G_{ie} + K_2 G_{ib} + K_3 s G_{ib}$$

$$= \frac{K_1 s(s + \omega_{c2})}{s^2 + \omega_{c2}s + \omega_{c1}\omega_{c2}} + \frac{K_2 \omega_{c1}\omega_{c2}}{s^2 + \omega_{c2}s + \omega_{c1}\omega_{c2}}$$

$$+ \frac{s K_3 \omega_{c1}\omega_{c2}}{s^2 + \omega_{c2}s + \omega_{c1}\omega_{c2}} \qquad (5.5\text{-}73)$$

This reduces to

$$\frac{Y}{X} = \frac{K_1 s^2 + (K_1 + K_3\omega_{c1})\omega_{c2}s + K_2\omega_{c1}\omega_{c2}}{s^2 + \omega_{c2}s + \omega_{c1}\omega_{c2}} \qquad (5.5\text{-}74)$$

The constants K_1, K_2, K_3 can have negative as well as positive values. By appropriate choice of the constants K_1, K_2, K_3, any desired transfer function can be implemented with diagrams c and d, or cascaded combinations of them.

5.5.6 Application of Serial-Simulation Procedure to ELF Transmitter Network

Now let us apply to the ELF transmitter network the simulation approach developed in Section 5.3. The Runge–Kutta integration procedure solves the system equations in a parallel manner, whereas the simulation method of Section 5.3 solves them in a serial manner. Therefore, the method of Section 5.3 is called *serial simulation*.

Figure 5.5-11a shows the computer signal-flow diagram of Fig 5.5-4a modified in accordance with the serial-simulation procedure. All of the $1/s$

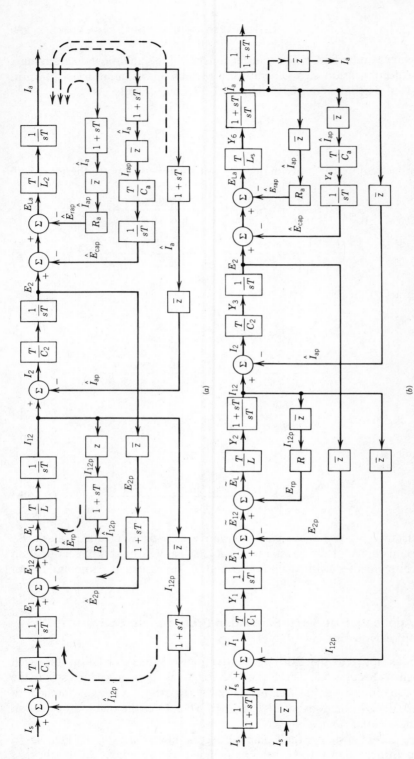

Figure 5.5-11 Signal-flow diagram of ELF transmitter circuit prepared for serial simulation: (*a*) modification of Fig 5.5-3, showing time delay, time-delay compensation, and normalization of transfer functions; (*b*) modification of *a* achieved by shifting $(1 + sT)$ transfer functions from feedback paths to forward paths.

integrator blocks are replaced with normalized integrations $1/sT$, and the element in the preceding block is appropriately multiplied by T. The transfer function $\bar{z}$ is placed in each feedback path to represent a one-cycle time delay, and a time-delay compensation factor $(1 + sT)$ is placed in series with each $\bar{z}$ transfer function.

To simplify Fig 5.5-11a, the compensation factors $(1 + sT)$ are shifted into the forward paths and combined with appropriate $1/sT$ integrations. Different signal-flow diagrams can result (each of which is valid) depending on how the factors $(1 + sT)$ are shifted. As shown by the dashed lines, the factors $(1 + sT)$ in the feedback paths of the antenna current I_a are shifted in the backward direction, and the other factors $(1 + sT)$ are shifted in the forward direction. To clarify this process, each compensation factor $(1 + sT)$ has been placed on the appropriate side of the time-delay factor $\bar{z}$.

Figure 5.5-11b shows the resultant simplified diagram. The source current I_s is fed through the lowpass transfer function $1/(1 + sT)$ to form the delayed signal $\tilde{I}_s$, which lags I_s by one sample period at low frequencies. If the signal I_s is not part of a feedback loop, the delayed signal $\tilde{I}_s$ can be approximated as the past value of I_s:

$$\tilde{I}_s \cong I_{sp} \tag{5.5-75}$$

Similarly, the antenna current I_a can be approximated as the past value of the future antenna current $\hat{I}_a$:

$$I_a \cong \hat{I}_{ap} \tag{5.5-76}$$

The computer program to simulate the signal-flow diagram of Fig 5.5-11b is outlined below. Remember that all of the variables in the computer equations represent the actual time-domain signals, not their transforms.

SPECIFY PARAMETERS:

R_a, L_a, C_a, C_1, C_2, L, $R = ?$ (Circuit parameters)
$L_{Max} = ?$ (Number of computation cycles per print cycle)
$TME_{Max} = ?$ (Maximum time of computation)
$T = ?$ (Sample period)

SET INITIAL VALUES TO ZERO:

$TME = 0$ (Initial time)
$L = 0$ (Print-cycle counter)
I_{sp}, I_{12p}, $\tilde{E}_{1p}$, Y_{1p}, E_{2p}, Y_{2p}, $\hat{I}_{ap}$, Y_{3p}, Y_{4p}, $\hat{E}_{capp}$, $Y_{6p} = 0$

SPECIFY INPUT FORCING FUNCTION(S) VERSUS TIME (TME):

100   $TME = TME + T$

101   $I_s = $ Function[TME]

SYSTEM EQUATIONS:

200 $\tilde{I}_s = I_{sp}$
201 $\tilde{I}_1 = \tilde{I}_s - I_{12p}$
202 $Y_1 = (T/C_1) * \tilde{I}_1$
203 $\tilde{E}_1 = \tilde{E}_{1p} + (1/2) * (Y_1 + Y_{1p})$
204 $\tilde{E}_{12} = \tilde{E}_1 - E_{2p}$
205 $E_{rp} = R * I_{12p}$
206 $\tilde{E}_L = \tilde{E}_{12} - E_{rp}$
207 $Y_2 = (T/L) * \tilde{E}_L$
208 $I_{12} = I_{12p} + (3/2) * Y_2 - (1/2) * Y_{2p}$
209 $I_2 = I_{12} - \hat{I}_{ap}$
210 $Y_3 = (T/C_2) * I_2$
211 $E_2 = E_{2p} + (1/2) * (Y_3 + Y_{3p})$
212 $Y_4 = (T/C_a) * \hat{I}_{ap}$
213 $\hat{E}_{cap} = \hat{E}_{capp} + (1/2) * (Y_4 + Y_{4p})$
214 $Y_5 = E_2 - \hat{E}_{cap}$
215 $\hat{E}_{rap} = R_a * \hat{I}_{ap}$
216 $E_{La} = Y_5 - \hat{E}_{rap}$
217 $Y_6 = (T/L_a) * E_{La}$
218 $\hat{I}_a = \hat{I}_{ap} + (3/2) * Y_6 - (1/2) * Y_{6p}$
219 $I_a = \hat{I}_{ap}$

CHECK PRINT COUNT, PRINT DATA:

300 $L = L + 1$
301 IF $(L < L_{Max})$ GOTO 400
302 $L = 0$
303 PRINT data (TME, X, X_e, $\hat{X}_v$, etc.)

SET PAST VALUES EQUAL TO PRESENT VALUES:

400 $I_{sp} = I_s$
401 $I_{12p} = I_{12}$
402 $\tilde{E}_{1p} = \tilde{E}_1$
403 $Y_{1p} = Y_1$
404 $E_{2p} = E_2$
405 $Y_{2p} = Y_2$
406 $\hat{I}_{ap} = \hat{I}_a$
407 $Y_{3p} = Y_3$
408 $Y_{4p} = Y_4$
409 $\hat{E}_{capp} = \hat{E}_{cap}$
410 $Y_{6p} = Y_6$

CHECK TIME LIMIT, REPEAT:

500   IF (TME < TME$_{\text{Max}}$) GOTO 100
501   END

As shown below, the variables Y_1, Y_2, Y_3, Y_4, and Y_6 are proportional to derivatives of the voltage and current variables:

$$Y_1 = T\frac{d\tilde{E}_1}{dt} \tag{5.5-77}$$

$$Y_2 = T\frac{d\tilde{I}_{12}}{dt} \tag{5.5-78}$$

$$Y_3 = T\frac{dE_2}{dt} \tag{5.5-79}$$

$$Y_4 = T\frac{d\hat{E}_{\text{cap}}}{dt} \tag{5.5-80}$$

$$Y_6 = T\frac{dI_a}{dt} \tag{5.5-81}$$

5.6   APPROXIMATE EFFECT OF SAMPLING WITHIN A CONTINUOUS-DATA FEEDBACK LOOP

The performance of an analog feedback control loop that contains a section employing digital data can be analyzed by means of the z-transform. In Ref [1.2], Section 9.5 describes the z-transform and shows how to use it in transient analysis and in frequency response analysis. That section also shows how the serial-simulation procedure, described in Section 5.3, is applied to simulate analog control loops containing digital data. The following summarizes the general conclusions derived in Ref [1.2] (Section 9.5).

The primary dynamic effects of sampling within a feedback loop carrying continuous data can be explained by reexamining Fig 5.3-7, which shows the time response of a conventional simple-hold D/A converter for a typical input signal. The response shows the following two major characteristics of the demodulated sampled-data signal:

1. There is a time delay in the digital demodulation process equal to half a sample period.
2. When the input signal is varying, there are appreciable components at the sampling frequency and its harmonics.

Time delay in the demodulation process causes phase lag that is proportional to frequency. Time delay of half a sample period corresponds to phase lag of half a cycle (180°) at the sampling frequency. Hence, the phase lag ϕ at any frequency f can be expressed as

$$\phi = 180(f/F_s) \quad \text{deg} \tag{5.6-1}$$

where F_s is the sampling frequency.

Harmonic components in the demodulated signal tend to cause instability if they are transmitted around a feedback loop. Additional filtering is required to attenuate the sampling harmonics, and this filtering adds additional phase lag to the loop. A certain amount of filtering is usually provided by the lowpass effect of elements within the loop, but additional filtering is usually needed to reduce sampling harmonics.

As shown in Ref [1.2] (Section 9.5), the dynamic effect of sampling with a simple-hold demodulator can be approximated by the transfer function of $(1 - sT/2)$, where T is the sample period. At low frequencies, the phase lag of this transfer function is the same as that of Eq 5.6-1. The right-half-plane zero of this transfer function provides high-frequency peaking in the frequency response, which approximates the destabilizing effect of sampling harmonics being transmitted around a feedback loop.

The analysis in Ref [1.2] (Section 9.5) shows that it is theoretically possible to have good stability in a sampled-data feedback control loop with a gain crossover frequency as high as $1/8$ of the sampling frequency. However, in most systems the gain crossover frequency is significantly lower than this, because stronger filtering is usually required to reduce sampling harmonics to acceptable levels. Under such conditions, the dynamic degradation of the sampling process is primarily the phase lag due to the half-cycle delay of the D/A converter (which was given in Eq 5.6-1) plus the phase lag of the filtering required to attenuate the sampling harmonics.

There are a number of reasons why strong filtering of sampling harmonics is usually required. For example, as is explained in Ref [1.2] (Chapter 10), mechanical control systems usually have poorly damped high-frequency resonance peaks caused by the dynamics of the mechanical structure. If a sampling harmonic of appreciable magnitude falls at the same frequency as a structural resonance peak, instability may result. Therefore, mechanical control systems generally require strong filtering of sampling harmonics.

Besides, in high-performance control systems, sampling harmonics may saturate the power amplifier unless they are strongly filtered. For example, in a tachometer feedback servomechanism, the tachometer feedback signal cancels most of the low-frequency portion of the amplified position error signal. Therefore, after the summation point of the tachometer signal, the control signal in a sampled-data system often consists largely of sampling harmonics. If these harmonics are not strongly filtered, they may saturate the power amplifier, and the resultant nonlinearities may severely degrade system performance.

Usually, the gain crossover frequency is less than $1/10$ of the sampling frequency. As shown in Eq 5.6-1, the time delay of the D/A conversion process produces a phase lag of $18°$ at $1/10$ of the sampling frequency. The total phase lag including filtering is often at least twice the phase lag of the D/A converter alone. Hence, the additional filtering required to attenuate harmonics may result in a total phase lag of $36°$ at $1/10$ of the sampling frequency.

Section 5.2 described the linear-interpolator hold circuit, which constructs straight-line interpolations between sampled data points. This process causes twice the phase lag of the simple-hold D/A converter, because it requires a time delay of one sample period to construct the interpolations. The phase lag ϕ of this circuit at any frequency f is equal to

$$\phi = 360(f/F_s)   \text{deg} \tag{5.6-2}$$

On the other hand, even though the linear-interpolator hold circuit has twice the phase lag as the simple-hold circuit, it may be dynamically superior because the harmonics of the demodulation process are very much lower. Very little additional filtering of harmonics is usually needed. Hence, the total phase lag of the demodulation process plus the harmonic filter may actually be less when a linear-interpolator hold circuit is used. The effect of a linear-interpolator hold circuit can be approximated by an over-sampled hold circuit, explained in Section 5.2.5, which is easier to build.

Chapter 6

Response to a Random Signal

Certain inputs to control systems are random, and can only be characterized quantitatively by statistical means. These inputs include in particular the disturbances caused by receiver noise and by wind gusts. This chapter summarizes the application of statistical techniques for handling such random input signals.

The chapter describes thermal noise and shot noise, the concept of noise bandwidth, and the Gaussian distribution. Autocorrelation and cross-correlation functions are presented, along with a general method of computing noise bandwidth.

These concepts are applied in Chapter 7 to calculate the effect of receiver noise in a tracking radar system, and in Section 6.4 of this chapter to calculate the tracking error produced by wind forces on a large antenna.

6.1 SUMMARY OF CHAPTER

The variable that is most commonly handled by statistical techniques is amplifier noise. As shown in Section 6.2, there are two major types of amplifier noise: (1) thermal noise, which is due to thermal agitation in resistive elements, and (2) shot noise, which is due to the discrete nature of electron flow, or the quantum nature of received radiation.

Noise transmission from these noise sources is characterized by the noise bandwidth of the amplifier. The noise bandwidths of common filters are listed in Section 6.2. An effective method for computing the noise bandwidth of a transfer function is given in Section 6.3, which provides direct expressions for the noise bandwidths of transfer functions of up to 10th order of s.

Most noise sources have an amplitude distribution that is approximately Gaussian. Section 6.2 explains the Gaussian, or normal, distribution and shows how it is applied.

Section 6.2 derives an approximate relation between the noise bandwidth of a lowpass filter and the rise time of the step response. This is extended to relate the noise bandwidth of a bandpass filter to the rise time of the envelope of the response to a step-modulated carrier.

Section 6.3 describes the convolution integral, a tool that is particularly useful in statistical analysis. Autocorrelation and cross-correlation functions are explained along with the Fourier transforms of these functions, which are the spectral densities.

Section 6.4 gives a practical application of these statistical analysis techniques. It shows how wind forces can be characterized statistically to calculate the wind-induced tracking error of a large satellite communication antenna.

The principles of this chapter are also used in Chapter 7, which applies statistical analysis to target detection and tracking in a tracking radar system.

6.2   CHARACTERISTICS OF RECEIVER NOISE SIGNALS

This section derives the basic equations for characterizing the major types of receiver noise: thermal (or Johnson) noise, which is due to thermal agitation in resistive elements, and shot (or Schottky) noise, which is due to the discrete nature of electron flow.

6.2.1   Thermal (or Johnson) Noise

Thermal agitation in a resistor generates electrical noise power that is proportional to the absolute temperature. If, as shown in Fig 6.2-1a, a resistor is connected to the input terminals of an amplifier having an input resistance matched to that of the resistor, the resistor delivers to the amplifier a noise power, per hertz of frequency, that is given by

$$\frac{dP}{df} = KT \tag{6.2-1}$$

where T is the absolute temperature in °K (Kelvin) and K is Boltzmann's constant given by

$$K = 1.374 \times 10^{-23} \text{ joule}/°\text{K} \tag{6.2-2}$$

At room temperature (25°C), $T = 298$ °K, so that

$$\frac{dP}{df} = 4.10 \times 10^{-21} \text{ Watt/Hz} \tag{6.2-3}$$

At the amplifier input, there is a signal to be amplified. What is important is the ratio of signal to noise in the amplified output. To evaluate this ratio, it is convenient to express the signal and noise outputs from the amplifier relative to the input, by normalizing the voltage gain and power gain of the amplifier

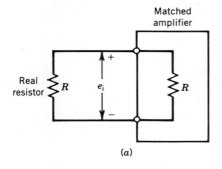

(a)

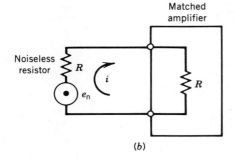

(b)

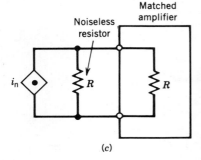

(c)

Figure 6.2-1 Definition of thermal noise: (a) basis for thermal noise definition; (b) representation of thermal noise by equivalent voltage noise source e_n; (c) representation of thermal noise by equivalent current noise source i_n.

to unity at the signal frequency. Consider an ideal amplifier with a frequency response that is flat between frequencies f_1 and f_2 and is zero outside that region, as shown in diagram a of Fig 6.2-2. The output noise power, normalized relative to the input, is

$$P_o = KT\Delta f \qquad (6.2\text{-}4)$$

$$\Delta f = f_2 - f_1 \qquad (6.2\text{-}5)$$

Frequency responses of practical amplifiers are illustrated in diagrams b and c. As shown, the amplifier gain is normalized to unity at zero frequency for a lowpass response, and it is normalized to unity at the carrier frequency for a

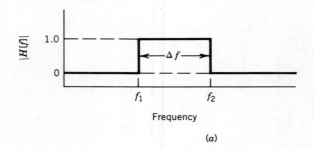

(a)

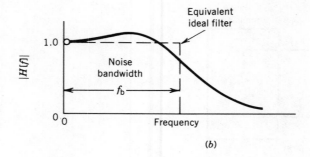

(b)

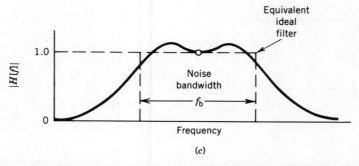

(c)

Figure 6.2-2 Definition of noise bandwidth: (*a*) magnitude of frequency response for ideal filter; (*b*) normalized magnitude of frequency response for practical lowpass filter; (*c*) normalized magnitude of frequency response for practical bandpass filter.

bandpass response. The following are defined:

E_i = input voltage to amplifier

E_o = normalized output voltage from amplifier

$|H| = |E_o/E_i|$ = normalized magnitude of frequency response

The normalized amplifier output power ΔP_o within a small frequency band Δf is related as follows to the input power ΔP_i in that band:

$$\frac{\Delta P_o}{\Delta P_i} = \left|\frac{E_o}{E_i}\right|^2 = |H|^2 \qquad (6.2\text{-}6)$$

Since $\Delta P_i = KT\Delta f$, the output power in the band Δf is

$$\Delta P_o = KT|H|^2\,\Delta f \qquad (6.2\text{-}7)$$

Hence the total output power is

$$P_o = KT\int_0^\infty |H[f]|^2\,df \qquad (6.2\text{-}8)$$

It is convenient to relate this power to the power from an ideal amplifier with a flat response. Setting Eq 6.2-8 equal to Eq 6.2-4 gives

$$\Delta f = \int_0^\infty |H[f]|^2\,df = f_b = \text{noise bandwidth} \qquad (6.2\text{-}9)$$

This represents the bandwidth of an ideal flat-response amplifier that has the same normalized output noise power as the practical amplifier. It is called the *noise bandwidth* f_b of the practical amplifier.

As shown in Fig 6.2-1b, the noise power generated in a resistor can be expressed in terms of a noise voltage of RMS value e_n in series with an ideal noiseless resistor. The RMS current flowing into the matched amplifier is

$$i = e_n/2R \qquad (6.2\text{-}10)$$

Hence, the power delivered to the amplifier is

$$P = i^2R = e_n^2/4R \qquad (6.2\text{-}11)$$

By Eq 6.2-4, this power P is set equal to KTf_b. Hence the noise voltage is

$$e_n = \sqrt{4KTRf_b} \qquad (6.2\text{-}12)$$

At room temperature (25°C) this is equal to

$$e_n = 0.128 \times 10^{-9}\sqrt{Rf_b} \ \ \text{V}/\sqrt{\Omega\text{-Hz}} \qquad (6.2\text{-}13)$$

This can be approximated within an error of 2% by

$$e_n \cong \tfrac{1}{8}\sqrt{Rf_b} \ \ \text{nV}/\sqrt{\Omega\text{-Hz}} \qquad (6.2\text{-}14)$$

As shown in Fig 6.2-1b, the noise can also be expressed in terms of a noise current of RMS value i_n in parallel with the resistor, where

$$i_n R = e_n \qquad (6.2\text{-}15)$$

The RMS noise current at room temperature is

$$i_n = \sqrt{(4KT/R)f_b} \cong \tfrac{1}{8}\sqrt{f_b/R}\ \ \text{nA-}\sqrt{\Omega/\text{Hz}} \qquad (6.2\text{-}16)$$

6.2.2 Noise-Bandwidth Examples

6.2.2.1 *Noise Bandwidths of Simple Lowpass Filters.* Table 6.2-1 shows the values of noise bandwidth in rad/sec, denoted ω_b, for some simple lowpass filters. Frequency-response plots for these filters are shown in Fig 6.2-3. A general method for calculating the noise bandwidth is presented in Section 6.3.4. The following is a derivation of the noise bandwidth of filter A. This is a single-order lowpass filter of break frequency $\omega_f = 2\pi f_f$, or time constant $\tau_f = 1/\omega_f$. Its transfer function in terms of angular frequency ω is

$$H[\omega] = \frac{1}{1 + j\omega/\omega_f} = \frac{\omega_f}{j\omega + \omega_f} \qquad (6.2\text{-}17)$$

The square of the magnitude of any transfer function $H[\omega]$ can be obtained from

$$|H[\omega]|^2 = H[\omega]H[\omega]^* = H[\omega]H[-\omega] \qquad (6.2\text{-}18)$$

where $H[\omega]^*$ is the complex conjugate of $H[\omega]$. Applying Eq 6.2-18 to Eq

TABLE 6.2-1 Noise Bandwidths of Simple Filters

Filter	Transfer Function $H[\omega]$	Noise Bandwidth ω_b (rad/sec)
A	$\dfrac{\omega_1}{s + \omega_1}$	$\dfrac{\pi}{2}\omega_1$
B	$\dfrac{\omega_1\omega_2}{(s + \omega_1)(s + \omega_2)}$	$\dfrac{(\pi/2)\,\omega_1\omega_2}{\omega_1 + \omega_2}$
C	$\dfrac{\omega_n^2}{s^2 + 2\zeta\omega_n s + \omega_n}$	$\dfrac{\pi}{2}\dfrac{\omega_n}{2\zeta}$

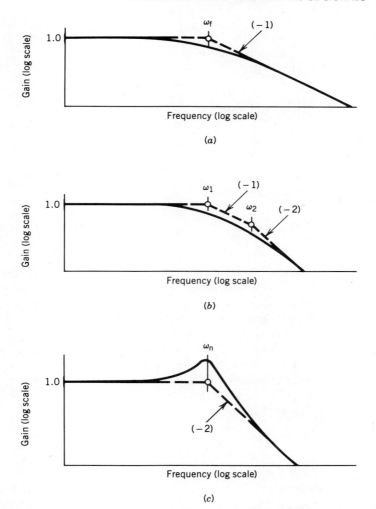

Figure 6.2-3 Frequency-response magnitude plots for filters considered in Table 6.2-1: (*a*) filter *A*; (*b*) filter *B*; (*c*) filter *C*.

6.2-17 gives for filter *A*

$$|H[\omega]|^2 = \frac{\omega_f}{(\omega_f + j\omega)}\frac{\omega_f}{(\omega_f - j\omega)} = \frac{\omega_f^2}{\omega_f^2 + \omega^2} \qquad (6.2\text{-}19)$$

From Eq 6.2-9, the noise bandwidth ω_b expressed in rad/sec is

$$\omega_b = \int_0^\infty |H[\omega]|^2 \, d\omega = \int_0^\infty \frac{\omega_f^2}{\omega_f^2 + \omega^2} \, d\omega \qquad (6.2\text{-}20)$$

Now,

$$\int_0^\infty \frac{d\omega}{\omega^2 + \omega_f^2} = \frac{1}{\omega_f} \arctan \frac{\omega}{\omega_f} \bigg|_{\omega=0}^\infty = \frac{1}{\omega_f} \frac{\pi}{2} \qquad (6.2\text{-}21)$$

Hence the noise bandwidth in rad/sec is

$$\omega_b = \omega_f^2 \frac{1}{\omega_f} \frac{\pi}{2} = \frac{\pi}{2} \omega_f \qquad (6.2\text{-}22)$$

The noise bandwidth in hertz is

$$f_b = \frac{\pi}{2} f_f = \frac{\pi}{2} \frac{\omega_f}{2\pi} = \frac{\omega_f}{4} = \frac{1}{4\tau_f} \qquad (6.2\text{-}23)$$

6.2.2.2   Noise Bandwidth of Ideal Averaging Integrator.

The next example is an ideal averaging integrator, which integrates the input signal over a fixed time interval T_i, to obtain the average of the input over that interval. Figure 6.2-4 shows its unit-impulse response, which has a constant value over the time interval T_i, and is zero outside that interval. Since the output is equal to the average of the input, the zero-frequency gain of the process must be unity. Hence the area under the impulse response is unity, and so the amplitude of the unit-impulse response is $1/T_i$.

To calculate the noise bandwidth of this filter, it is convenient to use the following general Fourier transform relation:

$$\int_{-\infty}^\infty h[t]^2 \, dt = \int_{-\infty}^\infty |H[f]|^2 \, df \qquad (6.2\text{-}24)$$

where $H[f]$ is the frequency response of the filter and $h[t]$ is the unit impulse

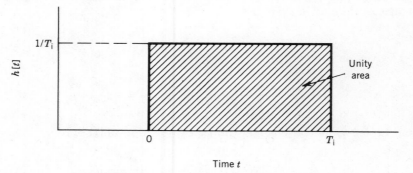

Figure 6.2-4   Impulse response of ideal integrator.

response. Since $|H[f]| = |H[-f]|$, the frequency-response integral from $-\infty$ to $+\infty$ is twice the integral from 0 to $+\infty$:

$$\int_{-\infty}^{\infty} |H[f]|^2 \, df = 2\int_{0}^{\infty} |H[f]|^2 \, df \qquad (6.2\text{-}25)$$

Applying Eqs 6.2-24, -25 to the definition of noise bandwidth in Eq 6.2-9 gives

$$f_b = \int_{0}^{\infty} |H[f]|^2 \, df = \frac{1}{2}\int_{-\infty}^{\infty} h[t]^2 \, dt \qquad (6.2\text{-}26)$$

Equation 6.2-26 allows the noise bandwidth to be calculated either in the frequency domain or in the time domain. For the ideal integrator, the time-domain calculation is simpler. Since the impulse response $h[t]$ is equal to $1/T_i$ over the integration interval T_i, and is zero elsewhere, the noise bandwidth is

$$f_b = \frac{1}{2}\int_{-\infty}^{\infty} h[t]^2 \, dt = \frac{1}{2}\int_{0}^{T_i}\left(\frac{1}{T_i}\right)^2 \, dt = \frac{1}{2T_i} \qquad (6.2\text{-}27)$$

6.2.2.3 *Noise Bandwidths of Simple Feedback Loops.*

The noise bandwidth of a feedback loop is defined in terms of its transfer function G_{ib}:

$$f_b = \int_{0}^{\infty} |G_{ib}|^2 \, df \qquad (6.2\text{-}28)$$

Table 6.2-2 shows the noise-bandwidth values for three useful loop transfer functions. Asymptotic frequency-response plots of loop gain are shown in Fig 6.2-5.

For loop D, the noise bandwidth, expressed in terms of the parameters of the loop transfer function, is independent of the value of the break frequency

TABLE 6.2-2   Noise Bandwidths of Feedback Loops in Fig 6.2-5

Loop	Transfer Function G	G_{ib}	Noise Bandwidth ω_b (rad/sec)
D	$\dfrac{\omega_c}{s(1 + s/\omega_f)}$	$\dfrac{\omega_c \omega_f}{s^2 + \omega_f s + \omega_c \omega_f}$	$\dfrac{\pi}{2}\omega_c$
E	$\dfrac{\omega_c(1 + \omega_i/s)}{s}$	$\dfrac{\omega_c(s + \omega_i)}{s^2 + \omega_c s + \omega_c \omega_i}$	$\dfrac{\pi}{2}(\omega_c + \omega_i)$
F	$\dfrac{\omega_c(1 + \omega_i/s)}{s(1 + s/\omega_f)}$	$\dfrac{\omega_c \omega_f(s + \omega_i)}{s^3 + \omega_f s^2 + \omega_c \omega_f s + \omega_c \omega_f \omega_i}$	$\dfrac{\pi\omega_f(\omega_c + \omega_i)}{2(\omega_f - \omega_i)}$

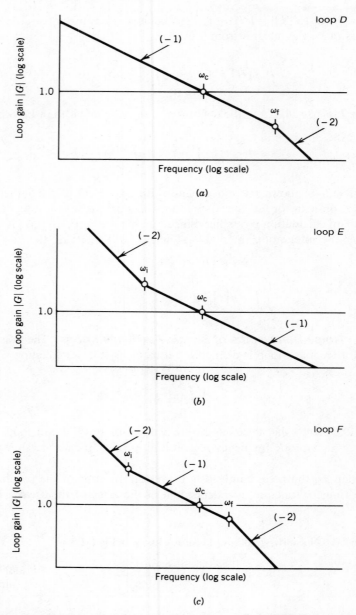

Figure 6.2-5 Magnitude asymptote plots of loop transfer functions considered in Table 6.2-2: (*a*) loop *D*; (*b*) loop *E*; (*c*) loop *F*.

ω_f. The corresponding G_{ib} transfer function of this loop is the same as the transfer function for filters A, B, and C shown in Table 6.2-1. If ω_f is infinite, G_{ib} is the same as $H[\omega]$ for filter A; if $4\omega_c < \omega_f < \infty$, G_{ib} has two real poles and is the same as H for filter B; and if $\omega_f < 4\omega_c$, G_{ib} is underdamped and the same as H for filter C.

6.2.3 Shot (or Schottky) Noise

6.2.3.1 *Coin-Flipping Probability Experiment.* Another major type of noise is shot (or Schottky) noise, which is caused by the discrete nature of electron flow in an electric current. The statistical characteristics of this noise can be derived from simple probability considerations. Let us consider a coin-flipping experiment.

If one flips a coin $2N$ times, on the average there are N heads, but the actual number of heads deviates from N in a random manner. To a good approximation, the RMS value of this deviation, σN, is as follows for reasonably large values of N:

$$\sigma N = \sqrt{N} \qquad (6.2\text{-}29)$$

Assume that the coin is flipped 200 times; the average number of heads is $N = 100$. The RMS deviation (or standard deviation) from this average is $\sqrt{N} = \sqrt{100} = 10$.

The probability of achieving a specific number of heads is described approximately by the Gaussian, or normal, distribution. The derivative of that distribution is called the Gaussian (or normal) probability density function, which is defined as

$$\frac{dP_g[x]}{dx} = \frac{1}{\sqrt{2\pi}} e^{-x^2/2\sigma^2} \qquad (6.2\text{-}30)$$

where σ is the RMS value (or standard deviation), and x is the deviation from the mean value. The normalized deviation x/σ is denoted z. A plot of the Gaussian probability density function $dP_g[z]/dz$ is shown in Fig 6.2-6, versus the normalized deviation z:

$$z = \frac{x}{\sigma}. \qquad (6.2\text{-}31)$$

The total area under the curve is unity. The probability that the number of heads falls within a given range of x is the area under the curve over the corresponding normalized range of x/σ. When normalized in terms of the variable z, the Gaussian probability density is

$$\frac{dP_g[z]}{dz} = \frac{1}{\sqrt{2\pi}} e^{-z^2/2} \qquad (6.2\text{-}32)$$

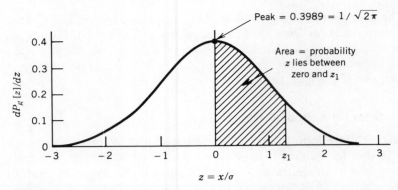

Figure 6.2-6 Plot of normal (or Gaussian) probability density function.

This Gaussian probability density $dP_g[z]/dz$ cannot be integrated analytically, and so its integral, the probability distribution, is commonly expressed in tabular form. Values of this integral, obtained from Feller [6.1], are shown in Table 6.2-3, where the limits of the integral are from zero to z. We denote this integral from zero to z as $P_g[z]$. The integral of the Gaussian density function from $-\infty$ to z is commonly denoted $\Phi[z]$. This is related as follows to the function $P_g[z]$ given in Table 6.2-3:

$$\Phi[z] = 0.5 + \text{sgn}[z]\, P_g[|z|] \qquad (6.2\text{-}33)$$

where $\text{sgn}[z]$ is the sign of z, which is $+1$ for $z > 0$, and -1 for $z < 0$.

TABLE 6.2-3 **Values of Gaussian Probability Integral from 0 to z, Denoted $P_g[z]$, where $z = x/\sigma$**

$z = x/\sigma$	$P_g[z]$	$z = x/\sigma$	$P_g[z]$	$z = x/\sigma$	$P_g[z]$
0.0	0.000 000	1.5	0.433 193	3.0	0.498 650
0.1	0.039 828	1.6	0.445 201	3.1	0.499 032
0.2	0.079 260	1.7	0.455 439	3.2	0.499 313
0.3	0.117 911	1.8	0.464 070	3.3	0.499 517
0.4	0.155 422	1.9	0.471 283	3.4	0.499 663
0.5	0.191 462	2.0	0.477 250	3.5	0.499 767
0.6	0.225 747	2.1	0.482 136	3.6	0.499 841
0.7	0.258 036	2.2	0.486 097	3.7	0.499 892
0.8	0.288 145	2.3	0.489 276	3.8	0.499 928
0.9	0.315 940	2.4	0.491 802	3.9	0.499 952
1.0	0.341 345	2.5	0.493 790	4.0	0.499 968
1.1	0.364 334	2.6	0.495 339	4.1	0.499 979
1.2	0.384 930	2.7	0.496 533	4.2	0.499 987
1.3	0.403 200	2.8	0.497 445	4.3	0.499 991
1.4	0.419 243	2.9	0.498 134	4.4	0.499 995

For large values of z, it is more convenient to consider the integral from z to infinity, which is equal to $(0.5 - P_g[z])$. This integral can be approximated by

$$\left(0.5 - P_g[z]\right) \cong \frac{1}{\sqrt{2\pi}\, z} e^{-z^2/2} \qquad \text{(for large } z) \qquad (6.2\text{-}34)$$

As shown by Feller [6.1], the exact value of $(0.5 - P_g[z])$ is less than this approximation, but is greater than the approximation multiplied by $(1 - 1/z^2)$. To solve for z as a function of $P_g[z]$, Eq 6.2-34 can be expressed as

$$z \cong \sqrt{-2\ln\left[\sqrt{2\pi}\, z\left(0.5 - P_g[z]\right)\right]} \qquad \text{(for large } z) \qquad (6.2\text{-}35)$$

This expression is a weak function of the value of z on the right-hand side, and so can be solved iteratively by guessing the value of z on the right-hand side of the equation, and using the calculated value of z to update the guess. For example, assume that $(0.5 - P_g[z])$ is 10^{-6}, and guess that $z = 3$ (a poor guess). The first calculation is

$$z = \sqrt{-2\ln\left[\sqrt{2\pi}\,(3)(10^{-6})\right]} = 4.86 \qquad (6.2\text{-}36)$$

The value $z = 4.86$ is used as the next guess. The iteration proceeds as follows:

$$
\begin{array}{llll}
\text{guess} & z = 4.86; & \text{calculate} & z = 4.757 \\
\text{guess} & z = 4.757; & \text{calculate} & z = 4.7617 \\
\text{guess} & z = 4.7617; & \text{calculate} & z = 4.7615 \qquad (6.2\text{-}37)
\end{array}
$$

Let us apply Table 6.2-3 to the coin-flipping example. What is the probability that the number of heads lies between 80 and 115? The mean value is 100, and so the deviation from the mean is

$$x = -20 \text{ to } +15 \qquad (6.2\text{-}38)$$

Since the RMS deviation from the mean (the standard deviation) σ is 10, the normalized deviation x/σ from the mean is

$$z = \frac{x}{\sigma} = -2.0 \text{ to } +1.5 \qquad (6.2\text{-}39)$$

The integral from $z = 0$ to $z = 1.5$ is obtained by reading from Table 6.2-3 the value for $z = 1.5$, which is

$$P_g[z] = P_g[1.5] = 0.4332 \qquad (6.2\text{-}40)$$

Since the normal probability curve is symmetric about $z = 0$, the integral from $z = -2.0$ to 0 is the same as the integral from $z = 0$ to $+2.0$. This probability

TABLE 6.2-4 Approximate Probability that the Number of Heads Lies within a Particular Range

Range	Probability
100 to 115	43.32%
80 to 100	47.73%
80 to 115	91.05%
Less than 80	50% − 47.73% = 2.27%
Greater than 115	50% − 43.32% = 6.68%

is read from Table 6.2-3 at $z = 2.0$, which is

$$P_g[-2.0] = P_g[2.0] = 0.4773 \tag{6.2-41}$$

The probability that the number of heads lies between 80 and 115 is the sum of Eqs 6.2-40, -41, which is 0.9105, or 91.05%. From these data, the probabilities shown in Table 6.2-4 can be obtained. These results are only approximate, because the distribution of number of heads is quantized, and so is only approximately Gaussian. On the other hand, the greater the average number of heads N, the more closely the distribution approximates a Gaussian distribution.

6.2.3.2 Derivation of Shot-Noise Formula. Let us relate this to the noise associated with the flow of electrons, which is called shot noise. A constant current I has the following average rate of electron flow:

$$\langle \dot{N} \rangle = \frac{I}{q} \tag{6.2-42}$$

The dot (·) above a variable represents differentiation, and the angular brackets $\langle \ \rangle$ around a variable represent average value. The parameter q is the charge on an electron. If this current is fed into an ideal averaging integrator of integration time T_i, the average number of electrons detected by the integrator is

$$\langle N \rangle = \langle \dot{N} \rangle T_i \tag{6.2-43}$$

The actual number that is detected deviates in a random manner from this average value. The RMS value of this deviation is

$$\sigma N = \sqrt{\langle N \rangle} = \sqrt{\langle \dot{N} \rangle T_i} = \sqrt{\left(\frac{I}{q}\right) T_i} \tag{6.2-44}$$

This deviation of electron count can be interpreted as a variation of the rate of flow of electrons. The equivalent RMS variation of the electron flow rate is

$$\sigma \dot{N} = \frac{\sigma N}{T_i} = \sqrt{\frac{I}{qT_i}} \tag{6.2-45}$$

The apparent RMS variation of the electric current is then

$$\sigma i = q \, \sigma \dot{N} = \sqrt{\frac{qI}{T_i}} \tag{6.2-46}$$

From Eq 6.2-27, the noise bandwidth f_b of the averaging integrator is $1/2T_i$. Replacing $1/T_i$ by $2f_b$ in Eq 6.2-46 gives the basic equation for shot noise, which expresses the RMS noise current in terms of the noise bandwidth of the integrator:

shot noise: $$\sigma i = \sqrt{2qIf_b} \tag{6.2-47}$$

Although this shot-noise formula was derived for an ideal integrator, it applies to any kind of filter. The value of the charge on an electron is

$$q = 1.59 \times 10^{-19} \, \text{coulomb} \tag{6.2-48}$$

Substituting this into Eq 6.2-47 gives the following quantitative formula for shot noise:

$$\sigma i = 0.566\sqrt{I_{(\mu A)}f_b} \;\; \text{pA}/\sqrt{\mu\text{A-Hz}} \tag{6.2-49}$$

where $I_{(\mu A)}$ represents the average current in microamperes (μA).

A noisy signal displayed on an oscilloscope exhibits fluctuations about an average value. It is convenient to characterize the magnitude of the fluctuations in terms of a "peak-to-peak" value for the deviation of the noise. However, what one calls "peak-to-peak" for a noise signal depends on the criterion that one sets to define peak-to-peak limits.

Assume that one chooses limits that are exceeded 1% of the time (the signal is within these limits 99% of the time). Table 6.2-3 shows that this corresponds to the range $\pm 2.6\sigma$. Hence for this 99% criterion, the peak-to-peak deviation is 5.2 times the RMS value σ. Alternatively, one might choose limits that include 95% of the signal. Since this corresponds to $\pm 2.0\sigma$, the peak-to-peak value for this criterion is 4 times the RMS value σ. The peak-to-peak value that one measures from observing a noisy signal is usually between 4 and 6 times the RMS value, depending on what criterion the observer selects in making his judgment.

This point is illustrated in Fig 6.2-7, which shows a photograph of a noisy waveform, compared with the corresponding plot of the Gaussian amplitude

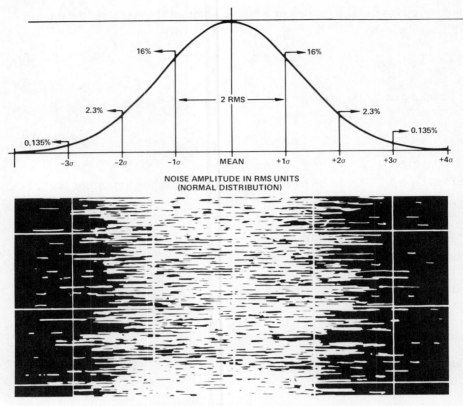

Figure 6.2-7   Gaussian amplitude distribution compared with random noise waveform. Reprinted from Ref [6.2] with permission from Analog Devices.

distribution. (This is obtained from an excellent practical discussion of noise in opamp circuits presented by Ryan and Scranton [6.2].)

6.2.4 Other Types of Noise

Thermal noise and shot noise have a flat noise spectrum at the point where they are generated, and so are called "white" noise. Besides thermal noise and shot noise, amplifiers also exhibit "flicker" noise and sometimes "popcorn" noise, which are particularly important at low frequencies.

Flicker (or $1/f$) noise has a spectrum that decreases with increasing frequency. The power density (watt/Hz) is inversely proportional to frequency, and so the noise voltage density (volt/$\sqrt{\text{Hz}}$) is inversely proportional to the square root of frequency. Flicker noise in semiconductors is due to random fluctuations in the number of surface recombinations. The contact between granules in carbon resistors also generates flicker noise as well as shot noise. For this reason, low-noise wirewound or metal-film resistors are used instead of carbon resistors in low-noise circuits. Note that a "low-noise" resistor generates just as much thermal noise as a carbon resistor.

Popcorn noise is characterized by random jumps between two or more levels. In semiconductors, it is due to random on–off recombination action in the semiconductor material, leading to erratic switching of the gain of the device. This noise has the sound of popcorn when displayed audibly. Figure 6.2-8 shows typical waveforms produced by white noise, flicker noise, and popcorn noise, which was presented by Ryan and Scranton [6.2].

6.2.5 Application to Error Components of a Control System

Suppose there are a number of independent noise sources contributing to the noise in a particular voltage E. Since the noise sources are independent, the noise powers add. Noise power is proportional to the square of the voltage, and so the total RMS voltage e is related as follows to the separate noise-voltage components e_1, e_2:

$$e^2 = e_1^2 + e_2^2 \qquad (6.2\text{-}50)$$

$$e = \sqrt{e_1^2 + e_2^2} \qquad (6.2\text{-}51)$$

This principle can be applied to any number of noise components. The combined RMS noise voltage is the RSS (root-sum-squared) combination of the individual noise-voltage components.

The noise contribution in a controlled variable is generally only one of many error components, produced by effects that are independent of one another. These can include errors due to static friction, tachometer imperfections, encoder quantization and inaccuracy, wind torque, etc. It is convenient to treat all of these error components as if they were noise components. The times at which they occur are independent of one another, and so they add in a random fashion. Hence, the combined error is the square root of the sum of the squares of the individual error components.

Many error components are characterized by peak values rather than RMS values. How does one combine the error of a position sensor, which is usually characterized by its peak value, with error components due to thermal and shot noise, which have Gaussian amplitude distributions? By convention, peak value is generally considered to be equivalent to an error of $\pm 3\sigma$. Hence $\frac{1}{3}$ of the peak error of the position sensor is considered to be its effective RMS value σ (or *sigma*).

As was shown in Fig 6.2-7, the value that one regards as the peak (or peak-to-peak) noise is somewhat arbitrary. Nevertheless, limits of ± 3 sigma (σ) are commonly used to define the peak value of noise for the following

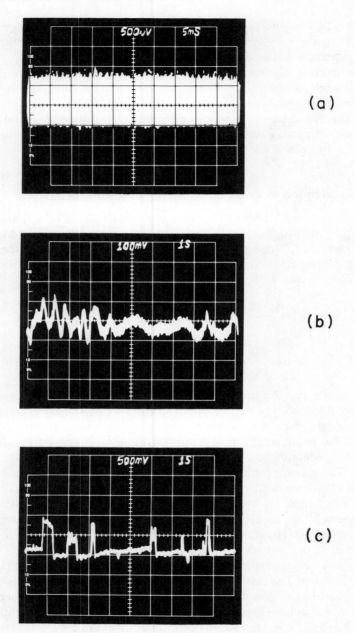

Figure 6.2-8   Noise signatures: (*a*) white noise; (*b*) flicker (or $1/f$) noise; (*c*) popcorn noise. Reprinted from Ref [6.2] with permission from Analog Devices.

reasons:

1. The probability of exceeding ± 3 sigma is only 0.37%.
2. The probability of exceeding ± 3 sigma by 10% (i.e., the probability of exceeding ± 3.3 sigma) is only 0.1%.
3. The probability of exceeding ± 3 sigma by 30% (i.e., the probability of exceeding ± 3.9 sigma) is only 0.01%.

Thus the limits of ± 3 sigma are not exceeded very often, and it is rare that these limits are exceeded by an appreciable factor. As a result, ± 3 sigma has been found to be a good practical definition of peak value in many engineering applications.

6.2.6 Relation between Rise Time and Noise Bandwidth

Chapter 2 showed that the rise time T_r of a lowpass filter is approximately related as follows to the half-power frequency ω_{hp}:

$$T_R \cong \frac{2}{\omega_{hp}} = \frac{2}{2\pi f_{hp}} = \frac{1}{\pi f_{hp}} \qquad (6.2\text{-}52)$$

where f_{hp} is the half-power frequency in hertz. Table 6.2-1 shows that for a single-order lowpass filter (A) the noise bandwidth f_b is related as follows to the half-power frequency f_{hp}:

$$f_b = \frac{\pi}{2} f_{hp} \qquad (6.2\text{-}53)$$

This same relation holds approximately for all lowpass filters that are reasonably well damped. Hence Eqs 6.2-52, -53 can be combined to give the following general approximation relating the rise time of a lowpass filter to its

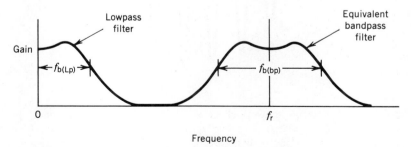

Figure 6.2-9 Relation between frequency-response magnitude plots of equivalent lowpass and bandpass filters.

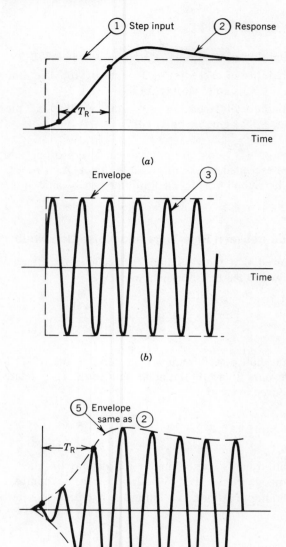

Figure 6.2-10 Waveforms showing effect of a bandpass filter on the envelope of a pulse-modulated carrier signal: (*a*) step response of equivalent lowpass filter; (*b*) step-modulated carrier input to bandpass filter; (*c*) response of bandpass filter to step-modulated carrier.

noise bandwidth:

$$T_R \cong \frac{1}{2f_b} \qquad (6.2\text{-}54)$$

As was shown in Chapter 4, the bandpass frequency response that is equivalent to a given lowpass frequency response is obtained by shifting the frequency response of the lowpass filter (including its response at negative frequencies) upward in frequency by the carrier (or reference) frequency f_r. This is illustrated in Fig 6.2-9. The figure shows that the noise bandwidth of the bandpass filter $f_{b(bp)}$ is twice the noise bandwidth of the equivalent lowpass filter $f_{b(Lp)}$:

$$f_{b(bp)} = 2f_{b(Lp)} \qquad (6.2\text{-}55)$$

Figure 6.2-10 compares the transient responses of equivalent lowpass and bandpass filters. Diagram a shows the step response of the lowpass filter. Diagram b shows a step-modulated carrier signal of frequency f_r that is fed into the equivalent bandpass filter, and diagram c shows the resultant response. The envelope of the response of the bandpass filter in diagram c is the same as the step response in diagram a of the equivalent lowpass filter. Hence, the rise time T_R of the step response of the lowpass filter is the same as that of the envelope of the response of the equivalent bandpass filter to a step-modulated carrier signal. This rise time is related approximately as follows to the noise bandwidth of the bandpass filter:

$$T_R \cong \frac{1}{f_{b(bp)}} \qquad (6.2\text{-}56)$$

This is obtained by combining Eqs 6.2-54, -55.

6.3 BASIC MATHEMATICAL TOOLS FOR STATISTICAL ANALYSIS OF SIGNALS

This section presents a summary of basic mathematical tools used in the statistical analysis of signals. Detailed derivations of these concepts are given by Truxal [6.3] (Chapters 7, 8), by Phillips [6.4], and by Newton, Gould, and Kaiser [6.5].

6.3.1 The Convolution Integral

The convolution integral is an equation for calculating the response of a system in the time domain. It is not a statistical analysis tool per se, but is very important in statistical analysis. It can be derived by physical reasoning in the following manner.

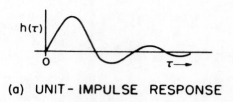

(a) UNIT - IMPULSE RESPONSE

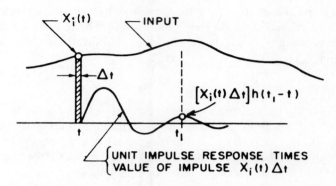

(b) CALCULATION OF RESPONSE TO INPUT

Figure 6.3-1   Waveforms for deriving the convolution integral. Reprinted with permission from Ref [6.6] © 1957 IRE (now IEEE).

As shown in Fig 6.3-1, assume that an input signal $x_i[t]$ is composed of a large number of narrow pulses of width Δt and magnitude $x_i[t]$. The time interval Δt is made sufficiently small, relative to the response time of the system, for each pulse to have the effect of an impulse of the same area $(x_i[t]\,\Delta t)$. Diagram a shows the unit-impulse response of the system, which is denoted $h[t]$. Diagram b shows the input $x_i[t]$ fed into the system, and one of the pulses into which $x_i[t]$ is divided. The response of the pulse is obtained by multiplying the unit impulse response $h[t]$ by $x_i[t]\,\Delta t$ to obtain the response curve shown in diagram b. At time t_1 the response to this pulse has the value

$$(x_i[t]\,\Delta t)h[t_1 - t] \tag{6.3-1}$$

The total response $x_o[t_1]$ to the input $x_t[t]$ at time t_1 is obtained by adding the responses for all of the pulses into which $x_i[t]$ is divided:

$$x_o[t_1] = \sum (x_i[t]\,\Delta t)h[t_1 - t] \tag{6.3-2}$$

This summation includes all of the pulses between $t = -\infty$ and $t = t_1$.

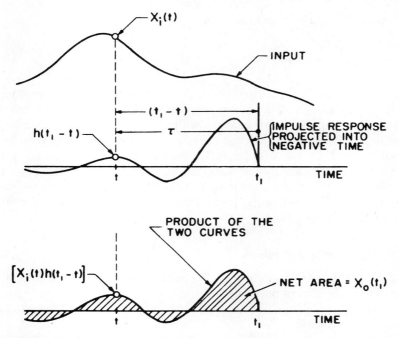

Figure 6.3-2   Alternative interpretation of convolution process. Reprinted with permission from Ref [6.6] © 1957 IRE (now IEEE).

Passing to the limit converts the summation to an integration:

$$x_o[t_1] = \int_{-\infty}^{t_1} dt \, x_i[h] h[t_1 - t] \qquad (6.3\text{-}3)$$

The convolution process is illustrated in Fig 6.3-2 in a different manner. The time t_1 at which the output is being calculated can be considered to represent the time of observation. The unit-impulse response $h[\tau]$ is shown projected from the time of observation t_1 in the direction of past (or negative) time, and hence represents $h[t_1 - t]$. Multiplying the input curve $x_i[t]$, point by point, by the impulse response $h[t_1 - t]$ gives the product curve $x_i[t]h[t_1 - t]$. By Eq 6.3-3, the response x_o at present time t_1 is the integral of this product curve from $-\infty$ up to the present time t_1, and hence represents the net area under the product curve.

As shown in Fig 6.3-2, the time interval $(t_1 - t)$ is denoted τ, which represents elapsed time: a time variable projected from present time t_1 into the past. To express the convolution integral in terms of τ, note that $d\tau$ is equal to $-dt$. Changing the variable from t to τ in Eq 6.3-3 gives

$$x_o[t_1] = \int_{t_1 - \tau = -\infty}^{t_1 - \tau = t_1} (-d\tau) x_i[t_1 - \tau] h[\tau] \qquad (6.3\text{-}4)$$

Expressing the limits in terms of τ gives

$$x_o[t_1] = \int_{\tau=0}^{\infty} d\tau\, h[\tau] x_i[t_1 - \tau] \qquad (6.3\text{-}5)$$

This is a more convenient form of the convolution integral. The construction of Fig 6.3-2 shows why the unit-impulse response $h[\tau]$ is termed a *weighting function*. The impulse response $h[\tau]$ multiplies, or "weights," the past values of the input. The impulse response $h[\tau]$ decays to zero at infinite τ, and hence the values of the input $x_i[t]$ far in the past have little effect on the present output $x_o[t_1]$. The more recent values of the input are weighted much more than long-past values. Since the impulse response is zero for negative values of time, the part of the input that has not yet occurred cannot affect the present value of the output.

6.3.2 Correlation Functions

6.3.2.1 *Definition.* The correlation functions are fundamental tools for measuring the statistical properties of signals. Consider a system with an input $x_i[t]$ and an output $x_o[t]$. There are four correlation functions pertaining to these signals, which are

$$\phi_{ii}[\tau] = \text{autocorrelation function of input } x_i[t]$$
$$\phi_{oo}[\tau] = \text{autocorrelation function of output } x_o[t]$$
$$\phi_{io}[\tau], \phi_{oi}[\tau] = \text{cross-correlation functions between input } x_i[t] \text{ and output } x_o[t]$$

These correlation functions are defined by

$$\phi_{ii}[\tau] = \lim_{T \to \infty} \frac{1}{2T} \int_{-T}^{+T} dt\, x_i[t] x_i[t + \tau] \qquad (6.3\text{-}6)$$

$$\phi_{oo}[\tau] = \lim_{T \to \infty} \frac{1}{2T} \int_{-T}^{+T} dt\, x_o[t] x_o[t + \tau] \qquad (6.3\text{-}7)$$

$$\phi_{io}[\tau] = \lim_{T \to \infty} \frac{1}{2T} \int_{-T}^{+T} dt\, x_i[t] x_o[t + \tau] \qquad (6.3\text{-}8)$$

$$\phi_{oi}[\tau] = \lim_{T \to \infty} \frac{1}{2T} \int_{-T}^{+T} dt\, x_o[t] x_i[t + \tau] \qquad (6.3\text{-}9)$$

Some important characteristics of the correlation functions can be obtained by inspection. The autocorrelation functions have the same value if τ is replaced by $-\tau$. Hence

$$\phi_{ii}[\tau] = \phi_{ii}[-\tau] \qquad (6.3\text{-}10)$$

and similarly for $\phi_{oo}[\tau]$. Thus, an autocorrelation function is an even function of τ, but the cross-correlation functions are generally not. In Eq 6.3-8, if the variable τ in $\phi_{io}[\tau]$ is replaced by $-\tau$, the relation is the same as that for $\phi_{oi}[\tau]$ of Eq 6.3-9. Thus,

$$\phi_{io}[\tau] = \phi_{oi}[-\tau] \tag{6.3-11}$$

The value of the autocorrelation function $\phi_{ii}[\tau]$ of Eq 6.3-6 at $\tau = 0$ is

$$\phi_{ii}[0] = \lim_{T \to \infty} \frac{1}{2T} \int_{-T}^{+T} dt \, x_i[t]^2 \tag{6.3-12}$$

By definition, this is the average of x_i^2, which is the mean square value of the input, denoted $\langle x_i^2 \rangle$. Thus

$$\phi_{ii}[0] = \langle x_i^2 \rangle \tag{6.3-13}$$

6.3.2.2 *Calculation of Correlation Functions.* The process of computing an autocorrelation function is illustrated in Fig 6.3-3. Diagram a shows a plot of an input function $x_i[t]$ and the function $x_i[t + \tau_1]$, which is the same plot shifted to the right by the time shift τ_1. These two curves, $x_i[t]$ and $x_i[t + \tau_1]$, are multiplied together, point by point, to obtain the product curve $x_i[t]x_i[t + \tau_1]$ shown in diagram b. The average value of this product curve is the autocorrelation function value $\phi_{ii}[\tau_1]$. Thus, $\phi_{ii}[\tau_1]$ is

$$\phi_{ii}[\tau_1] = \langle x_i[t]x_i[t + \tau_1] \rangle$$
$$= \lim_{t_2 - t_1 \to \infty} \frac{1}{t_2 - t_1} \int_{t_1}^{t_2} dt \, x_i[t]x_i[t + \tau_1] \tag{6.3-14}$$

In practice the averaging is performed over a finite interval of time $(t_2 - t_1)$. However, if this time interval is much greater than the maximum time shift τ_1, the value $\phi_{ii}[\tau_1]$ is approximated with high accuracy. The expression for $\phi_{ii}[\tau_1]$ given in Eq 6.3-14 is equivalent to the more common form given in Eq 6.3-6.

The averaging process of the product curve described in diagram b gives one point $\phi_{ii}[\tau_1]$ on the convolution curve, as shown in diagram c. If the interval $(t_2 - t_1)$ can be considered infinite, it makes no difference in diagram a if the second curve is shifted in the positive or negative direction by the amount τ_1. In either case there are two identical curves shifted from one another by the amount τ_1. Thus, the autocorrelation function $\phi_{ii}[\tau]$ must have the same value at $-\tau$ that it has at $+\tau$, and so $\phi_{ii}[\tau]$ is symmetric about the zero-τ axis.

If the input $x_i[t]$ does not have any periodic components, the variations in $x_i[t]$ are uncorrelated for large time shifts. Consequently, as τ approaches infinity the autocorrelation function $\phi_{ii}[\tau]$ approaches a constant equal to the square of the average value of the input x_i, which is denoted $\langle x_i \rangle^2$.

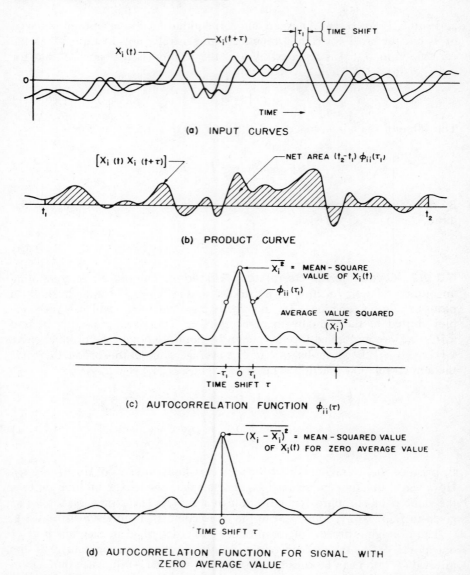

Figure 6.3-3 Description of autocorrelation process. Reprinted with permission from Ref [6.6] © 1957 IRE (now IEEE).

The final value of $\phi_{ii}[\tau]$ is almost always of no interest. Therefore it is generally removed, either by subtracting from the original input curve $x_i[t]$ its average value $\langle x_i \rangle$, or by subtracting from the autocorrelation function the square of the average value of the input, $\langle x_i \rangle^2$. Consequently, the autocorrelation function that is actually used has the form of Fig 6.3-3d, which has a final value of zero.

The cross-correlation function $\phi_{io}[\tau]$ given in Eq 6.3-8 is calculated in the same manner as the autocorrelation function, except that the input $x_i[t]$ is multiplied by the shifted output $x_o[t + \tau]$, and the average of the resultant product is calculated. To calculate the other cross-correlation function, $\phi_{oi}[\tau]$, the output curve $x_o[t]$ is held fixed, and the input $x_i[t]$ is shifted forward to obtain $x_i[t + \tau]$. The average of the resultant product curve $x_o[t]x_i[t + \tau]$ is taken to obtain the value of the cross-correlation function.

The same value results by shifting the output curve forward in time by τ_1 or by shifting the input curve backward in time by τ_1. Hence $\phi_{io}[\tau]$ is equal to $\phi_{oi}[-\tau]$, as shown in Eq 6.3-6. On the other hand, the sense of the relative shift between the two curves usually makes a difference. Therefore, $\phi_{io}[\tau]$ is generally not equal to $\phi_{io}[-\tau]$.

In calculating cross-correlation functions, the average values of the input $x_i[t]$ and output $x_o[t]$ are generally removed by subtraction. Consequently, the final values of the cross-correlation functions are generally zero.

6.3.2.3 Transient Approach for Relating Correlation Functions.

Figure 6.3-4 shows that the correlation functions can be related to one another in a simple fashion by regarding them as transient input and output signals of the system. (See Ref [6.6].) The input autocorrelation function $\phi_{ii}[\tau]$ is assumed to be a transient function of time, which is fed into the system $H[\omega]$. As shown, the resultant transient output from the system is the cross-correlation function $\phi_{io}[\tau]$. From this, the other cross-correlation function $\phi_{oi}[\tau]$ is formed by rotating the $\phi_{io}[\tau]$ transient about the zero-time axis. This $\phi_{oi}[\tau]$ transient is fed into the system, and the resultant system output is the output autocorrelation function $\phi_{oo}[\tau]$.

The relations illustrated in Fig 6.3-4 can be readily derived by using the convolution integral along with the definitions of the correlation functions. The following is a derivation of the transient relationship shown in diagram a. The convolution integral of Eq 6.3-5 gives the following for the output response $y[t]$ of the system to the transient input $\phi_{ii}[t]$:

$$y[t] = \int_0^\infty d\tau\, h[\tau]\phi_{ii}[t - \tau]$$

$$= \int_0^\infty d\tau\, h[\tau]\left\{ \lim_{T \to \infty} \frac{1}{2T} \int_{-T}^{T} dt_1\, x_i[t_1]x_i[t_1 + (t - \tau)] \right\} \quad (6.3\text{-}15)$$

where $h[\tau]$ is the impulse response of the system, and the function within the braces { } is the expression for $\phi_{ii}[t - \tau]$ obtained from Eq 6.3-6. Note that t_1 is the time variable of the autocorrelation integral. The order of integration of Eq 6.3-15 can be reversed, which gives

$$y[t] = \lim_{T \to \infty} \frac{1}{2T} \int_{-T}^{T} dt_1\, x_i[t_1]\left\{ \int_0^\infty d\tau\, h[\tau]x_i[(t_1 + t) - \tau] \right\} \quad (6.3\text{-}16)$$

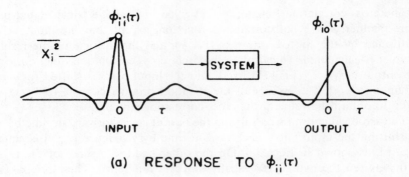

(a) RESPONSE TO $\phi_{ii}(\tau)$

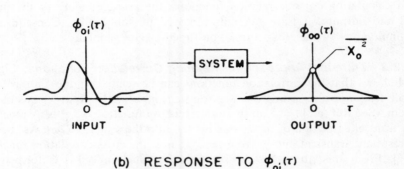

(b) RESPONSE TO $\phi_{oi}(\tau)$

Figure 6.3-4 Transient response of system to correlation-function input. Reprinted with permission from Ref [6.6] © 1957 IRE (now IEEE).

The expression within the braces is a convolution of the input $x_i[t_1 + t]$ with the system impulse response $h[\tau]$, and so represents the output $x_o[t_1 + t]$. Thus, the convolution integral of Eq 6.3-5 gives

$$x_0[t_1 + t] = \int_0^\infty d\tau\, h[\tau] x_i\big[(t_1 + t) - \tau\big] \tag{6.3-17}$$

Substituting Eq 6.3-17 into Eq 6.3-16 gives

$$y[t] = \lim_{T \to \infty} \frac{1}{2T} \int_{-T}^{T} dt_1\, x_i[t_1] x_o[t_1 + t] = \phi_{io}[t] \tag{6.3-18}$$

As shown, this represents $\phi_{io}[t]$, in accordance with the definition of the cross-correlation function in Eq 6.3-8.

6.3.3 Spectral Densities

Since a correlation function is not zero at negative values of time, it does not have a Laplace transform. However, if the average values of the time functions are subtracted, so that the final values of the correlation functions are zero, the correlation functions do have Fourier transforms. The Fourier transform of the autocorrelation function is called the *spectral density*, or sometimes the *power spectrum*. It is denoted $\Phi_{ii}[\omega]$ and $\Phi_{oo}[\omega]$ for the input and output signals. Applying the basic Fourier-transform equations yields

$$\Phi_{ii}[\omega] = \int_{-\infty}^{\infty} \phi_{ii}[\tau] e^{-j\omega\tau} \, d\tau \tag{6.3-19}$$

$$\phi_{ii}[\tau] = \frac{1}{2\pi} \int_{-\infty}^{\infty} \Phi_{ii}[\omega] e^{j\omega\tau} \, d\omega \tag{6.3-20}$$

Since an autocorrelation function is an even function of time (e.g., $\phi_{ii}[\tau]$ is equal to $\phi_{ii}[-\tau]$), the spectral densities $\Phi_{ii}[\omega]$ and $\Phi_{oo}[\omega]$ have zero phase. The Fourier transforms of the cross-correlation functions $\phi_{io}[\tau]$ and $\phi_{oi}[\tau]$ are denoted $\Phi_{io}[\omega]$ and $\Phi_{oi}[\omega]$. Unlike the spectral densities, these transforms generally have a phase response.

The spectral density can also be calculated by transforming the time function directly, rather than by transforming its autocorrelation function. Denote the Fourier transform of the time function $x_i[t]$ measured over the time interval $-T < t < +T$ as $X_{i(T)}[\omega]$. The spectral density of $x_i[t]$ can be shown to be

$$\Phi_{ii}[\omega] = \lim_{T \to \infty} \frac{1}{2T} X_{i(T)}[\omega] X_{i(T)}[\omega]^{*}$$

$$= \lim_{T \to \infty} \frac{1}{2T} \left| X_{i(T)}[\omega] \right|^{2} \tag{6.3-21}$$

The spectral density can be measured directly from the signal by means of a narrow-bandwidth variable-frequency filter. If the input signal $x_i[t]$ is passed through a narrow-bandwidth filter centered at the frequency ω_1, the average value of the output from the filter, divided by the bandwidth of the filter, approaches $\Phi_{ii}[\omega_1]$ as the bandwidth approaches zero.

As shown in Fig 6.3-4, the time response of the system to the transient input $\phi_{ii}[t]$ is the function $\phi_{io}[t]$. Hence, the Fourier transforms of these correlation functions are related by

$$\Phi_{io}[\omega] = H[\omega] \Phi_{ii}[\omega] \tag{6.3-22}$$

where $H[\omega]$ is the frequency response of the system. Similarly, the time response of the system to the transient input $\phi_{oi}[t]$ is $\phi_{oo}[t]$. Hence, the transforms of these correlation functions are related by

$$\Phi_{oo}[\omega] = H[\omega]\Phi_{oi}[\omega] \tag{6.3-23}$$

If a time function $f_2[t]$ is equal to $f_1[-t]$, their Fourier transforms are related by

$$F_2[\omega] = F_1[-\omega] = F_1[\omega]^* \tag{6.3-24}$$

Remember that $F[-\omega]$ is equal to $F[\omega]*$, which is the complex conjugate of $F[\omega]$. Thus, the cross-correlation transforms are related to each other by

$$\Phi_{oi}[\omega] = \Phi_{io}[-\omega] = \Phi_{io}[\omega]^* \tag{6.3-25}$$

Substitute Eq 6.3-25 into Eq 6.3-23:

$$\Phi_{oo}[\omega] = H[\omega]\Phi_{io}[\omega]^* \tag{6.3-26}$$

Substitute for $\Phi_{io}[\omega]*$ the conjugate of Eq 6.3-22:

$$\Phi_{oo}[\omega] = H[\omega]H[\omega]^*\Phi_{ii}[\omega]^*$$
$$= H[\omega]H[\omega]^*\Phi_{ii}[\omega] = |H[\omega]|^2\Phi_{ii}[\omega] \tag{6.3-27}$$

Since $\Phi_{ii}[\omega]$ has no phase, it is equal to $\Phi_{ii}[\omega]*$. This equation shows that the output spectral density is equal to the input spectral density multiplied by the square of the magnitude of the system frequency response. This principle was derived earlier in Section 6.2 from physical reasoning to develop the definition of noise bandwidth.

By Eq 6.3-20, the value of $\phi_{oo}[\tau]$ for $\tau = 0$ can be calculated as follows from the spectral density $\Phi_{oo}[\omega]$:

$$\phi_{oo}[0] = \frac{1}{2\pi}\int_{-\infty}^{\infty}\Phi_{oo}[\omega]\,d\omega = \int_{-\infty}^{\infty}\Phi_{oo}[f]\,df \tag{6.3-28}$$

where $d\omega$ is replaced by $2\pi\,df$. Since $\Phi_{oo}[-f]$ is equal to $\Phi_{oo}[f]$, the integral of $\Phi_{oo}[f]$ from $-\infty$ to 0 is equal to the integral from 0 to $+\infty$. Also, by Eq 6.3-13, $\phi_{oo}[0]$ is equal to $\langle x_o^2 \rangle$, the mean square value of the output x_o. Hence

$$\langle x_o^2 \rangle = \phi_{oo}[0] = 2\int_0^{\infty}\Phi_{oo}[f]\,df \tag{6.3-29}$$

Substituting Eq 6.3-29 into Eq 6.3-27 gives the following expression for the mean square value of the output in terms of the spectral density of the input:

$$\langle x_o^2 \rangle = 2 \int_0^\infty \Phi_{ii}[f] |H[f]|^2 \, df$$

$$= \frac{1}{\pi} \int_0^\infty \Phi_{ii}[\omega] |H[\omega]|^2 \, d\omega \tag{6.3-30}$$

6.3.4 Calculation of Noise Bandwidth

Frequency-response calculations of noise bandwidth and mean square error can be greatly simplified by using tables of integrals of the general form

$$I_n = \frac{1}{2\pi j} \int_{-j\infty}^{+j\infty} H[s] H[-s] \, ds \tag{6.3-31}$$

Tables of this integral I_n are shown in Appendix D. Setting $s = j\omega$ gives

$$I_n = \frac{1}{2\pi j} \int_{-j\infty}^{+j\infty} H[j\omega] H[-j\omega] \, d(j\omega)$$

$$= \frac{1}{2\pi} \int_{\omega=-\infty}^{+\infty} |H[\omega]|^2 \, d\omega$$

$$= \frac{1}{\pi} \int_0^{+\infty} |H[\omega]|^2 \, d\omega \tag{6.3-32}$$

The noise bandwidth of the filter $H[\omega]$ in rad/sec is defined as

$$\omega_b = \int_0^\infty |H[\omega]|^2 \, d\omega \tag{6.3-33}$$

Comparing Eqs 6.3-32, -33 gives

$$\omega_b = \pi I_n \tag{6.3-34}$$

The noise bandwidth in hertz, which is denoted f_b, is equal to

$$f_b = \omega_b / 2\pi = I_n / 2 \tag{6.3-35}$$

The transfer function $H[s]$ is defined as follows in terms of a ratio of two polynomials in s, where the order of the numerator is less than that of the denominator:

$$H[s] = \frac{c_{n-1} s^{n-1} + \cdots + c_1 s + c_0}{d_n s^n + d_{n-1} s^{n-1} + \cdots + d_1 s + d_0} \tag{6.3-36}$$

The parameter n is the order of s in the denominator of $H[s]$. Appendix D gives equations for the integral I_n for values of n from 1 to 6, as functions of the coefficients c_m, d_m of the numerator and denominator of $H[s]$. These were obtained from Appendix E of Newton, Gould, and Kaiser [6.5]. That reference shows how these integrals are derived and gives expressions for the integrals for values of n from 1 to 10. Somewhat different and more cumbersome forms of these integrals were given earlier in the Appendix of James, Nichols, and Phillips [6.4], which listed the integrals for values of n from 1 to 7.

Thus, the noise bandwidth of any transfer function $H[s]$ that does not exceed 6th order of s in the denominator can be obtained from the integral expressions of Appendix D. The noise bandwidth for a higher order of s in the denominator (up to 10th order) can be obtained from Ref [6.4].

Let us illustrate the use of this table of I_n integrals with an example. Consider the G_{ib} transfer function for loop F in Table 6.2-2, which is

$$G_{ib} = \frac{s\omega_c\omega_f + \omega_c\omega_f\omega_i}{s^3 + \omega_f s^2 + \omega_c\omega_f s + \omega_c\omega_f\omega_i} \qquad (6.3\text{-}37)$$

In accordance with Eq 6.3-36, the polynomial coefficients are

$$d_3 = 1, \qquad d_2 = \omega_f, \qquad d_1 = \omega_c\omega_f, \qquad d_0 = \omega_c\omega_f\omega_i \qquad (6.3\text{-}38)$$
$$c_2 = 0, \qquad c_1 = \omega_c\omega_f, \qquad c_0 = \omega_c\omega_f\omega_i \qquad (6.3\text{-}39)$$

Substitute $c_2 = 0$, $d_3 = 1$ in the expression for I_3 shown in Eqs D-5, -6 of Appendix D. This gives

$$I_3 = \frac{c_2^2 d_0 d_1 + \left(c_1^2 - 2c_0c_2\right)d_0 d_3 + c_0^2 d_2 d_3}{2d_0 d_3(d_1 d_2 - d_0 d_3)}$$

$$= \frac{c_1^2 d_0 + c_0^2 d_2}{2d_0(d_1 d_2 - d_0)} \qquad (6.3\text{-}40)$$

Substituting into this the remaining coefficients of Eqs 6.3-38, -39 gives

$$I_3 = \frac{(\omega_c\omega_f)^2(\omega_c\omega_f\omega_i) + (\omega_c\omega_f\omega_i)^2(\omega_f)}{2(\omega_c\omega_f\omega_i)[(\omega_c\omega_f)(\omega_f) - (\omega_c\omega_f\omega_i)]}$$

$$= \frac{\omega_c\omega_f + \omega_i\omega_f}{2(\omega_f - \omega_i)} = \frac{\omega_f(\omega_c + \omega_i)}{2(\omega_f - \omega_i)} \qquad (6.3\text{-}41)$$

Hence, for G_{ib} of Eq 6.3-37, the noise bandwidth in hertz is

$$f_b = \frac{I_3}{2} = \frac{1}{4}\frac{\omega_f(\omega_c + \omega_i)}{(\omega_f - \omega_i)} \qquad (6.3\text{-}42)$$

The noise bandwidth in rad/sec is

$$\omega_b = 2\pi f_b = \frac{\pi}{2} \frac{\omega_f(\omega_c + \omega_i)}{(\omega_f - \omega_i)} \qquad (6.3\text{-}43)$$

6.4 STATISTICAL ANALYSIS OF TRACKING ERROR DUE TO WIND TORQUE

6.4.1 General Analysis

This section presents a practical application of statistical signal analysis. It shows how wind forces can be characterized statistically, and uses this statistical model to calculate the tracking error produced by wind forces on a large parabolic reflector antenna. The antenna angular velocities are assumed to be sufficiently low that wind pressure due to antenna rotation is negligible.

The general expression for the pressure p exerted by wind is

$$p = \tfrac{1}{2} C_d \rho_a V^2 \qquad (6.4\text{-}1)$$

where V is the wind velocity, ρ_a is the mass density of air, and C_d is the drag coefficient of the structure, which varies with shape. For a solid structure, it is of the order of unity. The mass density of air is

$$\rho_a = 0.0024 \text{ lb-sec}^2/\text{ft}^4 \qquad (6.4\text{-}2)$$

For $C_d = 1.0$, the pressure at a wind velocity V of 60 mph (88 ft/sec) is

$$p = 9.29 \text{ lb/ft}^2 \quad (\text{at } 60 \text{ mph}) \qquad (6.4\text{-}3)$$

The torque produced by wind on an antenna is proportional to the wind pressure, and so is proportional to the square of the wind velocity. Since the tracking error due to wind is proportional to wind torque, the statistical properties of the wind torque, or wind pressure, are needed. These are not the same as the statistical properties of the wind velocity. If the wind pressure has a Gaussian distribution, the wind velocity cannot have a Gaussian distribution. The spectrum of the wind velocity is different from the spectrum of the wind torque.

On the other hand, if the relative deviation of wind velocity from mean wind velocity is small, the deviation of wind torque (or wind pressure) from mean wind torque (or mean wind pressure) is approximately proportional to the deviation of wind velocity from mean wind velocity. For this condition, the wind torque (or wind pressure) has approximately the same statistical properties as the wind velocity. Let us represent the wind velocity V and wind torque

T as the sum of an average value plus a deviation:

$$V = V_0 + \Delta V \tag{6.4-4}$$

$$T = T_0 + \Delta T \tag{6.4-5}$$

where V_0 is the mean wind velocity, T_0 is the wind torque at mean wind velocity, and $\Delta V, \Delta T$ are the variations from these values. Since the wind torque T is proportional to V^2, then

$$\frac{T}{T_0} = \left(\frac{V}{V_0}\right)^2 = \left(\frac{V + \Delta V}{V_0}\right)^2$$

$$= 1 + 2\frac{\Delta V}{V_0} + \left(\frac{\Delta V}{V_0}\right)^2 \tag{6.4-6}$$

By Eq 6.4-5, T/T_0 is equal to $(1 + \Delta T/T_0)$. Hence

$$\frac{\Delta T}{T_0} = 2\frac{\Delta V}{V_0} + \left(\frac{\Delta V}{V_0}\right)^2 \cong 2\frac{\Delta V}{V_0} \tag{6.4-7}$$

This approximation holds for $\Delta V/V_0 \ll 1$. Thus, for a small relative variation of wind velocity from mean wind velocity V_0, the relative variation of wind torque, $\Delta T/T_0$, is twice the relative variation of wind velocity, $\Delta V/V_0$. For example, a 10% variation of wind velocity from the mean velocity results in a 20% variation of wind torque from the mean wind torque.

The wind spectrum commonly used to specify wind torque on antennas is based on statistical measurements performed by Titus [6.7] of the wind torque exerted on a 60-ft antenna. He found that the following equation fitted the measured spectral data to reasonable accuracy:

$$\Phi_{\text{wt}}[\omega] = \frac{K\omega_1^2\omega_2^2}{(\omega^2 + \omega_1^2)(\omega^2 + \omega_2^2)} \tag{6.4-8}$$

where

$$\omega_1 = 0.12 \text{ rad/sec}, \qquad \omega_2 = 2.0 \text{ rad/sec} \tag{6.4-9}$$

and K is a constant. It is convenient to choose the value of the constant K to normalize the spectrum so that its integral over all angular frequencies ω is unity. If the constant K were unity, this spectrum would be the same as $|H[\omega]|^2$ for the lowpass filter B of Table 6.2-1, and so the integral of Eq 6.4-8 would be equal to the noise bandwidth of filter B. This is

$$\int_0^\infty \frac{\omega_1^2\omega_2^2}{(\omega^2 + \omega_1^2)(\omega^2 + \omega_2^2)} \, d\omega = \frac{\pi}{2}\frac{\omega_1\omega_2}{\omega_1 + \omega_2} \tag{6.4-10}$$

Hence, to make the integral of the spectrum unity, the constant K should be set equal to the reciprocal of the right-hand expression, which is

$$K = \frac{\omega_1 + \omega_2}{(\pi/2)\,\omega_1\omega_2} \tag{6.4-11}$$

Substituting Eq 6.4-11 into -8 gives the normalized wind-torque spectrum, the integral of which is unity:

$$\Phi_{\text{wt(n)}}[\omega] = \frac{(2/\pi)\,\omega_1\omega_2(\omega_1 + \omega_2)}{(\omega^2 + \omega_1^2)(\omega^2 + \omega_2^2)} \tag{6.4-12}$$

A statistical wind specification frequency often used by U.S. government agencies gives the wind-torque spectrum of Eq 6.4-8 along with the statement that the standard deviation (sigma) of wind gust velocity is equal to 25% of the mean wind velocity. For such a large standard deviation, the torque deviation is not approximately proportional to the velocity deviation, particularly for a 3-sigma variation. To have a practical analysis tool, the wind torque must be assumed to have a Gaussian distribution, and consequently the wind velocity does not. Nevertheless, the approximation of Eq 6.4-7 can still be used to calculate the standard deviation of the torque. If the standard deviation of the wind velocity, σV, is 25% of the mean wind velocity V_0, then the standard deviation of the torque, σT, is 50% of the torque T_0 at mean wind velocity:

$$\frac{\sigma T}{T_0} = 2\frac{\sigma V}{V_0} = 2(0.25) = 0.50 \tag{6.4-13}$$

On the other hand, this standard deviation is probably unrealistically high. A more reasonable specification is a standard deviation of wind velocity that is 15% of mean wind velocity. The standard deviation of torque is then 30% of the torque at mean wind velocity.

A general discussion of servo-system errors caused by load torque disturbances is given in Ref [1.1] (Chapter 6). It shows that the angle tracking error produced by a load torque T_d is given by

$$\Theta_e = T_d C^m G_{\text{ie(p)}} + \frac{T_d G_{\text{ie(r, p)}}}{s^2 J_{(m)}(1 + \omega_{\text{cm}}/s)} \tag{6.4-14}$$

The first term is due to compression of the mechanical compliance C^m of the antenna structure and the gear train, while the second term is due to coercion of the servo motor by the load torque. For wind torques, which are relatively low in frequency, the second term can usually be made appreciably smaller than the first and so is neglected in this analysis.

Equation 6.3-30 gave a general expression for the mean square output from a system, expressed in terms of the spectral density of the input signal and the system transfer function. In accordance with that equation, the RMS tracking error $\sigma\Theta$ due to wind torque is given by

$$(\sigma\Theta)^2 = \frac{1}{\pi}(C^m)^2 \int_0^\infty |G_{\text{ie(p)}}[\omega]|^2 \Phi_{\text{wt}}[\omega]\, d\omega \qquad (6.4\text{-}15)$$

where $\Phi_{\text{wt}}[\omega]$ is the spectral density of the wind torque. It is convenient to normalize the calculation by considering the RMS tracking error that would be produced under "locked-rotor" conditions, when the servo motor rotor is locked and the position loop is not operating. This locked-rotor error, which is denoted $\sigma\Theta_{\text{LR}}$, can be obtained by setting $|G_{\text{ie(p)}}|$ equal to unity in Eq 6.4-15, which gives

$$(\sigma\Theta_{\text{LR}})^2 = \frac{1}{\pi}(C^m)^2 \int_0^\infty \Phi_{\text{wt}}[\omega]\, d\omega \qquad (6.4\text{-}16)$$

Dividing Eq 6.4-16 by Eq 6.4-17 gives

$$\left(\frac{\sigma\Theta}{\sigma\Theta_{\text{LR}}}\right)^2 = \int_0^\infty |G_{\text{ie(p)}}[\omega]|^2 \Phi_{\text{wt(n)}}[\omega]\, d\omega \qquad (6.4\text{-}17)$$

The function $\Phi_{\text{wt(n)}}[\omega]$ is the normalized wind-torque spectrum given in Eq 6.4-12 and is defined by

$$\Phi_{\text{wt(n)}}[\omega] = \frac{\Phi_{\text{wt}}[\omega]}{\int_0^\infty \Phi_{\text{wt}}[\omega]\, d\omega} \qquad (6.4\text{-}18)$$

A general solution of Eq 6.4-17 can be obtained by assuming a tracking (position) loop having the loop transfer function of loop F, shown in Table 6.2-2 and Fig 6.2-5. The loop and error transfer functions are

$$G_{\text{(p)}} = \frac{\omega_c(1 - \omega_i/s)}{s(1 + s/\omega_f)} \qquad (6.4\text{-}19)$$

$$G_{\text{ie(p)}} = \frac{s^2(s + \omega_f)}{s^3 + \omega_f s^2 + \omega_c \omega_f s + \omega_c \omega_f \omega_i} \qquad (6.4\text{-}20)$$

The wind spectrum of Eq 6.4-12 can be related to the following transfer function:

$$H_{\text{wt}}[s] = \frac{1}{(s + \omega_1)(s + \omega_2)} \qquad (6.4\text{-}21)$$

For $s = j\omega$, the square of the magnitude of this is

$$|H_{wt}[\omega]|^2 = H_{wt}[\omega]H_{wt}[-\omega] = \frac{1}{(\omega^2 + \omega_1^2)(\omega^2 + \omega_2^2)} \quad (6.4\text{-}22)$$

Comparing this with Eq 6.4-12 shows that

$$\Phi_{wt(n)}[\omega] = \frac{2}{\pi}\omega_1\omega_2(\omega_1 + \omega_2)|H_{wt}[\omega]|^2 \quad (6.4\text{-}23)$$

Hence, Eq 6.4-17 can be expressed as

$$\left(\frac{\sigma\Theta}{\sigma\Theta_{LR}}\right)^2 = \frac{2}{\pi}\omega_1\omega_2(\omega_1 + \omega_2)\int_0^\infty |G_{ie(p)}[\omega]H_{wt}[\omega]|^2\,d\omega \quad (6.4\text{-}24)$$

Define the transfer function $H[s]$ as

$$H[s] = G_{ie(p)}[s]H_{wt}[s] \quad (6.4\text{-}25)$$

This is equal to

$$H[s] = \frac{s^2(s + \omega_f)}{(s^3 + \omega_f s^2 + \omega_c\omega_f s + \omega_c\omega_f\omega_i)(s + \omega_1)(s + \omega_2)}$$

$$= \frac{s^3 + \omega_f s^2}{D} \quad (6.4\text{-}26)$$

where the denominator D is

$$D = s^5 + s^4(\omega_f + \omega_1 + \omega_2) + s^3[\omega_c\omega_f + \omega_1\omega_2 + \omega_f(\omega_1 + \omega_2)]$$
$$+ s^2\omega_f[\omega_c\omega_i + \omega_1\omega_2 + \omega_c(\omega_1 + \omega_2)]$$
$$+ s\omega_c\omega_f[\omega_i(\omega_1 + \omega_2) + \omega_1\omega_2] + \omega_c\omega_f\omega_i\omega_1\omega_2 \quad (6.4\text{-}27)$$

Equation 6.4-24 can be expressed as follows in terms of the square of the magnitude of $H[s]$ for $s = j\omega$:

$$\left(\frac{\sigma\Theta}{\sigma\Theta_{LR}}\right)^2 = \omega_1\omega_2(\omega_1 + \omega_2)\frac{2}{\pi}\int_0^\infty |H[\omega]|^2\,d\omega \quad (6.4\text{-}28)$$

The integral in Eq 6.4-28 is equal to πI_5, where I_5 is the integral expression given in Appendix D for $n = 5$. From Eqs 6.4-26, -27, the coefficients of the numerator and denominator polynomials of $H[s]$ are

$$c_4 = c_1 = c_0 = 0 \tag{6.4-29}$$

$$c_3 = 1 \tag{6.4-30}$$

$$c_2 = \omega_f \tag{6.4-31}$$

$$d_5 = 1 \tag{6.4-32}$$

$$d_4 = \omega_f + \omega_1 + \omega_2 \tag{6.4-33}$$

$$d_3 = \omega_c \omega_f + \omega_1 \omega_2 + \omega_f(\omega_1 + \omega_2) \tag{6.4-34}$$

$$d_4 = \omega_f[\omega_c \omega_i + \omega_1 \omega_2 + \omega_c(\omega_1 + \omega_2)] \tag{6.4-35}$$

$$d_1 = \omega_c \omega_f[\omega_i(\omega_1 + \omega_2) + \omega_1 \omega_2] \tag{6.4-36}$$

$$d_0 = \omega_c \omega_f \omega_i \omega_1 \omega_2 \tag{6.4-37}$$

In the expression for I_5 of Appendix D, given in Eqs D-13 to -19, set $d_5 = 1$ and substitute the values of the numerator coefficients (c_0 to c_4) of Eqs 6.4-29 to -31. This gives

$$2I_5 = \frac{m_{51} + \omega_f^2 m_{52}}{D_5} \tag{6.4-38}$$

$$m_{51} = d_1 d_2 - d_0 d_3 \tag{6.4-39}$$

$$m_{52} = d_1 d_4 - d_0 \tag{6.4-40}$$

$$m_{53} = \frac{d_2 m_{52} - d_4 m_{51}}{d_0} \tag{6.4-41}$$

$$m_{54} = \frac{d_3 m_{53} - s_4 m_{52}}{d_0} \tag{6.4-42}$$

$$D_5 = d_0(d_1 m_{54} - d_3 m_{53} + m_{52}) \tag{6.4-43}$$

Setting the integral in Eq 6.4-24 equal to πI_5 gives

$$\left(\frac{\sigma\Theta}{\sigma\Theta_{LR}}\right)^2 = \omega_1 \omega_2(\omega_1 + \omega_2)\frac{2}{\pi}(\pi I_5) = \omega_1 \omega_2(\omega_1 + \omega_2)(2I_5) \tag{6.4-44}$$

The expression for $2I_5$ can be obtained from Eq 6.4-38 using the definitions of Eqs 6.4-39, to -43, along with the values for d_4 to d_0 in Eqs 6.4-33 to -37.

Let us apply this expression to three cases. As shown in Ref [1.1] (Section 3.4), for the loop transfer function of Eq 6.4-19, the value of Max$|G_{ib}|$ is minimized if ω_c is set equal to

$$\omega_c = \tfrac{1}{2}(\omega_i + \omega_f) \tag{6.4-45}$$

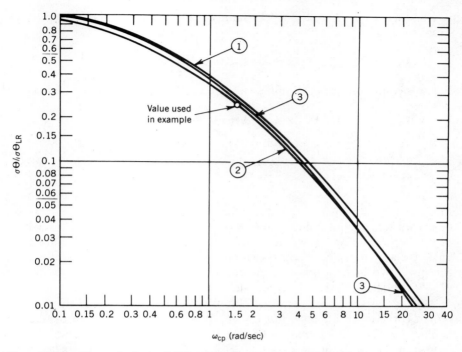

Figure 6.4-1   Plots of relative RMS tracking noise due to wind gusts on an antenna, as a function of the gain crossover frequency ω_{cp} of the tracking position loop.

This gain setting results in the best stability (the smallest value of $\text{Max}|G_{ib}|$) for a given ω_f/ω_i ratio. The value of this ratio is related as follows to $\text{Max}|G_{ib}|$:

$$\frac{\omega_f}{\omega_i} = \frac{\text{Max}|G_{ib}| + 1}{\text{Max}|G_{ib}| - 1} \tag{6.4-46}$$

Solving for ω_i, ω_f gives

$$\omega_i = \left(1 - \frac{1}{\text{Max}|G_{ib}|}\right)\omega_c \tag{6.4-47}$$

$$\omega_f = \left(1 + \frac{1}{\text{Max}|G_{ib}|}\right)\omega_c \tag{6.4-48}$$

Curve ① in Fig 6.4-1 is the plot of the square root of Eq 6.4-44 versus the position-loop asymptotic gain crossover frequency ω_c, using Eqs 6.4-47, -48 to relate ω_i, ω_f to ω_c. The value for $\text{Max}|G_{ib}|$ is 1.30. The parameters ω_1, ω_2 are set equal to 0.12 sec^{-1}, 2 sec^{-1} in accordance with Eq 6.4-9.

For curves ②, ③ in Fig 6.4-1, the ratio ω_f/ω_c, which is denoted β, is set equal to ω_c/ω_i:

$$\beta = \omega_f/\omega_c = \omega_c/\omega_i \qquad (6.4\text{-}49)$$

For this condition, the denominator of G_{ie} in Eq 6.4-20 can be factored to give

$$G_{ie} = \frac{s^2(s + \beta\omega_c)}{(s + \omega_c)(s^2 + 2\zeta\omega_c s + \omega_c^2)} \qquad (6.4\text{-}50)$$

$$\zeta = \frac{\beta - 1}{2} \qquad (6.4\text{-}51)$$

For curve ②, $\beta = 3$, $\zeta = 1$, and G_{ie} has a triple-order pole at $s = -\omega_c$. It can be shown that the maximum value of $|G_{ib}|$ is 1.30, and there is 25% overshoot of the step response. Of the three cases, this gives the best attenuation of wind gusts for a given value of ω_c. For curve ③, $\beta = \sqrt{6} = 2.45$ and $\zeta = 0.725$. It can be shown that for this case the maximum value of $|G_{ib}|$ is 1.42, and there is 33% overshoot in the step response.

This analysis was first performed by Briggs [6.8], who assumed the parameters of curve ③ of Fig 6.4-1 for $\beta = \sqrt{6}$. With such a low value of β, stability is marginal. The above analysis has extended his work to include more stable loop transfer functions.

6.4.2 Tracking Error of a Satellite Communication Antenna

Let us apply Fig 6.4-1 to calculate the tracking error due to wind gusts that is experienced by a large satellite communication antenna. Table 6.4-1 shows the steps for calculating the tracking error. The parameters are typical of a 32-m (105-ft) satellite communication antenna, which usually transmits at 6 GHz and receives at 4 GHz. The azimuth carriage (which weighs 500,000 lb) is carried on four railroad-car wheels, which roll on a circular track.

It is assumed that the antenna has a monopulse tracking feed which allows continuous tracking of the received signal from the satellite. The signal strength is assumed to be sufficiently high to achieve negligible tracking error due to receiver noise. The wind is assumed to have a mean velocity of 45 mph, with gusts, where the standard deviation of the wind-gust velocity is 15% of the mean wind velocity.

Item (1) of Table 6.4-1 shows the values of inertia J of the azimuth and elevation axes. As shown in (2), the servo structural resonant frequency in azimuth and elevation is 2.0 Hz, or 12.6 rad/sec. This 2-Hz servo structural resonant frequency is consistent with the data for a 105-ft antenna in Ref [1.1] (Fig 10.1-1 of Chapter 10). (The mean practical structural resonant frequency for this antenna diameter is 1.5 Hz.) The stiffness, shown in item (3), is given by

$$K = J\omega_n^2 \qquad (6.4\text{-}52)$$

TABLE 6.4-1   Calculation of Wind-Gust Tracking Error of 105-ft Satellite-Communication Antenna

Parameter	Azimuth	Elevation	Units
(1) Inertia J	4.0×10^6	2.0×10^6	ft-lb-sec^2
(2) Structural			
resonant	$f_n = 2.0$	2.0	Hz
frequency	$\omega_n = 12.6$	12.6	rad/sec
(3) Stiffness			
$K = J\omega_n^2$	6.35×10^8	3.175×10^8	ft-lb/rad
(4) Compliance	1.57×10^{-9}	3.15×10^{-9}	rad/ft-lb
$C^m = 1/K$	0.90×10^{-7}	1.80×10^{-7}	deg/ft-lb
(5) Wind torque:			
30 mph	427,000	313,000	ft-lb
45 mph	961,000	704,000	ft-lb
(6) Gust torque:			
45-mph mean,			
1-sigma	288,300	211,200	ft-lb
(7) Locked-rotor			
gust error,			
1-sigma	0.0259	0.0380	deg
(8) ω_{cp}	1.57	1.57	sec^{-1}
(9) $\sigma\Theta/\sigma\Theta_{LR}$	0.24	0.24	
(10) Actual			
gust error,			
1-sigma	0.0062	0.0091	deg
(11) 2-axis errors:			
1-sigma		0.0110	deg
3-sigma		0.0330	deg

where J is the antenna inertia and ω_n is the servo structural resonant frequency in rad/sec. Substituting the values of J and ω_n in items (1), (2) gives the antenna stiffness values shown in (3). The reciprocal of the stiffness is the compliance, in rad/ft-lb, shown in (4). This is multiplied by 57.3 deg/rad to obtain the antenna compliance expressed in deg/ft-lb.

Item (5) shows the calculated values of the wind torque for a 30-mph wind when the antenna is at the worst-case orientation relative to the wind. These torques are scaled to apply to 45 mph, by multiplying them by $[(45 \text{ mph})/(30 \text{ mph})]^2$. The standard deviation for the wind-gust velocity is assumed to be 15% of the mean wind velocity, and so the standard deviation of the wind-gust torque is 30% of the torque at mean wind velocity. Hence the 45-mph wind torque values in item (5) are multiplied by 0.3 to obtain the 1-sigma wind-gust torque values of item (6). These are multiplied by the antenna compliance values of (4) to obtain the 1-sigma values of the locked-rotor gust error shown in (7).

As was shown in Ref [1.1], (Section 10.5), with proper design the gain crossover frequency of the position loop can typically be $\frac{1}{8}$ of the structural resonant frequency. Since both axes have a structural resonant frequency of 2 Hz, the assumed value of position-loop gain crossover frequency for both axes is

$$\omega_{cp} = \omega_n/8 = (12.6 \text{ sec}^{-1})/8 = 1.57 \text{ sec}^{-1} \qquad (6.4\text{-}53)$$

Curve ② of Fig 6.4-1 is assumed. For this value of ω_c, the factor $\sigma\Theta/\sigma\Theta_{LR}$ for curve ② of Fig 6.4-1 is 0.24, shown in item (9). The 1-sigma locked-rotor gust error of item (7) is multiplied by 0.24 to obtain the actual 1-sigma gust error in item (10). The total two-axis deflection Θ of the antenna beam from the direction of the satellite is

$$\Theta = \sqrt{\Theta_d^2 + \Theta_{eL}^2} \qquad (6.4\text{-}54)$$

where Θ_d, Θ_{eL} are the deflection and elevation tracking errors. Taking the square root of the sum of the squares of the deflection and elevation gust components of error of item (10) gives the two-axis wind-gust tracking error shown in item (11). This 1-sigma two-axis error (0.01100) is multiplied by 3 to obtain the peak 3-sigma two-axis tracking error, which is 0.0330°.

The antenna aperture D is 32 m, or 3200 cm, and the wavelength at the transmit frequency, 6 GHz, is 5 cm. In Chapter 7, Eq 7.2-3 gives the following approximation for the half-power beamwidth Θ_b of the antenna at the transmit frequency:

$$\Theta_b = 65\frac{\lambda}{D} \text{ deg} = 65°\frac{5 \text{ cm}}{3200 \text{ cm}} = 0.102° \qquad (6.4\text{-}55)$$

In Chapter 7, Eq 7.2-6 gives the following for the transmit signal loss that corresponds to the peak 3-sigma wind-gust tracking error, 0.0330°:

$$\text{dB loss} = 12\left(\frac{\Theta}{\Theta_b}\right)^2 \text{ dB} = 12\left(\frac{0.0330°}{0.102°}\right)^2 \text{ dB} = 1.26 \text{ dB} \qquad (6.4\text{-}56)$$

Thus the peak signal loss due to wind-gust tracking error is 1.26 dB.

This statistical wind model assumes that the wind pressure (or wind torque) consists of a random gust component of wind pressure added to a constant mean wind pressure, where the gust component has a Gaussian distribution. Since the tracking loop has infinite gain at zero frequency, the constant average wind pressure (or wind torque) causes no tracking error. It is only the gust deviation from the mean wind torque that causes tracking error.

The concept of a "locked-rotor gust error" is somewhat fictitious. If the servo motor rotor were actually locked, there would be no tracking-loop compensation for the mean wind pressure, and so the mean wind pressure would produce error. Nevertheless, for analysis purposes it is convenient to consider a locked-rotor gust error that applies only to the random gust deviation of wind pressure, and not to the mean wind pressure.

Chapter 7

Target Detection and Tracking in a Tracking Radar

This chapter analyzes radar signal processing in a tracking radar. A specific system is assumed, and based on this model the calculations associated with target detection and tracking in range and angle are developed. The radar tracking equations derived in this chapter are applied later in Chapter 8 to characterize the radar data used in the Altair tracking system.

7.1 SUMMARY OF CHAPTER

Based on a specific model of a radar tracking system, Section 7.3 develops the radar signal-processing equations associated with target detection; Section 7.4 develops the equations for target tracking in range; and Section 7.5 develops the equations for target tracking in angle. As an introduction to this analysis, a simplified discussion of target tracking in range and angle is presented in Section 7.2, which summarizes the material in later sections.

This chapter applies the statistical analysis principles developed in Chapter 6 to calculate the effect of receiver noise. Setting the noise bandwidths of the range and angle tracking loops requires a compromise between tracking error due to receiver noise, and tracking error due to target motion. Target motion errors are characterized by the principles given in Chapter 3.

The signal-processing stages in a radar tracking system are essentially filtering operations, which separate the target-tracking information from the receiver noise in a nearly optimal manner. For example, the IF amplifier of a pulse-radar receiver narrows the signal bandwidth so as to minimize noise transmission, but this smears the radar pulse. Optimal design of the IF-amplifier frequency response requires a compromise between loss of signal information (pulse degradation) and the transmission of receiver noise.

Usually there is a noncoherent detection process at the output of the IF amplifier in a radar receiver. To achieve small tracking error, the signal-to-noise (or signal/noise) power ratio at the input to the noncoherent detector (the IF-amplifier output) should ideally be greater than unity. If this is not achieved, there is appreciable degradation of signal/noise ratio in the nonco-herent detection process.

Range and angle discrimination processes are incorporated into a tracking radar to provide range and angle tracking error signals. Range-tracking and angle-tracking feedback loops are closed, which use these error signals to keep the range gate centered around the target pulse, and to keep the antenna pointed at the target. The dynamic parameters of these feedback loops are chosen to provide a near optimal compromise between tracking error due to receiver noise and tracking error due to target motion.

To understand target tracking, one must also study target detection, be-cause the target detection requirements of a tracking radar strongly constrain its design. Section 7.3 describes the signal processing performed to detect a target in a tracking radar. The Rayleigh distribution is explained and applied, along with the Gaussian distribution, to develop equations for the probability of target detection and the probability of false alarm.

7.2   ERROR IN A RADAR TRACKING SYSTEM DUE TO RECEIVER NOISE

The detection and tracking processes of a tracking radar system are analyzed in Sections 7.3 to 7.5. The following is a summary of the equations for calculating the tracking errors in range and angle.

7.2.1   Transmission-Path Loss of Radar Example

Figure 7.2-1 is a simplified drawing of a tracking radar antenna, which has a parabolic reflector, in front of which is a feed supported on spars. Energy radiating from the feed is reflected by the paraboloid and projected in the direction of the antenna axis (the boresight), but the beam diverges somewhat because of diffraction. If there is a target in the transmitted beam, it reflects a small amount of energy back to the antenna, which is intercepted by the reflector and projected back to the feed. The direction of the antenna beam is controlled by servos, which drive the antenna gimbal structure in two axes, called azimuth and elevation. The azimuth servo rotates the whole structure about a vertical axis, thereby changing its heading relative to north. The elevation gimbal rides on the rotating azimuth carriage, and tilts the antenna boresight upward and downward relative to the horizontal plane.

The analysis in Sections 7.3–7.5 is based on a specific radar example, the basic parameters of which are listed in Table 7.2-1. The radar operates at a frequency f of 10 GHz [item (1)]. The wavelength λ is related to the frequency f by

$$\lambda = \frac{c}{f} \qquad (7.2\text{-}1)$$

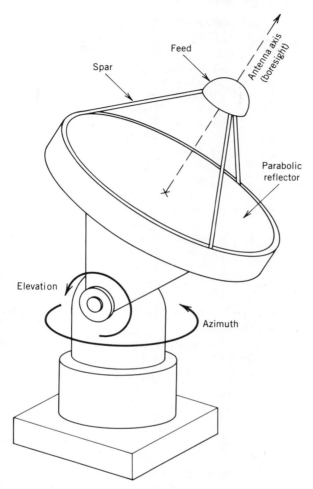

Figure 7.2-1 Tracking radar antenna.

where c is the speed of light (3×10^8 m/sec). As shown in item (2), the corresponding wavelength λ is 3 cm. The antenna diameter D is assumed to be 4 ft, or 122 cm [item (3)].

The peak (or maximum) gain G_m of the antenna gain pattern is equal to

$$G_m = \eta \left(\frac{\pi D}{\lambda} \right)^2 \tag{7.2-2}$$

where η is the antenna efficiency. A reasonable value for the antenna efficiency for a radar tracking antenna is 0.6, as shown in item (4). Applying the values

TABLE 7.2-1   Calculation of Signal Power into Receiver

Symbol		Parameter	Value
(1)	f	Frequency	10 GHz
(2)	λ	Wavelength	3.0 cm
(3)	D	Reflector diameter	4.0 ft = 122 cm
(4)	η	Antenna efficiency	0.60
(5)	G_m	Peak antenna gain	9390
(6)	Θ_b	Half-power bandwidth	1.60°
(7)	R	Range to target	50 km
(8)	A_t	Radar cross section	1.0 m^2
(9)	P_r/P_t	Transmission loss^{-1}	6.398×10^{-18}
(10)	$1/L_w$	Waveguide loss^{-1}	-2 dB $= 0.631$
(11)	P_{gen}	Transmitter power	100 kW
(12)	P_s	Receiver signal power	4.03×10^{-13} W
(13)	τ_p	Radar pulse width	0.1 μsec
(14)	E_s	Receiver signal energy	4.03×10^{-20} J
(15)	ΔR	Range resolution	15 m

of items (2), (3), (4) to Eq 7.2-2 gives the maximum antenna gain G_m shown in item (5), which is 9390.

The half-power beamwidth of an antenna gain pattern is denoted Θ_b. This is the angle between points 3 dB below peak gain. The approximate value for Θ_b is

$$\Theta_b \cong 65(\lambda/D) \text{ deg} \qquad (7.2\text{-}3)$$

Applying the values of items (2), (3) to Eq 7.2-3 gives an approximate 3-dB beamwidth Θ_b of 1.60°, as shown in item (6).

Near the peak of the antenna beam, any practical tracking antenna has a gain pattern that is approximately Gaussian, and so can be expressed as

$$\frac{G}{G_m} = \exp\left[-\left(\frac{\phi}{\Psi}\right)^2\right] \qquad (7.2\text{-}4)$$

where ϕ is the angle of the target signal from the peak of the antenna beam (the boresight), and Ψ is a constant that is proportional to the beamwidth. The ratio G/G_m should be $\frac{1}{2}$ when $\phi = \pm\Theta_b/2$. Setting this condition in Eq 7.2-4 gives

$$\Psi = 0.601\Theta_b \qquad (7.2\text{-}5)$$

Take 10 times the logarithm of the reciprocal of Eq 7.2-4. This gives the following expression for the loss in decibels (dB), relative to peak antenna gain, due to beam pointing error ϕ:

$$\text{dB loss} = 12(\phi/\Theta_b)^2 \text{ dB} \qquad (7.2\text{-}6)$$

Note that this loss is 3 dB when $\phi = \pm\Theta_b/2$.

It is assumed that the radar is tracking a target at a range R of 50 kilometers [item (7)]. The assumed radar cross section A_t of the target is 1.0 square meter [item (8)], which is a typical value for a small aircraft. The ratio of receive power P_r to transmit power P_t is given by the basic radar transmission equation:

$$\frac{P_r}{P_t} = \frac{G^2\lambda^2 A_t}{(4\pi)^3 R^4} \qquad (7.2\text{-}7)$$

where G is the antenna gain in the direction of the target. (This equation neglects atmospheric attenuation, which will be ignored.)

It is assumed that the angle tracking error is relatively small, so that the target is close to the peak of the radar beam. Hence antenna gain G in Eq 7.2-7 can be set equal to G_m given in item (5). Using the values of λ, R, and A_t of items (2), (7), (8) gives the transmission ratio P_r/P_t in item (9). The transmission loss is the reciprocal of this value, and so item (9) is called the reciprocal loss, which is designated loss^{-1}. Waveguide elements between the transmitter tube and the antenna, and between the antenna and the receiver, produce a signal loss that is called the waveguide loss L_w. The reciprocal waveguide loss L_w^{-1} is assumed to be -2 dB, which represents a ratio of 0.631 [item (10)]. The peak power generated in the transmitter is denoted P_{gen}, and is assumed to be 100 kilowatt [item (11)]. Multiplying items (9), (10), (11) gives the received signal power P_s at the input to the receiver. As shown in item (12), the signal power P_s is 4.03×10^{-13} watt.

The radar generates a pulsed waveform. The pulse width is denoted τ_p, and is assumed to be 0.1 microsecond, as shown in item (13). The received signal energy is denoted E_s, and is equal to the received signal power P_s multiplied by the pulse width τ_p:

$$E_s = P_s\tau_p \qquad (7.2\text{-}8)$$

The received signal power P_s of item (12) is multiplied by the pulse width τ_p of item (13) to obtain the received signal energy E_s shown in item (14), which is 4.03×10^{-20} joule.

The time t for the radar pulse to travel from the radar to the target at range R, and return, is equal to

$$t = \frac{2R}{c} \qquad (7.2\text{-}9)$$

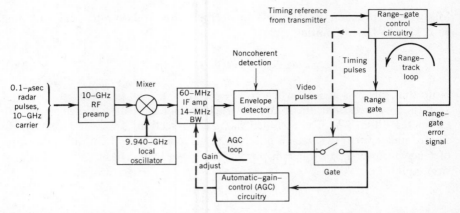

Figure 7.2-2   Block diagram of radar receiver, including range-gate and AGC circuitry.

where c is the speed of light (3×10^8 m/sec). Hence, the range increment ΔR corresponding to a time increment Δt is given by

$$\frac{\Delta R}{\Delta t} = \frac{c}{2} = 1.5 \times 10^8 \text{ m/sec} = 150 \text{ m/}\mu\text{sec} \qquad (7.2\text{-}10)$$

The 0.1-μsec pulse width τ_p of item (13) is multiplied by this factor 150 m/μsec to obtain the corresponding range increment ΔR shown in item (15), which is 15 m. This is the range resolution of the radar.

7.2.2 Calculating the RMS Range-Tracking Noise Error for a Single Radar Pulse

Figure 7.2-2 shows a block diagram of the radar receiver, including the range-tracking loop and the automatic-gain-control (AGC) loop. The receiver signal consists of a 10-GHz carrier, modulated with rectangular pulses of 0.1-μsec duration. This signal is fed through a 10-GHz RF preamplifier, and then to a mixer. (In early vacuum-tube radar receivers, RF amplification was not practical, and so the received radar signal was fed directly to the mixer.) The mixer multiplies (or heterodynes) the received radar signal with a local-oscillator signal at 9.940 GHz, to produce sum and difference frequency components at 19.940 GHz and 0.060 GHz. (This is mathematically the same as the modulation process discussed in Chapter 4, Section 4.2.) The sum signal at 19.940 GHz is rejected by filtering, and the difference signal at 0.060 GHz (or 60 MHz) is fed into the IF amplifier.

The 60-MHz IF amplifier has a 14-MHz bandwidth. As will be explained, this bandwidth is chosen to optimize the signal/noise (or signal-to-noise) ratio of the detected radar pulses. In accordance with Eq 6.2-56 in Section 6.2.6, the

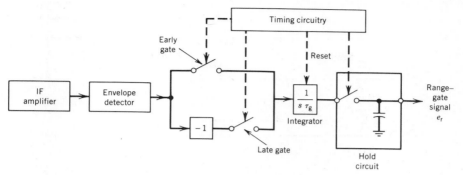

Figure 7.2-3 Basic design of early–late range-gate detector.

resultant rise time of the pulse envelope at the output of the IF amplifier is approximately

$$T_R \cong \frac{1}{f_{b(bp)}} = \frac{1}{14 \text{ MHz}} = 0.070 \ \mu\text{sec} \qquad (7.2\text{-}11)$$

Since this rise time is significant relative to the 0.1-μsec pulse width of the radar signal, the pulse modulation at the output of the IF amplifier is smoothed appreciably.

The output signal from the IF amplifier is fed through an envelope detector, which detects the peak value of each cycle of the waveform. This generates a video pulse that (except for noise) has the shape of the envelope of the IF signal. This video pulse is fed into a range-gate circuit, which generates a range gate that tracks the video pulse, and provides a signal proportional to the range-gate tracking error.

An early–late range-gate circuit is assumed, with the block diagram shown in Fig 7.2-3. The gating function provided by the circuit is shown in Fig 7.2-4. The parameter τ_g is defined as

τ_g = range-gate width = total duration of early gate plus late gate.

The early gate is closed for the time interval $\tau_g/2$ preceding the estimated time of the center of the received pulse, and the late gate is closed for an equal time after that estimated time. The range-gate circuit generates a range-error signal equal to the time difference between the actual center of the pulse and the estimated center of the pulse. This error signal is used in a feedback manner to vary the timing of the range gate, to keep the received pulse centered in the range gate.

Figure 7.2-5 shows the error response characteristic provided by the early–late gate circuit. This is a plot of the range-gate error signal, versus the timing error between the center of the radar pulse and the center of the range

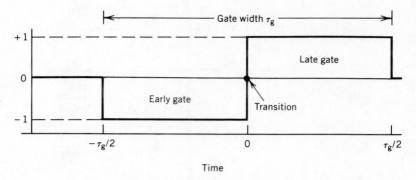

Figure 7.2-4 Range-gate function.

gate. The solid curve shows the error characteristic when the gate width τ_g is greater than $2\tau_p$. The dashed curves show the conditions for $\tau_g = 2\tau_p$ and for $\tau_g < 2\tau_p$.

The range tracker normally operates in the central (linear) region of the characteristic of Fig 7.2-5, between points 3 and 5. It is generally desirable that this central linear region be as wide as possible. Hence, τ_g should not be less than $2\tau_p$. When τ_g is greater than $2\tau_p$, the range-tracking noise error is increased. Therefore, a reasonable operating condition is a range-gate width τ_g that is twice the pulse width:

$$\tau_g = 2\tau_p \tag{7.2-12}$$

This condition is assumed in the example.

The IF amplifier generally consists of several stagger-tuned stages. The resultant bandpass response can be approximated quite well by an ideal

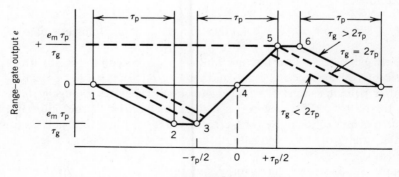

Figure 7.2-5 Plot of output voltage from range gate versus range-gate timing error.

TABLE 7.2-2 Calculation of RMS Range Tracking Error per Pulse, σR_1

	Symbol	Parameter	Value
(1)	T	Receiver noise temperature	700 K
(2)	K	Boltzmann's constant	1.374×10^{-23} J/°K
(3)	$p_n = KT$	Noise power density	9.62×10^{-21} W/Hz
(4)	E_s	Receiver signal energy	4.03×10^{-20} J
(5)	E_s/p_n	Matched-filter signal/noise	4.189
(6)	L_m	Receiver matching loss	1.216 = 0.85 dB
(7)	P_s/P_n	Signal/noise into detector	3.44
(8)	P_n/P_s	Noise/signal into detector	0.291
(9)	L_d	Noncoherent detection loss	1.291
(10)	p_{nd}	Detected noise power density	1.24×10^{-20} W/Hz
(11)	E_s/p_{nd}		3.25
(12)	$\sigma t_1/\tau_p$	Rel. RMS time error per pulse	0.261
(13)	τ_p	Pulse width	0.1 μsec
(14)	σt_1	RMS time error per pulse	0.0261 μsec
(15)	σR_1	RMS range error per pulse	3.92 m

rectangular bandpass filter. As will be shown in Section 7.4.3, the optimum noise bandwidth of a rectangular bandpass filter for the assumed range gate is

$$f_{b(bp)} = \frac{1.4}{\tau_p} \qquad (7.2\text{-}13)$$

For the assumed pulse width ($\tau_g = 0.1$ μsec), the optimum noise bandwidth of the bandpass IF amplifier is 14 MHz.

The RMS range-tracking error per pulse for the example is calculated in Table 7.2-2. The receiver noise power density at the input to the IF amplifier is denoted p_n and is equal to

$$p_n = KT \qquad (7.2\text{-}14)$$

where K is Boltzmann's constant, and T is the noise temperature of the receiver. As shown in item (1), the receiver noise temperature is assumed to be 700 °K. Item (2) gives the Boltzmann constant K, which was obtained from Chapter 6, Eq 6.2-2. By Eq 7.2-14, the temperature T in item (1) is multiplied by K in item (2) to obtain the noise power density p_n of item (3).

In calculating signal/noise ratios, the signal gain of the receiver is ignored, and all signal and noise levels are referenced to the input to the receiver. Hence the noise power density p_n of item (3) holds at the output of the IF amplifier as well as at the input.

As will be explained in Section 7.3, the ideal receiver for any signal is called the matched filter for that signal. For practical reasons, a matched filter is usually not implemented exactly. However, it provides an ideal reference against which actual receivers are compared. The signal/noise power ratio at the output of a matched-filter receiver is

Signal/noise at output of matched filter:

$$\text{Matched}\left[\frac{P_s}{P_n}\right] = \frac{E_s}{p_n} \qquad (7.2\text{-}15)$$

The signal/noise ratio P_s/P_n is the ratio of signal power to noise power at the time of the peak signal. In Table 7.2-2, item (4) is the received signal energy E_s, which was obtained from Table 7.2-1, item (14). This value for E_s in item (4) is divided by p_n in item (3), to obtain the ratio E_s/p_n of item (5). By Eq 7.2-15, this is the signal/noise power ratio at the output of an ideal matched-filter receiver. Note that this signal/noise power ratio, which is 4.189, is nondimensional.

As will be shown in Section 7.4, Eq 7.4-22, the IF amplifier for the assumed amplifier and range gate has a matching loss L_m of 0.85 dB, which is a factor of 1.216. This matching loss, shown in item (6), describes the reduction of actual signal/noise ratio relative to that of an ideal matched-filter receiver. Thus, the actual signal/noise power ratio at the output of the IF amplifier is

Signal/noise at IF amplifier output:

$$\frac{P_s}{P_n} = \frac{1}{L_m}\,\text{matched}\left[\frac{P_s}{P_n}\right] = \frac{E_s/p_n}{L_m} \qquad (7.2\text{-}16)$$

Therefore, the ratio E_s/p_n of item (5) is divided by the matching loss L_m of item (6) to obtain the signal/noise power ratio at the output of the actual IF amplifier, shown in item (7). This is the signal/noise power ratio at the input to the envelope detector.

There are two signal-detection processes in Fig 7.2-2: the mixing process, which converts the 10-GHz carrier to 60 MHz, and the envelope detector, which converts the pulse-modulated 60-MHz carrier to video pulses. The mixer cross-correlates the received signal with the local oscillator, and so provides coherent detection. This coherent detection process maintains phase information as well as amplitude information. In contrast, the envelope detector destroys phase information in the detection process, and so is a noncoherent detector.

With coherent detection, the signal/noise power ratio is unaffected by the detection process. However, with noncoherent detection, there is a loss of signal/noise ratio when the signal/noise ratio at the input to the noncoherent detector is not much greater than unity. The effect of noncoherent detection can be approximated by including in the analysis the following noncoherent detection loss L_d:

Noncoherent detection loss:

$$L_d = 1 + P_n/P_s \qquad (7.2\text{-}17)$$

The term P_n/P_s is the noise/signal power ratio at the input to the noncoherent detector, which is the reciprocal of the signal/noise power ratio.

For example, Eq 7.2-17 shows that if the signal/noise power ratio P_s/P_n at the input to the envelope detector (or any noncoherent detector) is unity, the noncoherent detection loss L_d is 2, and so there is a 3-dB loss in the detector If this signal/noise ratio is 10, the noncoherent detection loss L_d is 1.1 (or 0.4 dB). On the other hand, if the signal/noise ratio is 0.1, the noncoherent detection loss L_d is 11 (or 10.4 dB).

This illustrates the principle that the signal/noise ratio at the input to a noncoherent detection process should, if possible, be appreciably greater than unity. If this is achieved, the noncoherent detection loss is small, and the noncoherent detection process is essentially as good as coherent detection. When the signal/noise ratio is less than unity, there is considerable loss in noncoherent detection.

In Table 7.2-2, item (7) gives the signal/noise power ratio P_s/P_n at the input to the envelope detector. The reciprocal of this is the noise/signal ratio P_n/P_s shown in item (8), which is 0.291. Adding unity to this ratio gives the noncoherent detection loss L_d given in item (9). This loss L_d is 1.291, or 1.1 dB.

The noncoherent detection loss increases the effective noise density at the output of the envelope detector, which is denoted p_{nd}. This detected noise power density p_{nd} is related as follows to the noise power density p_n at the input to the envelope detector:

Effective noise power density at detector output:

$$p_{nd} = p_n L_d = p_n \left(1 + \frac{P_n}{P_s} \right) \qquad (7.2\text{-}18)$$

The noncoherent detection loss L_d is multiplied by the IF-amplifier noise power density p_n of item (3) to obtain the detected noise power density p_{nd}, given in item (10). The signal energy per pulse, E_s of item (4), is divided by p_{nd} of item (10) to obtain the ratio E_s/P_{nd}, shown in item (11), which is 3.25.

The RMS range-gate noise timing error per pulse is denoted σt_1, where the subscript 1 indicates that this is the error for a single pulse measurement. In

Section 7.4, Eq 7.4-23 shows that the relative RMS timing error is equal to

Relative RMS range-gate timing error per pulse:

$$\frac{\sigma t_1}{\tau_p} = \frac{1}{1.5\sqrt{2E_s/p_{nd}}} \tag{7.2-19}$$

This relation applies to the early–late range-gate circuit, for $\tau_g = 2\tau_p$ (as given by Eq 7.2-12), when the IF amplifier has an optimized rectangular bandpass of noise bandwidth given by Eq 7.2-13. Substitute into Eq 7.2-19 the value $E_s/p_{nd} = 3.25$ given in item (11). This gives $\sigma t_1/\tau_p = 0.261$, as shown in item (12).

The pulse width τ_p is 0.1 μsec, as was shown in Table 7.2-1, item (13). This is repeated in Table 7.2-2, item (13). Multiplying the ratio $\sigma t_1/\tau_p = 0.261$ in item (12) by this pulse width gives the RMS range-track timing error per pulse σt_1, shown in item (14). This is multiplied by 150 m/μsec (Eq 7.2-10) to obtain the RMS range error per pulse, σR_1, which is 3.92 m, as shown in item (15).

Thus, the range gate derives from each pulse a measure of the range to the target. Because of radar receiver noise, there is a random error in each pulse measurement, which has a Gaussian distribution with an RMS value of 3.92 m.

The timing of the range gate is controlled by digital range-gate control circuitry, which uses the range-gate error signal to keep the range gate centered over the received target pulse. The range-gate control loop is illustrated in the receiver block diagram that was shown in Fig 7.2-2. A timing reference pulse is received from the transmitter at the instant that the radar pulse is transmitted. The range-gate control circuitry counts the time following each transmitted pulse, for an interval equal to the estimated elapsed time between the transmitted pulse and the received target pulse, which is proportional to the estimated target range. By means of a digital feedback loop, this estimated elapsed time is controlled so that the range-gate error signal is minimized. The circuit provides a measure of the estimated target range, which is proportional to the estimated elapsed time.

In order for the range-gate control to work properly, the gain of the 60-MHz IF amplifier must be controlled to keep the video pulse level approximately constant at the output of the envelope detector. This is achieved by means of the automatic-gain-control (AGC) feedback loop illustrated in Fig 7.2-2.

To achieve automatic gain control, the video pulses from the envelope detector are fed through a gate, which is timed by the range-gate timing circuitry to be open for the full gate width of the range gate. The video pulses passing through this gate are integrated, and fed into the automatic gain control (AGC) circuitry. This AGC circuitry controls the IF amplifier gain to keep the pulse amplitudes approximately constant.

When the signal/noise power ratio at the output of the envelope detector is close to unity, or less than unity, this AGC loop is strongly affected by the noise level. Consequently, at low signal levels, gain control is degraded, which in turn degrades the dynamic performance of the range-gate tracking loop.

7.2.3   Response of Range-Tracking Feedback Loop

The digital range-gate tracking loop has a loop transfer function of the following form:

$$G_{(r)} = \frac{\omega_{c(r)}\left(1 + \dfrac{\omega_{i(r)}}{s}\right)}{s} \tag{7.2-20}$$

This transfer function ignores sampled-data effects. As was shown in Chapter 6, Table 6.2-2 (loop E), the noise bandwidth in rad/sec for this loop transfer function is

$$\omega_{b(r)} = \frac{\pi}{2}\left(\omega_{c(r)} + \omega_{i(r)}\right) \tag{7.2-21}$$

Hence, the noise bandwidth in hertz is

$$f_{b(r)} = \frac{\omega_{b(r)}}{2\pi} = \tfrac{1}{4}\left(\omega_{c(r)} + \omega_{i(r)}\right) \tag{7.2-22}$$

In Chapter 8, Eq 8.6-26 will show that the optimum value of the ratio ω_c/ω_i for a feedback loop of this type is

$$\text{Optimum}[\omega_c/\omega_i] = 1.84 \tag{7.2-23}$$

This condition yields the minimum peak error due to uncertainty of target acceleration, for a given noise bandwidth. It is convenient to round off this 1.84 ratio to 2.0, which gives the following for the range track loop:

$$\omega_{i(r)} = \tfrac{1}{2}\omega_{c(r)} \tag{7.2-24}$$

As shown in item (1) of Table 7.2-3, the range-tracking gain crossover frequency $\omega_{c(r)}$ is assumed to be 100 sec^{-1}. By Eq 7.2-24, $\omega_{i(r)}$ is 50 sec^{-1} [item (2)]. Substituting these values for $\omega_{c(r)}$, $\omega_{i(r)}$ of items (1),(2) into Eq 7.2-22 gives a noise bandwidth $f_{b(r)}$ of 37.5 Hz, as shown in item (3). By Eq 6.2-27 of Chapter 6, the equivalent integration time corresponding to this 37.5-Hz noise

TABLE 7.2-3   Calculation of Errors of Range-Tracking Feedback Loop

	Symbol	Parameter	Value
(1)	$\omega_{c(r)}$	Gain crossover frequency	100 sec^{-1}
(2)	$\omega_{i(r)}$	Integral break frequency	50 sec^{-1}
(3)	$f_{b(r)}$	Noise bandwidth	37.5 Hz
(4)	$T_{i(r)}$	Effective integration time	13.33 msec
(5)	F_p	Pulse repetition frequency	1.50 kHz
(6)	T_p	Pulse repetition period	0.667 msec
(7)	N	Number of pulses integrated	20
(8)	σR_1	RMS range noise error per pulse	3.92 m
(9)	σR	RMS range tracking noise error	0.877 m
(10)		Peak (3-sigma) range noise error	2.63 m
(11)	V_t	Maximum target velocity	600 m/sec
(12)		Maximum error at lock-on	± 6.0 m
(13)		Linear region of range gate	± 7.5 m

bandwidth is

$$T_{i(r)} = \frac{1}{2f_{b(r)}} = \frac{1}{2(37.5 \text{ Hz})} = 13.33 \text{ msec} \qquad (7.2\text{-}25)$$

This is shown in item (4).

The pulse repetition frequency F_p of the radar pulses is assumed to be 1.5 kHz, as shown in item (5). The reciprocal of this is the pulse repetition period T_p, which is 0.667 msec, as shown in item (6). The feedback action of the range-tracking loop effectively integrates the number of pulses that occur within the integration time $T_{i(r)}$ of the range-gate feedback loop. Hence, the effective number of pulses integrated is

$$N_{i(r)} = \frac{T_{i(r)}}{T_p} = \frac{13.33 \text{ msec}}{0.667 \text{ msec}} = 20.0 \qquad (7.2\text{-}26)$$

Item (8) shows the RMS range error per pulse, 3.92 m, obtained from Table 7.2-2, item (15). The averaging action of the range-tracking loop reduces the range error by the square root of the number of pulses integrated. Thus, the resultant RMS range noise error is

$$\sigma R = \frac{\sigma R_1}{\sqrt{N_{i(r)}}} = \frac{3.92 \text{ m}}{\sqrt{20}} = 0.877 \text{ m} \qquad (7.2\text{-}27)$$

This is shown in item (9). The peak limits of error are generally considered to be ± 3 sigma. Hence, the peak 3-sigma error is 3 times the 0.877-m RMS error, which is 2.63 m, as shown in item (10).

The range-tracking loop operates on sampled data, with a sampling frequency equal to the pulse repetition frequency, which is 1.5 kHz. This sampling frequency is nearly 100 times as large as the gain crossover frequency, 100 rad/sec or 16 Hz. Hence, the dynamic effects of sampling are small.

The range-tracking noise error can be reduced by decreasing the gain crossover frequency of the range-tracking loop, but this results in increased error due to target motion. Generally, the most important range-tracking error is the peak error occurring during lock-on. If this is too large, the signal may move outside the limits of the range gate and be lost. At lock-on, the range-tracking error experiences a ramp input equal to the radial velocity of the target. As shown in item (11) of Table 7.2-3, the maximum target velocity is assumed to be 600 m/sec, which is nearly Mach 2 at high altitudes. Assuming that this velocity is in the radial direction, the peak range-tracking error due to this ramp input is approximately

$$\text{Max}|R_e| = \frac{V_\tau}{\omega_{c(r)}} = \frac{600 \text{ m/sec}}{100 \text{ sec}^{-1}} = 6.0 \text{ m} \qquad (7.2\text{-}28)$$

As was shown in Fig 7.2-4, the error response characteristic of the range-tracking loop has a linear range equal to $\pm \tau_p/2$ in time, which is ± 0.05 μsec. Multiplying this by 150 m/μsec gives an equivalent variation of range of ± 7.5 m, which is shown in item (13). This is not much greater than the peak error ± 6.0 m, shown in item (12). Hence, the radar tracker may lose the target during lock-on if the initial range-gate error is excessive. To avoid this problem, the range gate may be widened during lock-on, and the gain crossover frequency of the range-gate tracking loop may be increased.

After the range gate has locked onto the target, the range-tracking error due to target motion is much smaller. The range-tracking error R_e after the lock-on transient has settled is approximately equal to

$$R_e \cong c_2 \frac{d^2R}{dt^2} = \frac{1}{\omega_{c(r)}\omega_{i(r)}} \frac{d^2R}{dt^2} \qquad (7.2\text{-}29)$$

where d^2R/dt^2 is the range acceleration of the target, and c_2 is the acceleration error coefficient. For the assumed parameters, this error reduces to

$$|R_e| \cong 0.0020 g_\tau \quad \text{meter} \qquad (7.2\text{-}30)$$

where g_t is the acceleration of the target in the radial direction expressed in g's. Even for a $10g$ acceleration, this error is only 0.02 m.

7.2.4 Measurement of Angular Tracking Error

7.2.4.1 Conical Scan. There are two primary methods for measuring angle-tracking error in a radar tracking system: conical scan and monopulse. Most early tracking radars used conical scan, in which the antenna beam is scanned in a conical pattern. If the target is at the center of the scan pattern, the amplitudes of the received radar pulses are constant during the scan. When the target is displaced from the scan center, the pulse amplitudes are modulated sinusoidally at the scan frequency. The sinusoid amplitude variation is proportional to the displacement of the target from the scan center, and the phase of the sinusoid characterizes the direction from the center.

Figure 7.2-6 shows a typical conical-scan pattern. The dashed circle is the path of the peak of the beam during the scan. The solid circles are the contours of 1 dB gain below peak gain, corresponding to four points on

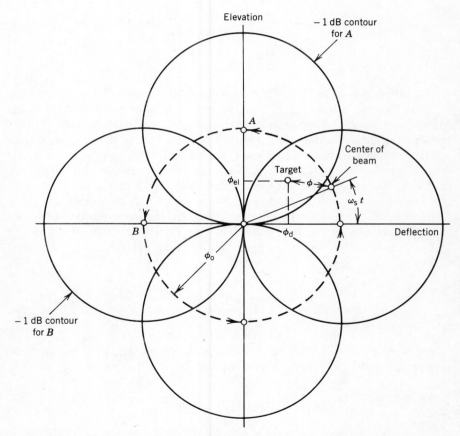

Figure 7.2-6 Typical conical-scan pattern, showing gain contours at four points in scan.

the scan separated by 90°. The angle ϕ_o is the offset angle of the scan, which is the amount that the center of the beam is offset from the center of the scan. For this case, the beam offset angle is $0.289\Theta_b$, where Θ_b is the 3-dB bandwidth of the antenna. With this offset, the gain at the center of the scan pattern is 1 dB below peak antenna gain.

As was shown in Fig 7.2-1, antenna position is controlled in terms of the azimuth and elevation angles of the antenna pedestal. However, target error is measured in terms of axes that rotate with the antenna dish. The coordinates of these tracking-error axes are designated *elevation* and *deflection*, but sometimes the term *traverse* is used in place of *deflection*. The conical angle scan provides signals proportional to the elevation and deflection components of target error, measured relative to the center of the scan pattern.

If the target lies at the center of the scan pattern (zero tracking error), the amplitudes of the radar pulses are constant during the scan. When the target is displaced from the center of the scan (as indicated in Fig 7.2-6), the amplitudes of the video radar pulses at the output of the envelope detector are modulated sinusoidally as follows:

$$E = E_0 \left(1 + k_s \frac{\phi_d}{\Theta_b} \cos[\omega_s t] + k_s \frac{\phi_{eL}}{\Theta_b} \sin[\omega_s t] \right) \qquad (7.2\text{-}31)$$

where

E_0 = pulse voltage amplitude when the target is at the center of the scan
k_s = conical-scan slope parameter
ϕ_d = deflection tracking error
ϕ_{eL} = elevation tracking error
Θ_b = 3-dB beamwidth of antenna gain pattern
ω_s = angular frequency of conical scan

As will be shown in Section 7.5, Eq 7.5-22, the conical-scan slope is equal to

$$k_s = 5.54 \frac{\phi_o}{\Theta_b} \qquad (7.2\text{-}32)$$

where

ϕ_o = offset angle of conical scan

In accordance with Eq 7.2-6, the conical-scan offset angle ϕ_o reduces the effective antenna gain, relative to peak gain. The gain reduction in decibels due to angle scan is

$$10 \log[L_{as}] = 12(\phi_o/\Theta_b)^2 \text{ dB} \qquad (7.2\text{-}33)$$

A typical conical-scan loss $L_{as} = 1$ dB. For this value, the corresponding offset angle ϕ_o and conical-scan slope k_s, obtained from Eqs 7.2-32, -33, are

$$\phi_o = 0.289\Theta_b \tag{7.2-34}$$

$$k_s = 1.60 \tag{7.2-35}$$

The radar system experiences angle-scan loss in both transmission and reception. Hence, there is a total angle-scan *crossover loss*, denoted L_k, which is equal to

$$L_k = L_{as}^2 \tag{7.2-36}$$

For this typical scan, which has a 1-dB angle-scan loss, the total angle-scan crossover loss L_k is 2 dB.

For small antennas, the conical angle scan may be achieved by rapidly oscillating the antenna dish, but for larger antennas the feed is oscillated. Figure 7.2-7 shows the block diagram of a conical-scan radar receiver showing the circuitry for demodulating the elevation and deflection error signals. This diagram also indicates the functions for range-gate control and automatic gain control (AGC) of the IF amplifier gain. Automatic gain control is needed to derive good angle error signals, just as it is needed to derive a good range-tracking error signal. (To simplify the diagram, the 10-GHz RF preamplifier stage is omitted.)

As shown in Fig 7.2-7, the signal at the output of the envelope detector is fed through the angle-tracking gate, of pulse width τ_{ga}. The angle-tracking gate is synchronized with the range-tracking gate. However, the angle-tracking gate width τ_{ga} may be smaller than the range-tracking gate width τ_g, in order to reduce angle-tracking noise error. The range gate can be narrower for angle-tracking, because the range-tracking loop is much faster than the angle-tracking loop, and generally keeps the target pulse close to the center of the range gate.

The signal that passes through the angle-tracking gate is fed to a pulse integrator, which integrates the energy occurring within the angle-tracking gate. The resultant integrated value is sampled by a sample-and-hold circuit, which holds the value fixed between radar pulses.

The conical-scan mechanism oscillates the feed of the antenna to produce the required conical scan pattern. Usually, the scan rate is no greater than $\frac{1}{10}$ of the pulse repetition frequency (PRF), although higher rates up to $\frac{1}{4}$ of the PRF can be used if the conical scan is synchronized with the pulse rate. For the example, the PRF is 1.5 kHz. Let us assume that the conical scan frequency is 125 Hz, which is $\frac{1}{12}$ of the PRF. For this scan rate, one radar pulse occurs every 30° of the conical-scan cycle.

Figure 7.2-8 shows the resultant video waveform. The circles (a) show the amplitudes of the radar pulses, which are obtained from the pulse integrator. The sample-and-hold circuit holds the signal constant between pulses, thereby

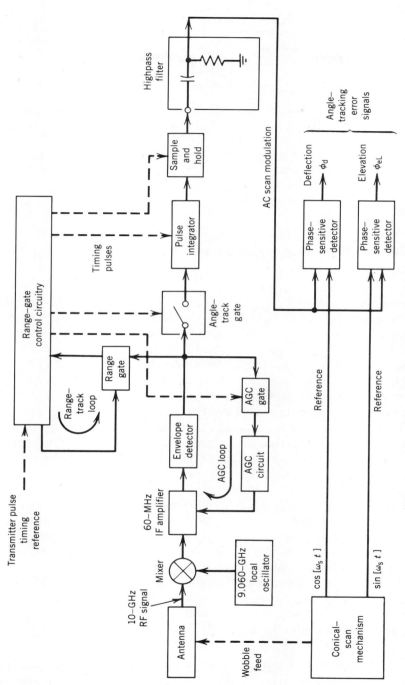

Figure 7.2-7 Receiver circuitry for conical-scan tracking.

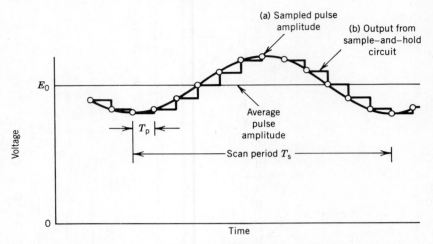

Figure 7.2-8   Derivation of conical-scan modulation from radar pulse amplitudes.

producing the waveform (b). This waveform (b) is fed through the highpass filter shown in Fig 7.2-7, which discards the average value E_0. The resultant AC scan modulation is fed to phase-sensitive detectors, which are excited by sine and cosine reference signals obtained from the conical-scan mechanism. (The operation of phase-sensitive detectors was discussed in Chapter 4, Section 4.2.3.) The outputs from these phase-sensitive detectors are DC voltages proportional to the deflection and elevation errors ϕ_d and ϕ_{eL}.

The radar signal consists of sampled data at the pulse repetition frequency. The circuit that holds the error signal fixed between radar pulses produces a time delay of the AC scan modulation equal to $\frac{1}{2}$ of a pulse repetition period T_p. This time delay is compensated for by applying a corresponding time shift of $T_p/2$ to the sine and cosine reference signals from the conical-scan mechanism.

The elevation error signal ϕ_{eL} is used as an error signal for the servo that drives the elevation axis of the antenna pedestal; and the deflection error signal ϕ_d is used as the error signal for the servo that drives the azimuth axis. To compensate for the fact that the deflection tracking error is not measured in the same coordinates as the azimuth pedestal error, the deflection error signal is fed through a gain compensation that is proportional to $1/\cos[\Theta_{eL}]$, where Θ_{eL} is the elevation angle of the pedestal. Since $1/\cos[\Theta_{eL}]$ is equal to $\sec[\Theta_{eL}]$, this gain variation is called *secant compensation*. Secant compensation is discussed further in Section 7.5.6.

7.2.4.2 Monopulse.
A serious deficiency with conical scan is that fluctuation of the radar reflectivity of the target (which is called *scintillation*) modulates the amplitudes of the target pulses, and so adds error to the

tracking signal. This problem is eliminated in a monopulse tracking system, which derives tracking information from a single pulse. Early tracking radars developed during World War II used conical scan. Modern high-performance radars generally use monopulse, but conical scan is sometimes used today because the system is much simpler.

Early monopulse tracking antennas used a four-horn feed to illuminate the antenna, but modern monopulse feeds often use more complicated structures to achieve better performance. With a four-horn feed, the signals from the horns illuminate four beams, which are squinted relative to one another in a manner similar to that indicated by the four solid beam contours in Fig 7.2-6. The signals received by the four horns are fed through microwave circuitry, which combines them to derive the following three signals: (1) the sum signal, (2) the elevation difference signal, and (3) the deflection difference signal. The transmitter power is fed into the channel for the sum signal, and excites the four horns equally.

A block diagram of the monopulse receiver circuitry is shown in Fig 7.2-9. The three signals from the feed (sum, elevation difference, and deflection difference) are fed into three matched receiver amplifiers. These consist of identical RF preamplifiers (not shown), identical mixers, and identical 60-MHz IF amplifiers. An important problem in the design of a monopulse receiver is the task of keeping the gains and phase shifts of the three amplifiers matched to one another.

The output from the sum-channel IF amplifier is fed to an envelope detector to derive video pulses. These video pulses are fed to a range gate to provide range tracking, just as for the conical-scan radar. The video pulses are also used for automatic gain control (AGC). In this case, the AGC circuitry must keep the gains of the three amplifiers matched as the radar signal level changes.

In Fig 7.2-10, the solid curve shows the amplitude of the IF signal in the elevation or deflection IF amplifier, plotted as a function of the angular tracking error in that axis. For positive error, the IF difference signal is in phase with the IF signal of the sum channel; while for negative error it is 180° out of phase.

The IF difference signal is equivalent to the position signal derived from an AC sensor, which was described in Chapter 4, Section 4.2. Hence, the corresponding DC error signal can be derived in the same manner, by feeding the AC signal from the IF amplifier into a phase-sensitive detector, which uses the sum-channel IF signal as a reference. By this means, the IF difference signal produces a positive DC error signal when it is in phase with the sum signal, and a negative DC error signal when it is out of phase with the sum signal. The dashed curve in Fig 7.2-10 shows the effective negative error signal for the latter case.

As shown in Fig 7.2-9, the signals from the elevation and deflection difference channels are fed to phase-sensitive detectors, which use the sum-channel signal as a reference. The output from each phase-sensitive detector is

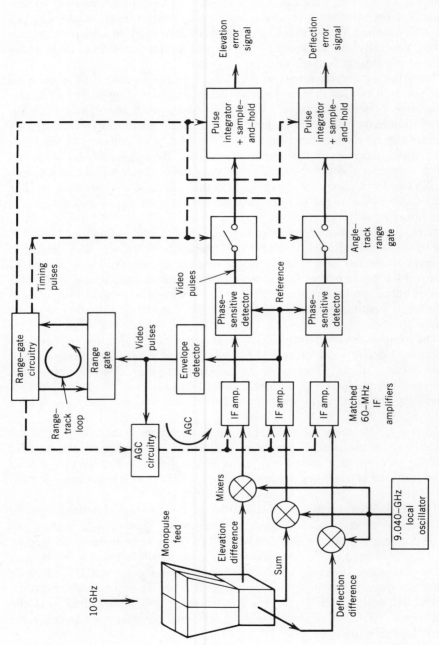

Figure 7.2-9 Receiver circuitry for monopulse tracking.

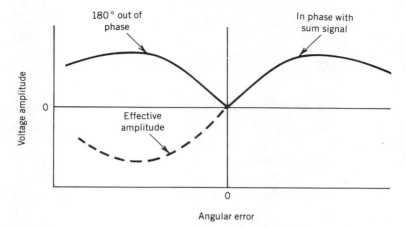

Figure 7.2-10 Amplitude of difference signal in monopulse receiver versus angle error.

a train of bipolar (positive and negative) video pulses. These video pulses are passed through an angle-scan range gate, a pulse integrator, and a sample-and-hold circuit, which are the same as those used for conical scan. The final outputs are bipolar DC voltages proportional to the elevation and deflection angular errors ϕ_{eL} and ϕ_d. These error signals are used to drive the elevation and azimuth antenna pedestal servos, in the same manner as the error signals derived from conical scan.

In the region near zero angle error, the effective voltage-amplitude plot of Fig 7.2-10 is nearly linear. Hence, the elevation and deflection IF amplifier voltages, E_{eL} and E_d, can be expressed as

$$E_{eL} = k_m \frac{\phi_{eL}}{\Theta_b} E_{sum} \qquad (7.2\text{-}37)$$

$$E_d = k_m \frac{\phi_d}{\Theta_b} E_{sum} \qquad (7.2\text{-}38)$$

where

E_{sum} = voltage amplitude in sum channel
k_m = monopulse slope

A monopulse feed causes a reduction of antenna gain in the sum channel relative to that for a simple feed. As will be explained in Section 7.5, Eq 7.5-49, the following are typical values of the monopulse slope k_s and the resultant sum-channel gain degradation of the antenna:

$$k_m = 1.6 \qquad (7.2\text{-}39)$$

$$\text{Gain degradation} = 1.0 \text{ dB} \qquad (7.2\text{-}40)$$

This 1-dB gain degradation for a monopulse feed results in a two-way signal loss of 2 dB, relative to that of a comparable antenna with a simple feed. Note that these values of tracking slope and gain degradation are the same as the corresponding typical parameters for conical scan given in Eq 7.2-35.

7.2.5 Equations for RMS Angle-Tracking Noise Error

The angle-tracking errors for conical-scan and monopulse tracking are analyzed in Section 7.5.4. The width of the range gate used to process the angle-tracking data is denoted τ_{ga}:

$$\tau_{\text{ga}} = \text{angle-track range-gate width}$$

This can be made appreciably narrower than τ_{g}, the range-gate width used for range tracking. It is assumed that the angle-tracking range-gate width τ_{ga} is greater than the radar pulse width τ_{p}. For this condition, the following are the effective RMS angle-tracking errors per pulse, $\sigma\Theta_1$, achieved by conical scan and monopulse:

Conical scan (for $\tau_{\text{ga}} > \tau_{\text{p}}$):

$$\frac{\sigma\Theta_1}{\Theta_{\text{b}}} = \frac{\sqrt{(1 + P_{\text{n}}/2P_{\text{s}})(\tau_{\text{ga}}/\tau_{\text{p}})}}{k_{\text{s}}\sqrt{E_{\text{s}}/P_{\text{n}}}} \tag{7.2-41}$$

Monopulse (for $\tau_{\text{ga}} > \tau_{\text{p}}$):

$$\frac{\sigma\Theta_1}{\Theta_{\text{b}}} = \frac{\sqrt{(1 + P_{\text{n}}/P_{\text{s}})(\tau_{\text{ga}}/\tau_{\text{p}})}}{k_{\text{s}}\sqrt{2E_{\text{s}}/P_{\text{n}}}} \tag{7.2-42}$$

These expressions are derived later in Eqs 7.5-71, -72. Equation 7.2-42 gives the actual RMS tracking error of the monopulse system for a single pulse. However, the conical-scan radar cannot derive a tracking error signal from a single pulse, and so Eq 7.2-41 merely gives an effective RMS tracking error per pulse for conical scan. For both cases, the actual angular-tracking error $\sigma\Theta$ is equal to

$$\sigma\Theta = \sigma\Theta_1/\sqrt{N_{\text{i(a)}}} \tag{7.2-43}$$

where $N_{\text{i(a)}}$ is the effective number of radar pulses integrated by the angle-tracking servo. The effective integration time of an angle-tracking servo is

$$T_{\text{i(a)}} = \frac{1}{2f_{\text{b(a)}}} \tag{7.2-44}$$

where $f_{b(a)}$ is the noise bandwidth of the angle-tracking servo. Hence, the number of radar pulses integrated by the angle-tracking servo is

$$N_{i(a)} = \frac{T_{i(a)}}{T_p} = \frac{1}{2f_{b(a)}T_p} \tag{7.2-45}$$

where T_p is the pulse repetition period of the radar.

In Eqs 7.2-41, -42 the ratio P_n/P_s is the noise/signal power ratio at the output of the IF amplifier for the conical-scan system, or at the output of the IF sum-channel amplifier for the monopulse system. This ratio is the reciprocal of the signal/noise power ratio, which is equal to

$$\frac{P_s}{P_n} = \frac{E_s/p_n}{L_m} \tag{7.2-46}$$

where L_m is the matching loss of the IF amplifier relative to that of an ideal matched-filter receiver. It is assumed that L_m is the same as that given for the range tracker, which is 0.85 dB [Table 7.2-2, item (6)].

Equations 7.2-41, -42 show that the monopulse and conical-scan radars have different noncoherent detection losses:

Conical-scan noncoherent detection loss:

$$L_{d(s)} = 1 + \frac{P_n}{2P_s} \tag{7.2-47}$$

Monopulse noncoherent detection loss:

$$L_{d(m)} = 1 + \frac{P_n}{P_s} \tag{7.2-48}$$

The monopulse noncoherent detection loss $(1 + P_n/P_s)$ is the same as that shown previously in Eq 7.2-17 for range-gate tracking. However, for conical scan, the noncoherent detection loss $(1 + P_n/2P_s)$ is lower. The reason for the reduced loss is that the conical-scan error signal is derived from the AC modulation of the video pulses, rather than from the pulse amplitude.

Note that there is an extra factor of $\sqrt{2}$ in the denominator of the monopulse tracking equation. Hence, at high values of the signal/noise ratio P_s/P_n, the monopulse error is lower than that for conical scan (assuming that $k_m = k_s$, and the tracking gain degradations of the systems are the same). However, when P_s/P_n drops below unity, the decreased noncoherent detection loss for conical scan compensates for this factor of $\sqrt{2}$, and so the RMS angular errors are nearly equal.

Thus, at high signal/noise ratios, the RMS tracking error for a monopulse system is typically $1/\sqrt{2}$, or 70%, of the error for a conical-scan system.

However, this is generally not an important consideration in the choice of monopulse over the much less complicated conical scan. The primary reason for using monopulse is that target scintillation can cause large tracking errors with conical scan, which do not occur with monopulse.

In the monopulse tracking radar, the phase-sensitive detectors that detect the IF signals in the difference channels are coherent detection processes. Hence, the signal/noise ratios at the outputs of the difference-signal IF amplifiers can be much less than unity without degrading the signal/noise ratio in the phase-sensitive detectors.

On the other hand, these phase-sensitive detectors are coherent detection processes only when the single/noise ratio of the reference signal derived from the IF sum channel is much greater than unity. When the sum-channel signal/noise ratio is low, the difference-channel signal/noise ratio is degraded in the phase-sensitive detectors. Hence, there is a noncoherent detection loss in the phase-sensitive detectors equal to $(1 + P_n/P_s)$, where P_n/P_s is the noise/signal ratio of the reference signal from the IF sum channel.

The effective single-pulse angle-tracking errors in Eqs 7.2-41, -42 can be expressed as follows in terms of the noncoherent detection loss values $L_{d(s)}$, $L_{d(m)}$ given in Eqs 7.2-47, -48:

Conical scan (for $\tau_{ga} > \tau_p$):

$$\frac{\sigma\Theta_1}{\Theta_b} = \frac{\sqrt{L_{d(s)}(\tau_{ga}/\tau_p)}}{k_s\sqrt{E_s/p_n}} \qquad (7.2\text{-}49)$$

Monopulse (for $\tau_{ga} > \tau_p$):

$$\frac{\sigma\Theta_1}{\Theta_b} = \frac{\sqrt{L_{d(m)}(\tau_{ga}/\tau_p)}}{k_m\sqrt{2E_s/p_n}} \qquad (7.2\text{-}50)$$

7.2.6 Computation of Angle-Tracking Errors for a Monopulse Example

The example assumes a monopulse tracking antenna. The 60% antenna efficiency given in Table 7.2-1, item (4), is consistent with a gain degradation of 1.0 dB in the monopulse tracking feed, as was assumed in Eq 7.2-40.

Table 7.2-4 shows the steps for calculating the relative RMS angle-tracking noise error per pulse. Item (1) is the ratio E_s/p_n for the monopulse sum channel, and item (2) is the sum-channel IF filter matching loss L_m. These are the same as the values given in items (5), (6) of Table 7.2-2. Dividing item (1) by item (2) gives the signal/noise ratio P_s/P_n shown in item (3). This signal/noise ratio occurs at the output of the sum-channel IF amplifier, which is the reference signal fed to the monopulse phase-sensitive detectors. This

TABLE 7.2-4 Calculation of RMS Angular Tracking Noise Error per Pulse

	Symbol	Parameter	Value
(1)	E_s/p_n	Sum-channel matched-filter signal/noise ratio	4.189
(2)	L_m	IF-amplifier matching loss	1.216 (0.85 dB)
(3)	P_s/P_n	Signal/noise ratio of sum channel	3.44
(4)	$L_{d(m)}$	Noncoherent detection loss in phase-sensitive detectors	1.291
(5)	k_m	Monopulse slope	1.6
(6)	τ_{ga}/τ_p	Relative angle-track gate width	1.5
(7)	$\sigma\Theta_1/\Theta_b$	Relative RMS angle noise error per pulse	0.300

ratio $P_s/P_n = 3.44$ of item (3) is the same as the P_s/P_n ratio of item (7) in Table 7.2-2.

In accordance with Eq 7.2-48, the noncoherent detection loss produced in the monopulse phase-sensitive detectors is

$$L_{d(m)} = 1 + \frac{P_n}{P_s} = 1 + \frac{1}{3.44} = 1.291 \qquad (7.2\text{-}51)$$

This is the same as the range-gate noncoherent-detection loss given in item (9) of Table 7.2-2 for range tracking.

In accordance with Eq 7.2-39, the monopulse slope k_m is assumed to be 1.6, as shown in item (5) of Table 7.2-4. Item (6) shows that the angle-track range-gate pulse width τ_{ga} is 1.5 times greater than the RF radar pulse width τ_p, which is 0.1 μsec. This width is sufficient to accommodate range-gate tracking error and smearing of the pulse by the IF amplifier, with only a small amount of energy falling outside the gate.

The values of items (1), (4), (5), (6) of Table 7.2-4 are substituted into Eq 7.2-50. This gives the value of $\sigma\Theta_1/\Theta_b$ shown in item (7). Thus, the RMS angle error per pulse, $\sigma\Theta_1$, is 0.300 times the antenna beamwidth Θ_b.

The actual RMS angle-tracking error depends on the noise bandwidths of the angle-tracking servos. This error is calculated in Table 7.2-5. As shown in item (1), the gain crossover frequency $\omega_{c(a)}$ of the angle-tracking loop is assumed to be equal to 15 rad/sec. This loop has an integral network with a break frequency $\omega_{i(a)}$ that is set equal to $\frac{1}{3}$ of the gain crossover frequency, or 5 rad/sec, as shown in item (2). For the range-tracking loop, the integral-network break frequency ω_i was set equal to $\frac{1}{2}$ of the gain crossover frequency, which is optimum for such a loop that has essentially ideal performance. However, a greater ratio ω_c/ω_i is needed in an angle-track loop, because of the dynamic limitations of the antenna drive. Although a ratio ω_c/ω_i of 3 is assumed for the angle-tracking loop, a somewhat larger ratio may sometimes be desirable.

TABLE 7.2-5   Calculation of Peak Two-Axis Angle-Tracking Error Due to Receiver Noise

Symbol		Parameter	Value
(1)	$\omega_{c(a)}$	Angle-tracking gain crossover frequency	15 sec^{-1}
(2)	$\omega_{i(a)}$	Angle-tracking integral break frequency	5 sec^{-1}
(3)	ω_b/ω_c	Noise bandwidth ratio	2.36
(4)	$f_{b(a)}$	Angle-tracking noise bandwidth	5.63 Hz
(5)	$T_{i(a)}$	Angle-tracking integration time	88.8 msec
(6)	T_p	Pulse repetition period	0.667 msec
(7)	$N_{i(a)}$	Number of pulses integrated by angle-tracking servos	133
(8)	$\sigma\Theta_1/\Theta_b$	Relative RMS angle noise error per pulse	0.300
(9)	$\sigma\Theta/\Theta_b$	Relative RMS angle noise error	0.0260
(10)	Θ_b	Half-power beamwidth	1.60°
(11)	$\sigma\Theta$	RMS one-axis noise error	0.0416°
(12)		Peak two-axis noise error	±0.176°
			±3.08 mrad

It is assumed that the loop transfer function of the angle-tracking loop can be approximated by

$$G_{(a)} = \frac{\omega_c(1 + \omega_i/s)}{s(1 + s/\omega_f)} \qquad (7.2\text{-}52)$$

The magnitude asymptote plot of this transfer function was shown in Chapter 6, Fig 6.2-5c (loop F). The noise bandwidth ω_b of loop F was given in Table 6.2-2 as

$$\frac{\omega_b}{\omega_c} = \frac{\pi\omega_f(\omega_c + \omega_i)}{2\omega_c(\omega_f - \omega_i)} = \frac{\pi(1 + \omega_i/\omega_c)}{2(1 - \omega_i/\omega_f)} \qquad (7.2\text{-}53)$$

It is assumed that $\omega_f/\omega_c = \omega_c/\omega_i = 3$. For this setting, the step response has 25% peak overshoot and the maximum value of $|G_{ib}|$ is 1.30. Hence the assumed loop parameters result in a good practical design. With this setting, the ratio ω_b/ω_c of Eq 7.2-53 is 2.36, as shown in item (3).

Thus the noise bandwidth ω_b for this case is 2.36 times greater than the gain crossover frequency ω_c. (This is actually an optimistic ratio for a practical antenna servo. When an antenna servo has strong structural resonance, the ratio ω_b/ω_c can be appreciably greater than 2.36.) The gain crossover frequency 15 rad/sec, shown in item (1), is multiplied by 2.36 to obtain the noise bandwidth ω_b of the angle tracking loop, which is 35.4 rad/sec. Dividing this by 2π gives the noise bandwidth f_b in hertz, which is 5.63 Hz, as shown in item (4).

Substitute the noise bandwidth $f_{b(a)} = 5.63$ Hz of item (4) into Eq 7.2-44. This gives the effective integration time $T_{i(a)}$ of the angle-tracking servos, shown in item (5) as 88.8 msec. The pulse repetition period T_p (0.667 msec) is shown in item (6), and was obtained from item (6) of Table 7.2-3. Dividing the integration time $T_{i(a)}$ by the pulse repetition period T_p gives the number of pulses integrated by the antenna servos, which is

$$N_{i(a)} = \frac{T_{i(a)}}{T_p} = \frac{88.8 \text{ msec}}{0.667 \text{ msec}} = 133 \qquad (7.2\text{-}54)$$

This is shown in item (7) of Table 7.2-5. Item (8) gives the relative RMS angle noise per pulse (0.300) obtained from item (7) of Table 7.2-4. This is divided by the square root of the number of pulses integrated (133) to obtain the relative RMS angle noise error:

$$\frac{\sigma\Theta}{\Theta_b} = \frac{\sigma\Theta_1/\Theta_b}{\sqrt{N_{i(a)}}} = \frac{0.300}{\sqrt{133}} = 0.0260 \qquad (7.2\text{-}55)$$

This is shown in item (9). The half-power beamwidth Θ_b of 1.60°, shown in item (10), was obtained from item (6) of Table 7.2-1. The values of items (9), (10) are multiplied together to obtain the actual RMS noise error $\sigma\Theta$, shown in item (11) to be 0.0416°.

There are two components of angle-tracking error: elevation and deflection. The total beam-tracking error ϕ_e is related as follows to the elevation error ϕ_{eL} and deflection error ϕ_d:

$$\phi_e^2 = \phi_{eL}^2 + \phi_d^2 \qquad (7.2\text{-}56)$$

Since the RMS error components for the two axes are equal, the total RMS 2-axis beam tracking error is $\sqrt{2}$ times the single-axis RMS error. The peak error is considered to be 3 times the RMS error. Hence, the peak 2-axis beam error $\text{Max}[\phi_e]$ is related as follows to the RMS single-axis error $\sigma\Theta$:

$$\text{Max}[\phi_e] = 3\sqrt{2}\,\sigma\Theta = 4.24\sigma\Theta \qquad (7.2\text{-}57)$$

Thus, the RMS single-axis noise error of item (11) in Table 7.2-5 is multiplied by 4.24 to obtain the peak two-axis error shown in item (12). The error in degrees is multiplied by 17.45 mrad/deg to obtain the error in milliradians, which is ± 3.08 mrad.

The angle error due to target motion is calculated in Table 7.2-6. It is assumed that the target is following a straight-line constant-velocity course. The tracking errors for this course were calculated in Chapter 3, Section 3.2, using error coefficients. The analysis showed that the maximum angular

TABLE 7.2-6   Calculation of Peak Angular Error Due to Target Motion

Symbol	Parameter	Value
(1) V_t	Maximum target velocity	600 m/sec
(2) R_{min}	Minimum range	800 m
(3) Ω_{Max}	Maximum angular velocity	0.75 rad/sec
(4) α_{Max}	Maximum angular acceleration	0.366 rad/sec^2
(5) c_2	Acceleration error coefficient	$\frac{1}{75}$ sec^2
(6)	Peak target-motion error	± 4.88 mrad

velocity Ω_{Max} and the maximum angular acceleration α_{Max} are

$$\Omega_{Max} = V_t / R_{min} \tag{7.2-58}$$

$$\alpha_{Max} = 0.65 \Omega_{Max}^2 \tag{7.2-59}$$

where V_t is the target velocity, and R_{min} is the minimum target range.

As was shown in item (11) of Table 7.2-3, the maximum target velocity V_t is assumed to be 600 m/sec, which is nearly Mach 2 at high altitudes. This is shown in item (1) of Table 7.2-6. The minimum range R_{min} is assumed to be 800 m, as shown in item (2). By Eq 7.2-58, the maximum angular velocity, shown in item (3), is

$$\Omega_{Max} = \frac{V}{R_{min}} = \frac{600 \text{ m/sec}}{800 \text{ m}} = 0.75 \text{ rad/sec} \tag{7.2-60}$$

Substituting this into Eq 7.2-59 gives the maximum angular acceleration, shown in item (4):

$$\alpha_{Max} = 0.65 \Omega_{Max}^2 = 0.65 (0.75 \text{ rad/sec})^2$$
$$= 0.366 \text{ rad/sec}^2 = 366 \text{ mrad/sec}^2 \tag{7.2-61}$$

It is assumed that the integral network has infinite gain at zero frequency, and so the velocity error coefficient is zero. Hence the maximum angle-tracking error due to target motion, after the initial lock-on transient has decayed, is approximately equal to the maximum angular acceleration multiplied by the acceleration error coefficient c_2. Item (5) gives the approximate acceleration error coefficient c_2, which is equal to $1/\omega_c \omega_i$:

$$c_2 = \frac{1}{\omega_{c(a)} \omega_{i(a)}} = \frac{1}{(15 \text{ sec}^{-1})(5 \text{ sec}^{-1})} = \frac{1}{75} \text{ sec}^2 \tag{7.2-62}$$

Hence the maximum error due to target motion is approximately

$$\text{Max}[\Theta_e] = c_2\alpha_{\text{Max}} = \left(\tfrac{1}{75}\text{ sec}^2\right)\left(366\text{ mrad/sec}^2\right)$$
$$= 4.88\text{ mrad} \tag{7.2-63}$$

This is shown in item (6).

Thus, the angle-tracking radar experiences a peak angle error due to target motion of ± 4.88 mrad, as shown in Table 7.2-6, and a peak two-axis error due to radar noise of ± 3.08 mrad, as shown in Table 7.2-5.

This example illustrates the statistical principles used in the calculation of radar tracking errors due to receiver noise. It shows that optimum setting of the dynamic parameters of the feedback loops require a compromise between radar noise error and error due to target motion.

7.3 ANALYSIS OF DETECTION PROCESSES IN A TRACKING RADAR SYSTEM

This section analyzes the target-detection processes of a pencil-beam tracking radar, such as might be used in a fire control system for directing guns or missiles against an attacking aircraft. It is assumed that the target has been detected by a continuously rotating search radar with a fan-shaped beam that is narrow in azimuth but broad in elevation. The tracking radar searches over the region of uncertainty of target location until it detects the target, and then tracks the target in range and angle. To perform angle track, the antenna has a monopulse tracking feed that provides angular error signals.

A simplified drawing of a tracking radar antenna was shown in Section 7.2, Fig 7.2-1. It has a parabolic reflector, in front of which is a feed supported on spars. Energy radiating from the feed is reflected by the paraboloid and projected in the direction of the antenna axis (the boresight), but the beam diverges somewhat because of diffraction. If there is a target in the transmitted beam, it reflects a small amount of energy back to the antenna, which is intercepted by the reflector and projected back to the feed. The direction of the antenna beam is controlled by servos that drive the antenna gimbal structure in two axes, called azimuth and elevation. The azimuth servo rotates the whole structure about a vertical axis, thereby changing its heading relative to north. The elevation gimbal rides on the rotating azimuth carriage and tilts the antenna boresight upward relative to the horizontal plane.

7.3.1 The Radar Transmission Equation

The following is a derivation of the radar transmission equation, which is similar to the derivation by Skolnick [7.1] (p. 3). If a radar antenna were isotropic, radiating its power uniformly in all directions, the transmitted power

P_t at the range R of the target would be spread uniformly over a spherical surface of area $4\pi R^2$. Therefore, the portion of the transmitted power reflected by the target would be the ratio of the effective radar cross-section area of the target, denoted A_t, divided by this spherical surface area $4\pi R^2$. The power reflected from the target would be

$$P_{ref} = P_t \frac{A_t}{4\pi R^2} \qquad (7.3\text{-}1)$$

where P_t is the power transmitted by the radar. With a directive radar antenna, the power radiated in the direction of the target is increased by the antenna gain G in that direction. Hence, the actual reflected power is

$$P_{ref} = \frac{P_t G A_t}{4\pi R^2} \qquad (7.3\text{-}2)$$

The target is assumed to act like an isotropic radiator. Hence the power P_r received by the radar antenna is related as follows to the power reflected from the target:

$$P_r = P_{ref} \frac{A_r}{4\pi R^2} \qquad (7.3\text{-}3)$$

where A_r is the effective receive (or capture) area of the radar antenna, for the particular direction to the target. Combining Eqs 7.3-2, -3 gives for the received power

$$P_r = \frac{P_t G A_r A_t}{\left(4\pi R^2\right)^2} \qquad (7.3\text{-}4)$$

Antenna theory shows that the antenna gain G in a particular direction is related as follows to the effective capture area A_r for a signal from that direction:

$$G = \frac{4\pi A_r}{\lambda^2} \qquad (7.3\text{-}5)$$

where λ is the wavelength. Solve Eq 7.3-5 for A_r and substitute this into Eq 7.3-4. This yields the following "radar equation", which gives the ratio of receive to transmit power for the radar:

$$\frac{P_r}{P_t} = \frac{G^2 \lambda^2 A_t}{\left(4\pi\right)^3 R^4} \qquad (7.3\text{-}6)$$

For the direction along the axis (or boresight) of the antenna, the antenna gain G and the capture area A_r are maximum. The maximum effective capture area is equal to the area of the antenna aperture multiplied by the antenna efficiency η. Let us assume a circular parabolic antenna with a reflector diameter D. The aperture area is the projected area of the circular reflector, which is $\pi D^2/4$. Hence the maximum effective capture area is

$$A_{r(max)} = \eta \frac{\pi D^2}{4} \qquad (7.3\text{-}7)$$

Substituting this into Eq 7.3-5 gives the following for the maximum antenna gain G_m:

$$G_m = \eta \left(\frac{\pi D}{\lambda} \right)^2 \qquad (7.3\text{-}8)$$

If the reflector were illuminated uniformly, the antenna efficiency η would be unity, but the sidelobes of the antenna gain pattern would be very high. Sidelobes are greatly reduced in practical antennas by tapering the illumination so that it is low at the edge of the dish (typically about 18 dB below the maximum). A typical value of the resultant antenna efficiency is 0.75. As will be shown in Section 7.5 (Eqs 7.5-47 and -49), the monopulse antenna tracking feed causes an additional loss that is typically 1 dB, which is a factor of 1.26. This results in a net antenna efficiency of $0.75/1.26 = 0.60$. Thus, an antenna efficiency of 60% is assumed for the antenna–feed combination.

As shown by Skolnick [7.1] (p. 60, Eq 2.49), the half-power (or -3-dB) beamwidth Θ_b of a practical radar antenna beam usually can be approximated roughly by

$$\Theta_b = 65(\lambda/D) \quad \text{deg} \qquad (7.3\text{-}9)$$

These equations are applied in Table 7.3-1 to calculate the minimum power received from the tracking radar during acquisition. It is assumed that the radar frequency f is 10 GHz (item 1). The wavelength λ (item 2) is calculated from

$$\lambda = \frac{c}{f} = \frac{3.0 \times 10^{10}\,\text{cm/sec}}{10 \times 10^9\,\text{Hz}} = 3.0\,\text{cm} \qquad (7.3\text{-}10)$$

where c is the speed of light. It is assumed that the antenna reflector diameter D is 4 ft (item 3). As explained in the preceding, the antenna efficiency η is set at 0.60 (item 4). From Eqs 7.3-8, -9 the values of peak antenna gain G_m and half-power beamwidth Θ_b shown in items (5, 6) are calculated using the parameters of items (2) to (4).

In the region within a few decibels of the peak antenna gain, the antenna gain pattern of any practical tracking antenna has approximately a Gaussian

TABLE 7.3-1   Calculation of Minimum Signal Power into Receiver

Symbol		Parameter	Value
(1)	f	Frequency	10 GHz $= 10 \times 10^9$ Hz
(2)	λ	Wavelength	3.0 cm
(3)	D	Reflector diameter	4.0 ft $= 122$ cm
(4)	η	Antenna efficiency	0.60
(5)	G_m	Peak antenna gain	$9390 = 39.9$ dB
(6)	Θ_b	Half-power beamwidth	$1.60°$
(7)		One-way search loss^{-1}	-2 dB $= 0.631$
(8)	G	Effective antenna gain	$5925 = 37.9$ dB
(9)	W	Width of angle scan	$0.567\Theta_b = 0.907°$
(10)	A_t	Target cross section	1.0 meter^2
(11)	R	Maximum range to target	50 km $= 31$ mile
(12)	P_r/P_t	Transmission loss^{-1}	2.55×10^{-18}
(13)		Waveguide loss^{-1}	-2 dB $= 0.631$
(14)	P_i/P_{gen}	Total loss^{-1}	1.61×10^{-18}
(15)	P_{gen}	Transmitter power	100 kW $= 10^5$ Watt
(16)	P_{si}	Signal power into receiver	1.61×10^{-13} Watt

shape and so can be expressed as

$$G = G_m \exp\left[-(\phi/\Psi)^2 \right] \qquad (7.3\text{-}11)$$

where ϕ is the angle measured from the center of the beam, and Ψ is a constant. The half-power beamwidth Θ_b is the width of the antenna beam between the points where G/G_m is $\frac{1}{2}$. Hence

$$\frac{G}{G_m} = \frac{1}{2} \quad \text{where } \phi = \pm\frac{1}{2}\Theta_b \qquad (7.3\text{-}12)$$

Combining Eqs 7.3-11, -12 shows that

$$\Psi = 0.601\Theta_b \cong 0.600\Theta_b \qquad (7.3\text{-}13)$$

Take 10 times the logarithm of the reciprocal of Eq 7.3-12. This gives the following expression for the antenna pointing loss in decibels relative to the peak antenna gain:

$$\text{dB loss} = 10 \log\left[\frac{G_m}{G} \right] = 12\left(\frac{\phi}{\Theta_b} \right)^2 \quad \text{decibel} \qquad (7.3\text{-}14)$$

The constant 12 is chosen so that this gives a loss of 3 dB at $\phi = \pm\Theta_b/2$. Figure 7.3-1a is a plot of the relative antenna gain G/G_m expressed in

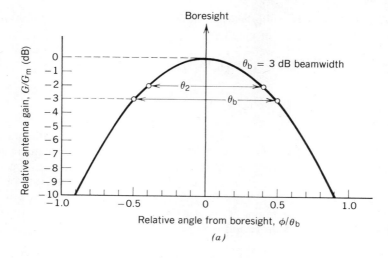

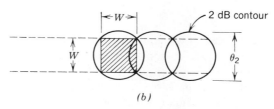

Figure 7.3-1 Antenna gain pattern (*a*) gain near peak of beam; (*b*) assumed angle scan pattern.

decibels, in the region near the peak of the beam. This is expressed as a function of the relative angle from boresight, ϕ/Θ_b.

When the search radar detects the target, it often has no measure of target elevation, and the uncertainty in azimuth is usually wider that $\pm \frac{1}{2}$ beamwidth of the tracking radar. The pencil beam of the tracking radar must search over the region of angular uncertainty until the target is acquired. It is assumed that during search the antenna beam patterns are overlapped at the -2-dB points. If the target falls at an overlap point of the beam, there is a 2-dB loss of antenna gain, relative to the peak antenna gain G_m, which represents the following factor:

$$\text{One-way search loss} = 2 \text{ dB} = 10^{0.2} = 1.585 = 1/0.631 \quad (7.3\text{-}15)$$

The antenna experiences this loss is both transmission and reception, and so the total two-way search loss in decibels is twice this value, or 4 dB. In item (7) of Table 7.3-1, the one-way angle-search loss is expressed in terms of the reciprocal loss, denoted loss^{-1}, which is -2 dB, or 0.631. The peak antenna gain G_m of item (5) is multiplied by the search loss^{-1} of item (7) to obtain the effective antenna gain G of item (8).

In accordance with Eq 7.3-14, the -2-dB beamwidth, denoted Θ_2, is related as follows to the -3 dB (half-power) beamwidth Θ_b:

$$\Theta_2 = \Theta_b\sqrt{(2 \text{ dB})/(3 \text{ dB})} = 0.816\Theta_b \qquad (7.3\text{-}16)$$

The 2-dB beamwidth Θ_2 and the 3-dB beamwidth Θ_b are illustrated in diagram a of Fig 7.3-1. Diagram b shows a portion of the angular search pattern. It is assumed that the antenna beam jumps from one location to the next, and remains fixed long enough to allow the radar to integrate over the required number of pulses. All of the points within the cross-hatched area are within 2 dB of the maximum antenna gain during the observation time. The width of this angular scan region is

$$W = \frac{\Theta_2}{\sqrt{2}} = \frac{0.816\Theta_b}{\sqrt{2}} = 0.567\Theta_b = 0.907° \qquad (7.3\text{-}17)$$

This is shown in item (9). The antenna beam is actually scanned in a continuous manner, but it is simpler to analyze the search process in terms of discrete observations.

As shown in item (10) of Table 7.3-1, the target is assumed to have a radar cross section A_t of 1.0 square meter, which is a typical value for a small aircraft. The maximum range to the target is assumed to be 50 kilometer (item 11). Applying the values of items (2, 8, 10, and 11) to the radar transmission relation Eq 7.3-6 gives the transmission loss^{-1} P_r/P_t shown in item (12). This represents the ratio of receive to transmit signal power at the ports of the antenna feed. Additional loss is experienced in the duplexer that isolates the receiver from the transmitted pulse, and in the waveguide runs (from the transmitter to the feed and from the feed to the receiver). This is represented as a total waveguide loss^{-1} of -2 dB, shown in item (13). Multiplying items (12) and (13) gives the total signal loss^{-1} of item (14) from the transmitter to the receiver. It is assumed that the transmitter generates a peak generated power P_{gen} of 100 kilowatt, as shown in item (15). Multiplying items (14) and (15) gives the power into the receiver at maximum range, shown in item (16), which is 1.6×10^{-13} watt.

7.3.2   Optimum Filtering of Radar Signal

The capability of the radar to detect the weak echo signal from the target in the presence of receiver noise is described in terms of the signal/noise power ratio at the envelope detector, which is calculated in Table 7.3-2. As shown in item (1) of Table 7.3-2, a receiver noise temperature of 700°K is assumed, which is typical of a modern receiver of wide dynamic range operating at 10

TABLE 7.3-2   Calculation of Signal-to-Noise Ratio at Output of IF Amplifier, Which is Input to Envelope Detector

Symbol	Parameter	Value
(1) T	Receiver noise temperature	700 °K
(2) K	Boltzmann constant	1.374×10^{-23} joule/°K
(3) $p_n = KT$	Noise power density (N_o)	9.62×10^{-21} watt/Hz
(4) τ_p	Pulse width	$0.1\ \mu\text{sec} = 10^{-7}$ sec
(5) ΔR	Range resolution	15 meter = 49 ft
(6) $f_{b(bp)}$	Optimum IF noise bandwidth for single-tuned stage	$0.63/\tau_p = 6.3 \times 10^6$ Hz
(7) P_{no}	Noise power at detector input	6.06×10^{-14} watt
(8) P_{si}	Signal power into receiver	1.61×10^{-13} watt
(9) P_{so}/P_{si}	Loss^{-1} of peak signal power in IF amplifier	$0.512 = -2.9$ dB
(10) P_{so}	Peak signal power at detector	8.24×10^{-14} watt
(11) P_{so}/P_{no}	Detector signal/noise ratio	$1.36 = 1.34$ dB
(12) $P_{si}\tau_p$	Received energy per pulse E_{si}	1.61×10^{-20} joule
(13) E_{si}/p_n	Detector signal/noise ratio for matched filter	$1.67 = 2.23$ dB
(14)	Receiver mismatch loss	$1.229 = 0.89$ dB
(15)	Mismatch loss of best practical filter	0.50 dB
(16)	Loss relative to best practical filter	0.39 dB

GHz. This is multiplied by Boltzmann's constant K shown in item (2) to obtain the noise power density $p_n = KT$ shown in item (3). The quantity p_n is commonly denoted N_0, the noise power per hertz of bandwidth.

It is assumed that the pulse width τ_p of the radar pulse is equal to 0.1 microsecond, as shown in item (4). The radar pulse travels a distance of $2R$ between transmission and reception, where R is the range to the target. Hence, the transmission time is $2R/c$, where c is the speed of light. The ratio of a range increment ΔR to a corresponding time increment Δt in the radar signal is then

$$\frac{\Delta R}{\Delta t} = \frac{c}{2} = \frac{3 \times 10^8 \text{ m/sec}}{2}$$

$$= 150 \text{ meter}/\mu\text{sec} = 492 \text{ ft}/\mu\text{sec} \qquad (7.3\text{-}18)$$

Thus the 0.1-μsec pulse width τ_p provides a range resolution ΔR of 15 meter, or 49 feet, as shown in item (5).

Figure 7.3-2 shows a block diagram of the portion of the radar receiver associated with detection of the target echo pulse. The received 10-GHz echo signal is fed into a balanced mixer, driven by a local oscillator of frequency f_{Lo}

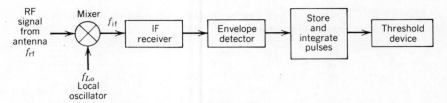

Figure 7.3-2   Block diagram of radar receiver, during search mode.

given by

$$f_{\text{Lo}} = f_{\text{rf}} - f_{\text{if}} \tag{7.3-19}$$

where f_{rf} is the RF carrier frequency (10 GHz), and f_{if} is the center frequency of the IF amplifier, which is assumed to be 60 MHz. Thus, the local-oscillator frequency f_{Lo} is (10,000 − 60) MHz, which is 9940 MHz. The mixer multiplies the received signal and the local oscillator signal, producing components at the sum frequency $(2f_{\text{rf}} + f_{\text{if}})$, which are filtered out, and components at the difference frequency f_{if}. The difference-frequency components are fed into the IF amplifier, which has a passband centered around the 60-MHz IF carrier frequency f_{if}. This is equivalent to the processing discussed in Chapter 4 relative to AC (carrier-frequency) servos. Diagram a of Fig 7.3-3 shows the spectra of the received signal and the local oscillator reference f_{Lo}; while diagram b shows the spectrum of the resultant IF signal.

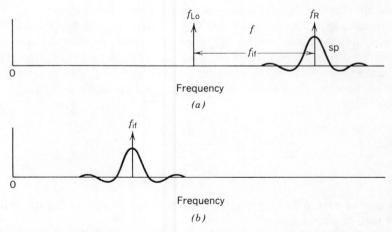

Figure 7.3-3   Spectra showing signal-frequency translation performed in receiver input mixer, which converts RF signal to IF signal: (a) spectrum of RF signal; (b) spectrum of IF signal.

As was shown in Chapter 6, Section 6.3, signal correlation is achieved by multiplying two signals together and integrating (or filtering) the result. The multiplication process implemented in the mixer performs correlation that cross-correlates the received signal with the local oscillator signal. Such a process is called *coherent detection*. It translates the signal spectrum from one band to another. The signal/noise power ratio does not change provided there is negligible noise in the local-oscillator signal.

It is assumed for simplicity that the IF amplifier has a passband defined by a single tuned circuit. This circuit narrows the frequency band, to limit the noise that is passed while transmitting the radar echo pulse without excessive attenuation. As will be shown, the optimum noise bandwidth of such a bandpass filter for a rectangular pulse is

$$f_{b(bp)} = 0.63/\tau_p = 6.3 \times 10^6 \text{ Hz} \qquad (7.3\text{-}20)$$

where τ_p is the pulse width, 0.1 μsec. This is shown in item (6) of Table 7.3-2. The noise power density p_n of item (3) is multiplied by the noise bandwidth in item (6) to obtain the noise power P_{no} shown in item (7). This is the noise power at the output of the IF amplifier, which is the input to the envelope detector. (Remember that all signal and noise power levels are expressed in terms of the equivalent signals at the input to the receiver, subsequent amplifier gains being normalized to unity.)

Item (8) shows the signal power input to the receiver, which was obtained from item (16) of Table 7.3-1. As will be shown, the IF amplifier attenuates the peak value of this 0.1-μsec pulse by the factor 0.716, and so it attenuates the peak power of the pulse by $(0.716)^2$, which is 0.512. Thus, as shown in item (8), there is a loss^{-1} of peak signal power in the IF amplifier equal to 0.512, or -2.9 dB. Multiplying items (8) and (9) gives item (10), which is the signal power of the detector input at the time of the peak signal. Dividing item (10) by item (7) gives the signal/noise power ratio at the detector input, which is a factor of 1.36 (equivalent to 1.34 dB).

The analysis justifying the calculations of items (6) to (11) of Table 7.3-2 is as follows. As explained in Chapter 4, a single-tuned bandpass IF stage is equivalent to a single-order lowpass filter of transfer function $1/(1 + \tau s)$. In Fig 7.3-4, curve (1) is the rectangular pulse modulation on the input signal to the IF amplifier. Curve (2) is the response of the equivalent lowpass filter to this modulation, and so is the envelope of the bandpass filter response to the IF pulse. Curve (2), which is denoted e_o, is the sum of the positive step response (3) plus the negative step response (4), which is delayed by the pulse width τ_p. Response (2) is the same as (3) in the region $0 < t < \tau_p$ and is equal to

$$e_o = (1 - e^{-t/\tau})e_{im} \quad \text{for } 0 < t < \tau_p \qquad (7.3\text{-}21)$$

where e_{im} is the amplitude of the input pulse. The maximum value of response

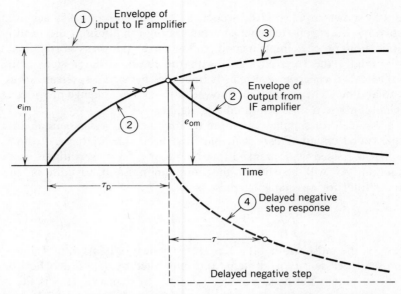

Figure 7.3-4   Calculation of envelope of output from IF amplifier having single-tuned stage, in response to IF signal with rectangular-pulse modulation.

e_o (curve 2) occurs at time $t = \tau_p$. Hence the maximum value is

$$\text{Max}[e_o] = e_{om} = (1 - e^{-\tau_p/\tau})e_{im} \qquad (7.3\text{-}22)$$

As was shown in Chapter 6, Eq 6.2-23, the noise bandwidth $f_{b(Lp)}$ of a single-order lowpass filter is equal to $1/4\tau$. Since the noise bandwidth $f_{b(bp)}$ of the equivalent bandpass filter is twice that of the lowpass filter, it is related as follows to the time constant τ:

$$f_{b(bp)} = 2f_{b(Lp)} = \frac{1}{2\tau} \qquad (7.3\text{-}23)$$

The square of the ratio of peak output signal to peak input signal is the signal power loss^{-1} of the IF amplifier that was shown as item (9) of Table 7.3-2. By Eq 7.3-22, this is

$$\frac{P_{so}}{P_{si}} = \left(\frac{e_{om}}{e_{im}}\right)^2 = (1 - e^{-\tau_p/\tau})^2 \qquad (7.3\text{-}24)$$

The output signal power at peak signal is proportional to the value of Eq 7.3-24. The output noise power is proportional to the noise bandwidth $f_{b(bp)}$ of the IF bandpass filter, which by Eq 7.3-23 is $1/2\tau$. Hence the output signal/noise power ratio at peak signal is proportional to the following ratio:

$$\frac{\left(1 - e^{-\tau_p/\tau}\right)^2}{f_{b(bp)}} = 2\tau\left(1 - e^{-\tau_p/\tau}\right)^2 \qquad (7.3\text{-}25)$$

To maximize this expression, differentiate it relative to the filter time constant τ, and set the derivative equal to zero. This yields

$$1 + \frac{2t_p}{\tau} = e^{-\tau_p/\tau} \qquad (7.3\text{-}26)$$

The solution of this is

$$\tau_p = 1.256\tau \qquad (7.3\text{-}27)$$

Combining Eqs 7.3-23, -27 gives the following for the optimum noise bandwidth of the IF filter:

$$f_{b(bp)} = \frac{1.256}{2\tau_p} = \frac{0.6325}{\tau_p} \qquad (7.3\text{-}28)$$

This is the optimum value used in item (6) of Table 7.3-2. Substituting 7.3-27 into Eq 7.3-24 gives the following for the loss^{-1} of peak signal in the IF amplifier when the noise bandwidth is optimum:

$$\frac{P_{so}}{P_{si}} = \left(1 - e^{-1.256}\right)^2 = (0.7152)^2 = 0.512 \qquad (7.3\text{-}29)$$

This was given as item (9) of Table 7.3-2.

This assumes a very simple IF amplifier: a single-tuned stage. If a more complicated filter is used, how much can the signal/noise power ratio be improved? The optimum receiver is called a *matched filter*. As explained by Skolnick [7.1] (Section 9.2), the matched filter for the signal $s[t]$ has an impulse response $h[t]$ related as follows to the received signal signal pulse $s[t]$:

$$h[t] = s[t_1 - t] \quad \text{(Matched Filter)} \qquad (7.3\text{-}30)$$

where t_1 is an arbitrary time interval that is chosen to make $h[t]$ realizable. In Fig 7.3-5, diagram *a* shows an input pulse signal $s[t]$. Diagram *b* shows this input pulse projected backward in time, which represents $s[-t]$. This theoretically is an optimum impulse response $h[t]$ of a filter for detecting the signal $s[t]$. However, it is physically unrealizable because it starts before the pulse occurs. Diagram *c* shows the pulse of diagram *b* delayed in time by a time interval t_1. This represents $s[t_1 - t]$, which is also an optimum impulse

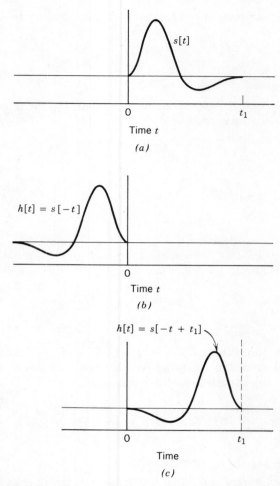

Figure 7.3-5 Waveforms illustrating matched-filter receiver: (a) assumed received pulse $s[t]$; (b) impulse response of ideal receiver; (c) impulse response of physically realizable ideal receiver.

response of a filter for detecting the signal $s[t]$. Since $s[t_1 - t]$ occurs completely in positive time, it can be physically realized. Thus the impulse response of the optimum matched filter for detecting a signal $s[t]$ is equal to the signal $s[t]$ projected backward in time, and delayed by an amount t_1 sufficient to achieve a realizable filter. The transfer function $H[s]$ of this optimum matched filter is related as follows to the transform $S[s]$ of the input signal:

$$H[s] = S^*[s]e^{-st_1} \qquad (7.3\text{-}31)$$

where $S^*[s]$ is the conjugate of the transform $S[s]$ of the signal $s[t]$ and e^{-st_1} is the transfer function of the t_1 time delay.

It can be shown that for a matched filter, the signal/noise power ratio at the output of the IF amplifier, at the time of the peak signal, is

$$\frac{P_{so}}{P_{no}} = \frac{E_{si}}{p_n} \qquad (7.3\text{-}32)$$

where E_{si} is the energy in the received pulse, and p_n is the noise power density. For a rectangular pulse, E_{si} is equal to $P_{si}\tau_p$, the product of the received signal power P_{si} and the pulse width τ_p. In Table 7.3-2, this is shown as item (12), which is obtained by multiplying the value of P_{si} of item (8) by the pulse width τ_p of item (4). Dividing E_{si} of item (12) by the noise power density p_n of item (3) gives the ratio E_{si}/p_n shown in item (13), which is 1.67. This is the optimum signal/noise power ratio at peak signal at the output of an ideal matched filter. Dividing the ideal signal/noise ratio of item (13) by the actual ratio of item (11) gives the value of item (14). This is the mismatch loss of the receiver, which is the loss of the receiver relative to an ideal matched receiver. This mismatch loss is 1.229, which is equivalent to 0.89 dB.

As shown by Skolnick [7.2] (pp. 5–25, Table 2), it generally is not practical to implement an ideal matched filter in a radar receiver for a rectangular modulated pulse. The best practical IF filter is one with a Gaussian frequency response given by

$$H[f] = \exp\left[-a(f - f_{if})^2\right] \qquad (7.3\text{-}33)$$

where a is a constant, f is the frequency, and f_{if} is the carrier frequency of the IF pulse. The optimum noise bandwidth of this filter is

$$f_{b(bp)} = 0.77/\tau_p \qquad (7.3\text{-}34)$$

For this case, the mismatch loss relative to a matched filter is 0.49 dB. This Gaussian filter response can be closely approximated by five identical tuned stages, synchronously tuned to the IF carrier frequency. This 5-stage filter has a mismatch loss of 0.50 dB, as shown in item (15). Subtracting this 0.50-dB loss from the mismatch loss relative to the matched filter, shown in item (14), gives the loss relative to the best practical filter, shown in item (16) as 0.39 dB. Thus, the receiver sensitivity can be improved by 0.39 dB by using a more complicated bandpass IF filter in place of the single-tuned stage.

7.3.3 Detection of Target

As we shown in Fig 7.3-2, the output of the IF amplifier is fed through an envelope detector, which extracts the envelope of the signal. The detected signal is usually fed to a storage element, which increases the signal/noise ratio by superimposing and averaging the returns from a number of radar pulses. The result is fed to a threshold device. When the threshold is exceeded, target detection is assumed. Angular search is stopped, and the beam is returned to the location of the target to take another look, with more pulses averaged. If the target is again detected, the tracking process is initiated, which centers the beam on the target. Otherwise, a false alarm is assumed, and angular search is continued.

Let us first consider the simple case of detecting a target from a single pulse. The noise at the output of the IF amplifier is narrow-band noise with a Gaussian distribution. When this is fed through an envelope detector, the resultant noise no longer has a Gaussian distribution. It has a Rayleigh distribution described by the following probability density function:

$$\frac{dP_r[z]}{dz} = z \exp[-z^2/2] \qquad (7.3\text{-}35)$$

The parameter z is equal to x/σ, where x is the value of the signal and σ is the standard deviation (RMS value) of the IF noise. This is defined for positive values of z, and the total integral for positive z is unity. Unlike the Gaussian distribution, the Rayleigh distribution can be integrated analytically. Let us define the integral from 0 to z as $P_r[z]$. Integrating Eq 7.3-35 gives

$$P_r[z] = \int_0^z \frac{dP_r[z]}{dz}\, dz = \int_0^z z \exp[-z^2/2]\, dz$$

$$= -\int_0^z \exp[-z^2/2]\, d[-z^2/2] = 1 - \exp[-z^2/2] \quad (7.3\text{-}36)$$

It can be shown that the mean and standard deviation of the Rayleigh distribution are

$$\text{Mean} = \sqrt{(\pi/2)}\,\sigma = 1.253\sigma \qquad (7.3\text{-}37)$$

$$\text{Standard deviation} = \sqrt{(2 - \pi/2)}\,\sigma = 0.655\sigma \qquad (7.3\text{-}38)$$

where σ is the RMS value of the IF signal prior to envelope detection. Remember that the standard deviation is the RMS deviation from the mean.

In Fig 7.3-6, the dashed curve (1) is the Gaussian distribution of the noise at the output of the IF amplifier, which is the input to the envelope detector. Solid curve (2) is the Rayleigh distribution of the envelope of this noise, which occurs at the output of the envelope detector. Curve (3) is the distribution of the signal-plus-noise at the output of the envelope detector when there is a constant-amplitude signal present. As shown by Schwartz, Bennett, and Stein [7.3] (p. 24), when the signal/noise ratio is much greater than unity, the detected signal-plus-noise has a distribution that is approximately Gaussian, with the same standard deviation as the noise at the detector input. The mean

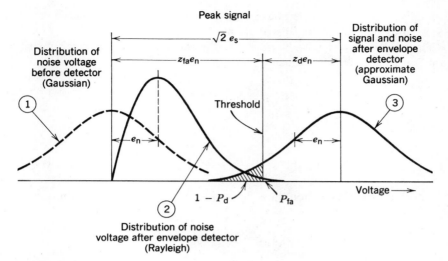

Figure 7.3-6 Distributions of noise and signal-plus-noise at threshold detector, showing probability of false alarm and probability of false detection.

value of this distribution is displaced from zero by the peak value of the signal. This peak value is shown as $\sqrt{2}\,e_s$, where e_s is the RMS value of the signal prior to detection.

To detect the presence of the target signal, a threshold is set as shown. Target detection occurs when the signal exceeds the threshold. Hence the area under curve (3) that exceeds the threshold is the probability of detection P_d. The area under curve (3) that falls below the threshold is the probability of false dismissal, equal to $(1 - P_d)$. The area under curve (2) that exceeds the threshold is the probability of false alarm P_{fa}, which is the probability that the noise will trigger the threshold when no target signal is present.

The RMS value of the noise is denoted e_n. The threshold level is denoted $z_{fa}e_n$; and the peak signal exceeds this threshold by $z_d e_n$. The parameters z_{fa}, z_d are the ratios x/σ of the corresponding probability distributions. Hence

$$\sqrt{2}\,e_s = z_{fa}e_n + z_d e_n = (z_{fa} + z_d)e_n \qquad (7.3\text{-}39)$$

Solving this for $(e_s/e_n)^2$ gives

$$\frac{P_s}{P_n} = \left(\frac{e_s}{e_n}\right)^2 = \frac{(z_{fa} + z_d)^2}{2} \qquad (7.3\text{-}40)$$

where

$$P_s = e_s^2 = \text{signal power at peak signal} \qquad (7.3\text{-}41)$$

$$P_n = e_n^2 = \text{noise power} \qquad (7.3\text{-}42)$$

This gives the signal/noise power ratio as a function of the relative variable $z = x/\sigma$ of the probability distribution.

The probability of false alarm P_{fa} is the integral of the Rayleigh distribution from z_{fa} to infinity, which (from Eq 7.3-36) is

$$P_{fa} = 1 - P_r[z_{fa}] = \exp[-z_{fa}^2/2] \tag{7.3-43}$$

Solving this for z_{fa} gives

$$z_{fa} = \sqrt{-2\ln[P_{fa}]} = \sqrt{2\ln[1/P_{fa}]} \tag{7.3-44}$$

The probability of detection P_d is derived from the Gaussian distribution, and is

$$P_d = 0.5 + P_g[z] \tag{7.3-45}$$

where $P_g[z]$ was defined in Eq 6.1-33 as the integral of the Gaussian distribution from 0 to z. Solving for $P_g[z]$ gives

$$P_g[z] = P_d - 0.5 \tag{7.3-46}$$

With this expression, the value of z_d can be derived from Table 6.1-3.

Let us calculate the required signal/noise power ratio at the threshold for single-pulse detection, to achieve a probability of detection P_d of 0.95 and a probability of false alarm P_{fa} of 10^{-6}. From Eq 7.3-44, z_{fa} is equal to

$$z_{fa} = \sqrt{-2\ln[P_{fa}]} = \sqrt{-2\ln[10^{-6}]} = 5.26 \tag{7.3-47}$$

From Eq 7.3-46 the value of z_d is derived from

$$P_g[z] = P_d - 0.5 = 0.95 - 0.5 = 0.45 \tag{7.3-48}$$

Table 6.2-3 in Section 6.2.3 gives the following:

$$P_g[1.6] = 0.445\,201 \tag{7.3-49}$$

$$P_g[1.7] = 0.455\,439 \tag{7.3-50}$$

Linear interpolation between these values for $z = 1.6, 1.7$ gives, for $P_g[z_d] = 0.450$,

$$z_d = 1.6 + 0.1(0.45 - 0.445\,201)/(0.455\,439 - 0.445\,701)$$
$$= 1.647 \cong 1.65 \tag{7.3-51}$$

TABLE 7.3-3 Approximate Signal/Noise Ratios for Single-pulse Detection for Different Values of Detection Probability P_d and False Alarm Probability P_{fa}

False Alarm Probability P_{fa}	$P_d = 0.90$		$P_d = 0.95$	
	Value	dB	Value	dB
10^{-4}	15.62	11.94	17.64	12.47
10^{-6}	21.52	13.33	23.87	13.78
10^{-8}	27.36	14.34	29.80	14.74
10^{-10}	32.72	15.15	35.62	15.52
10^{-12}	38.11	15.81	41.22	16.15

Substituting the values for z_{fa} and z_d of Eqs 7.3-47, -51 into Eq 7.3-40 gives

$$\frac{P_s}{P_n} = \frac{(z_{fa} + z_d)^2}{2} = \frac{(5.26 + 1.65)^2}{2}$$

$$= 23.87 = 13.78 \text{ dB} \qquad (7.3\text{-}52)$$

Table 7.3-3 shows the required signal/noise ratios at the threshold detector for 90% and 95% probability of detection P_d, and for several values of probability of false alarm P_{fa} from 10^{-4} to 10^{-12}.

When multiple pulses are integrated, the exact calculation of the required signal/noise ratio is very complicated. Nevertheless, a good estimate can be obtained by applying a simple approximation. As was stated earlier, for a coherent detection process (such as the superheterodyne conversion of the radar signal from RF to IF in the mixer), the signal/noise ratio does not change. However, a noncoherent detection process, such as envelope detection, degrades the signal/noise ratio. Barton [7.4] (pp. 27–29) has shown that the loss of signal/noise ratio in a noncoherent detector can be approximated by

$$\text{Noncoherent detection loss: } L_d \cong 1 + \frac{P_n}{P_s} \qquad (7.3\text{-}53)$$

where P_n/P_s is the noise/signal power ratio at the input of the noncoherent detector, which is the reciprocal of the signal/noise power ratio. (This equation was originally developed by Lamont V. Blake of the U.S. Naval Research Laboratory.)

By applying this principle, the signal/noise ratio P_{st}/P_{nt} at the threshold device, after integrating N pulses, is related approximately as follows to the

TABLE 7.3-4 Approximate Values of Signal/Noise Ratio at Input to Envelope Detector, Compared with Exact Values from Skolnick [7.2], for Noncoherent Integration of n Pulses for Different Probabilities of Detection and False Alarm

False Alarm Probability P_{fa}	Detection Probability P_d	Approximate dB Values				Skolnick dB Data Minus Approximate dB Values			
		$N = 1$	$N = 10$	$N = 100$	$N = 1000$	$N = 1$	$N = 10$	$N = 100$	$N = 1000$
10^{-4}	0.90	11.9	3.5	−3.2	−8.7	−0.2	0.5	0.9	1.0
	0.95	12.5	3.9	−2.9	−8.5	−0.3	0.5	0.9	1.2
10^{-12}	0.90	15.8	6.7	−0.8	−6.7	−0.1	0.6	1.2	1.4
	0.95	16.2	7.0	−0.6	−6.5	−0.2	0.5	1.3	1.4

signal/noise ratio P_s/P_n at the input to the envelope detector:

$$\frac{P_{st}}{P_{nt}} = \frac{N(P_s/P_n)}{1 + P_n/P_s}$$

$$= \frac{N(P_s/P_n)^2}{1 + P_s/P_n} \tag{7.3-54}$$

In this expression, the signal/noise ratio P_s/P_n at the input to the envelope detector is divided by the noncoherent detection loss of Eq 7.3-53, and the result is multiplied by the number of pulses N that are integrated to obtain the signal/noise ratio at the threshold device. Solving Eq 7.3-54 for P_s/P_n gives

$$\frac{P_s}{P_n} = \frac{P_{st}}{P_{nt}} \frac{1}{2N} \left(1 + \sqrt{1 + \frac{4NP_{nt}}{P_{st}}} \right) \tag{7.3-55}$$

Skolnick [7.2] (pp. 2–21) gives plots showing the exact values of signal/noise ratio at the input to the envelope detector required to satisfy detection probabilities of 90% and 95%, for the following false-alarm probabilities: 10^{-4}, 10^{-6}, 10^{-8}, 10^{-10}, and 10^{-12}. These are given for values of N from 1 to 10,000, where N is the number of pulses integrated. Table 7.3-4 shows samples of these data compared with the approximate values derived from the preceding approach. Values are shown for $P_d = 0.90$ and 0.95; for $P_{fa} = 10^{-4}$ and 10^{-12}; and for $N = 1$, 10, 100, and 1000.

Approximate values of signal/noise ratio in decibels (dB) are shown in Table 7.3-4, along with the differences between these approximate dB values and the exact dB values given by Skolnick. For example, for $P_{fa} = 10^{-4}$, $P_d = 0.9$, and $N = 1000$, the approximate signal/noise ratio at the envelope detector input is -8.7 dB; and the exact value given by Skolnick differs from this by 1.0 dB. Hence the exact value is $(-8.7 + 1.0)$ dB, which is -7.7 dB. Since the maximum approximation error for the cases shown in the table is only 1.4 dB, the approximation is adequate for preliminary analysis.

Table 7.3-5 shows the steps for calculating the number of radar pulses that must be integrated to achieve the specified values of probability of detection and probability of false alarm. Item (1) is the signal/noise ratio 1.36 at the input to the detector that was given in item (11) of Table 7.3-2. In accordance with Eq 7.3-53, the envelope detector causes the following noncoherent detection loss:

$$L_d = 1 + \frac{P_{no}}{P_{so}} = 1 + \frac{1}{1.36} = 1.735 \tag{7.3-56}$$

This is equivalent to 2.39 dB, as shown in item (2). Dividing item (1) by item

TABLE 7.3-5 Calculation of Number of Pulses Integrated Noncoherently Required for Detection of Target in Example

	Symbol	Parameter	Value
(1)	P_{so}/P_{no}	Signal/noise into detector	$1.36 = 1.34$ dB
(2)	$1 + P_{no}/P_{so}$	Noncoherent detector loss	$1.735 = 2.39$ dB
(3)	P_{sd}/P_{nd}	Signal/noise at detector output	$0.784 = -1.06$ dB
(4)	P_d	Probability of detection	0.95
(5)	P_{fa}	Probability of false alarm	10^{-4}
(6)	P_{st}/P_{nt}	Required threshold signal/noise	$17.6 = 12.5$ dB
(7)	N	Number of pulses integrated	$17.6/0.784 = 22.4$ (use 23)

(2) gives the following signal/noise ratio at the output of the detector:

$$\frac{P_{nd}}{P_{sd}} = \frac{P_{so}/P_{no}}{L_d} = \frac{1.36}{1.735} = 0.784 \qquad (7.3\text{-}57)$$

This is shown as item (3) and is equivalent to -1.06 dB.

Let us assume a probability of detection of 0.95 (item 4) and a probability of false alarm of 10^{-4} (item 5). For these parameters, Table 7.3-3 gives a required threshold signal/noise ratio P_{st}/P_{nt} of 17.6 (or 12.5 dB), which is given in item (6). Dividing item (6) by the detected single-pulse signal/noise ratio of item (3) gives the number of pulses N that must be integrated, which is

$$N = \frac{P_{st}/P_{nt}}{P_{sd}/P_{nd}} = \frac{17.6}{0.784} = 22.4 \qquad (7.3\text{-}58)$$

Since the number of pulses must be an integer, the actual number is 23, as shown in item (7).

7.3.4 Target Search in Range and Angle

The parameters of the search process are summarized in Table 7.3-6. The pulse repetition frequency f_p of the radar pulses is set at 1.5 kHz. The reciprocal for this is the pulse period T_p, which is 687 μsec. This is multiplied by 150 m/μsec, in accordance with Eq 7.3-19, to obtain the equivalent range interval, 100 kilometer, shown in item (3). This is called the ambiguous range interval, because echos from two targets separated by the ambiguous range interval appear in the same time slot. If the ambiguous range is sufficiently large, the echo from the more distant target is very low and so can be rejected. Multiply the pulse repetition period T_p of item (2) by the number of pulses integrated, $N = 23$, as shown in item (7) of Table 7.3-5. This gives the observation time T_{obs} shown in item (4), which is 15.3 millisecond.

TABLE 7.3-6 Parameters of Target Search

Symbol		Parameter	Value
(1)	f_p	Pulse repetition frequency	1.5 KHz
(2)	T_p	Pulse repetition period	667 μsec
(3)		Ambiguous range interval	100 km = 62 mile
(4)	T_{obs}	Observation time per cell	15.3 msec
(5)		Angular search region	
		Azimuth	2.7°
		Elevation	16.3°
(6)		Width of angle scan	0.907°
(7)		Number of angle search cells	3 × 18 = 54
(8)	T_{srch}	Search time	0.83 sec
(9)		Angular scan rate	63 deg/sec
(10)	$f_{b(if)}$	IF noise bandwidth	6.3 × 10^6
(11)	$T_{i(if)}$	Effective IF integration time	0.16 × 10^{-6} sec
(12)		Range correlation interval	24 meter
(13)		Range search region	1 mile = 1609 meter
(14)		Number of independent range cells	1609/24 = 67
(15)		Independent measurements during search	67 × 54 = 3618
(16)		Average false alarms per search	0.36

As shown in item (5), it is assumed that the radar searches over a region that is 2.7° in azimuth by 16.3° in elevation. The required elevation search is constrained by the possible variations of target altitude. At maximum range, 31 miles, this elevation search region represents an altitude variation of 48,000 ft. Item (6) gives the width of the angular scan (0.907°) obtained from item (9) of Table 7.3-1. Dividing the dimensions of the search region given in item (5) by this width (0.907°) gives the required number of angular search cells. As shown in item (7), the search region is 3 cells wide in azimuth by 18 cells high in elevation, which gives a total of 54 angular search cells. Item (8) shows that the total search time is 0.83 sec, which is obtained by multiplying the observation time per cell of item (4), which is 15.3 msec, by 54, the total number of angular search cells.

The antenna cannot be accelerated rapidly enough to allow the search to be performed by jumping the beam from one angular cell to another. Instead the antenna beam is scanned in a continuous manner over the angular region, probably as three separate scans in the elevation direction. The required angular scan rate is equal to the width of the scan (0.0963°) divided by the observation time per angular cell, T_{obs}, which is 15.3 msec. As shown in item (9), this scan rate is 63 deg/sec. The preceding approximate analysis is pessimistic. A more accurate analysis that considers the detailed variation with time of signal strength at each point would result in a greater probability of detection.

As was shown in item (6) of Table 7.3-2, the noise bandwidth of the IF amplifier is 6.3 MHz, which is repeated in item (10) of Table 7.3-6. In Chapter 6, Section 6.2.2.2 showed that an ideal averaging integrator of integration time T_i has a noise bandwidth f_b in hertz equal to $1/2T_i$. Accordingly, a low pass filter of noise bandwidth $f_{b(Lp)}$ has an equivalent integration time T_i equal to $1/2f_{b(Lp)}$. Since a bandpass filter is equivalent to a lowpass filter of half its noise bandwidth, a bandpass filter of noise bandwidth $f_{b(bp)}$ has an equivalent integration time T_i equal to $1/f_{b(bp)}$. Thus, the values of equivalent integration time for lowpass and highpass filters are

$$T_i = \frac{1}{2f_{b(Lp)}} = \frac{1}{f_{b(bp)}} \qquad (7.3\text{-}59)$$

In accordance with Eq 7.3-59, the reciprocal of the 6.3-MHz IF noise bandwidth of item (10) is the effective integration time T_i of the IF amplifier, shown in item (11) as 0.16 microsecond. This is multiplied by 150 m/μsec (Eq 7.3-18) to obtain the corresponding range interval, called the range correlation interval, which is 24 meter as shown in item (12). Thus, signals separated in range by 24 meters are uncorrelated.

Let us assume that the search radar provides range data that determines the target location within a range interval of 1 mile, including the effects of target motion. Hence, as shown in item (13), the tracking radar must search over 1 mile in range, or 1609 meter. Dividing this range search interval (item 13) by the range correlation interval, 24 meter (item 12), gives the number of independent range cells, which is 67 as shown in item (14). Multiplying the 67 independent range cells by the 54 angle search cells (item 7) gives the total number of independent measurements performed during search, which is 3618 (item 15). The probability of false alarm is 10^{-4}, as shown in item (5) of Table 7.3-5. The number of independent measurements, 3618, is multiplied by the 10^{-4} probability of false alarm to obtain the average number of false alarms during search, which is 0.36, as shown in item (16). Thus, there is a 36% probability that a false alarm will occur during search.

When a target signal is detected, the angular scan is stopped, and the antenna beam is returned to the location of the detected target. The beam is held fixed until the target is detected a second time, or until much more than 23 pulses have been integrated. If a second target detection does not occur, a false alarm is assumed, and the angular scan is continued. When detection occurs, the radar commences range and angle tracking. These tracking processes are described in Sections 7.4 and 7.5.

7.4 ANALYSIS OF RANGE TRACKING

After the tracking radar has detected the target, the radar begins to track the target signal in range and angle. This section analyzes range tracking, and Section 7.5 analyzes angle tracking.

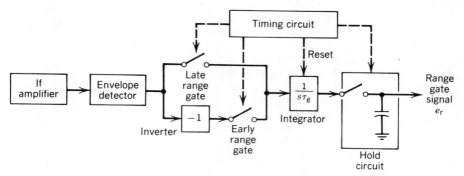

Figure 7.4-1 Basic design of early–late range-gate detector.

7.4.1 Early–Late Gate to Measure Range Error

Figure 7.4-1 shows a block diagram of an early–late range-gate circuit, which derives a range error signal. The signal from the IF amplifier is fed through an envelope detector, which detects the pulse modulation of the IF signal. This pulse is inverted and fed to a switch, called the *early gate*, and it is fed directly to a second switch, called the *late gate*. The early gate is closed for a fixed time interval, and then the late gate is closed for an equal interval. The sum of the output signals from the gates is fed to an integrator, which is reset to zero before the gates operate. The integrator output is fed to a sample-and-hold circuit, which samples the integrator output after the gates have opened and holds the sampled value until the next pulse.

The action of the early and late gates can be characterized by the range-gate function shown in Fig 7.4.-2a. Zero time t is defined as the transition between the early and late gates, and the total duration of the early and late gates is defined as the range gate width τ_g. Thus the early gate, which has a gain of -1, occurs between time $t = -\tau_g/2$ and $t = 0$; and the late gate, which has a gain of $+1$, occurs between $t = 0$ and $t = +\tau_g/2$.

It is assumed that the noise bandwidth of the IF amplifier is much greater than $1/\tau_p$, the reciprocal of the pulse width. Hence, the signal e_d at the output of the envelope detector is approximately a rectangular pulse of duration τ_p, as shown in Fig 7.4-2b. The center of the pulse (represented by the dashed line) is assumed to occur at a time Δt after the early–late gate transition. The portion of the pulse width occurring before the transition is denoted t_1, and the portion occurring after the transition is denoted t_2. These are related to Δt by

$$t_1 = \tfrac{1}{2}\tau_p - \Delta t \qquad (7.4\text{-}1)$$

$$t_2 = \tfrac{1}{2}\tau_p + \Delta t \qquad (7.4\text{-}2)$$

As shown in Fig 7.4-1, the integrator that follows the early and late gates has a gain factor of $1/\tau_g$. It is convenient to combine this gain factor with the

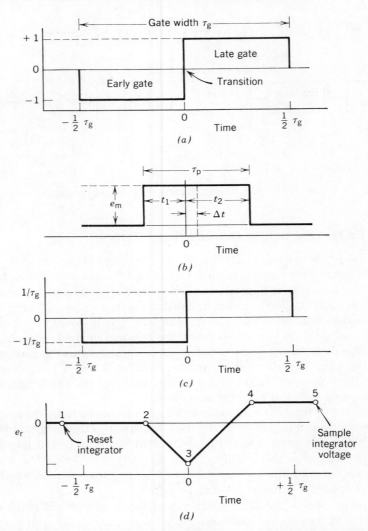

Figure 7.4-2 Waveforms for range-gate circuit: (*a*) range-gate function; (*b*) pulse at output of envelope detector; (*c*) effective range-gate function; (*d*) voltage at output of integrator.

range-gate function of diagram *a* in Fig 7.4-2 to form the effective range-gate function of diagram *c*. This effective range-gate function has amplitudes of $\pm 1/\tau_g$ rather than ± 1.0. To calculate the output e_r from the range-gate circuit, multiply the pulse of diagram *b* by the effective range-gate function of diagram *c* and integrate the product.

The voltage waveform at the output of the integrator is shown in Fig 7.4-2*d*. At point (1) the integrator voltage is reset to zero. Point (2) occurs at the leading edge of the pulse. The voltage being integrated after point (2) is $-e_m/\tau_g$, where e_m is the pulse amplitude and $-1/\tau_g$ is the magnitude of the effective range-gate function during the early gate. The integrator output drops negatively at a constant slope, reaching the following value at point (3):

$$E_3 = -\frac{e_m}{\tau_g}t_1 \qquad (7.4\text{-}3)$$

At point (3), the early gate is turned off and the late gate is turned on. Hence, the voltage being integrated changes from $-e_m/\tau_g$ to $+e_m/\tau_g$. The integrator output voltage now rises at a constant rate until point (4), which occurs at the trailing edge of the pulse. The integrator output voltage at point (4) is

$$E_4 = E_3 + \frac{e_m}{\tau_g}t_2 = \frac{e_m}{\tau_g}(t_2 - t_1) \qquad (7.4\text{-}4)$$

The integrator output stays constant after point (4). At point (5) this voltage is sampled by the sample-and-hold circuit to obtain the range-gate voltage e_r. Thus, the range-gate voltage is equal to the expression of Eq 7.4-4. Substitute into Eq 7.4-4 the values for t_1 and t_2 in Eqs 7.4-1, -2. This gives the following range-gate output voltage:

$$e_r = \frac{e_m}{\tau_g}(t_2 - t_1) = \frac{2e_m\,\Delta t}{\tau_g} \qquad (7.4\text{-}5)$$

The solid curve in Fig 7.4-3 shows a plot of the range-gate output voltage e_r versus the time interval Δt between the early–late gate transition and the center of the pulse. At point (4), the pulse is centered in the range gate, and so

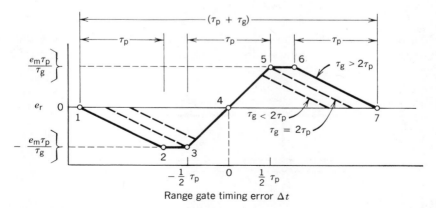

Figure 7.4-3 Plot of output voltage from range gate, versus range-gate timing error.

the range-gate output voltage e_r is zero. Between points (3) and (5) the pulse falls partly in the early gate and partly in the late gate, and so the range-gate voltage is described by Eq 7.4-5. Between points (5) and (6), the pulse is completely within the late gate, and so the range-gate output is $e_m(\tau_p/\tau_g)$, which is the product of the pulse area $e_m\tau_p$ multiplied by the effective amplitude $+1/\tau_g$ of the late gate. Similarly, between points (2) and (3) the pulse is completely within the early gate, and so the range-gate output is $-e_m(\tau_p/\tau_g)$. Between points (1) and (2), the pulse is partly outside the range gate and partly within the early gate, while between points (6) and (7) the pulse is partly outside the gate and partly within the late gate.

The solid curve in Fig 7.4-3 applies for the condition $\tau_g > 2\tau_p$. The dashed curves show plots for $\tau_g = 2\tau_p$ and for $\tau_g < 2\tau_p$.

7.4.2 Range-Tracking Error with Wide IF Bandwidth

Now let us consider the response of this range-gate circuit to noise. Since the noise bandwidth of the IF amplifier is wide, the noise that falls within the early gate is essentially uncorrelated with the noise falling within the late gate. Consequently, the sign reversal of the effective gating function of Fig 7.4-2c can be ignored in calculating the response to noise. Hence the effective range-gate function relative to noise can be approximated by the response shown in Fig 7.4-4, which has a constant amplitude equal to $1/\tau_g$ over the total duration τ_g of the range gate.

The noise component $e_{r(n)}$ in the range gate signal is equal to the integral of the product of the noise $e_{d(n)}[t]$ at the output of the detector, multiplied by the effective noise gating function of Fig 7.4-4. This integral is

$$e_{r(n)} = \frac{1}{\tau_g} \int_{-\tau_g/2}^{+\tau_g/2} e_{d(n)}[t]\, dt$$

$$= \text{average of } e_{d(n)}[t] \text{ over time interval } \tau_g \qquad (7.4\text{-}6)$$

Relative to noise, this shows that the range gate acts like an ideal averaging integrator, with an integration time T_i equal to τ_g. In Section 7.3, Eq 7.3-59

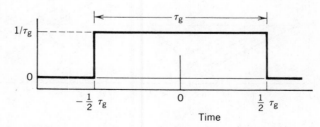

Figure 7.4-4 Effective response of range gate relative to noise.

showed that the equivalent bandpass noise bandwidth of an ideal averaging integrator is

$$f_{b(bp)} = \frac{1}{T_i} = \frac{1}{\tau_g} \qquad (7.4\text{-}7)$$

Let us assume that the signal/noise ratio at the input to the envelope detector is much greater than unity. Under this condition the detection process is equivalent to coherent detection, and so the signal/noise ratio of the range-gate signal is established completely by the filtering provided by the range gate. Hence, the noise power P_n at the output of the range gate is $P_n f_{b(bp)}$. The RMS noise at the output of the range gate is the square root of the noise power P_n, which is

$$\sigma e_r = \sqrt{P_n} = \sqrt{P_n f_{b(bp)}} = \sqrt{\frac{P_n}{\tau_g}} \qquad (7.4\text{-}8)$$

When the value of e_r of Eq 7.4-5 is set equal to this expression, the corresponding value for Δt is the RMS timing error, which is denoted σt_1. The subscript 1 indicates that this is the RMS timing error for a single-pulse measurement. Hence,

$$\sqrt{\frac{P_n}{\tau_g}} = \sigma e_r = 2e_m \frac{\sigma t_1}{\tau_g} \qquad (7.4\text{-}9)$$

Solving for σt_1 gives

$$\sigma t_1 = \left(\frac{\tau_g}{2e_m}\right)\sqrt{\left[\frac{P_n}{\tau_g}\right]} = \frac{\sqrt{P_n \tau_g}}{2e_m} \qquad (7.4\text{-}10)$$

As was shown in Fig 7.4-2b, e_m is the amplitude of the pulse at the output of the envelope detector, which is equal to the peak value of the IF signal. Since the peak value e_m of the signal is equal to $\sqrt{2}$ times the RMS value, the square of the peak signal value is twice the mean-square value of the signal, or twice the signal power P_s:

$$e_m^2 = 2P_s \qquad (7.4\text{-}11)$$

Substituting the square root of Eq 7.4-11 into Eq 7.4-10 gives

$$\sigma t_1 = \frac{1}{2}\sqrt{\frac{P_n \tau_g}{2P_s}} \qquad (7.4\text{-}12)$$

The signal energy per pulse is

$$E_s = P_s \tau_p \tag{7.4-13}$$

Solving this for P_s and substituting the result into Eq 7.4-12 gives

$$\sigma t_1 = \frac{1}{2} \sqrt{\frac{\tau_g \tau_p}{2 E_s / p_n}} \tag{7.4-14}$$

The quantity $2 E_s / p_n$ is a convenient nondimensional parameter for generalizing the performance of radar signal processing. Divide σt_1 by the pulse width τ_p to obtain the normalized timing error of the range gate:

$$\frac{\sigma t_1}{\tau_p} = \frac{1}{2} \sqrt{\frac{\tau_g / \tau_p}{2 E_s / p_n}} \tag{7.4-15}$$

If we assume that $\tau_g = 2 \tau_p$, this becomes

$$\frac{\sigma t_1}{\tau_p} = \frac{1}{\sqrt{2(2 E_s / p_n)}} \tag{7.4-16}$$

Equation 7.4-15 can be obtained from Barton and Ward [7.5] by combining their equations (3.33) and (3.49).

As was shown in Eq 7.1-53, when the signal/noise ratio P_s / P_n at the input to an envelope detector is not much greater than unity, the detector causes a noncoherent detection loss equal to $(1 + P_n / P_s)$. Barton and Ward [7.5] (pp. 39, 82) discuss this point relative to range and angle tracking. Thus, the effective noise power density p_{nd} at the output of the envelope detector is

$$p_{nd} = p_n (1 + P_n / P_s) \tag{7.4-17}$$

This should be used in Eqs 7.4-15, -16 instead of p_n.

7.4.3   Range-Tracking Error with Narrow IF Bandwidth

This range-gate design has the disadvantage that the IF noise bandwidth is wide, which results in a reduced signal/noise ratio at the envelope detector, and an increased noncoherent detection loss. A narrow IF bandwidth provides better performance. However, the analysis for this narrow-band case is much more complicated and so is not presented here. Instead, the results derived by Barton [7.4] (p. 363, Eq 11.8) are summarized. He gives the following general expression for the RMS range-gate noise for a single pulse measurement:

$$\frac{\sigma t_1}{\tau_p} = \frac{1}{M_r \sqrt{2 E_s / p_n}} \tag{7.4-18}$$

Plots of M_r are given by Barton [7.4] (Fig 11.8, p. 383) as a function of the relative range-gate width τ_g/τ_p for a number of different filters. This equation applies when the signal/noise power ratio at the input to the envelope detector is much greater than unity. To include the case of low signal/noise ratio, the noise density p_n in Eq 7.4-18 is replaced with p_{nd} given by Eq 7.4-17 to obtain

$$\frac{\sigma t_1}{\tau_p} = \frac{1}{M_r\sqrt{2E_s/p_{nd}}} \tag{7.4-19}$$

Most radar IF amplifiers are stagger-tuned to provide a frequency response with steep sides that is nearly flat over the passband. As shown by Barton [7.4] (p. 20), such a response can be approximated closely by the ideal *rectangular filter*. Plots of M_r are given by Barton (p. 383, Fig 11.8) for two rectangular filters corresponding to different values of IF noise bandwidth $f_{b(bp)}$. A gate width τ_g that is twice the pulse width τ_p is generally desirable because this provides a broad range-gate response characteristic, which is desirable during range-gate acquisition. For $\tau_g = 2\tau_p$, the values of M_r for these two noise bandwidths are

Rectangular filter ($\tau_g = 2\tau_p$):

$$M_r = 1.5 \quad \text{for} \quad f_{b(bp)} = 1/\tau_p \tag{7.4-20}$$

$$M_r = 1.6 \quad \text{for} \quad f_{b(bp)} = 2/\tau_p \tag{7.4-21}$$

In order to minimize the noncoherent detection loss described in Eq 7.4-17, the IF noise bandwidth should be chosen to minimize the filter matching loss L_m in the IF amplifier. Barton [7.4] (p. 21, Fig 1.11) gives a plot of the filter matching loss of a rectangular filter as a function of the noise bandwidth $f_{b(bp)}$ of the IF amplifier. This loss is a minimum for $f_{b(bp)} = 1.4/\tau_p$, and at that bandwidth it is

$$L_m = 0.85 \text{ dB} \quad \text{for} \quad f_{b(bp)} = 1.4/\tau_p \tag{7.4-22}$$

This indicates that for optimum bandwidth the peak signal/noise power ratio is 0.85 dB below that of a matched-filter receiver. The optimum IF amplifier bandwidth $1.4/\tau_p$ lies between the cases shown in Eqs 7.4-20, -21. To be conservative, the lower value $M_r = 1.5$ is assumed. Substituting this into Eq 7.4-19 gives

Optimum rectangular filter ($f_{b(bp)} = 1.4/\tau_p$, $\tau_g = 2\tau_p$):

$$\frac{\sigma t_1}{\tau_p} = \frac{1}{1.5\sqrt{2E_s/p_{nd}}} \tag{7.4-23}$$

7.4.4   Calculation of Range-tracking Errors

Table 7.4-1 shows the steps for calculating the range-gate noise error per pulse. Item (1) is the received signal power during search obtained from item (16) of Table 7.1-1. After the target is detected, the beam is centered on the target, and so the 2-dB one-way angle-search loss (item 7, Table 7.1-1) no longer occurs. Since this represents a two-way loss of 4 dB, during range tracking the received signal power is increased by 4 dB, as shown in item (2) of Table 7.4-1. There is still a small loss due to angular tracking error, but this is neglected. Since 4 dB is a factor of 2.5, the received signal power during search of item (1) is multiplied by 2.5 to obtain the received signal power during track, shown in item (3). The pulse width τ_p (given previously in Section 7.3, Table 7.3-2, item 4) is 0.1 μsec, as shown in item (4). Multiplying items (3) and (4) gives the received energy per pulse E_s shown in item (5). The noise power density p_n shown in item (6) is obtained from item (3) of Table 7.3-2. Divide E_s in item (5) by p_n in item (6) to obtain the ratio E_s/p_n shown in item (7). This is the signal/noise power ratio for a matched-filter receiver.

A rectangular IF filter is assumed with an optimized noise bandwidth of $f_{b(bp)} = 1.4/\tau_p$. As shown in Eq 7.4-22, the matching loss L_m of this filter is 0.85 dB, which represents a factor of $1/0.82$. Hence, the reciprocal matching loss L_m^{-1} is 0.82, as shown in item (8) of Table 7.4-1. The ratio E_s/p_n of item (7) (which is the signal/noise ratio of a matched filter) is multiplied by the reciprocal mismatch loss L_m^{-1} of item (8) to obtain the actual signal/noise power ratio at the input to the envelope detector, which is 3.44, as shown in item (10). The noncoherent detection loss in the envelope detector is then

$$L_d = 1 + \frac{P_n}{P_s} = 1 + \frac{1}{3.44} = 1.29 \qquad (7.4\text{-}24)$$

TABLE 7.4-1   Calculation of Range-Gate Noise Error Per Pulse

Symbol		Parameter	Value
(1)		Signal power during search	1.61×10^{-13} watt
(2)		Increase signal during track	4 dB = 2.5
(3)	P_s	Signal power during track	4.03×10^{-13} watt
(4)	τ_p	Pulse width	0.1×10^{-6} sec
(5)	E_s	Energy per pulse ($P_s\tau_p$)	4.03×10^{-20} joule
(6)	p_n	Noise power density	9.62×10^{-21} watt/Hz
(7)	E_s/p_n	Matched-filter signal/noise	4.189
(8)	L_m^{-1}	Mismatch Loss^{-1}	0.82 (-0.85 dB)
(9)	P_s/P_n	Signal/noise into detector	3.44
(10)	L_d	Noncoherent detection loss	1.29 (1.10 dB)
(11)	p_{nd}	Detected noise power density	1.24×10^{-20} watt/Hz
(12)	E_s/p_{nd}		3.25
(13)	$\sigma t_1/\tau_p$	Relative RMS time error per pulse	0.261
(14)	σt_1	RMS time error per pulse	0.0261 μsec
(15)	σR_1	RMS range error per pulse	3.92 meter

By Eq 7.4-17, the effective noise power density at the output of the envelope detector is

$$p_{nd} = p_n\left(1 + \frac{P_n}{P_s}\right) = p_n L_d = 1.29 p_n \qquad (7.4\text{-}25)$$

The value of p_{nd} is shown in item (11), which is obtained by multiplying L_d of item (10) by the input noise power density p_n of item (6). Item (12) gives the value of the nondimensional expression E_s/p_{nd}, which is obtained from the values given in items (5) and (11), and is

$$\frac{E_s}{p_{nd}} = 3.25 \qquad (7.4\text{-}26)$$

Substituting Eq 7.4-26 into Eq 7.4-23 gives the relative RMS timing error per pulse $\sigma t_1/\tau_p$, which is shown in item (13) to be 0.261. Multiplying this nondimensional value (0.261) by the pulse width τ_p (which is 0.1 μsec) gives the RMS timing error per pulse σt_1 shown in item (14). This is multiplied by 150 m/μsec (Eq 7.2-10) to obtain the RMS range error per pulse, which is 3.92 meter, as shown in item (15).

As was shown in Eq 7.4-5, the range-gate voltage e_r is proportional to the peak IF signal e_m. Hence, to use the range-gate voltage e_r as a tracking error signal, the system must compensate for variations of e_m caused by changes in the received signal power. To achieve this, the detected IF signal is fed through another gate, which is opened for the total gate period τ_g, and its output is fed through an integrator and a sample-and-hold circuit. The sampled output from this circuit, which is denoted e_{sig}, is

$$e_{sig} = \frac{1}{\tau_g} e_m \tau_p \qquad (7.4\text{-}27)$$

This voltage is used to close an automatic gain-control (AGC) feedback loop around the IF amplifier, which varies the IF amplifier gain so as to keep the signal e_m approximately constant. When the signal/noise ratio is high, the residual variation of e_m can be exactly compensated for by computing the ratio of the two signals e_r and e_{sig} for each pulse. Dividing Eq 7.4-5 by Eq 7.4-27 gives

$$\frac{e_r}{e_{sig}} = \frac{2\,\Delta t}{\tau_p} \qquad (7.4\text{-}28)$$

The range error R_e can be expressed as follows in terms of this ratio:

$$R_e = \frac{c}{2}\,\Delta t = (150 \text{ m}/\mu\text{sec})\,\frac{\tau_p}{2}\,\frac{e_r}{e_{sig}} \qquad (7.4\text{-}29)$$

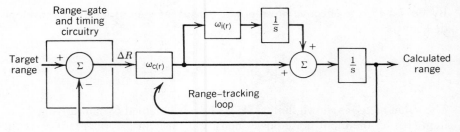

Figure 7.4-5   Range tracking loop.

This range error signal can be used to close a range-tracking loop, as shown in Fig 7.4-5. In modern radars this loop is generally implemented by digital processing. The loop transfer function of the range-tracking loop is

$$G_{(r)} = \frac{\omega_{c(r)}\left(1 + \omega_{i(r)}/s\right)}{s} \qquad (7.4\text{-}30)$$

The two integrations of this transfer function can be implemented with digital processing by applying the principles given in Chapter 5. Further discussion of digital range-gate control is given in Chapter 8. The resultant digital value of range R developed by this feedback loop controls a timing circuit. This timing circuit accurately sets the time of the range gate so that the center of the gate lags the center of the transmitted pulse by the time interval $2R/c$.

The digital range-gate loop is a sampled-data process having a sampling frequency equal to the pulse repetition frequency f_p of the radar, which is 1500 Hz. To have good stability, the range-tracking loop should have a gain crossover frequency that is no more than $1/10$ of this sampling frequency, or 150 Hz. However, an appreciably lower gain crossover frequency is actually used in order to minimize noise error in the range measurements. As shown in Table 7.4-2, item (1), the gain crossover frequency $\omega_{c(r)}$ of the range-tracking loop is set at 100 rad/sec (or 16 Hz). The integral-network break frequency $\omega_{i(r)}$ of the range-track loop is set at $1/2$ of this value, which is 50 rad/sec. (In Chapter 8, Section 8.6 will show that this setting is approximately optimum for such a loop.) In Chapter 6, Table 6.2-2 (loop E) showed that the noise bandwidth in hertz for the loop transfer function of the form of Eq 7.4-30 is

$$f_b = \frac{\omega_b}{2\pi} = \frac{(\pi/2)(\omega_c + \omega_i)}{2\pi} = \tfrac{1}{4}(\omega_c + \omega_i) \qquad (7.4\text{-}31)$$

Substitute into this the values of $\omega_{c(r)}$ and $\omega_{i(r)}$ of items (1) and (2) in Table 7.4-2. This gives the noise bandwidth $f_{b(r)}$ of the range-tracking loop shown in item (3), which is 37.5 Hz. The range tracking loop has an effective integration

TABLE 7.4-2   Calculation of Errors of Range-Tracking Feedback Loop

Symbol	Parameter	Value
(1) $\omega_{c(r)}$	Gain crossover frequency	100 sec^{-1}
(2) $\omega_{i(r)}$	Integral break frequency	50 sec^{-1}
(3) $f_{b(r)}$	Noise bandwidth	37.5 Hz
(4) $T_{i(r)}$	Effective integration time	13.33 msec
(5) T_p	Pulse repetition period	0.667 msec
(6) N	Number of pulses integrated	20
(7) σR_1	RMS range noise error per pulse	3.92 m
(8) σR	RMS tracking range noise error	0.877 m
(9)	Peak (3-sigma) range noise error	2.63 m
(10)	Maximum range rate (Mach 2)	600 m/sec
(11)	Maximum range acceleration (5G)	49.0 m/sec^2
(12)	Maximum error at lock-on	6.0 m
(13)	Linear region of range gate	± 7.5 m
(14)	Acceleration error coefficient	2.0×10^{-4} sec^2
(15)	Maximum acceleration error	0.0098 m

time $T_{i(r)}$ equal to

$$T_{i(r)} = \frac{1}{2f_{b(r)}} = \frac{1}{2 \times 37.5 \text{ Hz}} = 13.33 \times 10^{-3} \text{ sec} = 13.33 \text{ msec} \quad (7.4\text{-}32)$$

This is shown in item (4). By item (2) of Table 7.3-6, the pulse repetition period T_p is 667 μsec, or 0.667 msec, as shown in item (5). Hence the range-tracking loop effectively integrates the following number of pulses:

$$N = \frac{T_{i(r)}}{T_p} = \frac{13.33 \text{ msec}}{0.667 \text{ msec}} = 20.0 \quad (7.4\text{-}33)$$

This is shown in item (6).

The RMS range noise error per pulse σR_1, which was shown in item (15) of Table 7.4-1, is repeated here in item (7). The RMS noise error of the range-tracking feedback loop is

$$\sigma R = \frac{\sigma R_1}{\sqrt{N}} = \frac{\sigma R_1}{\sqrt{20}} \quad (7.4\text{-}34)$$

This value is shown in item (8), and is 0.877 meter. This RMS range error is multiplied by 3 to obtain the peak (3-sigma) range noise error, which is shown in item (9). This peak noise error is 2.63 meter.

There are also range-tracking errors due to motions of the target. It is assumed that the target aircraft has a maximum speed of 600 meter/sec, which is approximately Mach 2 at high altitudes. The assumed maximum acceleration is $5g$, which is $5(9.8 \text{ m/sec}^2)$, or 49 m/sec^2. These values of maximum velocity and acceleration are shown in items (10 and 11).

If the range-tracking loop acquires the target while it is moving at maximum velocity in the radial direction, the range-tracking loop experiences a ramp input of range rate $V = 600 \text{ m/sec}$. If the range gate is operating in the linear range, the maximum range error R_e due to this ramp is approximately the velocity of the ramp divided by the gain crossover frequency of the range-tracking loop:

$$\text{Max}|R_e| = \frac{V}{\omega_{c(r)}} = \frac{600 \text{ m/sec}}{100 \text{ sec}^{-1}} = 6.0 \text{ m} \qquad (7.4\text{-}35)$$

This is shown in item (12). As was shown in Fig 7.4-3, the central linear region of the range-gate error characteristic is $\pm \tau_p/2$. The pulse width ($\tau_p = 0.1 \ \mu\text{sec}$) corresponds to a range increment of 15 meter. Hence the linear region $\pm \tau_p/2$ of the range gate is equivalent to ± 7.5 meter, as shown in item (13). The 6.0-meter error of Eq 7.4-35 is within this linear region but is almost to the limit. Hence, the range-gate tracking loop will be barely able to acquire the target at maximum velocity with the specified gain crossover frequency of 100 sec^{-1}.

If the range-tracking loop can hold the target signal within the linear range of the range-gate detector, the integral network rapidly compensates for the target velocity, reducing the error due to a constant target velocity to zero. As was explained in Chapter 3, after the lock-on transient has settled, the error is approximately characterized by the acceleration error coefficient. The acceleration error coefficient for the range-tracking loop is

$$c_2 = \frac{1}{\omega_{c(r)}\omega_{i(r)}} = \frac{1}{(100 \text{ sec}^{-1})(50 \text{ sec}^{-1})} = 2 \times 10^{-4} \text{ sec}^2 \quad (7.4\text{-}36)$$

This is shown in item (14). Multiply this by the maximum target acceleration shown in item (11). This gives the acceleration component of range-tracking error, which is shown in item (15) to be only 0.0098 meter.

Thus, after the initial lock-on transient, there are two major components of range-tracking error: a peak noise error of ± 2.6 meter (item 9) and an acceleration error having a maximum value of ± 0.01 meter (item 15). Since the acceleration error is much lower than the noise error, an appreciably smaller range-tracking error could be achieved by reducing the gain crossover frequency of the range-tracking loop after lock-on.

7.5 ANALYSIS OF ANGLE TRACKING

7.5.1 Angle Tracking Coordinates

It is assumed that the antenna is positioned by means of an azimuth–elevation gimbal mount, which was illustrated in Section 7.2, Fig 7.2-1. Figure 7.5-1 shows the coordinates for such a mount. Vector OA is a unit vector along the boresight of the antenna, and vector OT is a unit vector aimed at the target. The monopulse tracking feed measures the elevation and deflection components of angular error between the target vector OT and the antenna vector OA. The elevation error $\Delta\theta_{eL}$ is an angular displacement measured in the same plane as elevation rotation, while the deflection error $\Delta\theta_d$ is an angular displacement measured in a plane that is perpendicular to the elevation plane at the antenna boresight. To correct for the deflection error $\Delta\theta_d$, the antenna gimbal is rotated through an azimuth angle $\Delta\theta_{az}$ equal to

$$\Delta\theta_{az} = \frac{\Delta\theta_d}{\cos[\theta_{eL}]} \qquad (7.5\text{-}1)$$

where θ_{eL} is the elevation angle of the antenna boresight. Note that when the antenna is pointed close to zenith, a very large azimuth angular rotation $\Delta\theta_{az}$

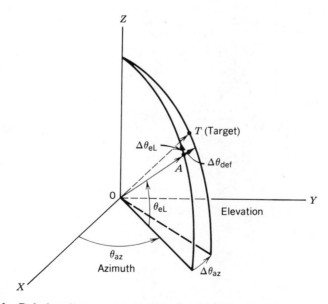

Figure 7.5-1 Relation between the azimuth–elevation coordinates of the antenna servo and the deflection–elevation coordinates of the tracking error, (A, antenna boresight; T, direction to target).

is required to compensate for a small deflection error $\Delta\theta_d$. An antenna that must move rapidly in the vicinity of zenith may require a more complicated antenna gimbal mount having three axes.

7.5.2   Measurement of Angular Tracking Error

7.5.2.1   *Conical Scan.*   Early tracking radars used conical scan to measure tracking error. Conical scan displaces the antenna beam slightly from boresight, and rotates it around the boresight in a conical pattern. This may be achieved by moving the antenna or the antenna feed. The following is an analysis of the conical-scan process.

Figure 7.5-2 shows a typical conical-scan pattern. The dashed circle shows the path of the center of the beam during the scan. The four circles show the

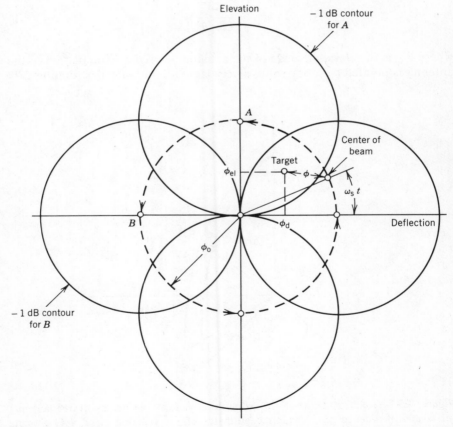

Figure 7.5-2   Typical conical scan pattern, showing gain contours at four points in scan.

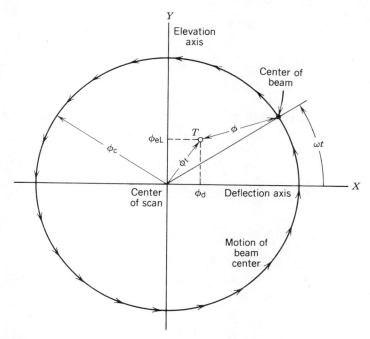

Figure 7.5-3   Coordinates of target relative to center of beam for conical scan pattern.

contours for 1 dB below peak gain, at points displaced by intervals of 90° along the scan. The angle ϕ_o is the offset angle of the scan, which is the amount that the center of the beam is offset from the center of the scan. For this particular scan pattern, the offset angle is chosen so that the beam patterns cross over at the -1-dB points.

The geometric relations of the scan pattern are shown in Fig 7.5-3. The deflection and elevation coordinates of the target (relative to the center of the scan pattern) are denoted ϕ_d and ϕ_{eL}, respectively. The deflection and elevation coordinates of the center of the beam are denoted $\phi_{d(c)}$ and $\phi_{eL(c)}$, which are equal to

$$\phi_{d(c)} = \phi_o \cos[\omega t] \tag{7.5-2}$$

$$\phi_{eL(c)} = \phi_o \sin[\omega t] \tag{7.5-3}$$

The angular displacement of the target from the center of the beam is denoted ϕ. This is computed from

$$\phi^2 = \left(\phi_{d(c)} - \phi_d\right)^2 + \left(\phi_{eL(c)} - \phi_{eL}\right)^2 \tag{7.5-4}$$

Combining Eqs 7.5-2 to -4 gives

$$\phi^2 = \left(\phi_o \cos[\omega t] - \phi_d \right)^2 + \left(\phi_o \sin[\omega t] - \phi_{eL} \right)^2$$
$$= \phi_o^2 + \left(\phi_d^2 + \phi_{eL}^2 \right) - 2\phi_d\phi_o \cos[\omega t] - 2\phi_{eL}\phi_o \sin[\omega t] \quad (7.5\text{-}5)$$

The angular displacement of the target from the center of the beam is denoted ϕ_t, which is given by

$$\phi_t^2 = \phi_d^2 + \phi_{eL}^2 \quad (7.5\text{-}6)$$

Combining Eqs 7.5-5, -6 gives

$$\phi^2 = \phi_o^2 + \phi_t^2 - 2\phi_d\phi_o \cos[\omega t] - 2\phi_{eL}\phi_o \sin[\omega t] \quad (7.5\text{-}7)$$

As was shown in Section 7.3, Eq 7.3-11, in the region near the center of the beam, the ratio of the antenna gain G divided by the maximum antenna gain G_m (at the center of the beam) is related approximately as follows to the angle ϕ:

$$\frac{G}{G_m} = \exp\left[-\left(\frac{\phi}{\Psi} \right)^2 \right] \quad (7.5\text{-}8)$$

The angle Ψ is proportional to the 3-dB beamwidth θ_b of the antenna gain pattern:

$$\Psi = 0.601\theta_b \quad (7.5\text{-}9)$$

The ratio of received power P divided by the power P_m received when the target at the center of the beam is

$$\frac{P}{P_m} = \left(\frac{G}{G_m} \right)^2 \quad (7.5\text{-}10)$$

The gain G is squared because it occurs in the radar transmission equation in both transmission and reception. The voltage level E of the IF amplifier is proportional to the square root of the received power. Hence, the ratio of the voltage level E divided by the voltage E_m when the target is at the center of the beam is

$$\frac{E}{E_m} = \sqrt{\frac{P}{P_m}} = \frac{G}{G_m} \quad (7.5\text{-}11)$$

Combining Eqs 7.5-7, -8, -11 gives

$$\frac{E}{E_m} = \exp\left[-(\phi/\Psi)^2\right]$$

$$= \exp\left[-\left(\frac{\phi_o}{\Psi}\right)^2 - \left(\frac{\phi_t}{\Psi}\right)^2 + 2\left(\frac{\phi_o}{\Psi^2}\right)(\phi_d \cos[\omega t] + \phi_{eL} \sin[\omega t])\right] \quad (7.5\text{-}12)$$

This can be expressed as

$$E/E_m = (F_1)(F_2)(F_3) \qquad (7.5\text{-}13)$$

where the functions F_1, F_2, F_3 are

$$F_1 = \exp\left[-\left(\frac{\phi_o}{\Psi}\right)^2\right] \qquad (7.5\text{-}14)$$

$$F_2 = \exp\left[-\left(\frac{\phi_t}{\Psi}\right)^2\right] \qquad (7.5\text{-}15)$$

$$F_3 = \exp\left[2\frac{\phi_o}{\Psi}\left(\frac{\phi_d}{\Psi}\cos[\omega t] + \frac{\phi_{eL}}{\Psi}\sin[\omega t]\right)\right] \qquad (7.5\text{-}16)$$

When the target is at the center of the scan pattern, the voltage E is a constant defined as E_0. Setting ϕ_t, ϕ_{eL}, and ϕ_d all equal to zero in Eq 7.5-12 gives the ratio E_0/E_m:

$$\frac{E_0}{E_m} = \exp\left[-\left(\frac{\phi_o}{\Psi}\right)^2\right] = F_1 \qquad (7.5\text{-}17)$$

Note that E_m and E_0 are defined as

E_m = receiver voltage when target is at center of antenna beam
E_0 = receiver voltage when target is at center of scan pattern

Combining Eqs 7.5-13, -17 gives

$$\frac{E}{E_0} = F_2 F_3 \qquad (7.5\text{-}18)$$

Remember that a function $\exp[x]$ is approximately equal to $(1 + x)$ for $x \ll 1$. Hence, for small target errors, the functions F_2, F_3 in Eqs 7.5-14, -15 can be approximated by

$$F_2 \cong 1 + \left(\frac{\phi_t}{\Psi}\right)^2 \tag{7.5-19}$$

$$F_3 \cong 1 + 2\frac{\phi_o}{\Psi}\left\{\frac{\phi_d}{\Psi}\cos[\omega t] + \frac{\phi_{eL}}{\Psi}\sin[\omega t]\right\} \tag{7.5-20}$$

Combining Eqs 7.5-18, -19, -20 gives

$$E = E_0\left\{1 - \left(\frac{\phi_t}{\Psi}\right)^2\right\}\left\{1 + 2\frac{\phi_o}{\Psi}\left(\frac{\phi_d}{\Psi}\cos[\omega t] + \frac{\phi_{eL}}{\Psi}\sin[\omega t]\right\} \tag{7.5-21}$$

Define the conical-scan slope k_s as

$$k_s = 2\frac{\phi_o}{\Psi}\frac{\theta_b}{\Psi} = 2\left(\frac{\theta_b}{\Psi}\right)^2\frac{\phi_o}{\theta_b}$$

$$= \frac{2}{(0.601)^2}\frac{\phi_o}{\theta_b} = 5.537\frac{\phi_o}{\theta_b} \tag{7.5-22}$$

Combining Eqs 7.5-21, -22 gives

$$E = E_0\left\{1 - \left(\frac{\phi_t}{\Psi}\right)^2\right\}\left(1 + k_s\frac{\phi_d}{\theta_b}\cos[\omega t] + k_s\frac{\phi_{eL}}{\theta_b}\sin[\omega t]\right) \tag{7.5-23}$$

The factor $[1 - (\phi_t/\Psi)^2]$ is a constant during the scan, and for small tracking errors is close to unity. Hence, it can generally be neglected, and Eq 7.5-23 becomes

$$E = E_0\left(1 + k_s\frac{\phi_d}{\theta_b}\cos[\omega t] + k_s\frac{\phi_{eL}}{\theta_b}\sin[\omega t]\right) \tag{7.5-24}$$

The ratio in decibels of the antenna gain G_0 when the target is at the center of the scan, divided by the antenna gain G_m when the target is at the center of the beam, is equal to

$$10\log\left[\frac{G_0}{G_m}\right] = -12\left(\frac{\phi_o}{\theta_b}\right)^2 \quad \text{dB} \tag{7.5-25}$$

Typically, G_0 is set 1 dB below peak gain G_m. Setting Eq 7.5-25 equal to -1 dB gives the following offset angle for this condition

$$\phi_o = 0.2887\theta_b \tag{7.5-26}$$

Substituting this into Eq 7.5-22 gives the corresponding conical-scan slope k_s, which is

$$k_s = 5.537(0.2887) = 1.598 \cong 1.600 \qquad (7.5\text{-}27)$$

Thus, a typical conical scan provides a slope of $k_s = 1.60$. The corresponding scan loss is 1 dB, which results in a two-way (transmit-plus-receive) loss of 2 dB.

Circuitry for deriving elevation and deflection tracking error signals from the conical-scan modulation of the received radar pluses was described in Section 7.2.4.1.

7.5.2.2 Monopulse.
With conical scan, the returns from many radar pulses must be compared to obtain a tracking error. Monopulse is a later development and was so named because it derives tracking error signals from a single radar pulse. A serious weakness of conical scan is that variations of the strength of the received signal due to target scintillation produce tracking error. Monopulse eliminates this problem but is much more complicated.

Figure 7.5-4 shows the basic method for deriving a monopulse tracking error signal for a single axis. To simplify the explanation, the diagram assumes that the radar antenna uses a refracting dielectric microwave lens rather than a parabolic reflector. The same principle applies with a parabolic reflector, but the geometry is more complicated. Two antenna feed horns (1) and (2) are placed side by side in the focal plane of the refracting lens. Feed horn (1) projects its ray in direction (1), as shown by the solid lines, while feed horn (2) projects its rays in direction (2) as shown by the dashed lines. These sets of rays are displaced by the offset angle $\mathscr{E}$ in opposite directions from the centerline boresight axis.

In Fig 7.5-4 the rays are shown projecting to precise focal points in the mouths of the two horns. However, because of diffraction, the image from a

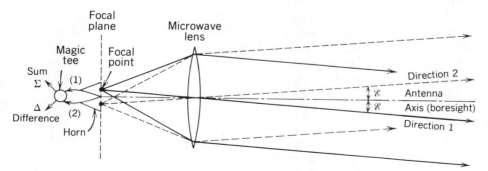

Figure 7.5-4 Geometric optics of single axis of monopulse tracking antenna, expressed in terms of microwave lens.

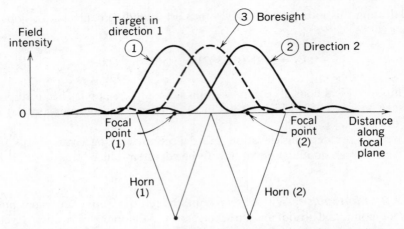

Figure 7.5-5 Plots of field intensity at focal plane for target signals received from different directions.

remote point target is projected as an extended diffraction pattern on the focal plane, called an Airy disk. This Airy disk pattern has the shape of the antenna angular gain pattern, multiplied by the focal distance F. Figure 7.5-5 shows typical plots of power versus displacement at the focal plane, along with the relative locations of the openings of feed horns (1) and (2). Solid curve (1) is the power from a point source in direction (1), solid curve (2) is the power from a point source in direction (2), and dashed curve (3) is the power from a point source on the boresight axis. Most of the energy from source (1) is intercepted by horn (1), and most of the energy from source (2) is intercepted by horn (2). The energy from source (3) on the boresight falls equally into both horns. Angular tracking error is measured by subtracting the signals received by the two horns.

As shown in Fig 7.5-4, the signals from the two horns are fed to a waveguide circuit device called a *magic Tee*, which has two output ports, one delivering a sum signal Σ proportional to the sum of the signals from the two horns, and the other delivering a difference signal Δ proportional to the difference of those signals. Figure 7.5-6a shows a magic-Tee waveguide device. The electric field intensities of the input signals at ports (1) and (2) are denoted by the complex quantities E_1, E_2. The intensities for the output signals at the sum and difference ports are

$$\text{Sum port:} \qquad E_s = \frac{E_1 + E_2}{\sqrt{2}} \qquad (7.5\text{-}28)$$

$$\text{Difference port:} \qquad E_d = \frac{E_1 - E_2}{\sqrt{2}} \qquad (7.5\text{-}29)$$

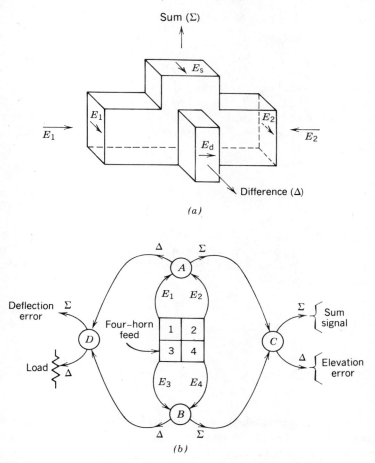

Figure 7.5-6 Microwave circuitry used in monopulse feeds: (*a*) magic Tee; (*b*) symbolic representation of microwave circuitry for 4-horn, 2-axis monopulse feed. (Magic Tees: *A*, *B*, *C*, and *D*.)

The factor $1/\sqrt{2}$ is required to maintain constancy of power. For example, if the two inputs are the same, then $E_1 = E_2 = E$. The difference signal E_d is zero, and the sum signal E_s is

$$E_s = 2E/\sqrt{2} = \sqrt{2}\,E \qquad (7.5\text{-}30)$$

Squaring this gives the power. Thus the sum signal power E_s^2 is equal to $2E^2$. This is twice the power in either input port, and so is equal to the total input power, as it should.

The following is an approximate analysis of the response of this single-axis monopulse tracking antenna. As was shown in Eq 7.3-11, the gain G of an antenna in the region near boresight can be approximated by

$$G = G_m \exp\left[-(\phi/\Psi)^2\right] \tag{7.5-31}$$

where G_m is the maximum antenna gain. The angle ϕ is measured from the axis of maximum gain, and ψ is a constant equal to $0.60\theta_b$, where θ_b is the antenna half-power beamwidth. The field intensity E of the signal received by the antenna is proportional to the square root of the antenna gain and so is equal to

$$E = E_m \sqrt{\frac{G}{G_m}} = E_m \exp\left[-\frac{\phi^2}{2\Psi^2}\right] \tag{7.5-32}$$

where E_m is the maximum electric field intensity, which occurs at maximum antenna gain. It is assumed that monopulse feed horns (1) and (2) intercept essentially all of the energy from their particular directions. Hence the signals from these two horns are

$$E_1 = E_m \exp\left[-\frac{(\theta - \mathscr{E})^2}{2\Psi^2}\right] \tag{7.5-33}$$

$$E_2 = E_m \exp\left[-\frac{(\theta + \mathscr{E})^2}{2\Psi^2}\right] \tag{7.5-34}$$

where θ is the angle measured from boresight, and $\mathscr{E}$ is the angular offset of the beams from boresight. For feed (1) the angle ϕ measured from the peak antenna gain is equal to $(\theta - \mathscr{E})$, while for feed (2), ϕ is equal to $(\theta + \mathscr{E})$. Equations 7.5-33, -34 can be expanded to give

$$\begin{aligned} E_1 &= E_m \exp\left[-\frac{\theta^2}{2\Psi^2} - \frac{\mathscr{E}^2}{2\Psi^2} + \frac{\mathscr{E}\theta}{\Psi^2}\right] \\ &= E_m \exp\left[-\frac{\theta^2}{2\Psi^2}\right]\exp\left[-\frac{\mathscr{E}^2}{2\Psi^2}\right]\exp\left[\frac{\mathscr{E}\theta}{\Psi^2}\right] \end{aligned} \tag{7.5-35}$$

$$\begin{aligned} E_2 &= E_m \exp\left[-\frac{\theta^2}{2\Psi^2} - \frac{\mathscr{E}^2}{2\Psi^2} - \frac{\mathscr{E}\theta}{\Psi^2}\right] \\ &= E_m \exp\left[-\frac{\theta^2}{2\Psi^2}\right]\exp\left[-\frac{\mathscr{E}^2}{2\Psi^2}\right]\exp\left[-\frac{\mathscr{E}\theta}{\Psi^2}\right] \end{aligned} \tag{7.5-36}$$

If the angle θ from boresight is small, the following approximations can be made:

$$\exp\left[-\frac{\theta^2}{2\Psi^2}\right] \cong 1 \tag{7.5-37}$$

$$\exp\left[\pm\frac{\mathscr{E}\theta}{\Psi^2}\right] \cong 1 \pm \frac{\mathscr{E}\theta}{\Psi^2} \tag{7.5-48}$$

Hence Eqs 7.5-35, -36 can be approximated as

$$E_1 \cong E_m \exp\left[-\frac{\mathscr{E}^2}{2\Psi^2}\right]\left(1 + \frac{\mathscr{E}\theta}{\Psi^2}\right) \tag{7.5-39}$$

$$E_2 \cong E_m \exp\left[-\frac{\mathscr{E}^2}{2\Psi^2}\right]\left(1 - \frac{\mathscr{E}\theta}{\Psi^2}\right) \tag{7.5-40}$$

Substituting these into Eqs 7.5-28, -29 gives the following signals at the sum and difference ports of the magic Tee:

$$E_s = \frac{E_1 + E_2}{\sqrt{2}} \cong \sqrt{2}\, E_m \exp\left[-\frac{\mathscr{E}^2}{2\Psi^2}\right] \tag{7.5-41}$$

$$E_d = \frac{E_1 - E_2}{\sqrt{2}} \cong \sqrt{2}\, E_m \exp\left[-\frac{\mathscr{E}^2}{2\Psi^2}\right]\frac{\mathscr{E}\theta}{\Psi^2} \tag{7.5-42}$$

The normalized monopulse slope constant k_m is defined as follows in terms of the sum and difference signals E_s and E_d from the feed:

$$\frac{E_d}{E_s} = k_m \frac{\theta}{\theta_b} \tag{7.5-43}$$

Substituting Eqs 7.5-41, -42 into this gives

$$k_m \frac{\theta}{\theta_b} = \frac{\mathscr{E}\theta}{\Psi^2} \tag{7.5-44}$$

Solving for the normalized monopulse slope k_m gives

$$k_m = \frac{\mathscr{E}\theta_b}{\Psi^2} \tag{7.5-45}$$

The maximum gain for the sum channel, denoted G_{sm}, is obtained by taking the square of Eq 7.5-41, which gives

$$\frac{G_{sm}}{G_m} = \left(\frac{E_s}{E_m}\right)^2 = 2\exp\left[-\left(\frac{\mathscr{E}}{\Psi}\right)^2\right] \tag{7.5-46}$$

The parameter G_m is the peak antenna gain that would be achieved if the antenna were illuminated with a single horn feed, while G_{sm} is the peak antenna gain of the sum channel when the monopulse feed is used. Typically a monopulse feed for a radar antenna causes a loss of about 1 dB. This is the loss that was assumed in determining the antenna efficiency $\nu = 0.60$ shown in Section 7.3, Table 7.3-1, item (4). For a 1-dB monopulse loss (which is a factor of 1.26), the reciprocal of Eq 7.5-46 is

$$\frac{G_m}{G_{sm}} = 1.26 = \tfrac{1}{2}\exp\left[\left(\frac{\mathscr{E}}{\Psi}\right)^2\right] \tag{7.5-47}$$

This gives

$$\frac{\mathscr{E}}{\psi} = \sqrt{\ln[2(1.26)]} = 0.961 \tag{7.5-48}$$

Substituting this into Eq 7.5-19 gives the following normalized monopulse gain:

$$k_m = \frac{\mathscr{E}}{\Psi}\frac{\theta_b}{\Psi} = \frac{0.961}{0.60} = 1.60 \tag{7.5-49}$$

The preceding analysis is approximate but yields values that are typical of well-designed monopulse feeds for radar antennas. A rigorous study is given by Hannan [7.6]. Discussions of this and other data are presented by Barton [7.4] (pp. 271–275) and by Barton and Ward [7.5] (pp. 24–32). Hannan shows that a simple four-horn monopulse feed typically has a monopulse sloped k_m of about 1.2 and experiences a loss of efficiency of about 2 dB. However, more sophisticated monopulse feeds usually achieve a monopulse slope k_m of about 1.6 with small loss of antenna efficiency. For our example, it is assumed that $k_m = 1.6$ and the efficiency loss of the feed is 1 dB relative to a non-monopulse feed.

The monopulse feed of Fig 7.5-4 provides angular tracking signals for only a single axis. By using four horns, tracking error signals can be derived for both elevation and deflection. The resultant microwave circuitry is shown symbolically in Fig 7.5-6b. At the focal plane, there is a square cluster of four feed horns labeled 1, 2, 3, 4. Elements A, B, C, D are magic-Tee waveguide devices. Four signals are derived from the microwave circuit: the sum signal, the elevation difference signal, the deflection difference signal, and a double-difference signal which is absorbed in a load.

7.5.3 General Equation for Monopulse Angle-Tracking Noise Error

Figure 7.5-7 shows a block diagram of the monopulse receiver circuitry commonly used to derive angular tracking error signals from the monopulse feed signals. The three RF signals from the monopulse feed (sum, elevation

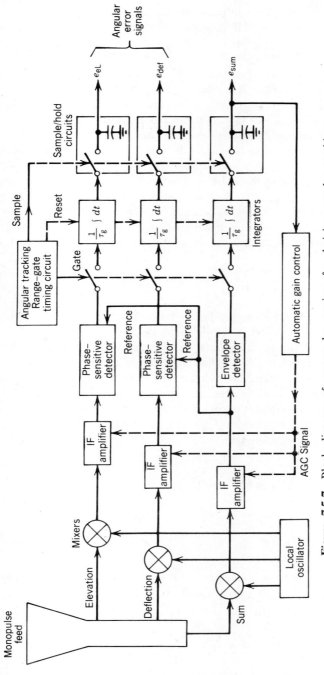

Figure 7.5-7 Block diagram of monopulse receiver for deriving angular tracking error signals.

difference, and deflection difference) are separately mixed with the same local-oscillator signal, which converts them to IF signals. These are fed through identical IF amplifiers. The phase shifts for the three channels, including the phase shifts at RF, are matched over the IF passband. The elevation and deflection difference signals are fed to phase-sensitive detectors, which use the sum-channel signal as a reference. (See the discussion of phase-sensitive detectors in Chapter 4, Section 4.1.3.) The outputs from the phase-sensitive detectors are passed through range gates to integrator circuits, which average the signals over the periods of the gates. These signals are then fed to sample-and-hold circuits, which hold the signals fixed between radar pulses.

If the signal/noise ratio of the sum channel is much greater than unity, the phase-sensitive detectors act like coherent detection processes, which preserve the signal/noise ratios in the difference channels. When this is not true, the sum channel adds noise to the difference channels. The resultant noise power density in a difference channel at the output of the phase-sensitive detector is approximately equal to

$$p_{nd} = p_n(1 + P_n/P_s) \qquad (7.5\text{-}50)$$

where P_n/P_s is the noise/signal power ratio of the sum channel, which is fed as a reference signal to the phase-sensitive detector.

It is assumed that the duration τ_{ga} of the angle-tracking range-gate is wider than the radar pulse width τ_p; that the pulse is completely within the range gate; and that the IF bandpass is relatively wide. Under these conditions, the noise bandwidth of the process is established by the range gate, which is an ideal averaging integrator of integration time T_i equal to τ_{ga}. Hence its equivalent IF noise bandwidth $f_{b(bp)}$ is equal to $1/\tau_{ga}$. The RMS noise in the range-gate output signal is then

$$e_n = \sqrt{p_{nd} f_{b(bp)}} = \sqrt{\frac{p_{nd}}{\tau_{ga}}}$$

$$= \sqrt{\frac{p_n(1 + P_n/P_s)}{\tau_{ga}}} \qquad (7.5\text{-}51)$$

By Eq 7.5-43, the RMS signal e_d in a difference channel is related as follows to the RMS signal e_s in the sum channel:

$$e_d = k_m \frac{\theta}{\theta_b} e_s = k_m \frac{\theta}{\theta_b} \sqrt{P_s} \qquad (7.5\text{-}52)$$

where θ is the angular error (elevation or deflection) for that channel. As shown, the RMS sum-channel signal e_s is set equal to the square root of the sum-channel signal power P_s. The phase-sensitive detector samples the difference signal at each peak value, and so generates a rectified pulse with an amplitude equal to the peak value of the IF difference signal, which is $\sqrt{2}\,e_d$. The range gate and integrator measure the average value of this pulse over the period τ_{ga} of the range gate, which is proportional to the ratio of the pulse width τ_p divided by the gate width τ_{ga}. Hence the detected angular error signal at the output of the integrator is

$$e_e = \sqrt{2}\,e_d \frac{\tau_p}{\tau_{ga}} \qquad (7.5\text{-}53)$$

Combining Eqs 7.5-52, -53 gives

$$e_e = k_m \frac{\theta}{\theta_b} \frac{\tau_p}{\tau_{ga}} \sqrt{2P_s} \qquad (7.5\text{-}54)$$

If the signal e_e is set equal to the RMS noise signal e_n given in Eq 7.5-51, the corresponding value for θ is the RMS angular error for a single-pulse measurement, denoted $\sigma\theta_1$. Thus,

$$k_m \frac{\sigma\theta_1}{\theta_b} \frac{\tau_p}{\tau_{ga}} \sqrt{2P_s} = \sqrt{\frac{P_n(1 + P_n/P_s)}{\tau_{ga}}} \qquad (7.5\text{-}55)$$

Solving for $\sigma\theta_1/\theta_b$ gives

$$\frac{\sigma\theta_1}{\theta_b} = \frac{\sqrt{(1 + P_n/P_s)(\tau_{ga}/\tau_p)}}{k_m\sqrt{2P_s\tau_p/P_n}} = \frac{\sqrt{(1 + P_n/P_s)(\tau_{ga}/\tau_p)}}{k_m\sqrt{2E_s/P_n}} \qquad (7.5\text{-}56)$$

This is the relative RMS angular error measured from a single pulse.

The angle-tracking servo has an effective integration time $T_{i(a)}$ equal to

$$T_{i(a)} = \frac{1}{2f_{b(a)}} \qquad (7.5\text{-}57)$$

where $f_{b(a)}$ is the noise bandwidth of the angle-tracking servo. The servo integrates the number of radar pulses that occur within its integration time. Hence, the number of pulses integrated are

$$N_{i(a)} = \frac{T_{i(a)}}{T_p} \frac{f_p}{2f_{b(a)}} \qquad (7.5\text{-}58)$$

where T_p is the pulse repetition period, and its reciprocal f_p is the pulse repetition frequency. The RMS angular error is

$$\sigma\theta = \frac{\sigma\theta_1}{\sqrt{N_{i(a)}}} \tag{7.5-59}$$

Combining Eqs 7.5-56, -58, -59 gives

$$\frac{\sigma\theta}{\theta_b} = \frac{\sqrt{(1 + P_n/P_s)(\tau_{ga}/\tau_p)}}{k_m\sqrt{(f_p/f_{b(a)})}\, E_s/P_n} \tag{7.5-60}$$

The elevation and deflection error signals e_{eL} and e_d at the outputs of the sample-and-hold circuits of Fig 7.5-6 vary in proportion to the received signal. To compensate for this, the signal from the sum channel is fed through an envelope detector, a range gate, an integrator, and a sample-and-hold circuit, just as for the error channel signals. This forms the signal e_{sum}, which is fed to an automatic gain-control (AGC) circuit. This AGC circuit varies the gains of the three amplifiers equally to maintain a constant signal level. It is difficult to keep the gains of the three amplifiers identical, and so there is usually some uncertainty in the sensitivities of the elevation and deflection error signals. This causes the gain crossover frequencies of the angle tracking loops to vary.

7.5.4 Tracking Error Expressions Given by Barton and Ward

Let us compare the expressions for angular tracking error that have been derived with those given by Barton and Ward [7.5] for the angular tracking errors of monopulse and conical-scan radars. Barton and Ward [7.5] (p. 46, Table 2.9) gave the following for the relative angular tracking error for monopulse and conical-scan radars:

Monopulse:

$$\frac{\sigma\theta}{\theta_b} = \frac{\sqrt{(P_s/P_n + 1)L_{mv}L_c}}{k_m(P_s/P_n)\sqrt{2N_{i(a)}L_m}} \tag{7.5-61}$$

Conical scan:

$$\frac{\sigma\theta}{\theta_b} = \frac{\sqrt{(2P_s/P_n + 1)L_{mv}L_c}}{k_s(P_s/P_n)\sqrt{N_{i(a)}L_m}} \tag{7.5-62}$$

Rearranging terms gives

Monopulse:

$$\frac{\sigma\theta}{\theta_b} = \frac{\sqrt{(1 + P_n/P_s)L_{mv}L_c}}{k_m\sqrt{2N_{i(a)}(P_s/P_n)L_m}} \tag{7.5-63}$$

Conical scan:

$$\frac{\sigma\theta}{\theta_b} = \frac{\sqrt{(1 + P_n/2P_s)L_{mv}L_c}}{k_s\sqrt{N_{i(a)}(P_s/P_n)L_m}} \qquad (7.5\text{-}64)$$

The parameters given in Eqs 7.5-61 to -64 are defined as

L_m = IF amplifier matching loss
L_{mv} = video matching loss
L_c = range-gate collapsing loss
P_s/P_n = signal/noise power ratio at output of IF amplifier
k_m = monopulse slope
k_s = conical-scan slope
$N_{i(a)}$ = effective number of pulses integrated by angle tracking servo

The expression for $N_{i(a)}$ was given in Eq 7.5-58.

Equations 7.5-63, -64 can be expressed in the form

$$\sigma\theta = \frac{\sigma\theta_1}{\sqrt{N_{i(a)}}} \qquad (7.5\text{-}65)$$

where $\sigma\theta_1$ is the effective RMS error for a single pulse. For the monopulse system, $\sigma\theta_1$ is the actual RMS error derived from the measurement made on a single pulse. However, the conical-scan system requires that the returns from several pulses be combined to measure angular error. Hence, for conical scan, $\sigma\theta_1$ is merely a ficticious effective signal-pulse error.

The signal/noise power ratio at the output of the IF amplifier is

$$\frac{P_s}{P_n} = \frac{E_s/p_n}{L_m} \qquad (7.5\text{-}66)$$

where E_s/p_n is the signal/noise ratio for an ideal matched-filter receiver, and L_m is the matching loss of the actual IF amplifier relative to the matched filter.

Combining Eqs 7.5-63 to -66 gives the following expressions for the effective single-pulse angular error:

Monopulse:

$$\frac{\sigma\theta_1}{\theta_b} = \frac{\sqrt{(1 + P_n/P_s)L_{mv}L_c}}{k_m\sqrt{2E_s/P_n}} \qquad (7.5\text{-}67)$$

Conical scan:

$$\frac{\sigma\theta_1}{\theta_b} = \frac{\sqrt{(1 + P_n/2P_s)L_{mv}L_c}}{k_s\sqrt{E_s/P_n}} \qquad (7.5\text{-}68)$$

The analysis of this chapter has assumed that the angle-scan range-gate width τ_{ga} is greater than the pulse width τ_p. For this case, Barton and Ward [7.5] (p. 42, Table 2.7) give several expressions for the range-gate collapsing loss for various processing techniques. For most of these cases, this loss is approximately equal to

$$L_c \cong \frac{\tau_{ga}}{\tau_p} \qquad \text{for } \tau_{ga} > \tau_p \tag{7.5-69}$$

The loss varies somewhat from this at high signal/noise ratios. Barton and Ward [7.5] (p. 41, Table 2.6) imply that the video matching loss for this condition is approximately unity:

$$L_{mv} \cong 1 \qquad \text{for } \tau_{ga} > \tau_p \tag{7.5-70}$$

Although the actual expression for L_{mv} for $\tau_{ga} > \tau_p$ is omitted in this reference, Eq 7.5-70 is implied in the discussion. Combining Eqs 7.5-67 to -70 gives the following for the effective single-pulse RMS angular error when the gate is wider than the pulse:

Monopulse ($\tau_{ga} > \tau_p$):

$$\frac{\sigma\theta_1}{\theta_b} = \frac{\sqrt{(1 + P_n/P_s)(\tau_{ga}/\tau_p)}}{k_m\sqrt{2E_s/P_n}} \tag{7.5-71}$$

Conical scan ($\tau_{ga} > \tau_p$):

$$\frac{\sigma\theta_1}{\theta_b} = \frac{\sqrt{(1 + P_n/2P_s)(\tau_{ga}/\tau_p)}}{k_s\sqrt{E_s/P_n}} \tag{7.5-72}$$

Note that the monopulse expression for $\sigma\theta_1/\theta_b$ in Eq 7.5-71, obtained from Barton and Ward, is the same as the expression given in Eq 7.5-56 derived from the analysis in Section 7.5.3.

The factors $(1 + P_n/P_s)$ and $(1 + P_n/2P_s)$ in Eqs 7.5-21, -22 are noncoherent detection losses, which are denoted

Monopulse noncoherent detection loss:

$$L_{d(m)} = 1 + \frac{P_n}{P_s} \tag{7.5-73}$$

Conical - scan noncoherent detection loss:

$$L_{d(s)} = 1 + \frac{P_n}{2P_s} \tag{7.5-74}$$

The monopulse noncoherent detection loss $L_{d(m)}$ in Eq 7.5-73 is the same as that shown in Eq 7.2-24 for range tracking. However, the conical-scan noncoherent detection loss $L_{d(s)}$ is smaller. The reason for the smaller noncoherent detection loss with conical scan is that the conical-scan information is derived from the AC variation of the detected pulses, rather than from the absolute values of the pulses.

7.5.5 Computation of Angular Tracking Errors for Monopulse Radar

Table 7.5-1 shows the steps for calculating the relative RMS angular tracking error per pulse. Item (1) is the ratio E_s/p_n for the monopulse sum channel, and item (2) is the sum-channel IF filter matching loss L_m. These are equivalent to the values given in items (7 and 8) of Table 7.4-1. Dividing item (1) by item (2) gives the signal/noise ratio P_s/P_n shown in item (3). This signal/noise ratio occurs at the output of the sum-channel IF amplifier, which is the reference signal fed to the monopulse phase-sensitive detectors. The ratio $P_s/P_n = 3.44$, given in item (3), is the same as the P_s/P_n ratio of item (9) in Table 7.4-1.

In accordance with Eq 7.5-73, the noncoherent detection loss produced in the monopulse phase-sensitive detectors is

$$L_{d(m)} = 1 + \frac{P_n}{P_s} = 1 + \frac{1}{3.44} = 1.291 \tag{7.5-75}$$

This is the same as the range-gate noncoherent detection loss given in item (10) of Table 7.4-1 for range tracking.

In accordance with Eq 7.5-49, the monopulse slope k_m is assumed to be 1.6, as shown in item (5) of Table 7.5-1. Item (6) shows that the angle-tracking

TABLE 7.5-1 Calculation of RMS Angular Noise Tracking Error Per Pulse

	Symbol	Parameter	Value
(1)	E_s/p_n	Sum-channel matched-filter signal/noise ratio	4.189
(2)	L_m	IF amplifier matching loss	1.216 (0.85 dB)
(3)	P_s/P_n	Signal/noise ratio of sum channel	3.44
(4)	$L_{d(m)}$	Non-coherent detection loss in phase-sensitive detectors	1.291
(5)	k_m	Monopulse slope	1.6
(6)	τ_{ga}/τ_p	Relative angle-tracking gate width	1.5
(7)	$\sigma\theta_1/\theta_b$	Relative RMS/angular noise error per pulse	0.300

TABLE 7.5-2 Calculation of Peak 2-Axis Angular Tracking Error Due to Receiver Noise

Symbol		Parameter	Value
(1)	$\omega_{c(a)}$	Angle-track gain crossover frequency	$15 \ \text{sec}^{-1}$
(2)	$\omega_{i(a)}$	Angle-track integral break frequency	$5 \ \text{sec}^{-1}$
(3)	ω_b / ω_c	Noise bandwidth ratio	2.36
(4)	$f_{b(a)}$	Angle-track noise bandwidth	5.63 Hz
(5)	$T_{i(a)}$	Angle-track integration time	88.8 msec
(6)	T_p	Pulse repetition period	0.667 msec
(7)	$N_{i(a)}$	Number of pulses integrated by angle-track servos	133
(8)	$\sigma\theta_1 / \theta_b$	Relative RMS angular noise error per pulse	0.300
(9)	$\sigma\theta / \theta_b$	Relative RMS angular noise error	0.0260
(10)	θ_b	Half-power beamwidth	1.60°
(11)	$\sigma\theta$	RMS 1-axis noise error	0.0416°
(12)		Peak 2-axis noise error	±0.176°
			±3.08 mrad

range-gate pulse width τ_{ga} it 1.5 times greater than the RF radar pulse width τ_p, which is 0.1 μsec. This width is sufficient to accommodate range-gate tracking error and smearing of the pulse by the IF amplifier, with only a small amount of energy falling outside the gate.

The values of items (1, 4, 5, 6) of Table 7.5-1 are substituted into Eq 7.5-71. This gives the value of $\sigma\theta_1 / \theta_b$ shown in item (7). Thus, the RMS angular error per pulse $\sigma\theta_1$ is 0.300 times the antenna beamwidth θ_b.

The actual RMS angular tracking error depends on the noise bandwidths of the angle tracking servos. This error is calculated in Table 7.5-2. As shown in item (1), the gain crossover frequency $\omega_{c(a)}$ of the angle-tracking loop is assumed to be 15 rad/sec. This loop has an integral network with a break frequency $\omega_{i(a)}$ that is set equal to 1/3 of the gain crossover frequency, or 5 rad/sec, as shown in item (2). For the range-tracking loop, the integral-network break frequency ω_i was set equal to 1/2 of the gain crossover frequency, which is an optimum setting for a loop that has essentially ideal performance. However, a greater ω_c / ω_i ratio is needed in an angle-tracking loop because of the dynamic limitations of the antenna drive. Although a ω_c / ω_i ratio of 3 is assumed for the angle-track loop, a somewhat larger ratio may sometimes be desirable.

It is assumed that the loop transfer function of the angle tracking loop can be approximated by

$$G_{(a)} = \frac{\omega_c (1 + \omega_i / s)}{s(1 + s/\omega_f)} \tag{7.5-76}$$

The magnitude asymptote plot of this transfer function was shown in Chapter

6, Fig 6.2-5 (loop F). The noise bandwidth ω_b of loop F was given in Table 6.2-2 as

$$\frac{\omega_b}{\omega_c} = \frac{\pi\omega_f(\omega_c + \omega_i)}{2\omega_c(\omega_f - \omega_i)} = \frac{\pi(1 + \omega_i/\omega_c)}{2(1 - \omega_i/\omega_f)} \qquad (7.5\text{-}77)$$

It is assumed that $\omega_f/\omega_c = \omega_c/\omega_i = 3$. For this setting, the loop is the same as loop D studied in Ref [1.1] (Section 2.4.2). Its transient responses are given in Fig 2.4-10 of that reference, which shows that the step response has 25% peak overshoot. The maximum value of $|G_{ib}|$ is 1.30. Hence the assumed loop parameters result in a good practical design. With this setting, the ω_b/ω_c ratio of Eq 7.5-77 is 2.36, as shown in item (7).

Thus the noise bandwidth ω_b for this case is 2.36 times greater than the gain crossover frequency ω_c. This is actually an optimistic ratio for a practical antenna servo. When an antenna servo has mechanical structural resonance, the ω_b/ω_c ratio can be appreciably greater than 2.36. The 15 rad/sec gain crossover frequency in item (1) is multiplied by 2.36 to obtain the noise bandwidth ω_b of the angle tracking loop, which is 35.4 rad/sec. Dividing this by 2π gives the noise bandwidth f_b in hertz, which is 5.63 Hz, as shown in item (4).

Substitute the noise bandwidth $f_{b(a)} = 5.63$ Hz of item (4) into Eq 7.5-57. This gives the effective integration time $T_{i(a)}$ of the angle-tracking servos, shown in item (5) as 88.8 msec. The pulse repetition period T_p (0.667 msec) shown in item (6) was obtained from item (5) of Table 7.4-2. Dividing the integration time $T_{i(a)}$ by the pulse repetition period T_p gives the number of pulses integrated by the antenna servos, which is

$$N_{i(a)} = \frac{T_{i(a)}}{T_p} = \frac{88.8 \text{ msec}}{0.667 \text{ msec}} = 133 \qquad (7.5\text{-}78)$$

This is shown in item (7) of Table 7.5-2. Item (8) gives the relative RMS angular noise per pulse (0.300) obtained from item (7) of Table 7.5-1. This is divided by the square root of the number of pulses integrated (133) to obtain the actual RMS angular noise error:

$$\frac{\sigma\theta}{\theta_b} = \frac{\sigma\theta_1/\theta_b}{\sqrt{N_{i(a)}}}$$

$$= \frac{0.300}{\sqrt{133}} = 0.0260 \qquad (7.5\text{-}79)$$

This is shown in item (9). The half-power beamwidth θ_b of 1.60° shown in item (10) was obtained from item (6) of Table 7.3-1. The values of items (9 and 10) are multiplied together to obtain the actual RMS noise error $\sigma\theta$, shown in item (11) to be 0.0416°.

There are two components of tracking error: elevation and deflection. The total beam-pointing error ϕ_e is related as follows to the elevation error ϕ_{eL} and deflection error ϕ_d:

$$\phi_e^2 = \phi_{eL}^2 + \phi_d^2 \tag{7.5-80}$$

Since the RMS error components for the two axes are equal, the total RMS beam-tracking error is $\sqrt{2}$ times the single-axis RMS error. The peak error is considered to be 3 times the RMS error. Hence, the peak two-axis beam error $\text{Max}[\phi_e]$ is related as follows to the RMS single-axis error $\sigma\theta$:

$$\text{Max}[\phi_e] = 3\sqrt{2}\,\sigma\theta = 4.24\sigma\theta \tag{7.5-81}$$

Thus, the RMS one-axis noise error of item (11) in Table 7.5-2 is multiplied by 4.24 to obtain the peak two-axis error shown in item (12). The error in degrees is multiplied by 17.45 mrad/deg to obtain the error in milliradians, which is ± 3.08 mrad.

The angular error due to target motion is calculated in Table 7.5-3. It is assumed that the target is following a straight-line constant-velocity course. The tracking errors for this course were calculated in Chapter 3 using error coefficients. The analysis showed that the maximum angular velocity Ω_{max} and the maximum angular acceleration α_{max} are

$$\Omega_{max} = \frac{V_t}{R_{min}} \tag{7.5-82}$$

$$\alpha_{max} = 0.65\Omega_{max}^2 \tag{7.5-83}$$

where V_t is the target velocity and R_{min} is the minimum target range.

As was shown in item (10) of Table 7.4-2, the maximum target velocity V_t is assumed to be 600 meter/sec. This is shown in item (1) of Table 7.5-3. The minimum range R_{min} is assumed to be 800 meter, as shown in item (2). By Eq

TABLE 7.5-3   Calculation of Peak Angular Error Due to Target Motion

Symbol		Parameter	Value
(1)	V_t	Maximum target velocity	600 meter/sec
(2)	R_{min}	Minimum range	800 meter
(3)	Ω_{max}	Maximum angular velocity	0.75 rad/sec
(4)	α_{max}	Maximum angular acceleration	0.366 rad/sec^2
(5)	c_2	Acceleration error coefficient	$\frac{1}{75}$ sec^2
(6)		Peak target motion error	± 4.88 mrad

7.5-82, the maximum angular velocity, shown in item (3), is

$$\Omega_{max} = \frac{V}{R_{min}} = \frac{600 \text{ m/sec}}{800 \text{ m}} = 0.75 \text{ rad/sec} \qquad (7.5\text{-}84)$$

Substituting this into Eq 7.5-83 gives the maximum angular acceleration, shown in item (4):

$$\alpha_{max} = 0.65\Omega_{max}^2 = 0.65(0.75 \text{ rad/sec})^2$$
$$= 0.366 \text{ rad/sec}^2 = 366 \text{ mrad/sec}^2 \qquad (7.5\text{-}85)$$

It is assumed that the integral network has infinite gain at zero frequency, and so the velocity error coefficient is zero. Hence the maximum angular tracking error due to target motion, after the initial lock-on transient has decayed, is approximately equal to the maximum angular acceleration multiplied by the acceleration error coefficient c_2. Item (5) gives the approximate acceleration error coefficient c_2, which is equal to $1/\omega_c\omega_i$:

$$c_2 = \frac{1}{\omega_{c(a)}\omega_{i(a)}} = \frac{1}{15 \times 5 \text{ sec}^{-1}} = \tfrac{1}{75} \text{ sec}^2 \qquad (7.5\text{-}86)$$

Hence the maximum error due to target motion is approximately

$$\text{Max}[\theta_e] = c_2\alpha_{max} = \left(\tfrac{1}{75} \text{ sec}^2 \right)(366 \text{ mrad/sec}^2)$$
$$= 4.88 \text{ mrad} \qquad (7.5\text{-}87)$$

This is shown in item (6).

Thus, the angle tracking radar experiences a peak angular error due to target motion of ± 4.88 mrad, as shown in Table 7.5-3, and a peak two-axis error due to radar noise of ± 3.08 mrad, as shown in Table 7.5-2.

This example illustrates the statistical principles used in the calculation of radar tracking errors due to receiver noise. It shows that optimum setting of the dynamic parameters of the feedback loops requires a compromise between radar noise error and error due to target motion.

7.5.6 Angle-Tracking in Two Axes

Figure 7.5-8 shows a signal-flow diagram for both axes of the angle-tracking servos, which position the antenna in elevation and azimuth. The details of the servo drives are not shown. Tachometers are coupled to the motor shafts, to provide feedback signals for the velocity feedback loops closed around the antenna drives. The gain crossover frequency of the velocity loop is assumed to be more than a factor of 3 times greater than the gain crossover frequency $\omega_{c(a)}$ of the tracking (position) loop.

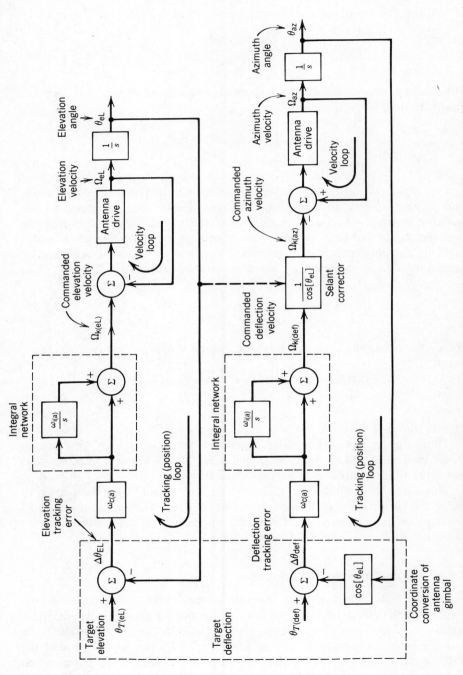

Figure 7.5-8 Condensed signal-flow diagram of a two-axis azimuth–elevation angle-tracking antenna system.

The antenna drive servos operate in terms of the elevation–azimuth gimbal axes of the antenna, but the radar monopulse tracker measures the tracking errors in terms of elevation–deflection axes. As was shown in Eq 7.5-1, an azimuth motion $\Delta\theta_{az}$ of the antenna gimbal produces the following deflection motion $\Delta\theta_d$ of the tracking line:

$$\Delta\theta_d = \Delta\theta_{az} \cos[\theta_{el}] \qquad (7.5\text{-}88)$$

where θ_{el} is the elevation angle of the antenna. The function $\cos[\theta_{eL}]$ is shown as a gain factor in Fig 7.5-8, within the dashed block that characterizes the gimbal coordinate conversion.

A *secant corrector* must be included in the azimuth tracking loop to keep the gain crossover frequency of that loop constant. The gain of the secant corrector is controlled by the elevation angle θ_{eL} to be proportional to the secant of the elevation angle, which is equal to $1/\cos[\theta_{eL}]$. In the system of Fig 7.5-8, the secant corrector is placed after the tracking-loop integral network, but in some radar systems it precedes the integral network. For either location, the tracking loop gain is kept constant. However, at high tracking rates, the tracking error can be much lower when the secant corrector follows the integral network.

The reason for this is as follows. Tracking error is approximately proportional to rate-of-change of integral-network output. When the secant corrector precedes the integral network, the integral network output is proportional to azimuth velocity, and the rate-of-change of integral network output is proportional to azimuth acceleration. When the secant corrector follows the integral network, the rate-of-change of integral network output is proportional to deflection acceleration. At high elevation angles, azimuth acceleration is much higher than deflection acceleration. Hence, the tracking error for a fast target can be much higher when the secant corrector precedes the integral network than when it follows the integral network (as in Fig 7.5-8).

The analysis of angular tracking error performed in Table 7.5-3 assumed that the secant corrector is placed after the tracking loop integral network. When this is not the case, the tracking error can be much larger.

Angle Sensor Cross-Coupling. Because of nonideal effects, radar angle-sensing circuitry usually has a small amount of cross-coupling between the two angle-tracking axes. The resultant angular error signals fed to the antenna servos can be expressed as

$$\phi'_{eL} = \phi_{eL} + k_{de}\phi_d \qquad (7.5\text{-}89)$$

$$\phi'_d = \phi_d + k_{ed}\phi_{eL} \qquad (7.5\text{-}90)$$

where ϕ'_{eL}, ϕ'_d are the elevation and deflection error signals fed to the servos and ϕ_{eL}, ϕ_d are the actual elevation and deflection angular errors. The coefficients k_{de} and k_{ed}, which can be either positive or negative, characterize the coupling between elevation and deflection axes in the angle error sensor. To achieve good tracking performance, the magnitudes of these cross-coupling coefficients must be much less than unity. Typically, specifications for angle error sensor circuitry require that the magnitudes of these coefficients should not exceed 0.1:

$$|k_{de}|, |k_{ed}| < 0.1 \qquad (7.5\text{-}91)$$

In other words, the cross-coupling between the elevation and deflection angle signals should not exceed -10 decilogs (dg), or -20 decibels (dB).

As shown by simulation studies performed by Pidhayny [7.7], when this -10-dg (or -20-dB) cross-coupling specification is exceeded, cross-coupling between the two axes can significantly degrade stability in the tracking system. On the other hand, greater cross coupling (as large as -5 dg or -10 dB) may be tolerated in certain tracking systems (such as satellite communication antennas) provided that the tracking loops (without cross coupling) have good stability margins.

7.5.7   Other Sources of Angular Tracking Error

Besides the angular tracking errors caused by receiver noise and target motion, there are a number of other sources of tracking error, which are discussed by Skolnick [7.2], Barton [7.4], and Barton and Ward [7.5]. The following is a summary of some of these errors.

Target Glint.   A target that is many wavelengths wide consists of a number of independent scatterers. The location of the apparent center of the reflected target signal depends on how these scatter signals add in phase as well as in amplitude. This apparent center of the target moves across the target as the orientation of the target changes. During 15% of the time, this apparent center actually falls outside the physical limits of the target.

Barton and Ward [7.5] (p. 170) show that the deviation of the apparent target center from the physical centroid of the target can be approximated by a Gaussian density function with the following standard deviation:

$$\sigma x = 0.35L \qquad (7.5\text{-}92)$$

where L is the physical width of the target. The constant 0.35 was chosen so that during 15% of the time, the apparent centroid falls outside the physical dimensions of the target (i.e., the deviation from the physical centroid exceeds $\pm L/2$). Assuming practical limits of ± 3 sigma, the peak-to-peak motion of the apparent center is $6(0.35\,L)$, which is $2.1L$. Thus, the practical limits of motion of the apparent target center are about twice the physical dimensions of the target.

Target Scintillation. The analysis has assumed that the effective radar cross section area A_t of the target is constant. However, the value of A_t often varies strongly with target orientation. This variation of target reflection is called scintillation. As shown by Skolnick [7.2] (pp. 2–18), when the detection probability is reasonably high, the probability of detection is appreciably less for a scintillating target than for a target of constant reflectivity having the same mean value of A_t. To compensate for scintillation, an effective value of target radar cross-section A_t may be specified that is lower than the mean value.

With a scintillating target, the tracking error of a conical-scan radar can greatly exceed that of a comparable monopulse radar. This is the primary reason for the wide use of monopulse despite its increased complexity.

Multipath Error. When the elevation of the target is low, the radar can receive two target signals: one from the direct path and a second from a path reflected by the ground. This ground-reflected signal interferes with the direct-path signal and can produce appreciable tracking error.

Polarization Shift. Sometimes the polarization of the signal reflected from the target is shifted appreciably in the reflection process. Some monopulse feeds are quite sensitive to polarization shift, and so may deliver poor error signals under such conditions, with strong angle-sensor cross-coupling. This may result in such a large tracking error that the radar loses the target signal.

7.6 OTHER TYPES OF RADAR SIGNALS

Besides the simple pulsed radar signal discussed in this chapter, modern radars often employ more complicated signals in order to improve target detection and tracking. These include the following.

Doppler Radar. A target moving in the radial direction causes a Doppler frequency shift Δf of the received signal given by

$$\Delta f = \frac{2 f_r}{c} V_r \tag{7.6-1}$$

where f_r is the RF transmitted frequency, c is the speed of light, and V_r is the radial velocity of the target. Many radars detect this Doppler shift and use it to discriminate the target echo from the clutter. Some Doppler radars use a continuous-wave (CW) signal, which discriminates only on the basis of Doppler shift. Other radars, called pulse Doppler, provide both range and Doppler discrimination.

Pulse-Compression Radar. The capability of a radar to detect a target, and the accuracy of target tracking, can be improved by lengthening the pulse while maintaining the same peak power, so that more energy is received from the target. However, this could decrease the range discrimination performance of the radar and limit its capability to discriminate a target from ground clutter. A "chirp" radar maintains accurate range discrimination with a long pulse by sweeping (or chirping) the RF frequency during the pulse. This allows the pulse to be compressed after it is received, to provide a time discrimination interval that is much shorter than the transmitted pulse width.

Coherent Pulse Integration. The examples that have been considered assume that the signal contributions from successive pulses are added in a noncoherent manner. Much better detectability at low signal/noise ratios can be achieved by integrating the radar pulses coherently. This requires that the phase of the IF signal be detected as well as the amplitude, so that the signals for successive pulses can be added in phase. With coherent integration, the noncoherent detection loss $L_d = (1 + P_n/P_s)$ given in Eq 7.4-24 can be reduced to unity.

Coherent pulse integration is typically performed in the following manner. The IF signal is fed to a pair of phase-sensitive detectors that are excited by sine and cosine reference signals. These reference signals are derived from oscillators that are phase-locked to the frequency of the transmitted pulse. The outputs from these phase-sensitive detectors are called the *in-phase* (I) and *quadrature* (Q) signals. These I and Q signals are sampled and converted to digital data. The I and Q signals are averaged over many radar pulses to provide coherent integration of the radar signal.

To allow integration over an appreciable time interval, the data processing must compensate for the phase change of the IF signal that is caused by the change of target range over the integration period. This can be achieved by phase shifting the sine and cosine reference signals or by appropriate processing of the digitized I and Q signals. During search, the range rate of the target is not accurately known, and so many independent channels of processing may be required to allow search over different range rates. For this and other reasons, the digital signal processing involved in coherent pulse integration can be very complicated.

Chapter 8

Kalman Optimal Filtering Applied to the Altair Radar Tracking System

This chapter examines the Kalman filter tracking equations used in the Altair radar, which provides optimal tracking of ballistic missile warheads. Based on this example, the principles of the Kalman filter are explained. This work was sponsored by the Ballistic Missile Defense Systems Command of the U.S. Army, under technical direction of the Massachusetts Institute of Technology, Lincoln Laboratory.

The aspects of the Altair radar that are specifically addressed in this chapter are the basic Kalman tracking filter equations given in Section 8.5, and the Kalman filter algorithms for setting optimal values for the smoothing constants of these tracking filters, which are presented in Section 8.4. Also, a brief qualitative description of the Altair antenna servos is given in Section 8.1.2. In Section 8.2, a digital radar tracking system is described that applies these Altair tracking filters. The coordinate transformations of this tracking system differ from those of Altair but are dynamically equivalent to them. Also, a radar model is presented in Section 8.1.3 to show how the tracking characteristics of such a radar are related to the radar system parameters.

8.1 INTRODUCTION TO OPTIMAL RADAR TRACKING SYSTEMS

8.1.1 General Discussion of Optimal Radar Tracking

In Chapter 3, error coefficients were used to calculate the tracking error of a radar antenna that is tracking a target moving along a straight-line constant-velocity course. However, this error can be compensated for because it is completely predictable from the range and angle radar measurements. When such prediction is performed, the tracking error that remains is due to unpredictable accelerations of the target, which cause the target path to vary from the straight-line, constant-velocity trajectory.

When a tracking radar such as the one described in Chapter 7 is used to direct an antiaircraft gun to fire at an attacking aircraft, a computer is employed to derive, from the radar tracking information, the commands for aiming the gun. The computer determines the angle by which the gun must lead the aircraft, so that the anti-aircraft shells are directed to the future position of the aircraft, when the shells reach its range. The computer corrects for the ballistic trajectory of the shell and the predicted motion of the aircraft. It determines the optimum angular commands to be fed to the antiaircraft gun so that the shells have maximum probability of hitting the target. In calculating the predicted trajectory of the target aircraft, the computer generally adds corrections to compensate for the tracking errors of the radar that are caused by geometry-induced angular accelerations. Thus, the computer usually corrects for the tracking errors that were calculated by error coefficients in Chapter 3.

There are various coordinate axes that can be used for the computer calculations that predict the target trajectory. In early fire-control systems, the choice of coordinates depended primarily on the accuracy limitations of the sensing instruments and the computing elements, rather than on the mathematical advantages of the coordinate system. However, it was always obvious that one would ideally like to perform tracking and prediction in terms of the coordinates of the target. Target tracking is essentially a filtering process, which smoothes the radar noise in the angle and range tracking signals.

With the development of digital technology, it became practical to track and predict target motions in more optimal coordinate systems. When the tracking and prediction problem was extended to exoatmospheric targets (ballistic missiles and other space vehicles), which follow highly predictable trajectories over long time periods, the advantages of tracking relative to target-centered coordinates increased greatly.

At this same time, a new general mathematical theory of optimum filtering and prediction was developed, the Kalman filter [8.1, 8.2]. Kalman filter theory, combined with advances in digital computer technology, was applied to the problem of tracking exoatmospheric vehicles to produce a number of very effective tracking and prediction systems. This practical success, combined with the confusing matrix equations of the Kalman filter, has created the feeling that there is a certain black magic in its mathematical abstractions.

One of the systems to which Kalman filter theory has been applied very successfully is the Altair radar tracking antenna. The Altair radar is located on Kwajalein Atoll in the Marshall Islands, 2100 miles west and south of Hawaii, just 9° north of the equator. This radar antenna is a mechanically steered parabolic dish with a diameter of 150 ft. It provides long-range tracking of the ballistic missiles launched from Vandenburg Air Force Base in California.

This chapter shows how Kalman filter theory is applied in the Altair radar system. An examination of this application provides a means of explaining Kalman theory in a simple, practical manner. It also allows one to separate those aspects of system performance that are due to the advantages of the Kalman theory from those that are the result of conventional tracking concepts.

8.1.2 Description of Altair Radar

The Altair radar system was built by GTE Sylvania Electronic Systems (now called GTE Government Systems). However, the antenna itself was built by Radiation, Inc. (now part of Harris Corp.). Since installation of the radar system in 1967, it has been operated by GTE under supervision from MIT Lincoln Laboratory.

The ALTAIR (ARPA Long-range Tracking And Instrumentation Radar) is a highly sensitive dual-frequency radar that operates at VHF and UHF. This is one of the radar sensors of the Kiernan REentry Measurement Site (KREMS) located on Kwajalein Atoll. The lagoon of Kwajalein Atoll is about 70 miles long, 10–15 miles wide, and 200 ft deep. Since the surrounding ocean is 2 miles deep, the Kwajalein lagoon is an ideal location for dropping and recovering ballistic warheads.

A picture of the 150-foot Altair radar antenna is shown in the frontispiece. The azimuth carriage weighs about one million pounds and is supported on four railroad-car wheel assemblies (bogies) that run on a circular railroad track, 118 ft in diameter. It is physically a unique antenna. Larger antennas have been built for astronomical observations, but move very slowly. The Altair antenna can rotate at a velocity of several degrees per second. To compare the size of antennas of different apertures, note that mass, volume, and torque requirements increase approximately as the cube of the antenna aperture diameter. Thus, an antenna with an aperture of 150 ft is about eight times as large as one with an aperture of 75 ft.

The elevation axis of the Altair antenna is driven by a single 150-hp DC motor, which is overdriven under transient conditions to deliver 300 hp. This motor is coupled to a geared speed reducer that drives a bull gear 83 ft in diameter. The frame holding the elevation motor and speed reducer is pivoted, and its weight preloads the elevation gear train to eliminate backlash. The azimuth axis is rotated by four 150-hp DC motors, with separate speed reducers that drive an internal bull gear mounted inside the circular railroad track. The azimuth motors operate in pairs preloaded against one another to eliminate backlash, where one pair delivers clockwise torque and the other pair delivers counterclockwise torque. Under transient conditions, either pair of motors can be overdriven to deliver 300 hp per motor.

Figure 8.1-1 is a block diagram of the present servo control system of the Altair antenna for the elevation axis. The control in azimuth is somewhat more complicated because there are four separate motors. The control angle θ_c of the antenna (azimuth or elevation) is measured by an optical encoder. In the computer, this digital signal θ_c is subtracted from the digital command angle θ_k to obtain the angular error θ_e. The error signal is appropriately scaled and converted to analog form in a digital-to-analog converter, to provide an error voltage equivalent to $\omega_{cp}\theta_e$. This is fed to an integral network with a transfer function approximately equal to $(1 + \omega_{ip}/s)$. The computer also generates a signal proportional to the desired velocity of the antenna. This desired velocity

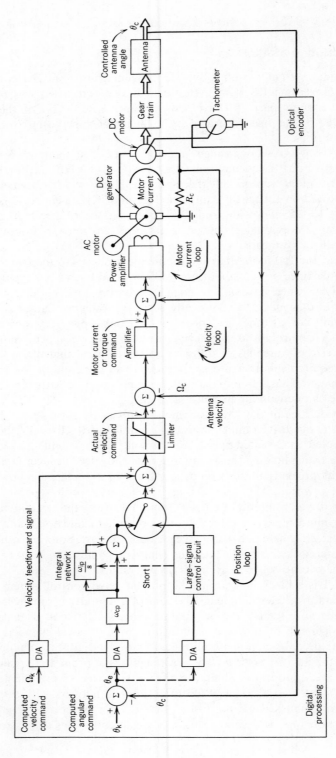

Figure 8.1-1 Simplified block diagram of servo control for Altair antenna.

signal is added to the output of the integral network, and the combined signal is fed as a velocity command to the tachometer feedback loop closed around the antenna drive.

The desired velocity signal is called a feedforward command signal, because it is inserted inside the position feedback loop to reduce the tracking error of that loop. The velocity feedforward signal is equivalent to the velocity correction signal that is generated in an integral network. However, it is dynamically superior because it is based on the predicted motion of the antenna. With this velocity feedforward command, the integral network must only compensate for errors due to load torques and errors due to inaccuracy in the calculation of the velocity feedforward signal.

When the angular error is large, the normal error signal path is bypassed by the large-signal control circuit. This circuit shorts out the integral network and provides an error channel of much lower gain and much wider dynamic range. This circuit provides the antenna control servo with an optimal saturated step response in a manner similar to that discussed in Ref [1.2] (Section 12.2).

The DC servo motor is controlled by a *Ward–Leonard* drive system, which is a field-controlled DC generator. The DC generator is driven at essentially constant speed by an AC induction motor. The voltage excited in the generator armature circuit is proportional to the generator magnetic field, which in turn is proportional to the generator field current. The armature current from the generator is fed to the DC motor on the antenna. A current-sense resistor in the armature circuit of the generator measures the current supplied by the generator, which is also the motor current. The voltage across this current-sense resistor is the feedback signal for a current feedback loop, which controls motor current and motor torque. A tachometer is coupled to the shaft of the DC motor to provide the feedback signal of a velocity feedback loop closed around the antenna drive system.

In 1980, the antenna servo system was modified under the author's direction to allow sustained acceleration capability without overheating the motors, and to reduce the mechanical shock and vibration applied to the antenna by the drive servos. Two 150-hp DC motors and gear boxes were added to the azimuth axis, along with DC generators, to provide a total of four DC servo motors in azimuth. Also, the obsolete thyristor (SCR) units that drove the generator fields were replaced with circuits having much better dynamic performance. This change allowed current feedback loops to be closed around the generators to provide direct, linear control of motor torque.

This modification illustrates an important aspect of nonlinear feedback control that is often overlooked: the effect of mechanical shock delivered by the drive system to the mechanical structure. In the original design, the two azimuth drive motors were preloaded against one another when moving the antenna slowly, but exerted torque in the same direction under conditions of high acceleration. When a motor reversed the direction of its torque, it rotated through the backlash before engaging the gearing in the opposite direction. This applied a shock to the antenna that sounded like the clanking of a railroad coupling when a railroad train starts to move. In the new design, the motors are continually preloaded and never rotate through the backlash. Another weakness of the original design was a nonlinear feedback loop that provided current and torque limiting. This nonlinear loop limited the current by oscillating at a frequency that varied with speed from 10 Hz at zero speed to about 5 Hz at maximum speed. The vibrations due to this nonlinear oscillation along with the backlash shocks were major causes of cracks in the antenna structure.

8.1.3 Model to Illustrate Radar Signal Calculations for an Altair-Type Radar

In order that the radar parameters may be described quantitatively, a radar model will now be developed. This radar model uses a commercially available UHF klystron transmitter tube designed for high-power radar applications, and an antenna of the same dimensions as the Altair antenna. The performance of the radar model is not intended to have any relation to that of the Altair radar. Nevertheless, the calculations applied to this model are equivalent to those that have been used to evaluate performance of the Altair radar.

The assumed UHF klystron transmitter tube is the Varian VA-812C pulse klystron amplifier, which is designed for "frequency-agile long-range radar service". Its characteristics are listed in Table 8.1-1. Since the frequency range of this tube is 400–450 MHz, it is assumed that the radar model operates at 425 MHz, the center of this range.

For this operating frequency, the estimated values of the antenna parameters are given in Table 8.1-2. As shown in item (1), the assumed antenna aperture D is 150 ft, which is the diameter of the Altair antenna. The assumed operating frequency (2) is 425 MHz, which is the center of the band of the VA-812C klystron. The corresponding wavelength λ is 70.6 cm, as shown in item (3). In accordance with Eq 7.3-9 of Chapter 7, the 3-dB beamwidth θ_b of

TABLE 8.1-1 Parameters of Varian VA-812c High-Power Klystron Assumed in Radar Model

Parameter	Value
(1) Tunable frequency range	400–450 MHz
(2) Peak output power	10 MWatt
(3) Average output power	30 kWatt
(4) Maximum pulse width	6 μsec
(5) 3-dB bandwidth	50 MHz
(6) Gain	30 dB
(7) Efficiency	40%
(8) Weight	1700 lb

TABLE 8.1-2　Estimated Values of Antenna Parameters for Radar Model

Parameter	Value
(1)　Antenna aperture D	150 ft
	45.7 m
(2)　Frequency	425 MHz
(3)　Wavelength λ	70.6 cm
(4)　Half-power beamwidth θ_b	1.00 °
(5)　Antenna efficiency η	48.9%
(6)　Peak antenna gain G_{max}	20,220 (43.06 dB)

an antenna gain pattern can be approximated by

$$\theta_b = 65\frac{\lambda}{D} \quad \text{degree} \tag{8.1-1}$$

where λ is the wavelength and D is the diameter of the antenna aperture. As shown in item (4), the 3-dB beamwidth θ_b calculated from Eq 8.1-1 is 1.00°.

The small monopulse-tracking radar antenna described in Chapter 7 had an antenna efficiency of 60% (Table 7.3-1, item 4). However, a radar antenna the size of Altair should have less efficiency because its relative surface accuracy should be lower.

Let us consider the effect of antenna surface inaccuracy on antenna efficiency. The reciprocal loss in antenna gain caused by the departure of the antenna surface from a true paraboloid is

$$\text{Loss}^{-1} = \exp\left[-(\sigma\phi)^2\right] \tag{8.1-2}$$

The parameter $\sigma\phi$ is the RMS phase error over the antenna aperture produced by the surface inaccuracy. This can be expressed as

$$\sigma\phi = \frac{2\pi\sigma L}{\lambda} \tag{8.1-3}$$

The parameter σL is the RMS path length error corresponding to the RMS phase error $\sigma\phi$. The RMS surface tolerance error is denoted ε. Since the wave is reflected from the antenna surface, any error of the surface location changes the path length by twice the surface error. Hence σL is twice the value of ε:

$$\sigma L = 2\varepsilon \tag{8.1-4}$$

Combining Eqs 8.1-2 to -4 gives

$$\text{Loss}^{-1} = \exp\left[-(4\pi\varepsilon/\lambda)^2\right] \tag{8.1-5}$$

Take 10 times the logarithm of this to obtain the corresponding loss in decibels. This yields

$$\text{dB loss} = \frac{10}{\ln[10]} \left(\frac{4\pi\varepsilon}{\lambda} \right)^2 = 686 \left(\frac{\varepsilon}{\lambda} \right)^2 \tag{8.1-6}$$

This is a convenient general formula for relating RMS antenna surface inaccuracy ε to the resultant degradation of antenna gain.

Let us assume that the RMS surface error of the radar model is 1.0 inch:

$$\varepsilon = 1.00 \text{ in.} = 2.54 \text{ cm} \tag{8.1-7}$$

From Eq 8.1-6, the corresponding loss value for the antenna model is

$$\text{dB loss} = 686 \left(\frac{\varepsilon}{\lambda} \right)^2 = 686 \left(\frac{2.54}{70.6} \right)^2 = 0.888 \text{ dB} \tag{8.1-8}$$

This decibel loss corresponds to a ratio of 1.227. Without this surface-tolerance loss, a reasonable value for antenna efficiency would be 60%, which is the value calculated for the small monopulse tracking antenna analyzed in Section 7.3. Hence, the assumed antenna efficiency for the antenna model is

$$\eta = 0.60/1.227 = 0.489 \tag{8.1-9}$$

This is indicated in item (5) of Table 8.1-2, which shows that the estimated efficiency for the antenna model is 48.9%. The corresponding peak antenna gain is

$$G_{\text{max}} = \eta \left(\frac{\pi D}{\lambda} \right)^2 = (0.489) \left[\frac{(\pi)(4570 \text{ cm})^2}{(70.6 \text{ cm})} \right]$$

$$= 20{,}220 = 43.06 \text{ dB} \tag{8.1-10}$$

This peak antenna gain is shown in item (6).

Table 8.1-3 shows the estimated parameters of the radar transmitter model, based on the characteristics of the VA-812C klystron given in Table 8.1-1. The peak power (item 1) and average power (2) are obtained from items (2) and (3) of Table 8.1-1. The maximum average power (2) is divided by the peak power (1) to obtain the maximum duty ratio, 0.003, shown in item (3). The maximum pulse width τ_p of item (4) is obtained from item (4) of Table 8.1-1. The pulse width (4) is multiplied by the peak power (1) to obtain the maximum energy per pulse (5), which is 60 joule. The pulse width (4) is divided by the duty ratio (3) to obtain the minimum pulse repetition period (6), which is 2000 μsec. The reciprocal of this minimum pulse repetition period (6) is the maximum pulse repetition frequency (7), which is 500 Hz. As shown in Chapter 7, Eq 7.3-18,

TABLE 8.1-3   Radar Transmitter Parameters of Model That Corresponds to Varian VA-812C Klystron

Parameter	Value
(1) Peak power P_t	10 MWatt
(2) Maximum average power	30 kWatt
(3) Maximum duty ratio	0.003
(4) Maximum pulse width τ_p	6 μsec
(5) Maximum energy per pulse	60 Joule
(6) Minimum pulse repetition period	2000 μsec
(7) Maximum pulse repetition frequency	500 Hz
(8) Minimum ambiguous range	300 km
(9) Chirp compression ratio	40 : 1
(10) Pulse compression loss L_{pc}	0.7 dB
(11) Compressed pulse width τ_{pc}	0.15 μsec

the ratio of range to time delay for a radar is 150 m/μsec. Hence, the minimum pulse repetition period (6) is multiplied by 150 m/μsec to obtain the minimum ambiguous range (8), which is 300 kilometer.

With a pulse width of 6 μsec, the range resolution would be 900 meter for a radar pulse of constant frequency. The range resolution can be improved by chirping the radar pulse. The radar frequency is swept (or chirped) during the pulse, which allows the radar pulse to be compressed after it is received. Pulse compression ratios somewhat higher than 100 : 1 have been achieved in chirp radars. Barton and Ward [7.5] (p. 230) give for illustration the parameters of a chirp pulse-compression radar, which has a compression ratio of 40 : 1 and a pulse-compression loss of 0.7 dB. As shown in items (9) and (10) of Table 8.1-3, these parameters are assumed in our radar model. The 6-μsec transmitted pulse width (4) is divided by 40 to obtain a compressed pulse width of 0.15 μsec (11). Thus, with chirp, the range resolution is reduced by a factor of 40, from 900 meter to 22.5 meter.

Table 8.1-4 shows the assumed loss factors for the radar model. The waveguide loss is defined as the total loss in the transmit waveguide and the receive waveguide, which includes the loss in the duplexer that separates the receive and transmit signals. Barton [7.4] (p. 122) reports that in most practical radars the waveguide transmission loss for transmit plus receive is typically about 1–2 dB, and the duplexer typically introduces an additional transmit-plus-receive loss of about 0.5 dB. As shown in items (2) and (5), our model assumes dissipative (resistive) waveguide losses of 1.0 dB each for transmit and receive because 1.0 dB is a convenient round number. The usual waveguide for use at 425 MHz is WR2100, which has a rectangular opening of 21 × 10.5 inch. As shown in Skolnick [7.2] (pp. 8–10), this waveguide operates from 350 to 530 MHz and has a loss of about 0.05 dB/(100 ft).

TABLE 8.1-4   Assumed Loss Factors for Radar Model

Parameter		Decibels	Ratio
(1)	Transmit waveguide loss (total) L_{tw}	1.18	1.312
(2)	Dissipative transmit waveguide loss L_{td}	1.00	1.259
(3)	Reflection transmit waveguide loss L_{tr}	0.18	1.042
(4)	Receive waveguide loss (total) L_{rw}	1.18	1.312
(5)	Dissipative receive waveguide loss L_{rd}	1.00	1.259
(6)	Reflection receive waveguide loss L_{rr}	0.18	1.042
(7)	Atmospheric propagation loss L_p	0.5	1.122
(8)	Total transmission loss L_t	1.68	1.472
(9)	Pulse compression loss L_{pc}	0.70	1.175
(10)	Receiver matching loss L_m	0.85	1.216
(11)	System loss relative to envelope detector input, L_s	3.23	2.104

Reflection loss values of 0.18 dB are assumed for both transmit and receive waveguides. This reflection loss is selected because it corresponds to a voltage standing-wave ratio (VSWR) of 1.5, which is the maximum VSWR that would be expected in a radar antenna. If the VSWR were greater than 1.5, the transmitter device and sometimes the receiver would experience operating difficulties.

The VSWR is related as follows to the voltage amplitudes of the reflected and incident waves in the waveguide:

$$\text{VSWR} = \frac{1 + |E_R|/|E_{in}|}{1 - |E_R|/|E_{in}|} \qquad (8.1\text{-}11)$$

where $|E_{in}|$ is the voltage amplitude of the incident wave and $|E_R|$ is the voltage amplitude of the reflected wave. Solving Eq 8.1-11 for the voltage amplitude ratio gives

$$\frac{|E_R|}{|E_{in}|} = \frac{\text{VSWR} - 1}{\text{VSWR} + 1} \qquad (8.1\text{-}12)$$

For VSWR = 1.5, this voltage ratio is 0.2. The corresponding power ratio is the square of the voltage ratio, which is

$$\frac{P_R}{P_{in}} = \left(\frac{|E_R|}{|E_{in}|}\right)^2 = (0.2)^2 = 0.04 \qquad (8.1\text{-}13)$$

Hence, the ratio of output power P_o to incident power P_{in} is

$$\frac{P_o}{P_{in}} = \frac{P_{in} - P_R}{P_{in}} = 1 - \frac{P_R}{P_{in}}$$

$$= 1 - 0.04 = 0.96 \qquad (8.1\text{-}14)$$

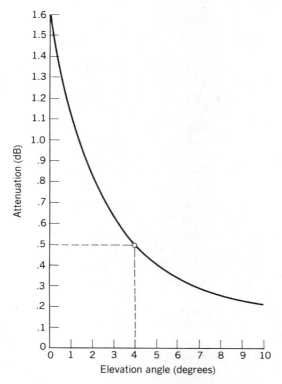

Figure 8.1-2   Two-way propagation loss through complete troposphere at 425 MHz, as a function of elevation angle.

The factor 0.96 corresponds to -0.18 dB. Thus, a VSWR of 1.5 is equivalent to a reflection loss of 0.18 dB.

Barton [7.4] (p. 470, Fig 15.2) gives a set of curves showing the two-way attenuation through the complete troposphere for frequencies from 100 MHz to 100 GHz. The individual curves show the attenuation for different elevation angles. Figure 8.1-2 is derived from these curves and shows the two-way tropospheric attenuation at 425 MHz as a function of elevation angle. The ionosphere also causes attenuation, but Barton [7.4] (p. 471) indicates that ionospheric attenuation is negligible at frequencies above 300 MHz.

With the strong variation of atmospheric propagation loss with elevation angle indicated in Fig 8.1-2, this propagation loss L_p should properly be considered a variable in the signal/noise ratio calculations. However, for simplicity the radar model assumes a fixed L_p loss of 0.5 dB, as indicated in item (7) of Table 8.1-4.

It is convenient to specify the noise temperature of the radar relative to the antenna aperture, rather than relative to the receiver input, because this simplifies the calculations. With this convention, the effect of the receive waveguide loss L_{rw} is incorporated into the system noise temperature, and so is not directly considered as a loss in calculating signal/noise ratio. Hence, the total transmission loss between transmit and receive (exclusive of the geometric signal loss of the radar transmission equation) is the sum of the transmit waveguide loss L_{tw} plus the atmospheric propagation loss L_p. This total signal transmission loss is denoted L_t. It is shown in item (8) of Table 8.1-4 and is the sum of the decibel values of items (1) and (7).

It is convenient to calculate the tracking performance in terms of the signal/noise power ratio at the input to the envelope detector, because the noncoherent detection loss caused by envelope detection depends on that signal/noise ratio. There are two signal-processing losses associated with this signal/noise ratio, which are shown in items (9) and (10) of Table 8.1-4. These are the pulse-compression loss L_{pc} of item (9), which is obtained from item (10) of Table 8.1-3, and the receiver matching loss L_m. The receiver matching loss L_m was obtained from Eq 7.4-22 of Chapter 7. The matching loss L_m is minimized when the noise bandwidth of the rectangular IF filter response is set equal to $1.4/\tau_{pc}$. A rectangular filter is a good approximation to the stagger-tuned frequency response that is typically used in pulsed-radar IF amplifiers.

Item (11) of Table 8.1-4 is the system loss L_s relative to the envelope detector input. The decibel value of this total system loss L_s (11) is the sum of the decibel values of the loss factors of items (8), (9), and (10). Thus, the system loss L_s is the combined effect of the transmission loss L_t, the pulse compression loss L_{pc}, and the receiver matching loss L_m.

The assumed values of noise temperature for the radar model are shown in Table 8.1-5. Barton [7.4] (p. 474, Fig 15.6) gives a family of plots of antenna noise temperature at different elevation angles for frequencies from 100 MHz to 10 GHz. These apply for typical conditions of cosmic, solar, atmospheric, and ground noise. At 425 MHz, the antenna noise temperature varies with elevation angle over a very small range, from 200 °K to 220 °K. Hence, the antenna noise temperature for the radar model is assumed to be 220 °K, as shown in item (1) of Table 8.1-5.

Skolnick [7.2] (pp. 5–10) gives a family of plots versus frequency showing the noise temperatures achievable in radar receivers as of 1970. Our radar model assumes a transistor amplifier. As shown in item (2) of Table 8.1-5, a

TABLE 8.1-5   Assumed Values of Noise Temperature for Radar Model

Element Characterized	Symbol	Noise Temperature (°K)
(1) Antenna	T_a	220
(2) Receiver	T_r	390
(3) System at antenna	T_s	806.8
(4) System at receiver	$T_{s(r)}$	615.9

transistor amplifier has a typical noise temperature of 390 °K at 425 MHz. A parametric amplifier has a much lower noise temperature (100 °K at 425 MHz). However, its dynamic range is so low it may not be adequate for our application.

Modern RF amplifiers in this frequency range generally use the gallium–arsenide field-effect transistor (GaAs FET). Such amplifiers can achieve a much lower noise temperature than a 1970 transistor amplifier, along with a wide dynamic range. However, our radar model is only intended to illustrate computation principles rather than to give precise radar performance. Therefore, the obsolete transistor amplifier considered by Skolnick is adequate for our purposes.

In evaluating receivers for this type of radar application, one must consider gain stability as well as dynamic range and noise temperature. The GaAs FET amplifiers having the lowest noise temperatures operate without feedback. Consequently, they probably cannot satisfy the very accurate gain stability that is required in the monopulse receivers of this radar application.

The noise performance of a receiver is often specified in terms of its noise figure, denoted F_r. Receiver noise temperature T_r is related as follows to receiver noise figure F_r:

$$T_r = (F_r - 1)T_0 \qquad (8.1\text{-}15)$$

where T_0 is the reference noise temperature defined as

$$T_0 = 290 \text{ K} \qquad (8.1\text{-}16)$$

This reference noise temperature T_0 corresponds to "room temperature" and is equivalent to 17 °C, or 62.8 °F. The noise figure F_r for the assumed 425-MHz transistor amplifier is 3.7 dB, which is a factor of 2.344. By Eq 8.1-15, the corresponding noise temperature is

$$T_r = (F_r - 1)T_0 = (2.344 - 1)(290 \text{ °K}) = 390 \text{ °K} \qquad (8.1\text{-}17)$$

The radar system noise temperature, reflected to the receiver input, is denoted $T_{s(r)}$ and is equal to

$$T_{s(r)} = T_r + \frac{T_a + (L_{rd} - 1)T_0}{L_{rw}} \qquad (8.1\text{-}18)$$

The term $(L_{rd} - 1)T_0$ is the thermal noise caused by the dissipative loss in the receive waveguide. In this term, the parameter L_{rd} is the dissipative receive waveguide loss (item 5 of Table 8.1-4) and T_0 is the reference noise temperature given in Eq 8.1-16. Both this dissipative loss noise and the antenna thermal noise (described by noise temperature T_a) are referenced to the antenna aperture. Hence the sum of these noise terms is attenuated by the total loss L_{rw} in the receive waveguide to obtain their noise contributions referenced to the receiver input. These attenuated noise temperatures are added to the receiver noise temperature T_r to obtain the total system noise temperature $T_{s(r)}$, referenced to the receiver input.

For computation purposes, it is more convenient to deal with the parameter T_s, which is the system noise temperature referenced to the antenna aperture. This is related to $T_{s(r)}$ by

$$T_s = L_{rw}T_{s(r)} \tag{8.1-19}$$

This shows that a noise of temperature T_s introduced at the antenna aperture, at the input to the loss L_{rw} of the receive waveguide, is equivalent to a noise of temperature $T_{s(r)}$ introduced at the receiver input, at the output of the L_{rw} receive waveguide loss. Combining Eqs 8.1-18, -19 gives for T_s

$$T_s = L_{rw}T_r + T_a + (L_{rd} - 1)T_0 \tag{8.1-20}$$

As shown in item (4) of Table 8.1-4, the total attenuation L_{rw} of the receive waveguide for the model is 1.18 dB, which represents a factor of 1.312. Item (5) shows that the assumed receive waveguide dissipation loss L_{rd} is 1.00 dB, which corresponds to a factor of 1.259. Hence, by Eq 8.1-20, the system noise temperature T_s, referenced to the antenna aperture, is equal to

$$\begin{aligned}
T_s &= L_{rw}T_r + T_a + (L_{rd} - 1)T_0 \\
&= 1.312(390 \text{ °K}) + 220 \text{ °K} + (1.259 - 1)(290 \text{ °K}) \\
&= 806.8 \text{ °K}
\end{aligned} \tag{8.1-21}$$

This is shown in Table 8.1-5, item (3). The corresponding noise temperature referenced to the receiver input is

$$T_{s(r)} = \frac{T_s}{L_{rw}} = \frac{806.8 \text{ °K}}{1.312} = 615.9 \text{ °K} \tag{8.1-22}$$

This is shown in item (4).

The ratio of received signal energy per pulse to noise power density is equal to

$$\frac{E_s}{P_n} = \frac{P_t\tau_p G^2\lambda^2 A_t}{(4\pi)^3 R^4 KT_s L_t} \tag{8.1-23}$$

where

E_s = received signal energy per pulse
p_n = noise power density (Watt/Hz) in receiver
P_t = peak transmitter power
τ_p = width of transmitter pulse
G = antenna gain in direction of target
λ = wavelength
A_t = effective radar cross section of target
R = range to target
K = Boltzmann constant = 1.38×10^{-23} Joule/°K   (8.1-24)
T_s = system noise temperature referenced to antenna aperture
L_t = transmission loss referenced to antenna aperture

This ratio given in Eq 8.1-23 is equal to the single-pulse signal/noise power ratio for an ideal matched-filter receiver. Thus,

$$\frac{P_{s1(m)}}{P_n} = \frac{E_s}{p_n}$$ (8.1-25)

where

$P_{s1(m)}/P_n$ = peak signal/noise power ratio for a single pulse of an ideal matched-filter receiver

The performance of the actual radar is calculated in terms of the signal/noise power ratio at the input to the envelope detector, which is denoted:

$P_{s1(d)}/P_n$ = peak signal/noise power ratio for a single pulse occurring at the input to the envelope detector

This is related as follows to the signal/noise power ratio of the matched-filter receiver:

$$\frac{P_{s1(d)}}{P_n} = \frac{P_{s1(m)}/P_n}{L_{pc}L_m} = \frac{E_s/p_n}{L_{pc}L_m}$$ (8.1-26)

where

L_{pc} = pulse compression loss
L_m = matching loss of IF receiver

Combining Eqs 8.1-23, -26 gives the following for the single-pulse signal/noise power ratio at the detector input for the actual receiver:

$$\frac{P_{\text{s1(d)}}}{P_{\text{n}}} = \frac{P_{\text{t}}\tau_{\text{p}}G^2\lambda^2 A_{\text{t}}}{(4\pi)^3 R^4 KT_{\text{s}}L_{\text{s}}} \tag{8.1-27}$$

where L_{s} is equal to

$$L_{\text{s}} = L_{\text{t}}L_{\text{pc}}L_{\text{m}} \tag{8.1-28}$$

The system loss factor L_{s} is defined as

L_{s} = system loss referenced to antenna aperture, defined relative to the signal at the input of the envelope detector

A convenient way to apply Eq 8.1-27 is to compute the zero-dB range R_0 for a 1 square-meter target. This parameter R_0 is the range at which the single-pulse signal/noise power ratio at the input to the envelope detector is 0 dB for a 1 square-meter target that is centered in the radar beam. In Eq 8.1-27, set $P_{\text{s1(d)}}/P_{\text{n}}$ equal to unity (0 dB), $A_{\text{t}} = 1.0$ m^2, $G = G_{\text{max}}$, and solve for R. This yields

$$R_0 = \left[\frac{P_{\text{t}}\tau_{\text{p}}G_{\text{max}}^2\lambda^2(1.0 \text{ m}^2)}{(4\pi)^3 KT_{\text{s}}L_{\text{s}}} \right]^{1/4} \tag{8.1-29}$$

The factor $P_{\text{t}}\tau_{\text{p}}$ is the transmitted energy per pulse given in item (5) of Table 8.1-3, which is

$$P_{\text{t}}\tau_{\text{p}} = 60 \text{ Joule} \tag{8.1-30}$$

The values $\lambda = 0.706$ meter and $G_{\text{max}} = 20,220$ are obtained from items (3) and (6) of Table 8.1-2; the value $L_{\text{s}} = 2.104$ is obtained from item (11) of Table 8.1-4; and the value $T_{\text{s}} = 806.8$ °K is obtained from item (3) of Table 8.1-5. From Eq 8.1-29, the resultant value for R_0 is

$$R_0 = 4.027 \times 10^6 \text{ meter} = 4027 \text{ km} \tag{8.2-31}$$

Using this value of zero-dB range R_0, one can calculate from the following equation the single-pulse signal/noise power ratio at the input to the envelope detector, for any given range R, target radar cross-section area A_{t}, and gain G in the direction of the target:

$$\frac{P_{\text{s1(d)}}}{P_{\text{n}}} = \left(\frac{R_0}{R}\right)^4 \frac{A_{\text{t}}}{(1 \text{ meter}^2)} \left(\frac{G}{G_{\text{max}}}\right)^2 \tag{8.1-32}$$

The parameter G_{max} is the peak antenna gain, while G is the antenna gain in the direction of the target. As was shown in Chapter 7, Eq 7.2-4, the ratio G/G_{max} is approximately equal to

$$\frac{G}{G_{max}} \cong \exp\left[-\left(\frac{\phi}{\Psi}\right)^2\right] \tag{8.1-33}$$

where ϕ is the angle of the target from the peak of the radar beam. The parameter Ψ is related as follows to the antenna half-power beamwidth θ_b:

$$\Psi = 0.60\theta_b \tag{8.1-34}$$

The estimated value of half-power beamwidth θ_b is 1.00 °, as was shown in Table 8.1-2, item (4).

From the single-pulse signal/noise ratio given in Eq 8.1-32, which occurs at the input to the envelope detector, one can obtain the signal/noise ratio P_s/P_n following the envelope detector, derived from integrating N pulses. This is

$$\frac{P_s}{P_n} = \frac{N\left(P_{s1(d)}/P_n\right)}{1 + P_n/P_{s1(d)}} \tag{8.1-35}$$

The factor $(1 + P_n/P_{s1(d)})$ is the noncoherent detection loss, which was explained in Chapter 7. It is caused by the noncoherent summation of the contributions from separate radar pulses. When the signal/noise ratio for a single pulse is much greater than unity, this noncoherent detection loss can be neglected.

It can be seen from Chapter 7, Eqs 7.4-17, -23, that the RMS range-gate timing error for a single pulse for the assumed optimized rectangular IF passband is as follows:

$$\frac{\sigma t_1}{\tau_{pc}} = \frac{\sqrt{1 + P_n/P_{s1(d)}}}{1.5\sqrt{2E_s/p_n}} \tag{8.1-36}$$

The pulse width τ_p has been replaced by the compressed pulse width τ_{pc}, and p_{nd} has been replaced by

$$p_{nd} = p_n\left(1 + \frac{P_n}{P_{s1(d)}}\right) \tag{8.1-37}$$

The parameter p_{nd} is the effective noise power density at the output of the envelope detector. As was shown in Eq 7.5-71, the RMS angular tracking error for a single pulse is as follows:

$$\frac{\sigma\theta_1}{\theta_b} = \frac{\sqrt{1 + P_n/P_{sl(d)}}\sqrt{\tau_{ga}/\tau_{pc}}}{k_m\sqrt{2E_s/P_n}} \tag{8.1-38}$$

Again, the pulse width τ_p has been replaced by the compressed pulse width τ_{pc}. The parameter τ_{ga} is the width of the range gate in the angle-tracking receiver. Applying Eq 8.1-26 to Eqs 8.1-36, -38 gives

$$\frac{\sigma t_1}{\tau_{pc}} = \frac{\sqrt{1 + P_n/P_{sl(d)}}}{1.5\sqrt{L_{pc}L_m}\sqrt{2P_{sl(d)}/P_n}} \tag{8.1-39}$$

$$\frac{\sigma\theta_1}{\theta_b} = \frac{\sqrt{1 + P_n/P_{sl(d)}}\sqrt{\tau_{ga}/\tau_{pc}}}{k_m\sqrt{L_{pc}L_m}\sqrt{2P_{sl(d)}/P_n}} \tag{8.1-40}$$

Set $N = 1$ in Eq 8.1-35 to obtain the signal/noise power ratio at the output of the envelope detector for a single pulse, which is

$$\frac{P_{sl}}{P_n} = \frac{P_{sl(d)}/P_n}{1 + P_n/P_{sl(d)}} \tag{8.1-41}$$

Substituting Eqs 8.1-41 into Eqs 8.1-39, -40 gives

$$\frac{\sigma t_1}{\tau_{pc}} = \frac{1}{1.5\sqrt{L_{pc}L_m}\sqrt{2P_{sl}/P_n}} \tag{8.1-42}$$

$$\frac{\sigma\theta_1}{\theta_b} = \frac{\sqrt{\tau_{ga}/\tau_{pc}}}{k_m\sqrt{L_{pc}L_m}\sqrt{2P_{sl}/P_n}} \tag{8.1-43}$$

The following parameter values are assumed for the radar model:

$$k_m = 1.60 \tag{8.1-44}$$

$$\frac{\tau_{ga}}{\tau_{pc}} = 1.50 \tag{8.1-45}$$

$$L_{pc}L_m = (1.175)(1.216) = 1.429 \quad \text{(or 1.55 dB)} \tag{8.1-46}$$

The monopulse slope value $k_m = 1.60$ was derived in Chapter 7, Eq 7.5-49, and is typical for a radar monopulse feed. The ratio $\tau_{ga}/\tau_{pc} = 1.50$ was obtained from Chapter 7, Table 7.5-1, item (6), and is a reasonable value for an angle-tracking radar receiver. The values for the losses L_{pc} and L_m were obtained from items (9) and (10) of Table 8.1-4. Substituting Eqs 8.1-44 to -46 into Eqs 8.1-42, -43 gives

$$\sigma t_1 = \frac{0.394\,\tau_{pc}}{\sqrt{P_{s1}/P_n}} \qquad (8.1\text{-}47)$$

$$\sigma\theta_1 = \frac{0.453\,\theta_b}{\sqrt{P_{s1}/P_n}} \qquad (8.1\text{-}48)$$

As was shown in item (11) of Table 8.1-3, the compressed pulse width τ_{pc} for the radar model is 0.15 μsec. Substituting this for τ_{pc} in Eq 8.1-47 gives the following for the RMS range-gate timing error for a single pulse:

$$\sigma t_1 = \frac{0.0592\ \mu\text{sec}}{\sqrt{P_{s1}/P_n}} \qquad (8.1\text{-}49)$$

This is multiplied by 150 m/μsec to obtain the corresponding RMS range error for a single pulse, which is

$$\sigma r_1 = \frac{8.88\ \text{meter}}{\sqrt{P_{s1}/P_n}} \qquad (8.1\text{-}50)$$

As was shown in item (4) of Table 8.1-2, the estimated half-power beamwidth θ_b of the radar model is 1.00 °. Hence, by Eq 8.1-48, the RMS angular tracking error for a single pulse in degrees and milliradians is

$$\sigma\theta_1 = \frac{0.453\ °}{\sqrt{P_{s1}/P_n}}$$

$$= \frac{7.90\ \text{mrad}}{\sqrt{P_{s1}/P_n}} \qquad (8.1\text{-}51)$$

Expressions for the RMS range and angle errors after integrating N pulses can be obtained from the following equations using the single-pulse errors computed in Eqs 8.1-50, -51:

$$\sigma_r = \sigma_{r1}/\sqrt{N} \qquad (8.1\text{-}52)$$

$$\sigma\theta = \sigma_{\theta 1}/\sqrt{N} \qquad (8.1\text{-}53)$$

Equations 8.1-50 to -53 can be used to calculate the noise errors in range and angle for the radar model. The single-pulse signal/noise power ratio P_{s1}/P_n is computed in Eq 8.1-41 from the ratio $P_{s1(d)}/P_n$, which is the signal/noise power ratio at the input to the envelope detector. This ratio $P_{s1(d)}/P_n$ can be obtained from Eqs 8.1-32, -33 for various values of range R

and angular tracking error ϕ, when the radar cross-section area A_t of the target is known. The value of R_0 for the radar model is obtained from Eq 8.1-31.

This procedure gives theoretical values for the RMS range and angle errors. As will be explained in Section 8.6.1, the Kalman filter optimizes the smoothing coefficients of the tracking system by using values of the single-pulse RMS range and angle errors $\sigma_{r1}, \sigma_{\theta 1}$ computed from the measured single-pulse signal/noise ratio $P_{s1(d)}/P_n$ at the input to the envelope detector. This signal/noise ratio is obtained by measuring the signal level in the receiver, along with the amplifier gain setting established by the automatic gain-control circuitry.

8.2 BASIC DESCRIPTION OF A KALMAN FILTER REAL-TIME TRACKING RADAR PROCESSOR

This section presents the equations of an optimal radar tracking model. The coordinate transformations of the model differ from those used in Altair, but are dynamically equivalent to them. Hence, the Altair tracking filter algorithms, which are analyzed in Section 8.5, can be applied to this model.

8.2.1 Coordinates and Block Diagram of Kalman Filter Tracking Model

A Kalman filter tracking radar such as Altair requires a monopulse tracking feed that can accurately measure the angular error of the target from boresight. The analysis of a monopulse tracking radar given in Chapter 7, Section 7.5, showed that the monopulse slope often varies with signal strength because the gains of the separate signal amplifiers are not precisely matched. In a Kalman filter tracking radar, great care must be taken in the control of amplifier gains to assure that the monopulse slope does not vary. With accurate calibration of monopulse slope, it is not necessary that the tracking antenna keep the target closely centered in the monopulse null of the radar pattern. The monopulse errors are added to the angular data from the optical encoders coupled to the antenna axes to obtain target-tracking information that is essentially independent of antenna pointing error. Consequently, rather large errors of antenna pointing can be tolerated.

An important consideration in the design of an optimal radar tracking system is the set of coordinates used in processing and smoothing the target-tracking data. Rectangular coordinates that follow the target are theoretically ideal for smoothing target-tracking information. However, it is much more convenient to use rectangular coordinates fixed at the radar site, and make these equivalent to target coordinates by including correction signals in the smoothing equations, which compensate for the predictable accelerations of the target relative to the radar site.

A critical issue affecting the choice of coordinates for smoothing radar signals is that range-tracking information is generally at least 100 times more accurate than target position information derived from the angular tracking data. To compensate for this accuracy difference, some Kalman filter radar systems process the tracking data in terms of rectangular coordinates that rotate with the antenna. The rectangular coordinates are located at the antenna aperture, with a range axis along the antenna boresight, a second axis parallel to the elevation axis, and a third axis perpendicular to the elevation axis. Effective target acceleration terms are included in the model to compensate for the rotation of these axes as the antenna moves.

A simpler approach for solving this great disparity between range and angle tracking accuracies is to use two tracking processors: (a) a three-dimensional target estimator operating in fixed coordinates at the radar site and (b) a separate range tracker. This simpler approach is used in the Altair radar.

Figure 8.2-1 shows a block diagram of an optimal Kalman-filter tracking radar model using separate target-estimator and range-tracking processors. The coordinate conversions of this tracking model are different from those of Altair, but, nevertheless, are dynamically equivalent to them. Hence the Altair tracking-filter equations can be studied in terms of this model.

The target estimator shown in Fig 8.2-1 performs three-dimensional smoothing of the target tracking data in terms of rectangular coordinates located at the radar site. The axes of these coordinates lie in the east (E), north (N), and vertical (V) directions. The horizontal east–north plane is located at the level of the elevation axis of the antenna, and the vertical axis passes through the center of azimuth rotation. The three-dimensional target estimator uses range as well as angle tracking data. However, its accuracy is established primarily by the errors in the angular tracking information, and so the *target estimator* is also called an *angle tracker*.

Although the coordinates of the target estimator are fixed at the radar site, the estimator acts as if it were using target-centered coordinates for smoothing the tracking data. This is achieved by including acceleration prediction in its computations, which accounts for the predictable motion of the target relative to radar-site coordinates. One of the outputs from the target estimator is the estimated range acceleration of the target. The estimated range acceleration is fed as a predicted acceleration signal to the range tracker, to compensate for the predictable motions of the target in range. This predicted acceleration signal allows the range tracker to operate as if it were tracking in terms of the coordinates of the target.

The optimal tracking system illustrated in Fig 8.2-1 operates as follows. The signals ΔR, $\Delta\theta_{eL}$, and $\Delta\theta_d$ are the error signals in range, elevation, and deflection derived from the range gate and from the monopulse tracking circuitry. The principles for deriving such signals were explained in Chapter 7.

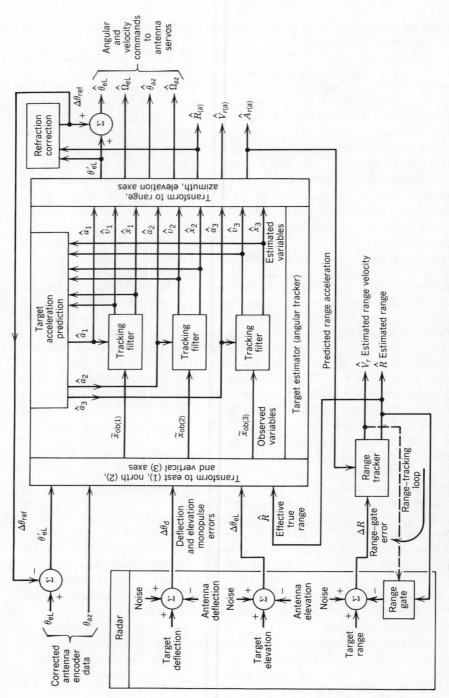

Figure 8.2-1 Block diagram of Kalman filter radar tracking model, which uses 3-axis target estimator plus separate range tracker.

However, sophisticated circuitry is required in this radar system to maintain fixed gains for these signals.

The variables θ_{eL}, θ_{az} are measures of the elevation and azimuth antenna angles, which are derived from optical encoders coupled to the antenna axes. As will be explained, the encoder data must be corrected to obtain the θ_{eL}, θ_{az} variables. The variable $\Delta\theta_{ref}$ is a refraction correction, to be described later, which corrects for refraction in the atmosphere. This refraction correction $\Delta\theta_{ref}$ is subtracted from the antenna elevation angle θ_{eL} to obtain the angle θ'_{eL}, which corresponds to the true elevation of the target.

Since the range tracker is much more accurate than the target estimator, the estimated range derived from the range tracker can be regarded as true range in the smoothing calculations of the target estimator. The target estimator uses the following five input signals: the estimated range $\hat{R}$ obtained from the range tracker; the corrected azimuth and elevation signals θ_{az}, θ'_{eL} derived from the encoder data, which describe the direction of the antenna boresight; and the elevation and deflection monopulse tracking error signals $\Delta\theta_{eL}$, $\Delta\theta_d$ obtained from the radar monopulse circuitry. In a manner to be described later, these five signals are transformed to rectangular radar-site axes to obtain the observed coordinates of the target. These coordinates are denoted $x_{ob(1)}$, $x_{ob(2)}$, $x_{ob(3)}$, which are the observed coordinates of the target measured at the radar site in the east, north, and vertical directions, respectively.

The observed coordinates of the target, $x_{ob(1)}$, $x_{ob(2)}$, $x_{ob(3)}$, are inaccurate, because of radar noise in the elevation and deflection monopulse signals. These observed variables are smoothed by the three tracking filters to obtain the estimated components of target position $\hat{x}_1$, $\hat{x}_2$, $\hat{x}_3$ and the estimated components of target velocity, $\hat{v}_1$, $\hat{v}_2$, $\hat{v}_3$.

The estimates of target position and velocity, characterized by the variables $\hat{x}_1$, $\hat{x}_2$, $\hat{x}_3$ and $\hat{v}_1$, $\hat{v}_2$, $\hat{v}_3$, are used in computer routines that calculate the predictable accelerations of the target due to gravity and coriolis effects, and atmospheric drag when the vehicle is within the atmosphere. The predicted accelerations of the target in the east, north, and vertical directions relative to the radar site are denoted $\hat{a}_1$, $\hat{a}_2$, $\hat{a}_3$. These are fed as acceleration command inputs to the tracking filters to compensate for the predictable motion of the target.

The estimated components of target position and velocity, $\hat{x}_1$, $\hat{x}_2$, $\hat{x}_3$, and $\hat{v}_1$, $\hat{v}_2$, $\hat{v}_3$, along with the predicted components of target acceleration $\hat{a}_1$, $\hat{a}_2$, $\hat{a}_3$, are transformed from east, north, and vertical coordinates at the radar site to range, azimuth, and elevation coordinates. This provides estimated position and velocity values for range, azimuth, and elevation, along with the estimated range acceleration. The estimated range $\hat{R}_{(a)}$ and the estimated target elevation angle $\hat{\theta}'_{eL}$ are used in a routine that computes the refraction correction $\Delta\theta_{ref}$. The refraction correction $\Delta\theta_{ref}$ is subtracted from the estimated true elevation angle $\hat{\theta}'_{eL}$ obtained from the target estimator to form the estimated antenna elevation angle $\hat{\theta}_{eL}$. The estimated azimuth and elevation antenna angles $\hat{\theta}_{az}$, $\hat{\theta}_{eL}$ and the estimated antenna azimuth and elevation angular velocities $\hat{\Omega}_{az}$, $\hat{\Omega}_{eL}$ are fed as command signals to the antenna servos. The estimated range acceleration $\hat{A}_{r(a)}$ is fed to the range tracker to act as a predicted range acceleration signal. This signal allows the range tracker to operate as if it were smoothing the data in terms of the coordinates of the target.

The range tracker is a feedback loop that derives its error signal ΔR from the range gate. The loop smoothes the tracking error data to obtain the estimated range $\hat{R}$ and the estimated range velocity $\hat{v}_r$. This smoothing process is assisted by the estimated range acceleration signal $\hat{a}_{r(a)}$ derived from the target estimator.

The target estimator and the range-tracker functions are performed digitally in a computer with sampled data. The sampled periods of these two processors are denoted:

T_a = angle-tracker (target estimator) sample period

T_r = range-tracker sample period

Since the range tracker has much higher accuracy than the angle tracker, its dynamic requirements are greater. Hence, it is generally desirable that the sample period T_r for the range tracker be smaller than the sample period T_a for the angle tracker.

The radar derives angle and range error signals from every radar pulse. The pulse repetition frequency of the radar is generally much higher than the sampling frequencies of the angle and range trackers. Hence the actual radar error signals $\Delta \theta_d$, $\Delta \theta_{eL}$, ΔR used for angle and range tracking are obtained by averaging the monopulse and range-gate error signals over the sample period T_a or T_r of the angle or range tracker.

The range gate is controlled by the range tracker to compensate for the changing range of the target between sampling instants of the range-tracking loop. As shown in Fig 8.2-1, this range-gate control is achieved by feeding to the range gate the estimated range velocity signal $\hat{v}_r$. The range gate is controlled to move at a rate proportional to this $\hat{v}_r$ signal.

As was stated previously, computations are required to correct the angle encoder data to obtain the antenna angle signals θ_{eL}, θ_{az}. This correction is illustrated in Fig 8.2-2. Bias corrections are added to the encoder signals to compensate for offset errors in setting the zeros of the encoders. A sag correction $\Delta \theta_{sag}$ is added to the indicated elevation angle to compensate for sag in the antenna. This sag correction is computed from an equation of the form

$$\Delta \theta_{sag} = \Delta \theta_0 \left(1 + K_1 \theta_{eL(u)} + K_2 \theta_{eL(u)}^2 \right) \qquad (8.2\text{-}1)$$

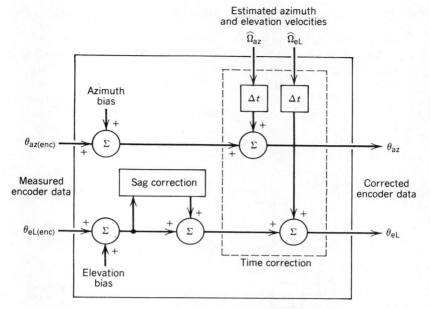

Figure 8.2-2 Block diagram of encoder signal-processing.

where $\theta_{eL(u)}$ is the uncorrected elevation angle and $\Delta\theta_0$, K_1, and K_2 are constants.

Except for antenna sag, the Altair calculations assume that the antenna moves as a rigid body, which is not exactly true. Structural antenna deflections caused by acceleration and wind torques result in small uncorrected tracking error components.

Finally, time corrections are added to both the azimuth and elevation angles to compensate for the difference between the valid time of the tracking computations performed by the target estimator and the time at which the encoders are sampled. These time corrections are equal to the amounts that the angles change during the time difference Δt between the two values of valid time. They are computed by multiplying the estimated azimuth and elevation angular rates $\hat{\Omega}_{az}$, $\hat{\Omega}_{eL}$ by this time difference Δt. Note that this time difference Δt may be either positive or negative.

The compensation for differences in valid time of the system variables is a very important part of a real-time process such as this. An uncompensated timing error adds inaccuracy to the data and may produce instability. The question of what represents the valid time of a variable can be very complicated and is beyond the scope of this book. Although the issue of valid time is ignored in most of the following discussion, it can be crucially important in making real-time process operate effectively.

8.2.2   Characteristics of Tracking Filters for Smoothing Radar Data

The tracking filter equations used in the Altair radar are analyzed in Section 8.5. By taking the Laplace transforms of these tracking equations, signal-flow diagrams of the computations are developed. When sampled-data effects are neglected, the signal-flow diagrams of the tracking computations simplify to the basic forms shown in Fig 8.2-3. The tracking filters used in the target estimator (angle tracker) have the form shown in diagram a, which has

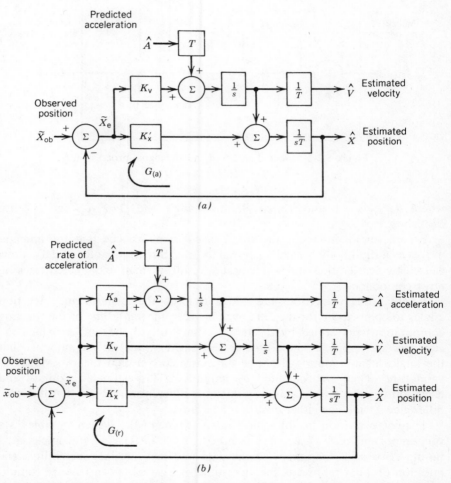

(a)

(b)

Figure 8.2-3 Signal-flow diagrams of Kalman tracking filters derived in Sections 8.4, not including sampled-data effects: (a) tracking filter with velocity memory used in angle tracker; (b) tracking filter with velocity and acceleration memory used in range tracker.

velocity memory. The range tracker uses the tracking filter shown in diagram *b*, which has both velocity and acceleration memory.

The parameters of the Kalman tracking filters are defined as follows:

T = sample period
K_x = Kalman position gain of tracking filter
K_v = Kalman velocity gain of tracking filter
K_a = Kalman acceleration gain of tracking filter

The parameter K'_x shown in Fig 8.2-3 is equal to

$$K'_x = K_x + \tfrac{1}{2}TK_v \qquad (8.2\text{-}2)$$

The term $\tfrac{1}{2}TK_v$ is much smaller than K_x, except immediately after lock-on, when K_x is close to unity. Hence, K'_x is generally nearly equal to K_x.

The loop transfer function of the tracking filter of diagram *a* in Fig 8.2-3 is denoted $G_{(a)}$ because that filter is used for angle tracking; and the loop transfer function for diagram *b* is denoted $G_{(r)}$ because that filter is used for range tracking. These loop transfer functions are equal to

$$G_{(a)} = \frac{K'_x}{Ts} + \frac{K_v}{Ts^2} \qquad (8.2\text{-}3)$$

$$G_{(r)} = \frac{K'_x}{Ts} + \frac{K_v}{Ts^2} + \frac{K_a}{Ts^3} \qquad (8.2\text{-}4)$$

These have the forms

$$G_{(a)} = \frac{\omega_c(1 + \omega_i/s)}{s}$$
$$= \frac{\omega_c}{s} + \frac{\omega_c\omega_i}{s^2} \qquad (8.2\text{-}5)$$

$$G_{(r)} = \frac{\omega_c(1 + \omega_{i1}/s)(1 + \omega_{i2}/s)}{s}$$
$$= \frac{\omega_c}{s} + \frac{\omega_c(\omega_{i1} + \omega_{i2})}{s^2} + \frac{\omega_c\omega_{i1}\omega_{i2}}{s^3} \qquad (8.2\text{-}6)$$

Magnitude asymptote plots of these loop transfer functions are shown in Fig 8.2-4. Setting the expressions for $G_{(a)}$ equal in Eqs 8.2-3, -5 gives:

Loop $G_{(a)}$ with velocity memory:

$$K'_x = \omega_c T \qquad (8.2\text{-}7)$$
$$K_v = \omega_c\omega_i T \qquad (8.2\text{-}8)$$

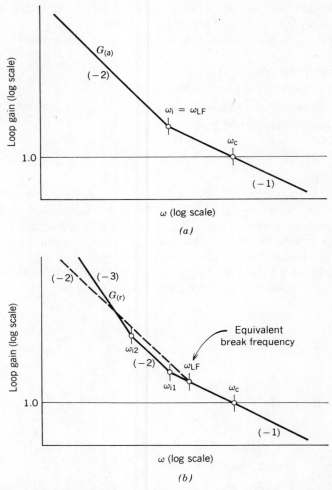

Figure 8.2-4   Magnitude asymptote plots of loop transfer functions $G_{(a)}$ and $G_{(r)}$ for angle and range tracking filters: (*a*) tracking filter with velocity memory used in angle tracker; (*b*) tracking filter with velocity and acceleration memory used in range tracker.

Setting the expressions for $G_{(r)}$ equal in Eqs 8.2-4, -6 gives:

Loop $G_{(r)}$ with velocity and acceleration memory:

$$K_x' = \omega_c T \qquad\qquad (8.2\text{-}9)$$

$$K_v = \omega_c(\omega_{i1} + \omega_{i2})T \qquad\qquad (8.2\text{-}10)$$

$$K_a = \omega_c\omega_{i1}\omega_{i2}T \qquad\qquad (8.2\text{-}11)$$

In accordance with the optimization techniques described in Ref [1.1] (Section 3.4), loop $G_{(r)}$ can be approximated by the following transfer function near gain crossover and at higher frequencies:

$$G_{(r)} \cong \frac{\omega_c (1 + \omega_{LF}/s)}{s} \qquad (8.2\text{-}12)$$

where ω_{LF} is the equivalent low-frequency break, which is equal to

$$\omega_{LF} = \sum_{\text{zeros}}^{LF} \omega_x - \sum_{\text{poles}}^{LF} \omega_x \qquad (8.2\text{-}13)$$

The first term is the sum of the break frequencies for the low-frequency (LF) zeros, and the second term is the sum for the low-frequency poles. The low-frequency zeros and poles are those with break frequencies less than ω_c. For loop $G_{(r)}$, ω_{LF} is equal to

$$\omega_{LF} = \omega_{i1} + \omega_{i2} \qquad (8.2\text{-}14)$$

Hence, Eq 8.2-10 becomes

$$\text{Loop } G_{(r)}: \quad K_v = \omega_c \omega_{LF} T \qquad (8.2\text{-}15)$$

This is equivalent to the expression for K_v of loop $G_{(a)}$ in Eq 8.2-8 because ω_{LF} for that loop is equal to ω_i. Hence, when the values of K_x and K_v are the same for the two tracking filters, the values for ω_c and ω_{LF} are the same. Therefore, the feedback loops for both filters have nearly the same values of $\text{Max}|G_{ib}|$, gain crossover frequency ω_{gc}, and noise bandwidth.

The range tracking loop $G_{(r)}$ has a low-frequency triple integration. As was shown in Chapter 3, a low-frequency triple integration usually does not reduce tracking error, and this is probably true in this application. The reason for using a triple integration in the range tracker is to improve *coasting* accuracy, not *tracking* accuracy.

During the initial part of reentry, a strong ionization wave is generated around the vehicle. This tends to obscure the radar echo from the vehicle, and so the radar echo may be lost momentarily. If the radar signal drops below a given threshold, the range gate error ΔR is set to zero, and the range tracking loop enters a *coast* mode. The estimated range acceleration and range velocity signals stored in the integrators keep the range gate moving at the proper velocity and acceleration for following the most likely target trajectory. The range acceleration memory, which is provided by the triple integration, allows the range tracker to follow the target with acceptable accuracy over a longer coast period than if a simple velocity-memory tracker were used.

In the Altair radar, optimal values of the tracking filter gains K_x and K_v are computed by an iterative Kalman filter routine to be described in Sections

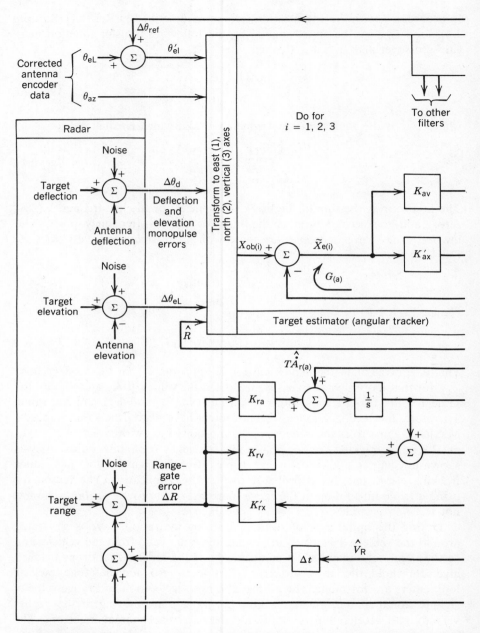

Figure 8.2-5 Kalman filter tracking radar model, which uses the same tracking filters as Altair.

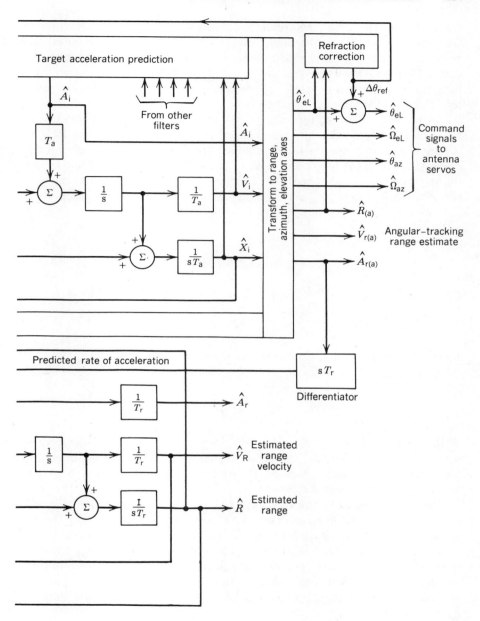

Figure 8.2-5 *(Continued)*

8.3 and 8.4. However, this routine cannot set an optimal value for the acceleration gain K_a of the range tracker because the routine is based on a covariance matrix of only second order. The value for K_a is set in accordance with a formula that is based on the values of the K_x and K_v parameters. This formula was designed to set K_a approximately at the maximum value consistent with good stability.

Let us derive our own formula for setting K_a. To assure good stability, the zeros of $G_{(r)}$ should not be underdamped because, as explained in Ref [1.2] (Chapter 13), low-frequency underdamped open-loop zeros generally produce underdamped closed-loop poles close to them. To maximize K_a while keeping the zeros of $G_{(a)}$ overdamped, the value for K_a should be chosen to make the two break frequencies ω_{i1}, ω_{i2} equal. Thus, by Eqs 8.2-14, the conditions for optimum K_a are

$$\omega_{i1} = \omega_{i2} \tag{8.2-16}$$

$$\omega_{LF} = \omega_{i1} + \omega_{i2} = 2\omega_{i1} \tag{8.2-17}$$

Applying Eqs 8.2-16, -17 to Eq 8.2-11 gives the optimum value for K_a:

$$K_a = \tfrac{1}{4}\omega_c\omega_{LF}^2 T \tag{8.2-18}$$

By combining Eqs 8.2-9, -15, -18, this can be expressed as

$$K_a = \frac{K_v^2}{4K_x'} = \frac{K_v^2}{4(K_x + K_v T/2)} \tag{8.2-19}$$

With this setting for K_a, the value for $\text{Max}|G_{ib}|$ of the $G_{(r)}$ triple-integration tracking filter is essentially the same as that for the $G_{(a)}$ double-integration filter having the same values of K_x and K_v. (Thus, the two loops have approximately the same stability.)

The velocity-memory tracking filter $G_{(a)}$ shown in diagram a of Fig 8.2-3 has a predicted acceleration input $\hat{A}$. This input compensates for the predictable motions of the target, and thereby allows the tracking data to be smoothed over a long time interval. For the tracking filter $G_{(r)}$ shown in diagram b, which has acceleration and velocity memory, it is more convenient to apply the predicted target motion information in terms of the predicted rate-of-acceleration signal $\dot{\hat{A}}$. This is the predicted input variable shown in diagram b.

When the tracking filters of Figs 8.2-3 are applied to the tracking system of Fig 8.2-1, the diagram of Fig 8.2-5 results. The $G_{(a)}$ tracking filter is used in the angle tracker (target estimator), and the $G_{(r)}$ tracking filter is used in the range tracker. To simplify the diagram, the signal-flow diagram for only one of the three $G_{(a)}$ tracking filters of the angle tracker is shown. The K_x, K_v gains for the angle tracker are represented as K_{ax} and K_{av}; and the K_x, K_v, K_a

gains of the range tracker are represented as K_{rx}, K_{rv}, K_{ra}. The sample period T is expressed as T_a for the angle tracker and as T_r for the range tracker.

In Fig 8.2-1, the estimated range acceleration $\hat{A}_{r(a)}$ derived from the angle tracker was fed as a command input to the range tracker. However, as was shown in Fig 8.2-3b, it is more convenient to use a rate-of-acceleration input to the range tracker because the range tracker has acceleration memory. Accordingly, the estimated range acceleration computed by the target estimator is differentiated to obtain the estimated rate of range acceleration. This is the command input to the range tracker that compensates for predicted motions of the target.

As was shown earlier, the estimated range velocity signal $\hat{V}_r$ is used to move the range gate at a constant velocity between sampling instants of the range tracker. This process is illustrated in the signal-flow diagram by multiplying the estimated range velocity $\hat{V}_r$ by the time increment Δt, to obtain a range correction signal. This correction is added to the estimated range $\hat{R}$ to obtain the actual range feedback signal applied to the range gate. This time interval Δt is a variable that changes from pulse to pulse.

8.2.3 Coordinate Conversion Equations for Radar Tracking Model

Now let us consider the coordinate conversion calculations that are implied in the block diagrams of Figs 8.2-1, -5 for the radar tracking model. As was stated previously, these coordinate conversions are not the same as those used in the Altair radar. However, they are dynamically equivalent and so are consistent with the tracking filters used in Altair for angle and range tracking.

8.2.3.1 *Conversion from East, North, Vertical Coordinates of Target Estimator to Range, Azimuth, Elevation Coordinates of Antenna.* Figure 8.2-6 shows the geometric relations for converting the estimated target position components $\hat{x}_1$, $\hat{x}_2$, $\hat{x}_3$ to the estimated values of range, azimuth, and elevation of the target. The projection of the target range vector in the horizontal plane is defined as $\hat{\rho}$, which is equal to

$$\hat{\rho} = \sqrt{\hat{x}_1^2 + \hat{x}_2^2} \qquad (8.2\text{-}20)$$

The estimated azimuth and elevation angles are

$$\hat{\theta}_{az} = \arctan[\hat{x}_1/\hat{x}_2] \qquad (8.2\text{-}21)$$

$$\hat{\theta}'_{eL} = \arctan[\hat{x}_3/\hat{\rho}] \qquad (8.2\text{-}22)$$

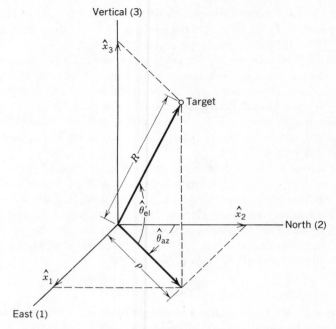

Figure 8.2-6 Conversion of rectangular components of estimated target position to estimated values of range, azimuth angle, and elevation angle.

The estimated elevation angle $\hat{\theta}'_{eL}$ is indicated with a prime because refraction correction must be added to this to obtain the actual elevation angle $\hat{\theta}_{eL}$ that commands the antenna elevation servo. The estimated range is

$$\hat{R}_{(a)} = \sqrt{\hat{\rho}^2 + \hat{x}_3^2} \qquad (8.2\text{-}23)$$

The subscript (a) indicates that this is the estimated range derived from the angle tracker, which is different from the estimated range $\hat{R}$ derived from the range tracker.

Note that the azimuth angle $\hat{\theta}_{az}$ for Altair is measured from north (axis 2) in the clockwise direction. In contrast, the azimuth–elevation tracking coordinates discussed in Chapter 7 measure azimuth angle in the counterclockwise direction, from an axis (1) equivalent to east in this reference frame. The approach of Chapter 7 is the convention generally used for geometric coordinate conversion. However, the Altair system measures azimuth in the clockwise direction from north, so that azimuth angle is consistent with the angle indicated on a compass.

Equations 8.2-20 to -23 can be rewritten as

$$\hat{\rho}^2 = \hat{x}_1^2 + \hat{x}_2^2 \tag{8.2-24}$$

$$\hat{R}_{(a)}^2 = \hat{x}_1^2 + \hat{x}_2^2 + \hat{x}_3^2 \tag{8.2-25}$$

$$\hat{x}_1 = \hat{x}_2 \tan\left[\hat{\theta}_{az}\right] \tag{8.2-26}$$

$$\hat{x}_3 = \hat{\rho} \tan\left[\hat{\theta}_{eL}\right] \tag{8.2-27}$$

These equations are differentiated to obtain expressions for transforming the rectangular components of target velocity $\hat{v}_1, \hat{v}_2, \hat{v}_3$ to obtain the target range rate $\hat{V}_{r(a)}$, the azimuth angular velocity $\hat{\Omega}_{az}$, and the elevation angular velocity $\hat{\Omega}_{eL}$. Differentiating Eqs 8.2-24, -25 gives

$$\hat{\rho}\frac{d\hat{\rho}}{dt} = \hat{x}_1 \hat{v}_1 + \hat{x}_2 \hat{v}_2 \tag{8.2-28}$$

$$\hat{R}_{(a)} \hat{v}_{r(a)} = \hat{x}_1 \hat{v}_1 + \hat{x}_2 \hat{v}_2 + \hat{x}_3 \hat{v}_3 \tag{8.2-29}$$

To differentiate Eqs 8.2-26, -27, remember that

$$d \tan[\theta] = \frac{1}{\cos^2[\theta]} d\theta \tag{8.2-30}$$

Hence the derivatives of Eqs 8.2-26, -27 are

$$\hat{v}_1 = \tan\left[\hat{\theta}_{az}\right] \hat{v}_2 + \frac{\hat{x}_2}{\cos^2\left[\hat{\theta}_{az}\right]} \hat{\Omega}_{az}$$

$$= \frac{\hat{x}_1}{\hat{x}_2} \hat{v}_2 + \frac{\hat{\rho}^2}{\hat{x}_2} \hat{\Omega}_{az} \tag{8.2-31}$$

$$\hat{v}_3 = \tan\left[\hat{\theta}_{eL}\right] \frac{d\hat{\rho}}{dt} + \frac{\hat{\rho}}{\cos^2\left[\hat{\theta}_{eL}\right]} \hat{\Omega}_{eL}$$

$$= \frac{\hat{x}_3}{\hat{\rho}} \frac{d\hat{\rho}}{dt} + \frac{\hat{R}^2}{\hat{\rho}} \hat{\Omega}_{eL} \tag{8.2-32}$$

Solving Eqs 8.2-29, -30 for the linear velocities $(d\hat{\rho}/dt)$, $\hat{v}_{r(a)}$ gives

$$\frac{d\hat{\rho}}{dt} = \hat{v}_1 \frac{\hat{x}_1}{\hat{\rho}} + \hat{v}_2 \frac{\hat{x}_2}{\hat{\rho}} \tag{8.2-33}$$

$$\hat{v}_{r(a)} = \hat{v}_1 \frac{\hat{x}_1}{\hat{R}_{(a)}} + \hat{v}_2 \frac{\hat{x}_2}{\hat{R}_{(a)}} + \hat{v}_3 \frac{\hat{x}_3}{\hat{R}_{(a)}} \tag{8.2-34}$$

Solving Eqs 8.2-31, -32 for the angular velocities $\hat{\Omega}_{az}$, $\hat{\Omega}_{eL}$ gives

$$\hat{\Omega}_{az} = \frac{1}{\hat{\rho}} \left(\frac{\hat{x}_2}{\hat{\rho}} \hat{v}_1 - \frac{\hat{x}_1}{\hat{\rho}} \hat{v}_2 \right) \tag{8.2-35}$$

$$\hat{\Omega}_{eL} = \frac{1}{\hat{R}_{(a)}} \left(\frac{\hat{\rho}}{\hat{R}_{(a)}} \hat{v}_3 - \frac{\hat{x}_3}{\hat{R}_{(a)}} \frac{d\hat{\rho}}{dt} \right) \tag{8.2-36}$$

The system does not require the values of acceleration in azimuth and elevation, but it does require the range acceleration. It can be shown by differentiating the expressions for $d\hat{\rho}/dt$ and $\hat{v}_{r(a)}$ that the range acceleration $\hat{a}_{r(a)}$ is given by

$$\frac{d^2\hat{\rho}}{dt^2} = \hat{a}_1 \frac{\hat{x}_1}{\hat{\rho}} + \hat{a}_2 \frac{\hat{x}_2}{\hat{\rho}} + \left(\hat{v}_1 \frac{\hat{x}_2}{\hat{\rho}} - \hat{v}_2 \frac{\hat{x}_1}{\hat{\rho}} \right) \hat{\Omega}_{az} \tag{8.2-37}$$

$$\hat{a}_{r(a)} = \frac{d^2\hat{\rho}}{dt^2} \frac{\hat{\rho}}{\hat{R}_a} + \hat{a}_3 \frac{\hat{x}_3}{\hat{R}_a} - \left(\hat{v}_3 \frac{\hat{\rho}}{\hat{R}_a} - \frac{d\hat{\rho}}{dt} \frac{\hat{x}_3}{\hat{R}_a} \right) \hat{\Omega}_{eL} \tag{8.2-38}$$

Equations 8.2-20 to -23 and Eqs 8.2-33 to -38 can be used to calculate the estimated values of azimuth and elevation angles and angular rates, and the estimated values of range, range rate, and range acceleration. These are derived from $\hat{x}_1$, $\hat{v}_i$, and $\hat{a}_i$, the estimated values of target position, velocity, and acceleration in radar-site east–north–vertical (ENV) rectangular coordinates.

8.2.3.2 Refraction Correction.
A refraction correction angle $\Delta\theta_{ref}$ is added to the elevation angle $\hat{\theta}'_{eL}$ derived from the target estimator to obtain the actual estimated elevation angle $\hat{\theta}_{eL}$ that commands the antenna servos in elevation:

$$\hat{\theta}_{eL} = \hat{\theta}'_{eL} + \Delta\theta_{ref} \tag{8.2-39}$$

The elevation error caused by atmospheric refraction is illustrated in Fig 8.2-7. The coordinates $\hat{\rho}$, $\hat{x}_3$ of the target estimator define the true position of the target (A). However, the wave reflected from the target follows the curved path, and so the target appears to be in position B. The elevation angle of the true target position A is $\hat{\theta}'_{eL}$, while the elevation angle of the apparent target position B is $\hat{\theta}_{eL}$. The refraction correction $\Delta\theta_{ref}$ is added to the true elevation angle $\hat{\theta}'_{eL}$ to obtain the apparent elevation angle $\hat{\theta}_{eL}$. The antenna is pointed in the direction of the apparent target at the elevation angle $\hat{\theta}_{eL}$.

The elevation refraction correction $\Delta\theta_{ref}$ can be calculated from:

$$\Delta\theta_{ref} = \theta_{inf} \left(1 - \frac{R_r^2}{\left(\hat{R}_{(a)} + R_r \right)^2} \right) \tag{8.2-40}$$

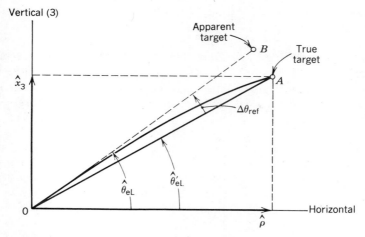

Figure 8.2-7 Geometry associated with refraction correction.

The parameters R_r and θ_{inf} are functions of elevation angle, and $\hat{R}_{(a)}$ is the estimated target range derived from the angle tracker. The equation shows that at infinite range the elevation refraction correction $\Delta\theta_{ref}$ is equal to the parameter θ_{inf}. The values of θ_{inf} and the characteristic refraction range R_r can be obtained from Table 8.2-1 by using the following interpolation formulas:

$$\theta_{inf} = \theta_{inf}[x] + \frac{d\theta_{inf}}{d\theta}(\theta - \theta_x) \qquad (8.2\text{-}41)$$

$$R_r = R_r[x] + \frac{dR_r}{d\theta}(\theta - \theta_x) \qquad (8.2\text{-}42)$$

The variable θ is the true elevation angle $\hat{\theta}'_{eL}$, and θ_x is the largest elevation angle given in the first column of Table 8.2-1 that does not exceed θ. The

TABLE 8.2-1   Table of Refraction Coefficients

Elevation, θ_x (deg)	Angle Coefficients		Range Coefficients	
	$\theta_{inf}[x]$ (deg)	$d\theta_{inf}/d\theta$ (deg/deg)	$R_r[x]$ kilometer	$dR_r/d\theta$ (km/deg)
− 3	0.725	−0.288	2130	−480
0.5	0.611	−0.172	1890	−500
1.0	0.525	−0.124	1640	−300
2.0	0.401	−0.0630	1340	−183.3
5.0	0.212	−0.0196	790	−66.8
10	0.114	−0.00567	456	−21.3
20	0.0573	−0.001625	243	−6.05
40	0.0248	−0.000496	122	−0.244

parameters $\theta_{inf}[x]$, $d\theta_{inf}/d\theta$, $R_r[x]$, $dR_r/d\theta$ are the values given in the row for that value of θ_x.

8.2.3.3 Conversion of Tracking Data to East, North, Vertical Coordinates of Target Estimator.

The following are equations that can be used to transform the antenna angle data, range data, and monopulse error data to radar-site east–north–vertical (ENV) coordinates to obtain the observed coor-

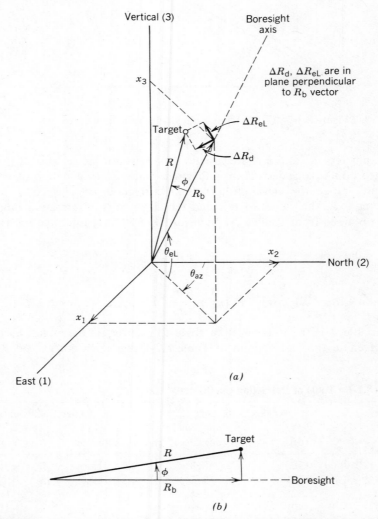

Figure 8.2-8 Geometry for computing observed coordinates of the target in east (1), north (2), and vertical (3) axes at the radar site: (a) geometry relating tracking error vectors to ENV coordinates at radar site ($\Delta \underline{R}_d$, $\Delta \underline{R}_{eL}$ are in plane perpendicular to $\underline{R}_b$ vector); (b) relation between range vector $\underline{R}$ and vector $\underline{R}_b$, which is the projection of the range vector along the antenna boresight.

dinates of the target used in the target estimator. The geometry for converting the tracking data to radar-site ENV coordinates is shown in Fig 8.2-8. For the moment, let us ignore refraction correction. The angles θ_{az}, θ_{eL} are the azimuth and elevation angles obtained from the corrected encoder data. The vector $\underline{R}$ is the range vector to the target, and the vector $\underline{R}_b$ is the projection of the range vector along the antenna boresight axis. The vectors $\Delta\underline{R}_d$ and $\Delta\underline{R}_{eL}$ are linear distances at the range of the target that correspond to the elevation and deflection monopulse tracking errors $\Delta\theta_{eL}$ and $\Delta\theta_d$.

Figure 8.2-8b shows the relation between the range vector $\underline{R}$ and the vector $\underline{R}_b$, which is the projection of $\underline{R}$ along the boresight. The angle between these two vectors is denoted ϕ. Since the angular tracking error is small, a very accurate approximation of the angle ϕ can be derived as follows from the elevation and deflection monopulse tracking error angles $\Delta\theta_{eL}$ and $\Delta\theta_d$:

$$\phi = \sqrt{\Delta\theta_{eL}^2 + \Delta\theta_d^2} \tag{8.2-43}$$

As shown in Fig 8.2-8b, the length of the vector R_b is equal to

$$R_b = R\cos\phi \tag{8.2-44}$$

Since ϕ is a small angle, Eq 8.2-44 can be approximated very accurately by

$$R_b = R\left(1 - \tfrac{1}{2}\phi^2\right) = R\left(1 - \tfrac{1}{2}\Delta\theta_{eL}^2 - \tfrac{1}{2}\Delta\theta_d^2\right) \tag{8.2-45}$$

The vectors $\Delta\underline{R}_d$, $\Delta\underline{R}_{eL}$ are positional errors that correspond to the deflection and elevation angular tracking errors. Their amplitudes are

$$\Delta R_d = R_b\tan[\Delta\theta_d] \tag{8.2-46}$$

$$\Delta R_{eL} = R_b\tan[\Delta\theta_{eL}] \tag{8.2-47}$$

Since the tracking errors are small, these equations can be approximated very accurately by

$$\Delta R_d = R_b\,\Delta\theta_d \tag{8.2-48}$$

$$\Delta R_{eL} = R_b\,\Delta\theta_{eL} \tag{8.2-49}$$

Figure 8.2-8a shows that the components of the vectors $\underline{R}_b$, $\Delta\underline{R}_d$, and $\Delta\underline{R}_{eL}$ along the ENV axes of the radar site are as follows:

Boresight range vector $\underline{R}_b$:

$$\text{East:} \qquad \tilde{x}_{rb(1)} = R_b\cos[\theta_{eL}]\sin[\theta_{az}] \tag{8.2-50}$$

$$\text{North:} \qquad \tilde{x}_{rb(2)} = R_b\cos[\theta_{eL}]\cos[\theta_{az}] \tag{8.2-51}$$

$$\text{Vertical:} \qquad \tilde{x}_{rb(3)} = R_b\sin[\theta_{eL}] \tag{8.2-52}$$

Deflection error vector $\Delta \underline{R}_d$:

$$\text{East:} \qquad \tilde{x}_{d(1)} = (R_b \, \Delta\theta_d)\cos[\theta_{az}] \qquad (8.2\text{-}53)$$

$$\text{North:} \qquad \tilde{x}_{d(2)} = -(R_b \, \Delta\theta_d)\sin[\theta_{az}] \qquad (8.2\text{-}54)$$

$$\text{Vertical:} \qquad \tilde{x}_{d(3)} = 0 \qquad (8.2\text{-}55)$$

Elevation error vector $\Delta \underline{R}_{eL}$:

$$\text{East:} \qquad \tilde{x}_{eL(1)} = -(R_b \, \Delta\theta_{eL})\sin[\theta_{eL}]\sin[\theta_{az}] \qquad (8.2\text{-}56)$$

$$\text{North:} \qquad \tilde{x}_{eL(2)} = -(R_b \, \Delta\theta_{eL})\sin[\theta_{eL}]\cos[\theta_{az}] \qquad (8.2\text{-}57)$$

$$\text{Vertical:} \qquad \tilde{x}_{eL(3)} = (R_b \, \Delta\theta_{eL})\cos[\theta_{eL}] \qquad (8.2\text{-}58)$$

The range vector to the target that is derived from the measured data is called the observed range vector $\tilde{\underline{R}}_{ob}$, and is equal to

$$\tilde{\underline{R}}_{ob} = \underline{R}_b + \Delta \underline{R}_d + \Delta \underline{R}_{eL} \qquad (8.2\text{-}59)$$

The components of this observed range vector $\tilde{\underline{R}}_{ob}$ in radar-site ENV coordinates are the observed components of target position, which are

$$\tilde{x}_{ob(i)} = \tilde{x}_{rb(i)} + \tilde{x}_{d(l)} + \tilde{x}_{eL(i)} \qquad (l = 1, 2, 3) \qquad (8.2\text{-}60)$$

Applying Eqs 8.2-50 to -58 to Eq 8.2-60 gives

$$\tilde{x}_{ob(1)} = \tilde{R}_b\{\sin[\theta_{az}](\cos[\theta_{eL}] - \Delta\theta_{eL}\sin[\theta_{eL}]) + \Delta\theta_d\cos[\theta_{az}]\} \qquad (8.2\text{-}61)$$

$$\tilde{x}_{ob(2)} = \tilde{R}_b\{\cos[\theta_{az}](\cos[\theta_{eL}] - \Delta\theta_{eL}\sin[\theta_{eL}]) - \Delta\theta_d\sin[\theta_{az}]\} \qquad (8.2\text{-}62)$$

$$\tilde{x}_{ob(3)} = \tilde{R}_b(\sin[\theta_{eL}] + \Delta\theta_{eL}\cos[\theta_{eL}]) \qquad (8.2\text{-}63)$$

$$\tilde{R}_b = \tilde{R}_{ob}\left(1 - \tfrac{1}{2}\Delta\theta_{eL}^2 - \tfrac{1}{2}\Delta\theta_d^2\right) \qquad (8.2\text{-}64)$$

The preceding analysis ignores the effect of refraction correction. Table 8.2-1 shows that ϕ_{inf}, which is the maximum possible refraction error, is less than 1.0 ° and so is a small correction. This small refraction correction can be included to high accuracy by replacing θ_{eL} in Eqs 8.2-61 to -63 with an angle θ'_{eL} equal to

$$\theta'_{eL} = \theta_{eL} + \Delta\theta_{ref} \qquad (8.2\text{-}65)$$

where $\Delta\theta_{ref}$ is the refraction correction angle computed by Eq 8.2-40.

8.2.4 Prediction of Target Acceleration in Tracking Model

As was shown in Figs 8.2-1, -5, the target estimator computes the predicted accelerations of the target $\hat{a}_1, \hat{a}_2, \hat{a}_3$ in radar-site ENV coordinates, based on the estimated values of target position $\hat{x}_1, \hat{x}_2, \hat{x}_3$ and the estimated values of target velocity $\hat{v}_1, \hat{v}_2, \hat{v}_3$. The following are equations that can provide target acceleration prediction for the radar model. The parameter values are expressed in terms of a radar model at the Altair site.

Assuming the vehicle does not have rocket propulsion, the only true force acting on the vehicle when it is above the atmosphere is gravity. For this application, gravity forces from the sun and moon can be ignored because they have negligible effect on the vehicle trajectory over the short time periods considered here. If the earth is assumed to have spherical mass symmetry, the vehicle stays within a plane that is fixed relative to inertial space, and travels along an elliptical orbit where the center of the earth is at one focus of the ellipse. If the position and the velocity vectors of the vehicle are known at any instant of time, the orbit of the vehicle is uniquely determined. The vehicle stays within the plane that passes through the velocity vector and the center of the earth. The elliptical orbit of the vehicle is fixed in that plane.

On the other hand, the trajectory of the vehicle appears to be much more complicated than this because the vehicle is being observed relative to a coordinate system fixed on the earth, which is rotating with respect to inertial space. To account for this rotation of the radar-site coordinates, additional effective vehicle acceleration terms are included. Sometimes these are simply called *coriolis acceleration* terms, but in the following they are separated into two classes called coriolis acceleration and *centripetal acceleration*.

When the vehicle is above the atmosphere, the effective accelerations experienced by the vehicle have three sets of components given by

$$\hat{a}_i = \hat{a}_{(g)i} + \hat{a}_{(cor)i} + \hat{a}_{(cnt)i} \qquad (i = 1, 2, 3) \qquad (8.2\text{-}66)$$

where

$\hat{a}_{(g)i}$ = gravitational components of effective vehicle acceleration $\hat{a}_i$
$\hat{a}_{(cor)i}$ = coriolis components of $\hat{a}_i$
$\hat{a}_{(cnt)i}$ = centripetal components of $\hat{a}_i$

As described previously, the index i has the values $1, 2, 3$ to describe the three components of acceleration relative to radar-site ENV coordinates. The position components of the effective center of the earth relative to the ENV coordinates of the radar site are denoted $x_{(c)i}$ and can be expressed as

$$\text{East:} \qquad x_{(c)1} = 0 \qquad (8.2\text{-}67)$$

$$\text{North:} \qquad x_{(c)2} = \Delta R_2 \qquad (8.2\text{-}68)$$

$$\text{Vertical:} \qquad x_{(c)3} = -(R_e - \Delta R_3) \qquad (8.2\text{-}69)$$

The constant R_e is the radius of the earth (see Eq 8.2-74). The constants ΔR_2, ΔR_3 are terms that are small relative to R_e, which account for the lack of spherical symmetry of the earth. These small terms can be derived for the particular location from standard geodetic data of the earth. For approximate analysis, they can be set to zero. Consider a set of axes that are parallel to the radar-site ENV axes but are located at the effective center of the earth. The coordinates of the target in these earth-centered axes are denoted $\hat{D}_i$ and are related as follows to the estimated position coordinates $\hat{x}_i$ of the target in radar-site coordinates:

$$\hat{D}_i = \hat{x}_i - \hat{x}_{(c)i} \qquad (i = 1, 2, 3) \tag{8.2-70}$$

where $\hat{x}_{(c)i}$ is given by Eqs 8.2-67 to -69. The distance of the vehicle from the effective center of the earth is denoted $\hat{D}$, and is equal to

$$\hat{D} = \sqrt{\hat{D}_1^2 + \hat{D}_2^2 + \hat{D}_3^2} \tag{8.2-71}$$

The gravitational components of effective target acceleration can be computed from

$$a_{(g)i} = -\frac{(gR_e^2)\hat{D}_i}{\hat{D}^3} \tag{8.2-72}$$

where g is the acceleration of gravity and R_e is the radius of the earth. As shown by Gray [8.3] (pp. 2–105, -106), the standard values for the acceleration of gravity and the radius of the earth at sea level on the equator are

$$g = 9.780\,490 \text{ meter/sec}^2 \tag{8.2-73}$$

$$R_e = 5,378,388 \text{ meter} \tag{8.2-74}$$

This gives a combined gR_e^2 constant of

$$gR_e^2 = 3.979\,08 \times 10^{14} \text{ meter}^3/\text{sec}^2 \tag{8.2-75}$$

This combined constant varies somewhat with latitude. The preceding reference by Gray gives data to correct the constant for latitude.

The coriolis and centripetal components of target acceleration can be computed as follows:

Coriolis acceleration:

$$\text{East:} \qquad \hat{a}_{(cor)1} = 2(\Omega_{e3}\hat{v}_2 - \Omega_{e2}\hat{v}_3) \tag{8.2-76}$$

$$\text{North:} \qquad \hat{a}_{(cor)2} = 2\Omega_{e3}\hat{v}_1 \tag{8.2-77}$$

$$\text{Vertical:} \qquad \hat{a}_{(cor)3} = 2\Omega_{e2}\hat{v}_1 \tag{8.2-78}$$

Centripetal acceleration:

East: $\hat{a}_{(cnt)1} = \left(\Omega_{e2}^2 + \Omega_{e3}^2 \right) \hat{R}_1$ (8.2-79)

North: $\hat{a}_{(cnt)2} = \Omega_{e3}^2 \hat{R}_2 - \Omega_{e2} \Omega_{e3} \hat{R}_3$ (8.2-80)

Vertical: $\hat{a}_{(cnt)3} = \Omega_{e2}^2 \hat{R}_3 - \Omega_{e2} \Omega_{e3} \hat{R}_2$ (8.2-81)

The parameters Ω_{e2}, Ω_{e3} are the components of rotation of the earth abut the north and vertical radar-site axes, which are equal to

$$\Omega_{e2} = \Omega_e \cos[\theta_L] \qquad (8.2\text{-}82)$$
$$\Omega_{e3} = \Omega_e \sin[\theta_L] \qquad (8.2\text{-}83)$$

The parameter Ω_e is the angular velocity of earth rotation about its axis relative to inertial space, and θ_L is the latitude of the site. The latitude of the Altair site is

$$\theta_L = 9\,^{\circ}23.5' = 9.392\,^{\circ} \qquad (8.2\text{-}84)$$

The period of rotation of the earth relative to the sun is 86,400 sec, the length of one day. However, the period of rotation of the earth relative to the fixed stars (inertial space) is less than this, because the earth rotates relative to the stars one extra turn per year as it rotates around the sun. The average period of rotation of the earth relative to inertial space is called a sidereal day, which is

$$T_e = 86,164.09 \text{ sec} \qquad (8.2\text{-}85)$$

The corresponding angular rate of rotation of the earth relative to inertial space is

$$\Omega_e = 2\pi/T_e = 7.292\,11 \times 10^{-5} \text{ rad/sec} \qquad (8.2\text{-}86)$$

Substituting Eqs 8.2-84, -86 into Eqs 8.2-82, -83 gives for a radar located at the Altair site

$$\Omega_{e2} = 7.194\,4 \times 10^{-5} \text{ rad/sec} \qquad (8.2\text{-}87)$$
$$\Omega_{e3} = 1.190\,0 \times 10^{-5} \text{ rad/sec} \qquad (8.2\text{-}88)$$

When the vehicle reenters the atmosphere, it experiences a very strong and much less predictable acceleration caused by atmospheric drag. As was shown in Chapter 6, Section 6.4, the drag pressure P_d due to wind is equal to

$$P_d = \tfrac{1}{2} C_d \rho_a v^2 \qquad (8.2\text{-}89)$$

where C_d is the drag coefficient of the vehicle, v is the vehicle velocity, and ρ_a is the mass density of the atmosphere. (Do not confuse ρ_a with $\hat{\rho}$ in Eq 8.2-20.) As shown by Hoerner [8.4], the drag coefficient for a 60° cone is 0.50, and that for a sphere is 0.47. Hence a reasonable estimate for the drag coefficient is

$$C_d = 0.5 \tag{8.2-90}$$

The mass density ρ_a of the atmosphere is approximately an exponential function of altitude. Hence, it can be expressed as

$$\rho_a = \rho_{ref} \exp\left[\frac{H_{ref} - H}{H_0}\right] \tag{8.2-91}$$

where H is the altitude above sea level, ρ_{ref} is the mass density of the atmosphere at the reference altitude H_{ref}, and H_0 characterizes the reduction of density with altitude. As shown by Gray [8.3] (pp. 2–137, Table 21-5), the standard ICAO atmosphere gives the following atmospheric mass density at a reference altitude of 20 kilometer:

$$\rho_{ref} = 0.08804 \text{ kg/m}^3 \quad \text{at } H_{ref} = 20.0 \text{ } km \tag{8.2-92}$$

The density at an altitude of 19 km is 0.10307 kg/m³. Comparing the density values at 19 and 20 km altitude gives the following for the parameter H_0:

$$H_0 = 6.34 \text{ kilometer} \tag{8.2-93}$$

Thus, in the vicinity of 20 km altitude, the mass density ρ_a of the atmosphere decreases by the factor $e = 2.72$ for every 6.34 km increase of altitude.

The atmospheric drag force F_d on the vehicle is equal to the atmospheric drag pressure P_d (the wind pressure) multiplied by the cross-section area of the vehicle:

$$F_d = (\text{Area}) \, P_d \tag{8.2-94}$$

where (Area) is the cross-section area of the vehicle in square meters. The drag acceleration a_d of the vehicle produced by this drag force is

$$a_d = \frac{F_d}{M} = \frac{(\text{Area})}{M} P_d = \tfrac{1}{2} C_d \frac{(\text{Area})}{M} \rho_a v^2$$

$$= \tfrac{1}{2} C_d \frac{(\text{Area})}{M} \rho_{ref} \exp\left[\frac{H_{ref} - H}{H_0}\right] v^2 \tag{8.2-95}$$

where M is the vehicle mass in kilograms. Thus, if one knows the ratio of cross section area divided by mass for the vehicle, one can estimate the drag acceleration of the vehicle as a function of altitude H and vehicle velocity v.

The drag coefficient C_d depends on the shape of the vehicle and the orientation of the vehicle relative to its velocity vector. The exact value of this coefficient is obtained from drag tables, which are unique for the particular vehicle. Equation 8.2-95 has the following form:

$$a_d = K_d[H]v^2 \tag{8.2-96}$$

where K_d is a drag parameter that varies with altitude H. This equation is expressed as follows to obtain the acceleration components along the three axes of the radar site:

$$\hat{a}_{d(i)} = -K_d[H]|\hat{v}|\hat{v}_i = -\hat{D}_d\hat{v}_i \quad (i = 1, 2, 3) \tag{8.2-97}$$

where $\hat{D}_d$ is the *drag function*. The variable $\hat{a}_{d(i)}$ is the computed drag acceleration along the i axis of radar-site ENV coordinates, $\hat{v}_i$ is the estimated target velocity along that axis, and $|\hat{v}|$ is the absolute value of target velocity, which is equal to

$$|\hat{v}| = \sqrt{\hat{v}_1^2 + \hat{v}_2^2 + \hat{v}_3^2} \tag{8.2-98}$$

Equation 8.2-97 is implemented in terms of the drag function $\hat{D}_d$, which by Eqs 8.2-95, -97 is

$$\hat{D}_d = K_d[H]|\hat{v}|$$

$$= \left(\frac{1}{2}C_d \frac{(\text{Area})}{M} \rho_{\text{ref}} \exp\left[\frac{H_{\text{ref}} - H}{H_0}\right]\right)|\hat{v}| \tag{8.2-99}$$

8.2.5 Sampled-Data Effects

The signal-flow diagram for the tracking filters in Figs 8.2-3, -5 are low-frequency approximations, which ignore sampled-data effects. Figures 8.2-9, -10 show the sampled-data signal-flow diagrams for the tracking filters used for angle and range tracking in Altair, which are derived in Section 8.5. These sampled-data signal-flow diagrams are expressed in terms of the complex pseudo-frequency p. as was explained in Chapter 5, the complex pseudo-frequency p is approximately equal to the Laplace complex frequency s for frequencies below $1/4$ of the sampling frequency. These sampled-data transfer functions were derived from the difference equations used in the Altair tracking algorithms by applying the following principles.

The major dynamic computations in the Altair tracking algorithms involve two types of integration formulas: simple integration and trapezoidal integration. The difference equations for these formulas used in the Altair software

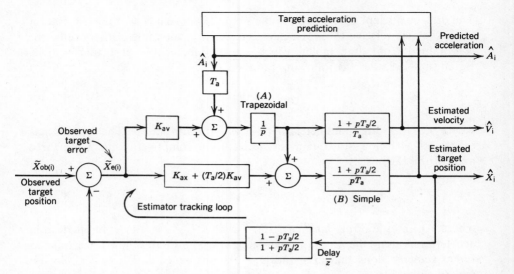

Figure 8.2-9 Sampled-data signal-flow diagram of angle (target estimator) tracking filter.

have the following general forms:

Simple integration:

$$y = y_p + Tx \tag{8.2-100}$$

Trapezoidal integration:

$$y = y_p + \tfrac{1}{2}T(x + x_p) \tag{8.2-101}$$

The variable x is the input signal for the computation step, and y is the output signal. The variables x_p, y_p are the past values of x, y computed in the previous computer cycle. The Laplace transforms of Eqs 8.2-100, -101 are

Simple integration:

$$Y = \bar{z}Y + Tx \tag{8.2-102}$$

Trapezoidal integration:

$$Y = \bar{z}Y + \tfrac{1}{2}T(X + \bar{z}X) \tag{8.2-103}$$

where $\bar{z}$ is equal to e^{-sT}, which is the transfer function for a one-cycle time delay. Solving for the transfer functions Y/X gives

Simple integration:

$$\frac{Y}{X} = \frac{T}{1 - \bar{z}} \tag{8.2-104}$$

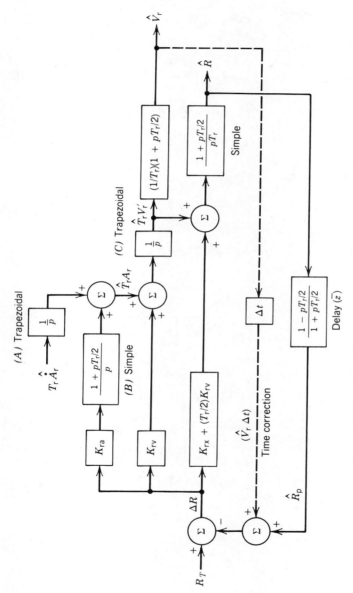

Figure 8.2-10 Sampled-data signal-flow diagram of range tracking filter.

Trapezoidal integration:

$$\frac{Y}{X} = \frac{T(1 + \bar{z})}{1 - \bar{z}} \tag{8.2-105}$$

The variable $\bar{z}$ is related to U by

$$\bar{z} = \frac{1 - U}{1 + U} \tag{8.2-106}$$

The variable U is proportional to the complex pseudo-frequency p:

$$U = \frac{pT}{2} \tag{8.2-107}$$

Combining Eqs 8.2-106, -107 gives

$$\bar{z} = \frac{1 - pT/2}{1 + pT/2} \tag{8.2-108}$$

Substituting Eq 8.2-106 into Eq 8.2-104, -105 gives

Simple integration:

$$\frac{Y}{X} = \frac{T}{2\left[1 - \dfrac{1 - U}{1 + U}\right]} = \frac{T(1 + U)}{2U} \tag{8.2-109}$$

Trapezoidal integration:

$$\frac{Y}{X} = \frac{T\left[1 + \dfrac{1 - U}{1 + U}\right]}{2\left[1 - \dfrac{1 - U}{1 + U}\right]} = \frac{T(2)}{2(2U)} = \frac{T}{2U} \tag{8.2-110}$$

Set U equal to $pT/2$, in accordance with Eq 8.2-107. Equations 8.2-109, -110 become

Simple integration:

$$\frac{Y}{X} = \frac{T(1 + U)}{2U} = \frac{1 + pT/2}{p} \tag{8.2-111}$$

Trapezoidal integration:

$$\frac{Y}{X} = \frac{T}{2U} = \frac{1}{p} \tag{8.2-112}$$

These are the transfer functions given in the signal-flow diagrams of Figs 8.2-9, -10 for the two integration formulas. At frequencies below $1/4$ of the sampling frequency, p is approximately equal to s, and these transfer functions become approximately

Simple integration:

$$\frac{Y}{X} \cong \frac{1 + sT/2}{s} \tag{8.2-113}$$

Trapezoidal integration:

$$\frac{Y}{X} \cong \frac{1}{s} \tag{8.2-114}$$

The signal-flow diagram of the radar tracking system, including sampled-data effects, is shown in Fig 8.2-11. The estimated rate of range-acceleration $\dot{\hat{A}}_{r(a)}$ is computed by differentiating the estimated range acceleration $\hat{A}_{r(a)}$, obtained from the target estimator. The difference equation for this differentiation is

$$\dot{\hat{a}}_{r(a)} = \frac{\hat{a}_{r(a)} - \hat{a}_{r(a)p}}{T_a} \tag{8.2-115}$$

where $\hat{a}_{r(a)p}$ is the past value of $\hat{a}_{r(a)}$ computed in the previous cycle. The Laplace transform of Eq 8.2-115 is

$$\dot{\hat{A}}_{r(a)} = \frac{\hat{A}_{r(a)} - \bar{z}\hat{A}_{r(a)}}{T_a} = \frac{\hat{A}_{r(a)}(1 - \bar{z})}{T_a} \tag{8.2-116}$$

where $\dot{\hat{A}}_r$, $\hat{A}_{r(a)}$ are the Laplace transforms of $\dot{\hat{a}}_r$, $\hat{a}_{r(a)}$. Solving for the transfer function $\dot{\hat{A}}_r/\hat{A}_{r(a)}$ yields

$$\frac{\dot{\hat{A}}_r}{\hat{A}_{r(a)}} = \frac{1 - \bar{z}}{T_a} = \frac{\left[1 - \dfrac{1 - U}{1 + U}\right]}{T_a} = \frac{2U}{T_a(1 + U)} \tag{8.2-117}$$

where $\bar{z}$ was replaced by $(1 - U)/(1 + U)$ in accordance with Eq 8.2-106. Setting U equal to $pT/2$, in accordance with Eq 8.2-107, gives

$$\frac{\dot{\hat{A}}_r}{\hat{A}_{r(a)}} = \frac{2U}{T_a(1 + U)} = \frac{p}{1 + pT_a/2} \tag{8.2-118}$$

Thus, this transfer function approximate $s/(1 + sT_a/2)$ at frequencies below $1/4$ of the sampling frequency.

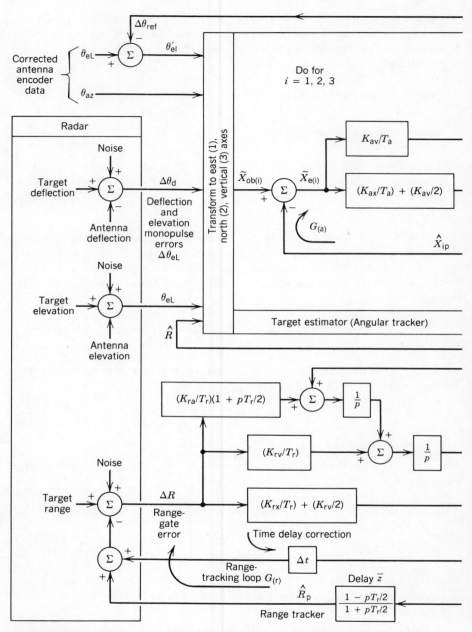

Figure 8.2-11 Sampled-data signal-flow diagram of radar tracking model, which uses Altair tracking filters.

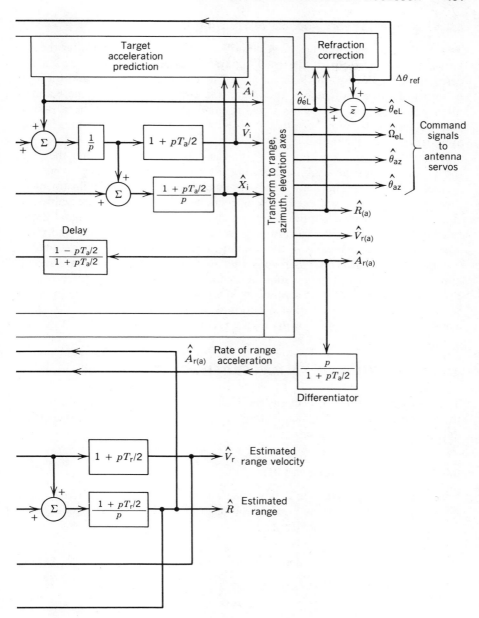

Figure 8.2-11   *(Continued)*

8.2.6   Kalman Theory Aspects of Tracking Configuration

The tracking system that has been described is based on Kalman filter theory. However, the portions of the tracking system presented in this section are certainly not unique to the Kalman filter. The concept of filtering and tracking in terms of the coordinates of the target was around long before the Kalman filter was invented. The same tracking configuration that has been shown could have evolved independently of Kalman filter theory, from a direct application of conventional concepts.

The aspect of this tracking system that is unique to Kalman filter theory is the optimal adjustment of the tracking filter gain (or smoothing) constants K_{ax}, K_{av}, K_{rx}, K_{rv}. The theory and the equations for optimal adjustment of these gain constants are presented in Sections 8.3 and 8.4. These gain constants are varied in an optimal manner as functions of (1) the signal/noise ratio in the radar receiver (determined from the strength of the received radar signal) and of (2) estimates of the unpredictable portion of target motion.

The fact that these filter gain constants are varied with radar signal strength and with estimates of target motion is not unique to Kalman filter theory. Prior to that theory, it was obvious to most individuals working in radar tracking systems that such gain variation is desirable, particularly for tracking exoatmospheric targets following highly predictable trajectories.

8.3   DESCRIPTION OF KALMAN FILTER EQUATIONS

The following discussion explains the Kalman filter equations and shows how they are applied. The reader is referred to Gelb [8.5] for a mathematical derivation of these equations.

8.3.1   Continuous Kalman Filter

Figure 8.3-1 is a matrix signal-flow diagram of the continuous Kalman filter system. This is the same as Fig 4.3-1 on page 123 of Gelb [8.5], except for the following changes of symbolism:

$$\underline{w} \rightarrow \underline{u} \qquad (8.3\text{-}1)$$

$$\underline{v} \rightarrow \underline{n} \qquad (8.3\text{-}2)$$

$$\underline{v} \rightarrow \underline{\mathscr{E}} \qquad (8.3\text{-}3)$$

The symbol $\hat{\underline{z}}$ in the diagram is defined by the author to represent a variable not specifically labeled by Gelb. The blocks labeled $\underline{G}$, $\underline{F}$, $\underline{H}$, and $\underline{K}$ are matrices. The matrices as well as the vector signals may vary with time, but for simplicity the time variation is ignored in the symbols (e.g., $\underline{F}$ is not represented as $\underline{F}[t]$).

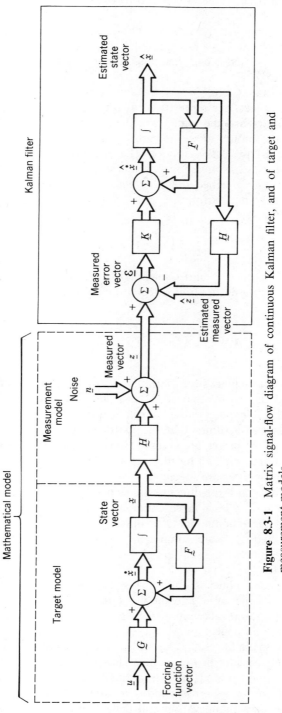

Figure 8.3-1 Matrix signal-flow diagram of continuous Kalman filter, and of target and measurement models.

483

It is assumed that the dynamic equations of the target are known. The target model shown in Fig 8.3-1 expresses these dynamic equations in state variable form. This model has the same configuration as that illustrated in Chapter 5, Fig 5.5-7.

To obtain the state equations, the dynamic system equations are written so that all dynamic effects are represented as integrators. The outputs from the resultant integrators are the state variables (indicated in Fig 8.3-1 by the vector $\underline{x}$), and the inputs to the integrators are the derivatives of the state variables (indicated by the vector $\underline{\dot{x}}$). The equations that express the state variable derivatives $\underline{\dot{x}}$ as functions of the state variables $\underline{x}$ and the input forcing functions $\underline{u}$ are called the *state equations*. These state equations are written as follows in matrix format:

$$\underline{\dot{x}} = \underline{F}\underline{x} + \underline{G}\underline{u} \qquad (8.3\text{-}4)$$

The elements in matrices $\underline{F}$ and $\underline{G}$ are the coefficients of the state equations. The integrations represent a series of equations of the form

$$x_n = \int \dot{x}_n \, dt \qquad (8.3\text{-}5)$$

where x_n is a particular state variable of the vector $\underline{x}$ and $\dot{x}_n$ is the corresponding derivative of that state variable. This method of representing the equations of a dynamic system is explained in Chapter 5, Section 5.5.

The Kalman filter assumes that the matrices $\underline{F}, \underline{G}$, which define the dynamic response of the target, are known. However, the forcing signal inputs to the target system, represented by the vector $\underline{u}$, are not known explicitly; only their statistical properties are specified.

The target moves in three dimensions. Hence, the state vector of the target model generally has at least three components to describe position in three dimensions plus three components to describe velocity in three dimensions, and it may have three more components to describe acceleration in three dimensions. These are the usual components of the state vector, but many more are possible.

The measurement model in Fig 8.3-1 shows the relationship between the state variables $\underline{x}$ of the target and the measurements $\underline{z}$ made on the target by the radar. The radar with its monopulse feed measures the range to the target and the elevation and deflection angular errors between the line-of-sight to the target and the antenna boresight axis. If the radar senses Doppler shift, it also provides a direct measure of range rate. Thus the measurement vector $\underline{z}$ has three or four components: (1) elevation angular error, (2) deflection angular error, (3) range (or range error), and sometimes (4) range rate. It is often convenient to express the elevation and deflection angular tracking errors in terms of equivalent positional errors as follows:

$$\Delta x_{eL} = R \, \Delta\phi_{eL} \tag{8.3-6}$$

$$\Delta x_{def} = R \, \Delta\phi_{def} \tag{8.3-7}$$

The variable R is the range to the target and $\Delta\phi_{eL}$ and $\Delta\phi_{def}$ are the elevation and deflection angular errors. The variables Δx_{eL} and Δx_{def} are positional errors of target location measured perpendicular to the range vector that are equivalent to the elevation and deflection angular errors.

The vector $\underline{n}$ describes the receiver noise contributions that corrupt the radar measurements of the target. The parameters of the target that are measured by the radar are functions of the target state vector $\underline{x}$. The matrix $\underline{H}$ describes the transformation relating the state-vector components $\underline{x}$ of the target to the parameters that are measured by the radar. Many of the elements of the $\underline{H}$ matrix are zero because the radar has no direct measurement of the corresponding target state variable. In particular, if the radar does not measure Doppler shift, it has no direct measure of target velocity. For that case, the coefficients of the $\underline{H}$ matrix related to the components of target velocity (or any higher derivatives) are zero.

(The preceding paragraph requires some qualification. The arguments presented are precisely correct only when the sampling period of the Kalman filter calculations is equal to the pulse repetition period of the radar. However, in most tracking systems the error signals used in the Kalman filter equations are derived from many radar pulses. If differential calculations are made, velocity information can be derived from the set of radar pulses received within a single Kalman filter sample even if target Doppler is not measured. On the other hand, this differential calculation is not performed in the Altair radar tracking system, and so the following discussion ignores this qualification.)

The solid block in Fig 8.3-1 describes the signal processing of the Kalman filter, which processes the radar measurement data $\underline{z}$ to obtain the estimated target state vector, which is denoted $\hat{\underline{x}}$. The Kalman filter incorporates the same feedback matrix $\underline{F}$ that characterizes the target model. The estimated state vector $\hat{\underline{x}}$ is measured by the same matrix $\underline{H}$ that characterizes the radar measurements of the target. If the estimated state vector $\hat{\underline{x}}$ were exactly the same as the target state vector $\underline{x}$ and the measurement noise $\underline{n}$ were zero, the estimated measurement vector $\hat{\underline{z}}$ would be exactly equal to the actual measurement vector $\underline{z}$. The difference between the actual measurement vector $\underline{z}$ and the estimated measurement vector $\hat{\underline{z}}$ is the measurement error $\underline{\mathscr{E}}$:

$$\underline{\mathscr{E}} = \underline{z} - \hat{\underline{z}} \tag{8.3-8}$$

Assuming the radar does not measure Doppler, there are three components of this error vector that correspond to errors in range, elevation angle, and deflection angle.

The error vector $\underset{\sim}{\mathscr{E}}$ is multiplied by the gain matrix $\underset{\sim}{K}$ to develop correction components. These are added to the derivatives of the estimated state vector represented by the vector $\dot{\underset{\sim}{x}}$. The corrections continually change the estimated state vector $\hat{\underset{\sim}{x}}$ so that its components ideally follow those of the actual target state vector $\underset{\sim}{x}$.

The elements of the gain matrix $\underset{\sim}{K}$ are varied optimally, based on a knowledge of the statistical properties of the forcing vector $\underset{\sim}{u}$ applied to the target and the statistical properties of the measurement noise $\underset{\sim}{n}$. Let us examine the manner in which the components of these vectors $\underset{\sim}{u}$ and $\underset{\sim}{n}$ are characterized statistically.

In Section 6.4 Eq 6.4-12 showed that the torque exerted by wind on an antenna can be described statistically by the following normalized spectrum:

$$\phi_{\text{wt(n)}}[\omega] = \frac{(2/\pi)\,\omega_1\omega_2(\omega_1 + \omega_2)}{\left(\omega^2 + \omega_1^2\right)\left(\omega^2 + \omega_2^2\right)} \tag{8.3-9}$$

where $\omega_i = 0.12 \text{ sec}^{-1}$ and $\omega_2 = 2 \text{ sec}^{-1}$. To simulate a wind torque disturbance with this spectrum, one can feed the signal from a white noise source through a lowpass filter having the following transfer function:

$$H[s] = \frac{1}{(1 + s/\omega_1)(1 + s/\omega_2)} \tag{8.3-10}$$

Thus, the wind torque disturbance can be described statistically in terms of this lowpass filter $H[s]$ and the amplitude of the input white noise.

This same philosophy is used to characterize statistically the input signals $\underset{\sim}{u}$ applied to the target model. Each component of the input signal vector $\underset{\sim}{u}$ represents the amplitude of a white noise signal. The dynamic properties of the filters (such as $H[s]$ of Eq 8.3-10) that establish the spectra of the forcing signals experienced by the target are incorporated into the state equations of the target. This is achieved by adding more components to the target state vector $\underset{\sim}{x}$ and by increasing the order of the $\underset{\sim}{F}$ matrix accordingly.

Pure white noise is conceptually difficult to describe because its RMS value is infinite. However, pure white noise can be accurately approximated by band-limited white noise having a bandwidth that is much wider than that of the system to which the noise is applied Obviously, practical white noise experiments always use band-limited white noise. If band-limited white noise has a noise bandwidth $f_{\text{b(w)}}$ and an RMS value σ_n, the noise power density is

$$\frac{dP}{df} = \frac{\sigma_n^2}{f_{\text{b(w)}}} \tag{8.3-11}$$

Thus, the components of the input white noise vector $\underset{\sim}{u}$ can be described in terms of the RMS values σ_n of the individual band-limited white noise signals and their noise bandwidth. If the target model is appropriately defined, it is always possible to describe the inputs in terms of white noise signals that are statistically independent. However, the model may sometimes be simplified if the white noise inputs contain components that are cross-correlated. In that case, the covariances between the white noise signals must be defined as well as the RMS values.

The statistical properties of the band-limited white noise signals that characterize the input forcing vector $\underline{u}$ are described by a covariance matrix, which is denoted Q. For example, if the forcing vector $\underline{u}$ has three components, the covariance matrix $\underset{\sim}{Q}$ has the form

$$\underset{\sim}{Q} = \begin{bmatrix} \langle u_1^2 \rangle & \langle u_1 u_2 \rangle & \langle u_1 u_3 \rangle \\ \langle u_2 u_1 \rangle & \langle u_2^2 \rangle & \langle u_2 u_3 \rangle \\ \langle u_3 u_1 \rangle & \langle u_3 u_2 \rangle & \langle u_3^2 \rangle \end{bmatrix} \tag{8.3-12}$$

The values $\langle u_1^2 \rangle, \langle u_2^2 \rangle, \langle u_3^2 \rangle$ are the mean-square values of the band-limited white noise signals u_1, u_2, u_3, which are the averages of the squares of the signals. The quantities $\langle u_1 u_2 \rangle, \langle u_1 u_3 \rangle, \langle u_2 u_3 \rangle$ are the covariances of these signals, which are the averages of the signal products: $u_1 u_2$, $u_1 u_3$, and $u_2 u_3$. For example, the parameters $\langle u_1^2 \rangle$ and $\langle u_1 u_2 \rangle$ are calculated by the following averaging integrations:

$$\langle u_1^2 \rangle = \frac{1}{T} \int_0^T u_1[t]^2 \, dt \tag{8.3-13}$$

$$\langle u_1 u_2 \rangle = \frac{1}{T} \int_0^T u_1[t] u_2[t] \, dt \tag{8.3-14}$$

where T is a long time period over which the averaging is performed. If the forcing inputs u_1, u_2, u_3 are statistically independent, the cross-correlation quantities, such as $\langle u_1 u_2 \rangle$, are all zero. In that case, the $\underset{\sim}{Q}$-matrix has elements only along the diagonal, which are equal to the mean-square values of the signals.

The system model for the Kalman filter has another input signal vector, denoted $\underline{n}$, which represents the white noise contributions in the radar measurements of target position. The measurement noise is characterized by a covariance matrix denoted $\underset{\sim}{R}$, which is equivalent to the matrix $\underset{\sim}{Q}$ that characterizes the target forcing signals. However, the components of the noise vector $\underline{n}$ are always uncorrelated in Kalman filter theory. Assuming three independent radar measurements (range, elevation error, and deflection error), the noise vector $\underline{n}$ has three components, and the covariance matrix $\underset{\sim}{R}$ that characterizes this noise has the form

$$\underset{\sim}{R} = \begin{bmatrix} \langle n_1^2 \rangle & 0 & 0 \\ 0 & \langle n_2^2 \rangle & 0 \\ 0 & 0 & \langle n_3^2 \rangle \end{bmatrix} \tag{8.3-15}$$

The elements $\langle n_1^2 \rangle, \langle n_2^2 \rangle, \langle n_3^2 \rangle$ are the squares of the RMS values (the mean-square values, or variances) of the band-limited white noise components for the three measurements.

To calculate the optimum parameters of the $\underset{\sim}{K}$ gain matrix in the Kalman filter, the Kalman filter derives an estimate of the statistical parameters of the errors between the components of the estimated state vector $\hat{x}$ and the corresponding components of the actual state vector $\underset{\sim}{x}$ of the target. These statistical parameters are described in terms of an error covariance matrix $\underset{\sim}{P}$, of the form

$$
\underset{\sim}{P} = \begin{bmatrix}
\sigma_{11}^2 & \sigma_{12}^2 & \sigma_{13}^2 & \cdots \\
\sigma_{21}^2 & \sigma_{22}^2 & \sigma_{23}^2 & \cdots \\
\sigma_{31}^2 & \sigma_{32}^2 & \sigma_{33}^2 & \cdots \\
\vdots & \vdots & \vdots & \vdots
\end{bmatrix} \tag{8.3-16}
$$

The elements $\sigma_{11}^2, \sigma_{22}^2, \sigma_{33}^2$ are the squares of the RMS values of error in the estimates of the components x_1, x_2, x_3 of the target state vector $\underset{\sim}{x}$. These are called the mean-square error values, or variances. The parameters $\sigma_{12}^2, \sigma_{13}^2, \ldots$ are the covariances of these estimated errors. For example, σ_{12}^2 is ideally equal to

$$
\sigma_{12}^2 = \frac{1}{T} \int_0^T (\hat{x}_1 - x_1)(\hat{x}_2 - x_2) \, dt \tag{8.3-17}
$$

These components of the $\underset{\sim}{P}$ error covariance matrix are not the true statistical parameters of the error; they are merely estimates of those parameters.

The $\underset{\sim}{P}$ error covariance matrix can become quite large if it is rigorously applied to all of the components of the state vector. As will be shown, the calculations that compute the gain matrix require two inversions of the $\underset{\sim}{P}$ matrix and so are extremely complicated when the $\underset{\sim}{P}$ matrix is large. In Altair this problem is circumvented by simplifying the $\underset{\sim}{P}$ matrix such that all of the position components of the state vector are reduced to a single position component in the $\underset{\sim}{P}$ matrix; and similarly for the velocity components. By this means, a 6×6 covariance $\underset{\sim}{P}$ matrix is reduced to a 2×2 covariance matrix.

Now let us examine the matrix equations that describe the continuous Kalman filter. The following equations (with some changes of symbolism) are obtained from Table 4.3-1 on page 123 of Gelb [8.5] except that the equation for $\dot{\underset{\sim}{P}}$ is obtained from Eq 4.3-8 on page 122 of Gelb.

System model:

$$
\dot{\underset{\sim}{x}} = \underset{\sim}{F}\underset{\sim}{x} + \underset{\sim}{G}\underset{\sim}{u} \tag{8.3-18}
$$

Measurement model:

$$
\underset{\sim}{z} = \underset{\sim}{H}\underset{\sim}{x} + \underset{\sim}{n} \tag{8.3-19}
$$

Calculation of estimated state vector:

$$\dot{\hat{x}} = F\hat{x} + K(z - H\hat{x}) \qquad (8.3\text{-}20)$$

Calculation of estimated error covariance:

$$\dot{P} = FP + PF^{\mathrm{T}} + GQG^{\mathrm{T}} - PHR^{-1}HP \qquad (8.3\text{-}21)$$

Calculation of Kalman gain matrix:

$$K = PH^{\mathrm{T}}R^{-1} \qquad (8.3\text{-}22)$$

All of the matrices and vectors may vary with time, but for simplicity the functional time relations are omitted from the symbols.

Appendix C summarizes the rules of matrix manipulation and explains the operations indicated in these equations, which include the following: matrix multiplication, taking the transpose of a matrix, and taking the inverse of a matrix. The matrix F^{T} is the transpose of the matrix F, which is formed by interchanging rows and columns in the F matrix. The matrix R^{-1} is the inverse of the matrix R. Inverting a matrix is a very complicated process when the order of the matrix is high. Appendix C gives rules for matrix inversion that are practical for inverting matrices up to fourth order.

Equations 8.3-18 to -20 can be derived directly from the signal-flow diagram of Fig 8.3-1. Equations 8.3-21, -22 show how the covariance matrix P of the estimation errors can be computed and how the Kalman gain matrix K can be derived from the covariance matrix P. These continuous equations help to explain the principles of the Kalman filter. On the other hand, to develop the actual equations for calculating the gain matrix by digital computation, it is more convenient to use the discrete (or sampled-data) form of the Kalman filter.

Equation 8.3-21 is a rather complicated matrix equation that expresses the $\dot{P}$ derivatives of the elements of the covariance matrix P as functions of the P elements and the elements of other matrices of the system. These include the system model matrices F and G, the measurement matrix H, and the covariance matrices Q and R, which describe the statistical properties of the target forcing signals and the measurement noise. The elements of the P covariance matrix can be calculated by integrating these $\dot{P}$ values. Thus, in principle, these equations can be solved in a feedback manner to obtain the elements of the P covariance matrix and the elements of the K gain matrix.

8.3.2 Sampled-Data (Discrete) Kalman Filter

Figure 8.3-2 is a signal-flow diagram of the sampled-data (or discrete) form of the Kalman filter calculations. As was explained in Chapter 5, Section 5.5.4,

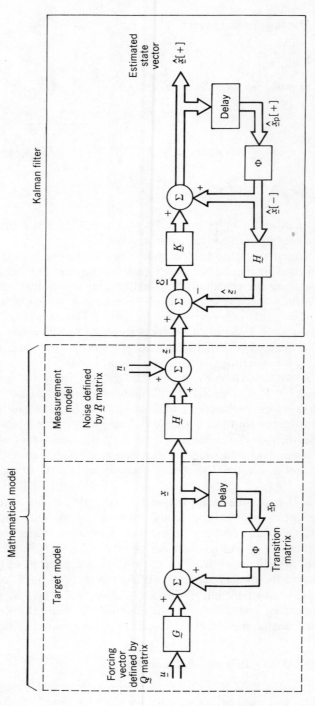

Figure 8.3-2 Matrix signal-flow diagram of sampled-data (discrete) Kalman filter, and of target and measurement models.

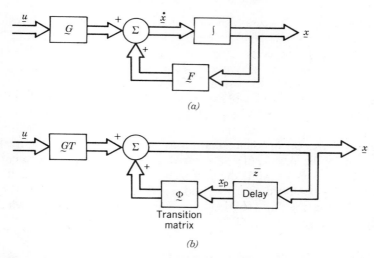

(a)

(b)

Figure 8.3-3 Conversion of continuous matrix model of a system to sampled-data form: (*a*) signal-flow diagram of model; (*b*) corresponding sampled-data matrix.

the matrix signal-flow diagram for a continuous system can be converted to sampled-data form as shown in Fig 8.3-3. Diagram *a* is the target model portion of Fig 8.3-1. Diagram *b* is the corresponding matrix signal-flow diagram for implementing these equations with sampled-data calculations. The matrix $\underset{\sim}{F}$ of the state equations is replaced by the transition matrix $\underset{\sim}{\Phi}$. By Eq 9.4-58, the transition matrix $\underset{\sim}{\Phi}$ is related to the state equation matrix $\underset{\sim}{F}$ by

$$\underset{\sim}{\Phi} = \underset{\sim}{I} + T\underset{\sim}{F} + \frac{(T\underset{\sim}{F})^2}{2!} + \frac{(T\underset{\sim}{F})^3}{3!} + \ldots \qquad (8.3\text{-}23)$$

where $\underset{\sim}{I}$ is an identity matrix and T is the sample period. In Kalman filter designs, the transition matrix $\underset{\sim}{\Phi}$ is often approximated by the first two terms of this expansion:

$$\underset{\sim}{\Phi} \cong \underset{\sim}{I} + \underset{\sim}{F}T \qquad (8.3\text{-}24)$$

When the principle of Fig 8.3-3 is applied to the continuous form of the Kalman filter system given in Fig 8.3-1, the sampled-data form shown in Fig 8.3-2 results. In making this conversion, the matrix block $\underset{\sim}{G}$ should be multiplied by the sample period T. However, this is avoided by redefining the covariance matrix $\underset{\sim}{Q}$ of the $\underset{\sim}{u}$ forcing signals to include the T constant.

In the continuous signal-flow diagram of Fig 8.3-1, the estimated measurement vector $\hat{\underset{\sim}{z}}$ is derived directly from the estimated state vector $\hat{\underset{\sim}{x}}$ by multiplying the $\hat{\underset{\sim}{x}}$ vector by the measurement matrix $\underset{\sim}{H}$. In the sampled-data design of Fig 8.3-2, the estimated state vector $\hat{\underset{\sim}{x}}$ is represented as $\hat{\underset{\sim}{x}}[+]$. The estimated measurement vector $\hat{\underline{z}}$ is not derived directly from $\hat{\underline{x}}[+]$. Instead, it is derived from the vector labeled $\hat{\underline{x}}[-]$, which is an approximation of $\hat{\underline{x}}[+]$. If the measurement error vector $\underline{\mathscr{E}}$ were zero, the vectors $\hat{\underline{x}}[+]$ and $\hat{\underline{x}}[-]$ would be equal. The reason for deriving $\hat{\underline{z}}$ from $\hat{\underline{x}}[-]$ rather than from $\hat{\underline{x}}[+]$ is beyond the scope of this discussion.

Except for changes of variable symbolism, Fig 8.3-2 is essentially the same as Fig 4.2-2 given in Gelb [8.5] (p. 117) for the discrete Kalman filter system. However, Gelb has omitted the $\underline{G}$ matrix by appropriately defining the parameters of the $\underline{Q}$ covariance matrix. In our diagram the subscript (p) is used to define the past value of a variable: for example, the present state vector is denoted $\hat{\underline{x}}[+]$, while the past value is denoted $\hat{\underline{x}}_p[+]$. In Gelb, these two vectors are expressed as $\hat{\underline{x}}_k[+]$ and $\hat{\underline{x}}_{k-1}[+]$, respectively.

The following are the sampled-data Kalman filter equations, which are equivalent to those given by Gelb:

System model:

$$\underline{x} = \underline{\Phi}\underline{x}_p + \underline{G}\underline{u} \qquad (8.3\text{-}25)$$

Measurement model:

$$\underline{z} = \underline{H}\underline{x} + \underline{n} \qquad (8.3\text{-}26)$$

Calculation of estimated state vector:

$$\hat{\underline{x}}[-] = \underline{\Phi}\hat{\underline{x}}_p[+] \qquad (8.3\text{-}27)$$

$$\hat{\underline{x}}[+] = \hat{\underline{x}}[-] + \underline{K}(\underline{z} - \underline{H}\hat{\underline{x}}[-]) \qquad (8.3\text{-}28)$$

Calculation of estimated error covariance:

$$\underline{P}[-] = \underline{\Phi}\underline{P}_p[+]\underline{\Phi}^\mathrm{T} + \underline{G}\underline{Q}\underline{G}^\mathrm{T} \qquad (8.3\text{-}29)$$

$$\underline{P}[+]^{-1} = \underline{P}[-]^{-1} + \underline{H}^\mathrm{T}\underline{R}^{-1}\underline{H} \qquad (8.3\text{-}30)$$

Calculation of Kalman gain matrix:

$$\underline{K} = \underline{P}[+]\underline{H}^\mathrm{T}\underline{R}^{-1} \qquad (8.3\text{-}31)$$

Equations 8.3-31, -32 are equivalent to Eqs 4.2-19, 4.2-20 given by Gelb [8.5] on pages 111, 112. The other equations are equivalent to those given by Gelb in Table 4.2-1. Note, however, that Gelb does not include the matrix $\underline{G}$ because $\underline{G}$ is defined as unity in the discrete Kalman model given by Gelb.

In the discrete (sampled-data) calculations, the white noise signals characterized by the Q and R matrices are replaced by random number sequences having a Gaussian distribution. The random numbers of the digital sampled-data calculations are equivalent to the corresponding white noise signals filtered with an integration time equal to the sample period T. (Remember that an integration time T corresponds to a noise bandwidth in hertz equal to $1/2T$.) The RMS value σ_s of a sampled-data digital variable is related as follows to the RMS value σ_w of the corresponding band-limited white noise:

$$\sigma_s^2 = \frac{\sigma_w^2(1/2T)}{f_{b(w)}} = \frac{\sigma_w^2}{2Tf_{b(w)}} \tag{8.3-32}$$

where $f_{b(w)}$ is the noise bandwidth of the band-limited white noise.

The implementation of these sampled-data filter equations is explained in Section 8.4 by showing how they are applied in the Altair radar tracking system.

8.4 CALCULATION OF KALMAN-FILTER SMOOTHING GAINS IN ALTAIR

This section derives the equations implemented in the Altair radar tracking system to calculate the K_x, K_v gains of the Kalman tracking filters. This analysis is based on an internal Altair memorandum by D. A. Cassell dated October 1967.

8.4.1 Equations for Target Model

A difficult practical problem encountered in implementing Kalman filter theory is the formulation of a realistic target model. The Altair Kalman tracker was designed by D. E. Johansen [8.6] and was based on a concept developed in his Ph.D. thesis research. He concluded that the best way to characterize target motion in the Altair application is to define an uncertainty bound for the estimate of target acceleration. Within the uncertainty bound, the acceleration is assumed to change unpredictably in a manner that maximizes tracking error.

The Altair tracking program calculates the accelerations of the target in free space that are due to coriolis and gravitational effects. During reentry, the program estimates the drag force on the target and calculates the resultant acceleration. There are errors in these computations of target acceleration. The bound Δa on these acceleration errors is calculated from the following equation:

$$\Delta a = \Delta a_{\text{g}} + \varepsilon \hat{D}_{\text{d}} |\hat{v}|$$ (8.4-1)

where

Δa = estimated bound on acceleration uncertainty
Δa_{g} = estimated error in the models of gravitation and coriolis acceleration
$\hat{D}_{\text{d}}$ = calculated drag factor of the vehicle, which varies with the estimated vehicle altitude above the earth and with target speed
$\hat{v}$ = estimated velocity of the vehicle
ε = assumed relative error bound on the drag factor

The acceleration uncertainty is represented as a_{e}. The quantity Δa is the bound on the magnitude of a_{e}.

The assumption of uncertainty in target acceleration yields the following sampled-data equations that describe the unpredictable portions of the position and velocity state variables of the target:

$$x = x_{\text{p}} + Tv_{\text{p}} + \tfrac{1}{2}T^2 a_{\text{e}}$$ (8.4-2)

$$v = v_{\text{p}} + Ta_{\text{e}}$$ (8.4-3)

where x_{p}, v_{p} are the past values of x, v. Equations 8.4-2, -3 can be expressed in matrix form as follows:

$$\begin{bmatrix} x \\ v \end{bmatrix} = \begin{bmatrix} 1 & T \\ 0 & 1 \end{bmatrix} \begin{bmatrix} x_{\text{p}} \\ v_{\text{p}} \end{bmatrix} + \begin{bmatrix} \tfrac{1}{2}T^2 \\ T \end{bmatrix} a_{\text{e}}$$ (8.4-4)

This is the assumed matrix equation of the Kalman target model, which by Eq 8.3-25 has the following general form:

$$\underline{x} = \underline{\Phi} \underline{x}_{\text{p}} + \underline{G} \underline{u}$$ (8.4-5)

Hence the matrices $\underline{\Phi}$ and $\underline{G}$ are

$$\underline{\Phi} = \begin{bmatrix} 1 & T \\ 0 & 1 \end{bmatrix}$$ (8.4-6)

$$\underline{G} = \begin{bmatrix} \tfrac{1}{2}T^2 \\ T \end{bmatrix}$$ (8.4-7)

The bound Δa on the acceleration uncertainty a_{e} is treated as if it were the

RMS value of the target forcing signal u. The covariance matrix Q of the target forcing signal has only a single element, which is a mean-square value equal to Δa^2:

$$Q = \Delta a^2 \qquad (8.4\text{-}8)$$

In the actual estimator equations, the estimated state vector $\hat{x}$ has six components: three position components and three velocity components. However, in computing the covariance matrix of estimation error, the state vector is regarded as having a single position component and a single velocity component, as was shown in the matrix expression of Eq 8.4-4. Separate routines are used to calculate the error covariance parameters and gain constants for the angle tracker and the range tracker.

Remember from Eq 8.3-24 that the transition matrix Φ is approximated by $(I + FT)$. To express Φ in this form, expand Eq 8.4-6 as

$$\Phi = \begin{bmatrix} 1 & T \\ 0 & 1 \end{bmatrix} = \begin{bmatrix} 1 & 0 \\ 0 & 1 \end{bmatrix} + \begin{bmatrix} 0 & T \\ 0 & 0 \end{bmatrix} = I + \begin{bmatrix} 0 & T \\ 0 & 0 \end{bmatrix} \qquad (8.4\text{-}9)$$

where I is the identity matrix. Hence the state matrix F for the equivalent continuous system is

$$F = \frac{1}{T} \begin{bmatrix} 0 & T \\ 0 & 0 \end{bmatrix} = \begin{bmatrix} 0 & 1 \\ 0 & 0 \end{bmatrix} \qquad (8.4\text{-}10)$$

When the forcing signal u is zero, the corresponding system equations are

$$\begin{bmatrix} \dot{x} \\ \dot{v} \end{bmatrix} = Fx = \begin{bmatrix} 0 & 1 \\ 0 & 0 \end{bmatrix} \begin{bmatrix} x \\ v \end{bmatrix} \qquad (8.4\text{-}11)$$

The state equations of this model are

$$\dot{x} = v \qquad (8.4\text{-}12)$$
$$\dot{v} = 0 \qquad (8.4\text{-}13)$$

Since the radar measures only the position of the target (not its velocity), the matrix equation for the measurement process is

$$z = Hx + n = \begin{bmatrix} 1 & 0 \end{bmatrix} \begin{bmatrix} x \\ v \end{bmatrix} + n \qquad (8.4\text{-}14)$$

The measurement matrix H is

$$H = \begin{bmatrix} 1 & 0 \end{bmatrix} \qquad (8.4\text{-}15)$$

Although the measurement vector z actually has three dimensions, it is approximated as a single position component when calculating the Kalman gains. Hence, the covariance matrix $\underset{\sim}{R}$ describing the measurement noise $\underset{\sim}{n}$ has a single element and is expressed as

$$\underset{\sim}{R} = \sigma_n^2 \qquad (8.4\text{-}16)$$

To calculate the Kalman gains for the range tracker, the variable σ_n is set equal to the RMS range error σR per computational sample of the range tracker, which is equal to

$$\sigma_n = \sigma R = \sigma R_1 \sqrt{T_r/T_p} \qquad (8.4\text{-}17)$$

The variable σR_1 is the RMS range-tracking error per radar pulse, T_p is the radar pulse repetition period, and T_r is the sample period of the Altair range tracker. As was described in Eqs 8.3-6, -7, the angular tracking errors are expressed in terms of equivalent positional errors by multiplying the angular errors by the range R. In calculating the Kalman gains for the angle tracker (or target estimator), the variable σ_n is set equal to $R\sigma\theta$, where $\sigma\theta$ is the RMS angular error per computation sample. Thus, for angle tracking σ_n is equal to

$$\sigma_n = R\sigma\theta = R\sigma\theta_1\sqrt{T_a/T_p} \qquad (8.4\text{-}18)$$

The variable $\sigma\theta_1$ is the RMS angular tracking error per pulse, and T_a is the computational sample period for angle tracking.

8.4.2 Calculation of Elements of Covariance Matrices

The covariance matrices $\underset{\sim}{P_0}$, $\underset{\sim}{P_1}$, and $\underset{\sim}{P_2}$ are defined as follows in terms of the matrix variables given in Eqs 8.3-29, -30:

$$\underset{\sim}{P_2} = \underset{\sim}{P}[+] \qquad (8.4\text{-}19)$$

$$\underset{\sim}{P_1} = \underset{\sim}{P}[-] \qquad (8.4\text{-}20)$$

$$\underset{\sim}{P_0} = \underset{\sim}{\Phi}\underset{\sim}{P_p}[+]\underset{\sim}{\Phi}^T = \underset{\sim}{\Phi}\underset{\sim}{P_{2p}}\underset{\sim}{\Phi}^T \qquad (8.4\text{-}21)$$

Remember that $\underset{\sim}{P_p}[+]$ is a matrix with elements that are the past values of the elements in $\underset{\sim}{P}[+]$. Similarly, matrix $\underset{\sim}{P_{2p}}$ contains the past value of matrix $\underset{\sim}{P_2}$. The matrix $\underset{\sim}{P_0}$ is the first term of Eq 8.3-29. Hence, Eqs 8.3-29, -30 become

$$\underset{\sim}{P_1} = \underset{\sim}{P_0} + \underset{\sim}{G}\underset{\sim}{Q}\underset{\sim}{G}^T \qquad (8.4\text{-}22)$$

$$\underset{\sim}{P_2}^{-1} = \underset{\sim}{P_1}^{-1} + \underset{\sim}{H}^T\underset{\sim}{R}^{-1}\underset{\sim}{H} \qquad (8.4\text{-}23)$$

In calculating the error covariance matrix, the state vector is assumed to have only a single position component and a single velocity component. Hence the error covariance matrices $\underset{\sim}{P_0}$, $\underset{\sim}{P_1}$, and $\underset{\sim}{P_2}$ are 2 × 2 matrices.

The elements of the covariance matrices have a large dynamic range. To reduce this range, the P matrices are factored in the following manner. The P_2 matrix is factored in terms of a matrix M defined as

$$M = \begin{bmatrix} M_1 & M_2 \\ 0 & M_3 \end{bmatrix} \qquad (8.4\text{-}24)$$

The P_2 matrix is defined as

$$P_2 = MM^T = \begin{bmatrix} M_1 & M_2 \\ 0 & M_3 \end{bmatrix} \begin{bmatrix} M_1 & 0 \\ M_2 & M_3 \end{bmatrix} = \begin{bmatrix} (M_1^2 + M_2^2) & M_2 M_3 \\ M_2 M_3 & M_3^2 \end{bmatrix} \qquad (8.4\text{-}25)$$

(Remember that M^T is the transpose of the matrix M. The transpose is formed by interchanging the rows and columns of the matrix, which is equivalent to rotating the matrix about its diagonal. The rules for taking the transpose of a matrix and for multiplying matrices are summarized in Appendix C.) In like manner the P_1 matrix is factored in terms of a matrix N:

$$N = \begin{bmatrix} N_1 & N_2 \\ 0 & N_3 \end{bmatrix} \qquad (8.4\text{-}26)$$

The P_1 matrix is defined as

$$P_1 = NN^T = \begin{bmatrix} N_1 & N_2 \\ 0 & N_3 \end{bmatrix} \begin{bmatrix} N_1 & 0 \\ N_2 & N_3 \end{bmatrix} = \begin{bmatrix} (N_1^2 + N_2^2) & N_2 N_3 \\ N_2 N_3 & N_3^2 \end{bmatrix} \qquad (8.4\text{-}27)$$

The P_0 matrix is factored in terms of a matrix M_0:

$$M_0 = \begin{bmatrix} M_{01} & M_{02} \\ 0 & M_{03} \end{bmatrix} \qquad (8.4\text{-}28)$$

The P_0 matrix is defined as

$$P_0 = M_0 M_0^T = \begin{bmatrix} M_{01} & M_{02} \\ 0 & M_{03} \end{bmatrix} \begin{bmatrix} M_{01} & 0 \\ M_{02} & M_{03} \end{bmatrix}$$

$$= \begin{bmatrix} (M_{01}^2 + M_{02}^2) & M_{02} M_{03} \\ M_{02} M_{03} & M_{03}^2 \end{bmatrix} \qquad (8.4\text{-}29)$$

In Eq 8.4-21, the P_0 matrix is expressed in terms of the past value of the P_2 matrix. Substitute into Eq 8.4-21 the expression for P_2 in Eq 8.4-25 and the expression for the transition matrix $\underset{\sim}{\Phi}$ given in Eq 8.4-9. This gives

$$\underset{\sim}{P}_0 = \underset{\sim}{\Phi}\underset{\sim}{P}_{2p}\underset{\sim}{\Phi}^T$$

$$= \begin{bmatrix} 1 & T \\ 0 & 1 \end{bmatrix} \begin{bmatrix} M_{1p} & M_{2p} \\ 0 & M_{3p} \end{bmatrix} \begin{bmatrix} M_{1p} & 0 \\ M_{2p} & M_{3p} \end{bmatrix} \begin{bmatrix} 1 & 0 \\ T & 1 \end{bmatrix} \qquad (8.4\text{-}30)$$

The subscript p was added to all of the M constants of the $\underset{\sim}{P}_2$ matrix to obtain the $\underset{\sim}{P}_{2p}$ matrix. The p subscripts indicate that these M constants are past values calculated in the previous computation cycle. Apply to the last two matrices of Eq 8.4-30 the following matrix equation:

$$(\underset{\sim}{A}\underset{\sim}{B})^T = \underset{\sim}{B}^T\underset{\sim}{A}^T \qquad (8.4\text{-}31)$$

Equation 8.4-30 becomes

$$\underset{\sim}{P}_0 = \begin{bmatrix} 1 & T \\ 0 & 1 \end{bmatrix} \begin{bmatrix} M_{1p} & M_{2p} \\ 0 & M_{3p} \end{bmatrix} \left[\begin{bmatrix} 1 & T \\ 0 & 1 \end{bmatrix} \begin{bmatrix} M_{1p} & M_{2p} \\ 0 & M_{3p} \end{bmatrix} \right]^T \qquad (8.4\text{-}32)$$

This has the same form given in Eq 8.4-29:

$$\underset{\sim}{P}_0 = \underset{\sim}{M}_0 \underset{\sim}{M}_0^T \qquad (8.4\text{-}33)$$

Comparing Eqs 8.4-32, -33 gives for the matrix $\underset{\sim}{M}_0$

$$\underset{\sim}{M}_0 = \begin{bmatrix} 1 & T \\ 0 & 1 \end{bmatrix} \begin{bmatrix} M_{1p} & M_{2p} \\ 0 & M_{3p} \end{bmatrix} = \begin{bmatrix} M_{1p} & (M_{2p} + TM_{3p}) \\ 0 & M_{3p} \end{bmatrix} \qquad (8.4\text{-}34)$$

Set this equal to the expression for $\underset{\sim}{M}_0$ in Eq 8.4-28. This yields

$$M_{01} = M_{1p} \qquad (8.4\text{-}35)$$

$$M_{02} = M_{2p} + TM_{3p} \qquad (8.4\text{-}36)$$

$$M_{03} = M_{3p} \qquad (8.4\text{-}37)$$

Substitute into the expression for $\underset{\sim}{P}_1$ in Eq 8.4-22 the expressions for the matrices $\underset{\sim}{G}$ and $\underset{\sim}{Q}$ in Eqs 8.4-7, -8. This gives

$$\underset{\sim}{P}_1 = \underset{\sim}{P}_0 + \underset{\sim}{G}\underset{\sim}{Q}\underset{\sim}{G}^T = \underset{\sim}{P}_0 + \begin{bmatrix} \frac{1}{2}T^2 \\ T \end{bmatrix} \Delta a^2 \begin{bmatrix} \frac{1}{2}T^2 & T \end{bmatrix}$$

$$= \underset{\sim}{P}_0 + \begin{bmatrix} \frac{1}{2}T^2 \\ T \end{bmatrix} \begin{bmatrix} \frac{1}{2}T^2 & T \end{bmatrix} \Delta a^2 = \underset{\sim}{P}_0 + \begin{bmatrix} \frac{1}{4}T^4 & \frac{1}{2}T^3 \\ \frac{1}{2}T^3 & T^2 \end{bmatrix} \Delta a^2$$

$$= \underset{\sim}{P}_0 + \begin{bmatrix} \frac{1}{4}\Delta a^2 T^4 & \frac{1}{2}\Delta a^2 T^3 \\ \frac{1}{2}\Delta a^2 T^3 & \Delta a^2 (T^2) \end{bmatrix} \qquad (8.4\text{-}38)$$

The coefficients for the P_0 matrix were defined in Eq 8.4-29. For the moment, let us simplify the matrix to the form

$$P_0 = \begin{bmatrix} P_{01} & P_{02} \\ P_{02} & P_{03} \end{bmatrix} \tag{8.4-39}$$

Substituting Eq 8.4-39 into Eq 8.4-38 gives for the P_1 matrix

$$P_1 = \begin{bmatrix} \left(P_{01} + \tfrac{1}{4}\Delta a^2 T^4\right) & \left(P_{02} + \tfrac{1}{2}\Delta a^2 T^3\right) \\ \left(P_{02} + \tfrac{1}{2}\Delta a^2 T^3\right) & \left(P_{03} + \Delta a^2 T^2\right) \end{bmatrix} \tag{8.4-40}$$

The terms $\tfrac{1}{4}\Delta a^2 T^4$ and $\Delta a^2 T^2$ are denoted as

$$\delta x^2 = \tfrac{1}{4}\Delta a^2 T^4 = \left(\tfrac{1}{2}\Delta a\, T^2\right)^2 \tag{8.4-41}$$

$$\delta v^2 = \Delta a^2 T^2 = (\Delta a\, T)^2 \tag{8.4-42}$$

The variable δx, which is equal to $\tfrac{1}{2}\Delta a\, T^2$, is a position bias produced by the acceleration uncertainty; and δv, which is equal to $\Delta a\, T$, is an equivalent velocity bias. The off-diagonal elements $\tfrac{1}{2}\Delta a^2 T^3$ in the matrix of Eq 8.4-40 can be considered to be covariances between these position and velocity biases. However, the acceleration uncertainty is actually deterministic and so should have no effect on the off-diagonal elements of the covariance matrix. Therefore, the $\tfrac{1}{2}\Delta a^2 T^3$ terms are discarded, and the P_1 matrix of Eq 8.4-40 reduces to

$$P_1 = \begin{bmatrix} \left(P_{01} + \delta x^2\right) & P_{02} \\ P_{02} & \left(P_{03} + \delta v^2\right) \end{bmatrix} \tag{8.4-43}$$

Since the acceleration uncertainty is deterministic, the corresponding δx and δv position and velocity bias values should affect the square roots of the covariance matrix elements in a linear manner. Equation 8.4-43 does not exhibit this behavior. To correct this deficiency, the diagonal terms of the matrix are modified as follows so that the effects of the bias terms are linear:

$$P_1 = \begin{bmatrix} \left(\sqrt{P_{01}} + \delta x\right)^2 & P_{02} \\ P_{02} & \left(\sqrt{P_{03}} + \delta v\right)^2 \end{bmatrix} \tag{8.4-44}$$

This is the form of the P_1 matrix that is actually used in the Altair tracking equations.

Substitute for P_{01}, P_{02}, P_{03} in Eq 8.4-44 the actual coefficients of the $\underset{\sim}{P}_0$ matrix given in Eq 8.4-29. The resultant $\underset{\sim}{P}_1$ matrix is

$$\underset{\sim}{P}_1 = \begin{bmatrix} \left(\sqrt{[M_{01}^2 + M_{02}^2]} + \delta x\right)^2 & M_{02}M_{03} \\ M_{02}M_{03} & (M_{03} + \delta v)^2 \end{bmatrix} \tag{8.4-45}$$

Set this equal to the expression for the $\underset{\sim}{P}_1$ matrix in Eq 8.4-27 to obtain

$$N_1^2 + N_2^2 = \left(\sqrt{M_{01}^2 + M_{02}^2} + \delta x\right)^2 \tag{8.4-46}$$

$$N_2 N_3 = M_{02}M_{03} \tag{8.4-47}$$

$$N_3 = M_{03} + \delta v \tag{8.4-48}$$

Equation 8.4-23 gave an expression for the $\underset{\sim}{P}_2$ matrix. Substitute into this the relations of Eqs 8.4-15, -16 for the $\underset{\sim}{H}$ and $\underset{\sim}{R}$ matrices. This yields

$$
\begin{aligned}
\underset{\sim}{P}_2^{-1} &= \underset{\sim}{P}_1^{-1} + H^T R^{-1} H \\
&= \underset{\sim}{P}_1^{-1} + \begin{bmatrix} 1 \\ 0 \end{bmatrix} \sigma_n^{-2} [1 \quad 0] = \underset{\sim}{P}_1^{-1} + \begin{bmatrix} 1 \\ 0 \end{bmatrix} [1 \quad 0] \sigma_n^{-2} \\
&= \underset{\sim}{P}_1^{-1} + \begin{bmatrix} 1 & 0 \\ 0 & 0 \end{bmatrix} \sigma_n^{-2} = \underset{\sim}{P}_1^{-1} = \begin{bmatrix} \sigma_n^{-2} & 0 \\ 0 & 0 \end{bmatrix}
\end{aligned} \tag{8.4-49}
$$

Equation 8.4-16 shows that the measurement noise matrix $\underset{\sim}{R}$ consists of a single element σ_n^2 and so is a scalar equal to σ_n^2. Hence the inverse of this matrix, denoted $\underset{\sim}{R}^{-1}$, is equal to $1/\sigma_n^2 = \sigma_n^{-2}$.

The term $\underset{\sim}{P}_1^{-1}$ of Eq 8.4-49 is the inverse of the matrix $\underset{\sim}{P}_1$ given in Eq 8.4-45. To simplify the calculation of this inverse matrix, matrix $\underset{\sim}{P}_1$ is defined in the following general form:

$$\underset{\sim}{P}_1 = \begin{bmatrix} a & b \\ b & c \end{bmatrix} \tag{8.4-50}$$

As shown in Appendix C, the inverse of a matrix is given by

$$\underset{\sim}{P}_1^{-1} = \frac{\text{adj}[\underset{\sim}{P}_1]}{\det[\underset{\sim}{P}_1]} \tag{8.4-51}$$

The formulas given in Appendix C show that the determinant and adjoint of the matrix $\underset{\sim}{P}_1$ of Eq 8.4-50 are

$$\det[\underset{\sim}{P}_1] = ac - b^2 \tag{8.4-52}$$

$$\text{adj}[\underset{\sim}{P}_1] = \begin{bmatrix} c & -b \\ -b & a \end{bmatrix} \tag{8.4-53}$$

Hence, by Eq 8.4-51, the inverse of matrix $\underset{\sim}{P}_1$ is

$$\underset{\sim}{P}_1^{-1} = \frac{1}{ac - b^2}\begin{bmatrix} c & -b \\ -b & a \end{bmatrix} \qquad (8.4\text{-}54)$$

Substituting Eq 8.4-54 into Eq 8.4-49 gives

$$\underset{\sim}{P}_2^{-1} = \underset{\sim}{P}_1^{-1} + \begin{bmatrix} \sigma_n^{-2} & 0 \\ 0 & 0 \end{bmatrix}$$

$$= \frac{1}{ac - b^2}\begin{bmatrix} c & -b \\ -b & a \end{bmatrix} + \begin{bmatrix} \sigma_n^{-2} & 0 \\ 0 & 0 \end{bmatrix}$$

$$= \frac{1}{ac - b^2}\begin{bmatrix} [c + (ac - b^2)\sigma_n^{-2}] & -b \\ -b & a \end{bmatrix} \qquad (8.4\text{-}55)$$

The inverse of this is $\underset{\sim}{P}_2$, which can be expressed as follows by defining the matrix $\underset{\sim}{A}$:

$$\underset{\sim}{P}_2 = (ac - b^2)\begin{bmatrix} [c + (ac - b^2)\sigma_n^{-2}] & -b \\ -b & a \end{bmatrix}^{-1} = (ac - b^2)\underset{\sim}{A}^{-1} \quad (8.4\text{-}56)$$

The determinant of the matrix $\underset{\sim}{A}$ defined by Eq 8.4-56 is

$$\det[\underset{\sim}{A}] = a\{c + (ac - b^2)\sigma_n^{-2}\} - b^2$$

$$= ac + a(ac - b^2)\sigma_n^{-2} - b^2$$

$$= (ac - b^2)(1 + a\sigma_n^{-2}) \qquad (8.4\text{-}57)$$

The adjoint of the matrix $\underset{\sim}{A}$ defined by Eq 8.4-56 is

$$\text{adj}[\underset{\sim}{A}] = \begin{bmatrix} a & b \\ b & [c + (ac - b^2)\sigma_n^{-2}] \end{bmatrix} \qquad (8.4\text{-}58)$$

By Eq 8.4-51, the inverse of the matrix $\underset{\sim}{A}$ is

$$\underset{\sim}{A}^{-1} = \frac{\text{adj}[\underset{\sim}{A}]}{\det[\underset{\sim}{A}]} \qquad (8.4\text{-}59)$$

Combining Eqs 8.4-56 to -59 gives

$$\underset{\sim}{P}_2 = \frac{ac - b^2}{(ac - b^2)(1 + a\sigma_n^{-2})} \begin{bmatrix} a & b \\ b & [c + (ac - b^2)\sigma_n^{-2}] \end{bmatrix}$$

$$= \frac{\sigma_n^2}{\sigma_n^2 + a} \begin{bmatrix} a & b \\ b & [c + (ac - b^2)\sigma_n^{-2}] \end{bmatrix}$$

$$= \frac{1}{\sigma_n^2 + a} \begin{bmatrix} a\sigma_n^2 & b\sigma_n^2 \\ b\sigma_n^2 & c(\sigma_n^2 + a - b^2/c) \end{bmatrix} \tag{8.4-60}$$

Set Eq 8.4-60 equal to the general expression for $\underset{\sim}{P}_2$ given in Eq 8.4-25 to obtain

$$M_1^2 + M_2^2 = \frac{a\sigma_n^2}{\sigma_n^2 + a} \tag{8.4-61}$$

$$M_2 M_3 = \frac{b\sigma_n^2}{\sigma_n^2 + a} \tag{8.4-62}$$

$$M_3^2 = \frac{c(\sigma_n^2 + a - b^2/c)}{\sigma_n^2 + a} \tag{8.4-63}$$

Comparing the expressions for $\underset{\sim}{P}_1$ in Eqs 8.4-27, -50 gives the values of the a, b, c coefficients:

$$a = N_1^2 + N_2^2 \tag{8.4-64}$$

$$b = N_2 N_3 \tag{8.4-65}$$

$$c = N_3^2 \tag{8.4-66}$$

Substituting Eqs 8.4-64 to -66 into Eqs 8.4-61 to -63 gives

$$M_1^2 + M_2^2 = \frac{(N_1^2 + N_2^2)\sigma_n^2}{\sigma_n^2 + N_1^2 + N_2^2} \tag{8.4-67}$$

$$M_2 M_3 = \frac{N_2 N_3 \sigma_n^2}{\sigma_n^2 + N_1^2 + N_2^2} \tag{8.4-68}$$

$$M_3^2 = \frac{N_3^2 [\sigma_n^2 + N_1^2 + N_2^2 - (N_2 N_3)^2/N_3^2]}{\sigma_n^2 + N_1^2 + N_2^2}$$

$$= \frac{N_3^2 (\sigma_n^2 + N_1^2)}{\sigma_n^2 + N_1^2 + N_2^2} \tag{8.4-69}$$

By Eq 8.3-32 of Section 8.3, the equation for the Kalman gain matrix is

$$K = P[+] H^T R^{-1} = P_2 H^T \frac{1}{\sigma_n^2}$$

$$= \begin{bmatrix} (M_1^2 + M_2^2) & M_2 M_3 \\ M_2 M_3 & M_3^2 \end{bmatrix} \begin{bmatrix} 1 \\ 0 \end{bmatrix} \frac{1}{\sigma_n^2}$$

$$= \begin{bmatrix} (M_1^2 + M_2^2) \\ M_2 M_3 \end{bmatrix} \frac{1}{\sigma_n^2} = \begin{bmatrix} (M_1^2 + M_2^2)/\sigma_n^2 \\ M_2 M_3/\sigma_n^2 \end{bmatrix} \qquad (8.4\text{-}70)$$

The expressions for the matrices H, R, and P_2 were obtained from Eqs 8.4-15, -16, and -25. The coefficients for the K matrix are defined as

$$K = \begin{bmatrix} K_x \\ K_v \end{bmatrix} \qquad (8.4\text{-}71)$$

Relating this to Eq 8.4-70 gives

$$K_x = \frac{M_1^2 + M_2^2}{\sigma_n^2} = \frac{N_1^2 + N_2^2}{\sigma_n^2 + N_1^2 + N_2^2} \qquad (8.4\text{-}72)$$

$$K_v = \frac{M_2 M_3}{\sigma_n^2} = \frac{N_2 N_3}{\sigma_n^2 + N_1^2 + N_2^2} \qquad (8.4\text{-}73)$$

Equations 8.4-67, -68 were used to obtain the right-hand expressions for K_x, K_v. By Eqs 8.4-41, -42, the equations for δx and δv are

$$\delta x = \tfrac{1}{2} \Delta a \, T^2 \qquad (8.4\text{-}74)$$

$$\delta v = \Delta a \, T \qquad (8.4\text{-}75)$$

8.4.3 Summary of Gain Subroutine Equations

Let us now summarize the equations, expressing them in the proper sequence for computer simulation. To simplify the expressions, the following are defined

$$N = \sqrt{N_1^2 + N_2^2} \qquad (8.4\text{-}76)$$

$$M = \sqrt{M_1^2 + M_2^2} \qquad (8.4\text{-}77)$$

$$\sigma = \sqrt{\sigma_n^2 + N^2} = \sqrt{\sigma_n^2 + N_1^2 + N_2^2} \qquad (8.4\text{-}78)$$

$$\sigma_1 = \sqrt{\sigma_n^2 + N_1^2} \qquad (8.4\text{-}79)$$

The equations for calculating the Kalman gains are as follows. On the left are the numbers of the equations from which they were derived.

(8.4-35) $$M_{01} = M_{1p}$$ (8.4-80)

(8.4-36) $$M_{02} = M_{2p} + TM_{3p}$$ (8.4-81)

(8.4-37) $$M_{03} = M_{3p}$$ (8.4-82)

(8.4-46) $$N = \sqrt{M_{01}^2 + M_{02}^2} + \delta x$$ (8.4-83)

(8.4-48) $$N_3 = M_{03} + \delta v$$ (8.4-84)

(8.4-47) $$N_2 = \frac{M_{02} M_{03}}{N_3}$$ (8.4-85)

(8.4-76) $$N_1 = \sqrt{N^2 - N_2^2}$$ (8.4-86)

(8.4-67) $$M = N\sigma_n / \sigma$$ (8.4-87)

(8.4-69) $$M_3 = N_3 \sigma_1 / \sigma$$ (8.4-88)

(8.4-68, -88) $$M_2 = N_2 N_3 \frac{\sigma_n^2}{\sigma^2 M_3} = N_2 \frac{\sigma_n^2}{\sigma \sigma_1}$$ (8.4-89)

(8.4-77) $$M_1 = \sqrt{M^2 - M_2^2}$$ (8.4-90)

(8.4-72) $$K_x = N^2 / \sigma^2$$ (8.4-91)

(8.4-73) $$K_v = N_2 N_3 / \sigma^2$$ (8.4-92)

$$M_{1p} = M_1$$ (8.4-93)

$$M_{2p} = M_2$$ (8.4-94)

$$M_{3p} = M_3$$ (8.4-95)

Equation 8.4-90 can be simplified as follows:

$$
\begin{aligned}
M_1 = M^2 - M_2^2 &= \frac{N^2 \sigma_n^2}{\sigma^2} - \frac{N_2^2 \sigma_n^4}{\sigma^2 \sigma_1^2} \\
&= \left(\frac{\sigma_n}{\sigma \sigma_1}\right)^2 \left(N^2 \sigma_1^2 - N_2^2 \sigma_n^2\right) \\
&= \left(\frac{\sigma_n}{\sigma \sigma_1}\right)^2 \left[\left(N_1^2 + N_2^2\right)\left(\sigma_n^2 + N_1^2\right) - N_2^2 \sigma_n^2\right] \\
&= \left(\frac{\sigma_n}{\sigma \sigma_1}\right)^2 N_1^2 \left(\sigma_n^2 + N_1^2 + N_2^2\right) = \left(\frac{N_1 \sigma_n}{\sigma_1}\right)^2 \quad (8.4\text{-}96)
\end{aligned}
$$

The following are the equations for calculating the K_x, K_v Kalman gains that are actually implemented in the Altair tracking software, with some changes of symbolism:

$$N_3 = M_3 + \delta v \tag{8.4-97}$$

$$N_2 = (M_3/N_3) M_2 \tag{8.4-98}$$

$$N = \sqrt{M_1^2 + M_2^2} + \delta x \tag{8.4-99}$$

$$N_1 = \sqrt{N^2 - N_2^2} \tag{8.4-100}$$

$$\sigma_1 = \sqrt{\sigma_n^2 + N_1^2} \tag{8.4-101}$$

$$\sigma = \sqrt{\sigma_1^2 + N_2^2} = \sqrt{\sigma_n^2 + N^2} \tag{8.4-102}$$

$$M_1 = (\sigma_n/\sigma_1) N_1 \tag{8.4-103}$$

$$M_2 = (\sigma_n/\sigma)(\sigma_n/\sigma_1) N_2 \tag{8.4-104}$$

$$M_3 = (\sigma_1/\sigma) N_3 \tag{8.4-105}$$

$$K_x = (N/\sigma)^2 \tag{8.4-106}$$

$$K_v = (N_2/\sigma)(N_3/\sigma) \tag{8.4-107}$$

$$M_2 = M_2 + TM_3 \tag{8.4-108}$$

The Altair software specifications use the symbols η, ζ, and R, which are related as follows to the preceding symbols:

$$\eta = N \tag{8.4-109}$$

$$\zeta = \sigma_1 \tag{8.4-110}$$

$$R = \sigma_n \tag{8.4-111}$$

Equations 8.4-97 to -108 are equivalent to those given previously in Eqs 8.4-76 to -96 except that the Altair equations are expressed in the sequential form for FORTRAN calculations.

8.4.4 Simplified Form of Gain Subroutine Equations

The gain subroutine expressions of Eqs 8.4-76 to -96 can be simplified in the following manner, with changes in symbolism to put them in a form that is more meaningful physically. Combine Eqs 8.4-80, -81, -83:

$$
\begin{aligned}
N &= \sqrt{M_{ip}^2 + (M_{2p} + TM_{3p})^2} + \delta x \\
&= \sqrt{M_{1p}^2 + M_{2p}^2 + 2TM_{2p}M_{3p} + T^2M_{3p}^2} + \delta x \\
&= \sqrt{M_p^2 + 2TM_{2p}M_{3p} + T^2M_{3p}^2} + \delta x
\end{aligned}
\tag{8.4-112}
$$

Combine Eqs 8.4-82, -84:

$$N_3 = M_{3p} + \delta v \tag{8.4-113}$$

Combine Eqs 8.4-81, -82, -85:

$$N_2 N_3 = \left(M_{2p} + TM_{3p} \right) M_{3p} = M_{2p} M_{3p} + TM_{3p}^2 \qquad (8.4\text{-}114)$$

From Eqs 8.4-78, -79, -86,

$$\sigma_1^2 = \sigma_n^2 + N_1^2 = \sigma_n^2 + N^2 - N_2^2 = \sigma^2 - N_2^2 \qquad (8.4\text{-}115)$$

By Eqs 8.4-25, -27, the matrices P_1, P_2 can be expressed as

$$P_1 = \begin{bmatrix} N^2 & N_2 N_3 \\ N_2 N_3 & N_3^2 \end{bmatrix} \qquad (8.4\text{-}116)$$

$$P_2 = \begin{bmatrix} M^2 & M_2 M_3 \\ M_2 M_3 & M_3^2 \end{bmatrix} \qquad (8.4\text{-}117)$$

It is convenient to denote the elements of the covariance matrix P_2 as follows:

$$P_2 = \begin{bmatrix} \sigma_x^2 & \sigma_{xv}^2/T \\ \sigma_{xv}^2/T & (\sigma_v/T)^2 \end{bmatrix} \qquad (8.4\text{-}118)$$

The parameter σ_x is the estimated RMS error of position, and σ_v/T is the estimated RMS error of velocity. The parameter σ_v is the RMS velocity error normalized in terms of position, which is obtained by multiplying the RMS velocity error by the sampling period T. Thus, σ_v is the RMS displacement over a sampling period that would be produced by a velocity equal to the RMS velocity error. The covariance σ_{xv}^2 is also normalized in terms of position, and so all three terms $\sigma_x^2, \sigma_v^2, \sigma_{xv}^2$ have the units of position squared. The parameter δv is normalized in terms of position to give the following parameter δx_v:

$$\delta x_v = T \delta v = \Delta A T^2 = 2 \delta x \qquad (8.4\text{-}119)$$

The advantages of normalizing all signals in terms of equivalent position variables was demonstrated in the discussion of the serial simulation procedure in Chapter 5.

The computation process alternates between the covariance matrices P_1 and P_2. Hence it is convenient to denote the elements of the "alternate" covariance matrix P_1 in the same manner as those of P_2 given in Eq 8.4-118, with the subscript a for alternate added to each symbol. Thus the matrix P_1 is defined as

$$P_1 = \begin{bmatrix} \sigma_{xa}^2 & \sigma_{xva}^2/T \\ \sigma_{xva}^2/T & (\sigma_{va}/T)^2 \end{bmatrix} \qquad (8.4\text{-}120)$$

Relating Eqs 8.4-117, -118 gives

$$M = \sigma_x \tag{8.4-121}$$

$$M_3 T = \sigma_v \tag{8.4-122}$$

$$M_2 M_3 T = \sigma_{xv}^2 \tag{8.4-123}$$

Combining Eqs 8.4-122, -123 gives

$$M_2 = \sigma_{xv}^2 / \sigma_v \tag{8.4-124}$$

Relating Eqs 8.4-116, -120 gives

$$N = \sigma_{xa} \tag{8.4-125}$$

$$N_3 T = \sigma_{va} \tag{8.4-126}$$

$$N_2 N_3 T = \sigma_{xva}^2 \tag{8.4-127}$$

Combining Eqs 8.4-126, -127 gives

$$N_2 = \sigma_{xva}^2 / \sigma_{va} \tag{8.4-128}$$

By Eqs 8.4-78, -115, -125, and -128 the variables σ and σ_1 become

$$\sigma^2 = \sigma_n^2 + N^2 = \sigma_n^2 + \sigma_{xa}^2 \tag{8.4-129}$$

$$\sigma_1^2 = \sigma^2 - N_2^2 = \sigma_n^2 + \sigma_{xa}^2 - \sigma_{xva}^4 / \sigma_{va}^2 \tag{8.4-130}$$

Using the values of Eqs 8.4-119 and -121 to -130 allows the computation routine to be reduced to the following:

(8.4-112)
$$\sigma_{xa} = \sqrt{\sigma_{x(p)}^2 + 2\sigma_{xv(p)}^2 + \sigma_{v(p)}^2} + \delta x \tag{8.4-131}$$

(8.4-113)
$$\sigma_{va} = \sigma_{v(p)} + \delta x_v = \sigma_{v(p)} + 2\,\delta x \tag{8.4-132}$$

(8.4-114)
$$\sigma_{xva}^2 = \sigma_{xv(p)}^2 + \sigma_{v(p)}^2 \tag{8.4-133}$$

(8.4-87)
$$\sigma_x^2 = \frac{\sigma_{xa}^2 \sigma_n^2}{\left(\sigma_n^2 + \sigma_{xa}^2\right)} \tag{8.4-134}$$

(8.4-88)
$$\sigma_v^2 = \sigma_{va}^2 \left(1 - \frac{\sigma_{xva}^4}{\sigma_{va}^2\left(\sigma_n^2 + \sigma_{xa}^2\right)}\right)$$

$$= \sigma_{va}^2 - \frac{\sigma_{xva}^4}{\sigma_n^2 + \sigma_{xa}^2} \tag{8.4-135}$$

(8.4-89)
$$\sigma_{xv}^2 = \sigma_{xva}^2 \frac{\sigma_n^2}{\sigma^2} = \frac{\sigma_{xva}^2 \sigma_n^2}{\sigma_n^2 + \sigma_{xa}^2} \tag{8.4-136}$$

(8.4-91)
$$K_x = \frac{\sigma_{xa}^2}{\sigma_n^2 + \sigma_{xa}^2} \tag{8.4-137}$$

(8.4-92)
$$K_v T = \frac{\sigma_{xva}^2}{\sigma_n^2 + \sigma_{xa}^2} \tag{8.4-138}$$

The subscript (p) refers to the past value of the variable. Further simplification is achieved as follows. Substituting Eq 8.4-134 into Eq 8.4-136 gives

$$\sigma_{xv} = \sigma_{xva}(\sigma_x / \sigma_{xa}) \qquad (8.4\text{-}139)$$

Substituting Eq 8.4-134 into Eq 8.4-137 gives

$$K_x = \left(\frac{\sigma_x}{\sigma_n}\right)^2 \qquad (8.4\text{-}140)$$

Substituting Eqs 8.4-134, -139 into Eq 8.4-138 gives

$$K_v T = \left(\frac{\sigma_{xva}\sigma_x}{\sigma_{xa}\sigma_n}\right)^2 = \left(\frac{\sigma_{xv}}{\sigma_n}\right)^2 = \left(\frac{\sigma_{xv}}{\sigma_x}\right)^2 K_x \qquad (8.4\text{-}141)$$

Substituting Eqs 8.4-134, -138 into Eq 8.4-135 gives

$$\sigma_v^2 = \sigma_{va}^2 - \frac{\sigma_{xva}^2 \sigma_{xv}^2}{\sigma_n^2} \qquad (8.4\text{-}142)$$

A physical discussion of these iterative equations for calculating the Kalman filter gains is given in Section 8.6.

8.5 DERIVATION OF SIGNAL-FLOW DIAGRAMS FOR ALTAIR KALMAN TRACKING FILTERS

8.5.1 Basic Angle-Tracker and Range-Tracker Filter Equations

This section analyzes the signal-filtering equations of the Altair target estimator and the range tracker, and reduces them to the signal-flow diagrams presented in Section 8.2. In the equations for the target estimator, the index i takes on the values of 1, 2, and 3, which correspond to motions in the east, north, and vertical directions relative to coordinates at the Altair site. The following are the computer equations solved by the Altair tracker in the target estimator (or angle tracker) and in the range tracker. These equations are solved iteratively in the order given:

Target Estimator (Angle Tracker) Equations ($i = 1, 2, 3$)

Angle-updator routine:

$$\hat{x}_i = \hat{x}_i + K_{ax}\tilde{z}_i \qquad (8.5\text{-}1)$$

$$\hat{v}_i = \hat{v}_i + K_{av}\tilde{z}_i \qquad (8.5\text{-}2)$$

Angle-extrapolator routine:

$$\hat{a}_i = a_{gi} - \hat{D}_d \hat{v}_i \tag{8.5-3}$$

$$\hat{v}_i^{(0)} = \hat{v}_i \tag{8.5-4}$$

$$\hat{v}_i = \hat{v}_i + T_a \hat{a}_i \tag{8.5-5}$$

$$\hat{x}_i = \hat{x}_i + \tfrac{1}{2} T_a \left(\hat{v}_i + \hat{v}_i^{(0)} \right) \tag{8.5-6}$$

Range-tracker Equations

Range-updator routine

$$\hat{r} = \hat{r} + K_{rx} \tilde{z}_r \tag{8.5-7}$$

$$\hat{v}_r = \hat{v}_r + K_{rv} \tilde{z}_r \tag{8.5-8}$$

$$\hat{a}_e = \hat{a}_e + K_{ra} \tilde{z}_r \tag{8.5-9}$$

Range-extrapolator routine

$$\hat{a}_r^{(0)} = \hat{a}_r \tag{8.5-10}$$

$$\hat{v}_r^{(0)} = \hat{v}_r \tag{8.5-11}$$

$$\hat{a}_r = \hat{a}_r + T_r \hat{a}_r \tag{8.5-12}$$

$$\hat{v}_r = \hat{v}_r + \tfrac{1}{2} T_r \left(\hat{a}_r + \hat{a}_r^{(0)} + 2 \hat{a}_e \right) \tag{8.5-13}$$

$$\hat{r} = \hat{r} + \tfrac{1}{2} T_r \left(\hat{v}_r + \hat{v}_r^{(0)} - \tfrac{1}{6} T_r^2 \hat{a}_r \right) \tag{8.5-14}$$

The variables $\hat{v}_i^{(0)}$, $\hat{a}_r^{(0)}$, and $\hat{v}_r^{(0)}$ are temporary calculation variables. The symbols in these equations are defined as follows:

$\tilde{z}_i$ = measurements of estimator error ($\tilde{z}_1$, $\tilde{z}_2$, $\tilde{z}_3$) in Altair ENV coordinates.

$\hat{x}_i$ = estimated target position ($\hat{x}_1$, $\hat{x}_2$, $\hat{x}_3$) in Altair ENV coordinates

$\hat{v}_i$ = estimated target velocity ($\hat{v}_1$, $\hat{v}_2$, $\hat{v}_3$) in Altair ENV coordinates

K_{ax}, K_{av} = position and velocity Kalman gains for angle tracker

T_a = sampling period for angle tracker (or target estimator)

a_{gi} = gravity and coriolis acceleration (a_{g1}, a_{g2}, a_{g3}) in Altair ENV coordinates (represented by a_i' in the Altair software specification)

$\hat{D}_d$ = atmospheric drag factor, which varies with target altitude and speed; target speed is equal to $\sqrt{\left[\hat{v}_1^2 + \hat{v}_2^2 + \hat{v}_3^2\right]}$

$\hat{a}_i$ = total predicted acceleration ($\hat{a}_1$, $\hat{a}_2$, $\hat{a}_3$) of the vehicle in Altair ENV coordinates

$\tilde{z}_r$ = equivalent error measurement of range tracker

$\hat{r}$ = estimated target range

$\hat{v}_r$ = estimated range velocity of target (represented by $\hat{u}$ in the Altair software specifications)

$\hat{a}_r$ = predicted target range acceleration

$\dot{\hat{a}}_r$ = rate-of-change of predicted target range acceleration

$\hat{a}_e$ = estimated error in predicted range acceleration

T_r = sample period for range tracker

To simplify the presentation, the symbols $\tilde{z}_i$ and $\tilde{z}_r$ for the estimator and range tracking error signals are replaced by the symbols $\tilde{x}_{ei}$ and $\tilde{r}_e$:

$$\tilde{x}_{ei} = \tilde{z}_i \tag{8.5-15}$$

$$\tilde{r}_e = \tilde{z}_r \tag{8.5-16}$$

This allows the symbol $\bar{z}$ to be used without confusion to represent $\exp[-sT]$, the transfer function of a one-cycle sampled-data time delay.

Equations 8.5-1 to -6 for the angle tracker are converted to the following equations in order to eliminate the sequential requirement of the calculations:

Angle tracker equations:

$$\hat{x}_{(a)i} = \hat{x}_{ip} + K_{ax}\tilde{x}_{ei} \tag{8.5-17}$$

$$\hat{v}_{(a)i} = \hat{v}_{ip} + K_{av}\tilde{x}_{ei} \tag{8.5-18}$$

$$\hat{a}_i = a_{gi} - \hat{D}_d\hat{v}_{(a)i} \tag{8.5-19}$$

$$\hat{v}_i = \hat{v}_{(a)i} + T_a\hat{a}_i \tag{8.5-20}$$

$$\hat{x}_i = \hat{x}_{(a)i} + \tfrac{1}{2}T_a\left(\hat{v}_i + \hat{v}_{(a)i}\right) \tag{8.5-21}$$

The symbols $\hat{x}_{ip}$, $\hat{v}_{ip}$ represent the past values of $\hat{x}_i$, $\hat{v}_i$ computed in the previous cycle. The symbols $\hat{x}_{(a)i}$, $\hat{v}_{(a)i}$ define *alternate* or *intermediate* variables that are related to $\hat{x}_i$, $\hat{v}_i$.

In similar fashion, Eqs 8.5-7 to -14 for the range tracker are converted to the following equations in order to eliminate the sequential requirement of the calculations:

Range-tracker equations:

$$\hat{r}_{(a)} = \hat{r}_p + K_{rx}\tilde{r}_e \tag{8.5-22}$$

$$\hat{v}_{r(a)} = \hat{v}_{rp} + K_{rv}\tilde{r}_e \tag{8.5-23}$$

$$\hat{a}_e = \hat{a}_{ep} + K_{ra}\tilde{r}_e \tag{8.5-24}$$

$$\hat{a}_r = \hat{a}_{rp} + T_r\hat{\dot{a}}_r \tag{8.5-25}$$

$$\hat{v}_r = \hat{v}_{r(a)} + \tfrac{1}{2}T_r\big(\hat{a}_r + \hat{a}_{rp} + 2\hat{a}_e\big) \tag{8.5-26}$$

$$\hat{r} = \hat{r}_{(a)} + \tfrac{1}{2}T_r\big(\hat{v}_r + \hat{v}_{r(a)} - \tfrac{1}{6}T_r^2\hat{\dot{a}}_r\big) \tag{8.5-27}$$

The symbols $\hat{r}_p$, $\hat{v}_{rp}$, $\hat{a}_{ep}$, $\hat{a}_{rp}$ represent the past values of $\hat{r}$, $\hat{v}_r$, $\hat{a}_e$, $\hat{a}_r$ computed in the previous cycle. The symbols $\hat{r}_{(a)}$, $\hat{v}_{r(a)}$ are *alternate* values related to $\hat{r}$, $\hat{v}_r$.

8.5.2 Laplace Transforms of Angle-Tracker Filter Equations

The angle tracker equations can be simplified by substituting Eqs 8.5-17, -18 into Eqs 8.5-19 to -21 to give

$$\hat{a}_i = a_{gi} - \hat{D}_d\big(\hat{v}_{ip} + K_{av}\tilde{x}_{ei}\big) \tag{8.5-28}$$

$$\hat{v}_i = \hat{v}_{ip} + K_{av}\tilde{x}_{ei} + T_a\hat{a}_i$$
$$= \hat{v}_{ip} + T_a(K_{av}/T_a)\tilde{x}_{ei} + T_a\hat{a}_i \tag{8.5-29}$$

$$\hat{x}_i = \hat{x}_{ip} + K_{ax}\tilde{x}_{ei} + \tfrac{1}{2}T_a\big(\hat{v}_i + \hat{v}_{ip} + K_{av}\hat{x}_i\big)$$
$$= \hat{x}_{ip} + T_a\big[(K_{ax}/T_a) + \tfrac{1}{2}K_{av}\big]\tilde{x}_{ei} + \tfrac{1}{2}T_a\big(\hat{v}_i + \hat{v}_{ip}\big) \tag{8.5-30}$$

Equations 8.5-29, -30 include the following general sampled-data integration formulas:

Simple integration:

$$y = y_p + Tx \tag{8.5-31}$$

Trapezoidal integration:

$$y = y_p + \tfrac{1}{2}T(x + x_p) \tag{8.5-32}$$

where T is the sampling period. As was shown in Section 8.2, Eqs 8.2-111, -112, the Laplace transforms of these integration formulas are

Simple integration:

$$\frac{Y}{X} = \frac{1 + pT/2}{p} \tag{8.5-33}$$

Trapezoidal integration:

$$\frac{Y}{X} = \frac{1}{p} \tag{8.5-34}$$

As was explained in Chapter 5, the complex pseudo-frequency p is approximately equal to the Laplace-transform complex-variable s for frequencies less than $\frac{1}{4}$ of the sampling frequency. Applying the relations of Eqs 8.5-31 to -34 gives the following for the Laplace transforms of Eqs 8.5-28 to -30:

$$\hat{A}_i = A_{gi} - \hat{D}_d\left(\hat{V}_{ip} + K_{av}\tilde{X}_{ei}\right) \tag{8.5-35}$$

$$\hat{V}_i = \frac{1 + pT/2}{p}\left(\frac{K_{av}}{T_a}\tilde{X}_{ei} + \hat{A}_i\right) \tag{8.5-36}$$

$$\hat{X}_i = \frac{1 + pT/2}{p}\left(\frac{K_{ax}}{T_a} + \frac{K_{av}}{2}\right)\tilde{X}_{ei} + \frac{1}{p}\hat{V}_i \tag{8.5-37}$$

where $\hat{X}_i$, $\hat{V}_i$, A_{gi}, $\tilde{X}_{ei}$, $\hat{A}_i$ are the transforms of the time-domain signals $\hat{x}_i$, $\hat{v}_i$, a_{gi}, $\tilde{x}_{ei}$, $\hat{a}_i$. (Note that Eq 8.5-35 has no integrations.) The variable $\hat{V}_{ip}$ is the transform of the past value of $\hat{v}_i$ and is related as follows to the transform of $\hat{v}_i$:

$$\hat{V}_{ip} = e^{-sT}\hat{V}_i = \bar{z}\hat{V}_i = \frac{1 - pT/2}{1 + pT/2}\hat{V}_i \tag{8.5-38}$$

Remember from Chapter 5 that the transfer function for a one-cycle time delay is

$$\bar{z} = e^{-sT} = \frac{1 - U}{1 + U} = \frac{1 - pT/2}{1 + pT/2} \tag{8.5-39}$$

8.5.3   Signal-Flow Diagram of Angle-Tracker Filter

Equations 8.5-35 to -39 are expressed in signal-flow diagram form in Fig 8.5-1. Remember from Section 8.2 that the variable $\tilde{X}_{ei}$ is the error between the observed target position $X_{(ob)i}$ and the past value of the estimated target

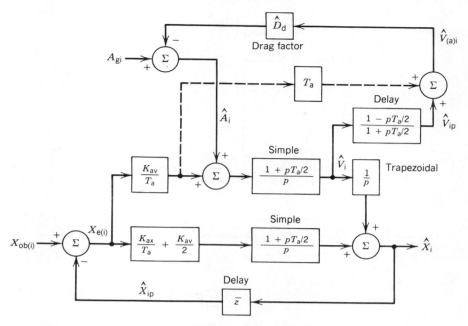

Figure 8.5-1 Signal-flow diagram for tracking filter used in Altair angle tracker (target estimator).

position $\hat{X}_{ip}$:

$$\tilde{X}_{ei} = X_{(ob)i} - \hat{X}_{ip} \qquad (8.5\text{-}40)$$

In the diagram, this equation is implemented to close the estimator feedback loop.

The dashed path has only a small effect on the system response and so will be ignored to simplify the discussion. When this is done, the drag component of target acceleration is derived from the past value of the estimated velocity $\hat{v}_{ip}$ rather than from the variable $\hat{v}_{(a)i}$. The Altair computations use the variable $\hat{v}_{(a)i}$ rather than $\hat{v}_{ip}$ to compute drag acceleration because $\hat{v}_{ip}$ is not available in a register at that time. However, $\hat{v}_{ip}$ is theoretically a better variable than $\hat{v}_{(a)i}$ for computing drag acceleration.

When the dashed path in Fig 8.5-1 is ignored, the signal-flow diagram can be modified to the form shown in Fig 8.5-2 by interchanging the $1/p$ and the $(1/p)(1 + pT/2)$ transfer blocks in the $\hat{V}_i$ signal path. Figure 8.5-2 is simplified further to form the final signal-flow diagram to Fig 8.5-3. Note that the transfer function $(1 - pT/2)/(1 + pT/2)$ is equal to $\bar{z}$, and represents a one-cycle time delay. Except for movement of the T_a constant, this is the same

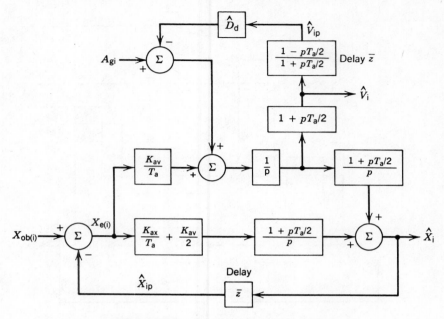

Figure 8.5-2   Simplification of Fig 8.5-1.

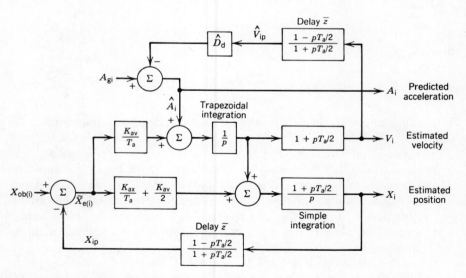

Figure 8.5-3   Final simplification of signal-flow diagram of angle tracking filter.

as the signal-flow diagram for the angle tracker was given in Fig 8.2-9 of Section 8.2.

8.5.4   Laplace Transforms of Range-Tracker Filter Equations

Now let us derive an equivalent signal-flow diagram for the range tracker. The range-tracker equations are simplified by substituting Eqs 8.5-22, -23 into Eqs 8.5-26, -27 to give

$$\hat{v}_r = \hat{v}_{rp} + K_{rv}\tilde{r}_e + \tfrac{1}{2}T_r(\hat{a}_r + \hat{a}_{rp}) + T_r\hat{a}_e$$

$$= \hat{v}_{rp} + T_r((K_{rv}/T_r)\tilde{r}_e + \hat{a}_e) + \tfrac{1}{2}T_r(\hat{a}_r + \hat{a}_{rp}) \tag{8.5-41}$$

$$\hat{r} = r_p + K_{rx}\tilde{r}_e + \tfrac{1}{2}T_r(\hat{u} + \hat{u}_p + K_{rv}\tilde{r}_e) - T_r(\tfrac{1}{12}T_r^2)\hat{a}_r$$

$$= r_p + T_r\left(\frac{K_{rx}}{T_r} + \tfrac{1}{2}K_{rv}\right)\tilde{r}_e + \tfrac{1}{2}T_r(\hat{u} + \hat{u}_p) - T_r(\tfrac{1}{12}T_r^2)\hat{a}_r \tag{8.5-42}$$

Equation 8.5-24 can be expressed as

$$\hat{a}_e = \hat{a}_{ep} + T_r(K_{ra}/T_r)\tilde{r}_e \tag{8.5-43}$$

Applying the integration relations of Eqs 8.5-30 to -33 gives the following for the Laplace transforms of Eqs 8.5-41 to -43:

$$\hat{V}_r = \frac{(1 + pT/2)}{p}\left(\frac{K_{rv}}{T_r}\tilde{R}_e + \tilde{A}_e\right) + \frac{1}{p}\hat{A}_r \tag{8.5-44}$$

$$\hat{R} = \frac{(1 + pT/2)}{p}\left(\frac{K_{rx}}{T_r} + \frac{K_{rv}}{2}\right)\tilde{R}_e + \frac{1}{p}\hat{U} - \frac{(1 + pT/2)}{p}\left(\frac{T_r^2}{12}\right)\hat{A}_r \tag{8.5-45}$$

$$\hat{A}_e = \frac{(1 + pT/2)}{p}\frac{K_{ra}}{T_r}\tilde{R}_e \tag{8.5-46}$$

Similarly, the transform of Eq 8.5-25 is

$$\tilde{A}_r = \frac{(1 + pT/2)}{p}\hat{A}_r \tag{8.5-47}$$

The variables $\hat{R}, \hat{V}_r, \hat{A}_e, \hat{A}_r, \dot{\hat{A}}_r, \tilde{R}_e$ are the transforms of the time-domain signals $\hat{r}, \hat{v}_r, \hat{a}_e, \hat{a}_r, \dot{\hat{a}}_r, \tilde{r}_e$.

8.5.5 Signal-Flow Diagram of Range-Tracker Filter

A signal-flow diagram of the range tracker incorporating Eqs 8.5-44 to -47 is shown in Fig 8.5-4. This diagram also includes the relation

$$\tilde{R}_e = R_t - \hat{R}_p \tag{8.5-48}$$

where $\tilde{R}_e$ is the range-tracker error, R_t is the actual range to the target, and $\hat{R}_p$ is the past value of the estimated target range $\hat{R}$.

The dashed path in Fig 8.4-4 provides a correction signal that is derived from the target estimator rate-of-acceleration signal. The purpose of this correction signal is not clear and is not explained in the Altair literature. Since its effect on the system response is small, it will be ignored.

The range-tracker signal-flow diagram of Fig 8.5-4 is simplified to form Fig 8.5-5 by ignoring the dashed path and placing the summation of the acceleration signals $\hat{A}_r, \hat{A}_e$ in front of the $(1/p)(1 + pT/2)$ integration. Figure 8.5-5 is

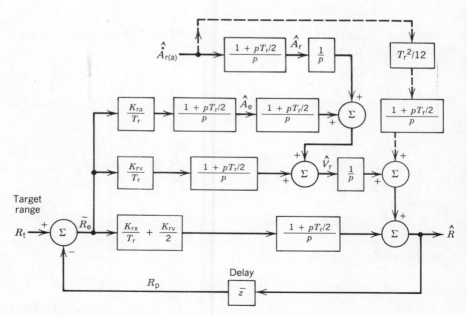

Figure 8.5-4 Signal-flow diagram for tracking filter used in Altair range tracker.

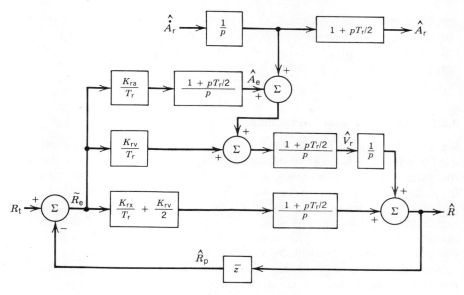

Figure 8.5-5 Simplification of Fig 8.5-4.

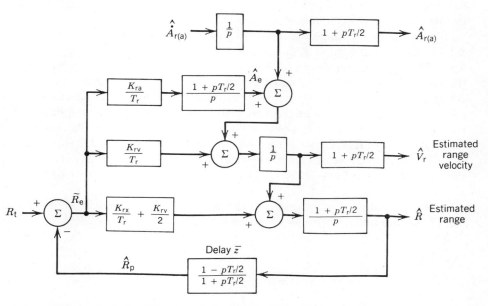

Figure 8.5-6 Final simplification of signal-flow diagram of the range tracking filter.

simplified further to form Fig 8.5-6 by placing the summation for the velocity signal prior to the $(1/p)(1 + pT/2)$ transfer block of the range feedback loop. Except for movement of the T_r constant, this is the signal-flow diagram for the range tracker that was given in Fig 8.2-10.

8.5.6 Implications of Signal-Flow Diagram Analysis

Section 8.5 has shown how the Kalman tracking-filter computer algorithms used in Altair can be represented as signal-flow diagrams expressed in terms of the pseudo complex frequency p. At frequencies below $\frac{1}{4}$ of the sampling frequency, the pseudo complex frequency p is approximately equal to the true complex frequency s. One can apply this approach to any Kalman filter system, to derive a signal-flow diagram representation of the tracking filter portion of the system. This can be done, even when the actual system computations are implemented directly in terms of matrix equations.

The signal-flow diagram representation enables one to understand the dynamic characteristics of the system much more clearly. It allows one to relate the Kalman filter algorithms to conventional tracking concepts, and thereby to gain insight into the application of Kalman filter theory to practical engineering problems. Section 8.6 will show how this insight can lead to changes for improving system performance.

The Kalman tracking-filter computer algorithms are derived from Eqs 8.3-27, -28 of the general Kalman-filter matrix expressions given in Section 8.3.2. The Laplace transforms of these equations can be taken by representing the one-cycle time delay (indicated by the vector x_p) as the transfer function $\bar{z} = \exp[-sT]$. Although the matrices may be time varying, the time variation is so slow the matrices can be considered to be constants in taking the Laplace transform. It is theoretically possible for the filter gain matrix K to vary too rapidly to allow this. However, as will be shown in Section 8.6.2, practical Kalman filters have velocity memory, and for a filter with velocity memory, the variation of the gain matrix K must be slow in order for the filter to reach a stable and accurate solution.

After the Laplace transforms of the Kalman filter expressions in Eqs 8.3-27, -28 are taken, the time-delay transfer function $\bar{z}$ is replaced by

$$\bar{z} = \frac{1 - pT/2}{1 + pT/2} \tag{8.5-49}$$

The resultant equations can then be represented in signal-flow diagram form, expressed in terms of the pseudo complex frequency p.

8.6 RESPONSES OF OPTIMAL ALGORITHMS FOR SETTING TRACKING FILTER GAINS

8.6.1 Summary of Equations for Calculating Optimal Kalman Filter Gains

The following is a summary of the simplified equations derived in Section 8.4 for calculating the Altair tracking filter gains:

(8.4-131) $$\sigma_{xa} = \sqrt{\sigma_{x(p)}^2 + 2\sigma_{xv(p)}^2 + \sigma_{v(p)}^2} + \delta x$$ (8.6-1)

(8.4-132) $$\sigma_{va} = \sigma_{v(p)} + 2\,\delta x$$ (8.6-2)

(8.4-133) $$\sigma_{xva}^2 = \sigma_{xv(p)}^2 + \sigma_{v(p)}^2$$ (8.6-3)

(8.4-134) $$\sigma_x^2 = \frac{\sigma_{xa}^2 \sigma_n^2}{\sigma_n^2 + \sigma_{xa}^2}$$ (8.6-4)

(8.4-139) $$\sigma_{xv} = \sigma_{xva}\frac{\sigma_x}{\sigma_{xa}}$$ (8.6-5)

(8.4-142) $$\sigma_v^2 = \sigma_{va}^2 - \frac{\sigma_{xva}^2 \sigma_{xv}^2}{\sigma_n^2}$$ (8.6-6)

(8.4-140) $$K_x = \left(\frac{\sigma_x}{\sigma_n}\right)^2$$ (8.6-7)

(8.4-141) $$K_v T = \left(\frac{\sigma_{xv}}{\sigma_n}\right)^2 = \left(\frac{\sigma_{xv}}{\sigma_x}\right)^2 K_x$$ (8.6-8)

(8.4-74) $$\delta x = \tfrac{1}{2}\Delta a\, T^2$$ (8.6-9)

The numbers on the left give the equations in Section 8.4 from which these were obtained.

The inputs to this program are the values of Δa and σ_n. The variable Δa is the estimated uncertainty in predicted acceleration, which is computed from Eq 8.4-1. The variable σ_n is the RMS position error per sample period. It has different values for the range tracker and the angle tracker (target estimator), which are as follows:

Range tracker:

$$\sigma_n = \sigma_{n(r)} = \sigma r_1\sqrt{T_p/T_r}$$ (8.6-10)

Angle tracker:

$$\sigma_n = \sigma_{n(a)} = \hat{R}\sigma\theta_1\sqrt{T_p/T_a}$$ (8.6-11)

The variables σr_1, $\sigma\theta_1$ are the RMS errors in range and angle for a single pulse measurement, which were given in Eqs 8.1-50, -51, T_r, T_a are the values of sample period T for range and angle tracking, and T_p is the pulse repetition period.

The Kalman filter gains for the range tracker are calculated by the iterative solution of Eqs 8.6-1 to -9, setting $T = T_r$, and $\sigma_n = \sigma_{n(r)}$ as given in Eq 8.6-10. The resultant values of K_x, K_v are the gains K_{rx}, K_{rv} used in the range tracker. The Kalman filter gains for the angle tracker are calculated by the iterative solution of Eqs 8.6-1 to -9, setting $T = T_a$, and $\sigma_n = \sigma_{n(a)}$ as given in Eq 8.6-11. The resultant values of K_x, K_v are the gains K_{ax}, K_{av} used in the angle tracker.

8.6.2 Simpler Optimal Gain Algorithm Based on Remainder Coefficient

To develop a clearer understanding of the responses of these iterative equations, let us consider a simpler method for optimally varying the tracking filter gains. The philosophy behind the Johansen implementation of Kalman filtering in the Altair system is that target motion is characterized by a limit on the uncertainty of target acceleration. Since the tracking system corrects for the predictable portion of target acceleration, the predictable acceleration does not cause tracking error. The Johansen target model assumes that the uncertainty of target acceleration varies in a worst-case manner between the limits of $\pm\Delta a$ such that it maximizes tracking error.

The theory presented in Chapter 3, Section 3.3, gives the following upper limit to the tracking error when the acceleration remains within the bounds $\pm\Delta a$, provided the velocity error coefficient of the tracking loop is zero:

$$|x_e| \leq \Delta x = r_2\,\Delta a \qquad (8.6\text{-}12)$$

The parameter r_2 is the acceleration remainder coefficient for the tracking loop. The factor $r_2\,\Delta a$ is denoted Δx, which is the absolute value of the worst-case tracking error due to target motion.

As explained in Section 3.3, the acceleration remainder coefficient can be calculated from the response of the tracking loop to a unit step of acceleration in the manner illustrated in Fig 8.6-1. The solid curve (A) is the error response of the tracking filter to a unit step of acceleration. The final value of this error response is the acceleration error coefficient c_2. The dashed curve (B) is derived from the solid curve (A) by inverting the segments of curve (A) having negative slope, and displacing all of the segments vertically to form the monotonically increasing continuous curve (B). The final value of the dashed curve (B) is the acceleration remainder coefficient, denoted r_2.

The following analysis applies to the tracking loops of the Altair target estimator, which have double integrations at low frequency. The Altair range tracker, which has a triple integration at low frequency, is examined later in Section 8.6.4. Each of the three tracking loops of the Altair target estimator

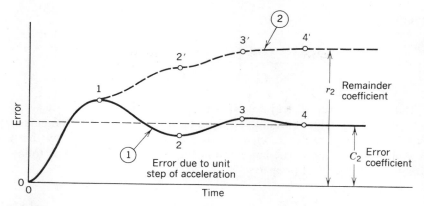

Figure 8.6-1   Calculation of the remainder coefficient for acceleration, from the error response to a unit step of acceleration.

(angle tracker) has a loop transfer function of the following form:

$$G = \frac{\omega_c (1 + \omega_i/s)}{s} = \frac{\omega_c (s + \omega_i)}{s^2} \qquad (8.6\text{-}13)$$

The corresponding error transfer function is

$$G_{ie} = \frac{1}{1 + G} = \frac{s^2}{s^2 + \omega_c s + \omega_c \omega_i}$$

$$= \frac{s^2}{s^2 + 2\zeta \omega_n s + \omega_n^2} \qquad (8.6\text{-}14)$$

where

$$\omega_n = \sqrt{\omega_c \omega_i} \qquad (8.6\text{-}15)$$

$$2\zeta = \frac{\omega_c}{\omega_n} = \sqrt{\frac{\omega_c}{\omega_i}} \qquad (8.6\text{-}16)$$

The acceleration error coefficient is

$$c_2 = \frac{1}{\omega_c \omega_i} = \frac{1}{\omega_n^2} \qquad (8.6\text{-}17)$$

Appendix E shows that for $\zeta > 1$ the acceleration remainder coefficient r_2 is equal to the acceleration error coefficient c_2; and for $\zeta < 1$ the coefficient r_2 is equal to

$$r_2 = \frac{c_2\left(1 + \exp\left[-\zeta\pi/\sqrt{1 - \zeta^2}\right]\right)}{1 - \exp\left[-\zeta\pi/\sqrt{1 - \zeta^2}\right]} \quad \text{(for } \zeta < 1) \qquad (8.6\text{-}18)$$

The remainder coefficient describes the error response of the tracking loop to unpredictable target motions. To minimize the tracking error, one must also consider the response to receiver noise, which is characterized by the noise bandwidth of the loop. The noise bandwidth in rad/sec is given as follows in Chapter 6, Table 6.1-2 (loop E):

$$\omega_b = \frac{\pi}{2}(\omega_c + \omega_i) \qquad (8.6\text{-}19)$$

Divide this by 2π to obtain the noise bandwidth f_b in hertz; then apply the relation for 2ζ given in Eq 8.6-16. This yields

$$f_b = \frac{\omega_b}{2\pi} = \tfrac{1}{4}(\omega_c + \omega_i) = \tfrac{1}{4}\omega_c\left(1 + \frac{\omega_i}{\omega_c}\right)$$

$$= \tfrac{1}{4}\omega_c\left[1 + \frac{1}{(2\zeta)^2}\right] \qquad (8.6\text{-}20)$$

By combining Eqs 8.6-16, -17, the acceleration error coefficient can be expressed as

$$c_2 = \frac{1}{\omega_c^2}\frac{\omega_c}{\omega_i} = \left(\frac{2\zeta}{\omega_c}\right)^2 \qquad (8.6\text{-}21)$$

Solve Eq 8.6-21 for $(2\zeta)^2$ and substitute the result into Eq 8.6-20. This yields

$$c_2 = \frac{\left(1 + 4\zeta^2\right)^2}{64\zeta^2 f_b^2} \qquad (8.6\text{-}22)$$

Substituting Eq 8.6-22 into Eq 8.6-18 gives the following expression for the acceleration remainder coefficient of the angle tracking loop:

$$r_2 = \frac{\left(1 + 4\zeta^2\right)^2\left\{1 + \exp\left[-\zeta\pi/\sqrt{1 - \zeta^2}\right]\right\}}{64\zeta^2 f_b^2\left\{1 - \exp\left[-\zeta\pi/\sqrt{1 - \zeta^2}\right]\right\}} \qquad (8.6\text{-}23)$$

This expresses the acceleration remainder coefficient r_2 in terms of the noise bandwidth f_b and the damping ratio ζ of the tracking loop.

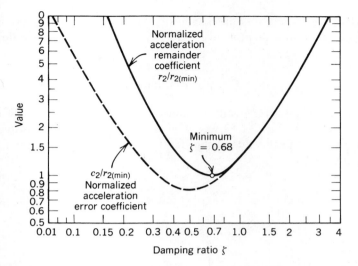

Figure 8.6-2 For a tracking filter, the variation (vs. damping ratio) of normalized acceleration remainder coefficient and error coefficient for constant noise bandwidth.

If the noise bandwidth f_b is constant, the tracking error due to receiver noise is constant. Hence the optimum damping ratio ζ can be determined by keeping f_b fixed in Eq 8.6-23 and varying ζ to minimize the remainder coefficient r_2. This yields the minimum error due to target motion for any given receiver noise error.

Differentiate Eq 8.6-23 relative to the damping ratio ζ, keeping the noise bandwidth f_b constant, and set the derivative equal to zero. This yields the following optimum damping ratio:

$$\text{Optimum}[\zeta] = 0.6786 \tag{8.6-24}$$

Substituting this value for ζ into Eq 8.6-23 gives the following for the minimum value of the acceleration remainder coefficient r_2 for a given noise bandwidth f_b:

$$r_{2(\text{min})} = \frac{0.30589}{f_b^2} \tag{8.6-25}$$

In Fig 8.6-2, the solid curve A is a plot of the ratio $r_2/r_{2(\text{min})}$ obtained by dividing Eq 8.6-23 by Eq 8.6-25. This ratio is called the *normalized acceleration remainder coefficient*. This plot shows how the acceleration remainder coefficient varies as a function of the damping ratio ζ when the noise bandwidth is held fixed. The dashed curve (B) shows for reference the *normalized acceleration error coefficient*, which is obtained by dividing the acceleration error coefficient in Eq 8.6-22 by the value of $r_{2(\text{min})}$ in Eq 8.6-25.

The damping ratio of the tracker feedback loop is established by the ratio ω_c/ω_i. The optimum value for this frequency ratio is obtained as follows from Eqs 8.6-16, -24:

$$\text{Optimum } [\omega_c/\omega_i] = \text{optimum } [2\zeta]^2 = 1.8420 \qquad (8.6\text{-}26)$$

The feedback loop corresponds to loop C, described in Ref [1.1] (Section 2.4.2). As shown in Eqs 2.4-52 to -54 of that reference, the following are the expressions of the peak value of the step response and the maximum value of $|G_{ib}|$ for this loop:

$$\text{Max}[x_b] = 1 + \exp\left[-\frac{\zeta}{\sqrt{1 - \zeta^2}}\left(\text{arc cos}[2\zeta^2 - 1]\right)\right] \qquad (8.6\text{-}27)$$

$$\left(\frac{1}{\text{Max}[G_{ib}]}\right)^2 = 1 - \frac{2}{\eta^2}\left(1 + \eta - \sqrt{1 + 2\eta}\right) \qquad (8.6\text{-}28)$$

where

$$\eta = (2\zeta)^2 = \frac{\omega_c}{\omega_i} \qquad (8.6\text{-}29)$$

Setting the damping ratio equal to its optimum value in Eq 8.6-24 gives the following for the optimum values of peak overshoot and maximum $|G_{ib}|$:

$$\text{Optimum peak overshoot} = 21.8\% \qquad (8.6\text{-}30)$$

$$\text{Optimum Max}|G_{ib}| = 1.290 \qquad (8.6\text{-}31)$$

This value for $\text{Max}|G_{ib}|$ is very close to the setting $\text{Max}|G_{ib}| = 1.30$, which we have used as a practical feedback-control design criterion.

It is convenient to express the acceleration bound Δa in terms of the variable δx used in the Altair calculation of optimal Kalman gains. Solving Eq 8.6-9 for Δa gives

$$\Delta a = 2\,\delta x/\text{T}^2 \qquad (8.6\text{-}32)$$

Substituting this into Eq 8.6-12 gives the following for the error bound Δx due to uncertainty of target motion:

$$\Delta x = r_2\,\Delta a = 2r_2\,\delta x/\text{T}^2 \qquad (8.6\text{-}33)$$

Substitute into this the value for r_2 in Eq 8.6-25. This gives the following error bound when the ratio ω_c/ω_i is set for optimum:

$$\Delta x = 2(0.30589)\,\delta x/(f_b T)^2 \qquad (8.6\text{-}34)$$

As was shown in Chapter 6, Section 6.1, the noise bandwidth f_b of a lowpass filter is equal to $1/2T_i$, where T_i is the effective integration time of the filter. It is convenient to define a parameter N_i, which is the effective number of samples integrated by the tracking loop. This is obtained by dividing the integration time T_i by the sampling period T:

$$N_i = \frac{T_i}{T} = \text{number of samples integrated} \qquad (8.6\text{-}35)$$

Since f_b is equal to $1/2T_i$, the factor f_bT is related to N_i by

$$f_bT = \frac{T}{2T_i} = \frac{1}{2N_i} \qquad (8.6\text{-}36)$$

Substituting this into Eq 8.6-34 gives

$$\Delta x = 8(0.30589)\,\delta x\,N_i^2 = 2.447\,\delta x\,N_i^2 \qquad (8.6\text{-}37)$$

The RMS error σ_x caused by tracking noise is related to N_i by

$$\sigma_x^2 = \frac{\sigma_n^2}{N_i} \qquad (8.6\text{-}38)$$

where σ_n is the RMS noise for one sample.

The more samples N_i that are integrated, the smaller is the error due to receiver noise, as characterized by σ_x, but the greater is the error due to uncertainty of target motion, as characterized by Δx. The parameter N_i should be set to minimize the total error, which is the sum of these two error components. Since the receiver noise is independent of the tracking error due to target motion, these two error components are statistically independent. Hence the effective RMS value of the combined error, which is denoted σ_{sum}, can be expressed as

$$\sigma_{sum}^2 = \sigma_x^2 + (\Delta x/b)^2 \qquad (8.6\text{-}39)$$

where b is a constant. The variable σ_x is the RMS tracking error due to receiver noise, while $\Delta x/b$ is the "effective RMS value" of the error due to uncertainty of target motion. Since the target motion error is deterministic, it does not have a Gaussian distribution. Therefore, judgment is required in choosing the constant b, which sets its *effective RMS value*.

The relation for Δx in Eq 8.6-35 can be expressed as

$$\Delta x = C_a N_i^2 \qquad (8.6\text{-}40)$$

where C_a is a constant when δx is constant. Squaring this gives

$$\Delta x^2 = C_a^2 N_i^4 \qquad (8.6\text{-}41)$$

Substitute into Eq 8.6-39 the expressions for σx^2 and Δx^2 given in Eqs 8.6-38, -41. This gives

$$\sigma_{\text{sum}}^2 = \frac{\sigma_n^2}{N_i} + \frac{C_a^2 N_i^4}{b^2} \tag{8.6-42}$$

We desire the value for N_i that minimizes the total RMS error σ_{sum}. This is obtained by differentiating Eq 8.6-42 relative to N_i and setting the derivative equal to zero. This gives

$$-\frac{\sigma_n^2}{N_i^2} + \frac{4C_a^2 N_i^3}{b^2} = 0 \tag{8.6-43}$$

Solving for N_i gives

$$N_i^5 = \left(\frac{\sigma_n}{C_a}\right)^2 \left(\frac{b}{2}\right)^2 \tag{8.6-44}$$

Divide the expression for σ_x^2 in Eq 8.6-38 by the expression for Δx^2 in Eq 8.6-41. This yields

$$\frac{\sigma_x^2}{\Delta x^2} = \frac{\sigma_n^2 / N_i}{C_a^2 N_i^4} = \frac{(\sigma_n / C_a)^2}{N_i^5} \tag{8.6-45}$$

Combining Eqs 8.6-44, -45 gives the following when N_i is optimum:

$$\sigma_x = \left(\frac{2}{b}\right) \Delta x \tag{8.6-46}$$

Square Eq 8.6-13 and substitute into this the relation for σ_x^2 in Eq 8.6-38 and the square of Δx in Eq 8.6-37. This gives

$$\frac{\sigma_n^2}{N_i} = \left(\frac{2}{b}\right)^2 (2.447\,\delta x)^2 N_i^4 \tag{8.6-47}$$

Solving Eq 8.6-47 for N_i gives

$$N_i = \left(\frac{b\sigma_n}{4.894\,\delta x}\right)^{2/5} = 0.5298 b^{0.4} \left(\frac{\sigma_n}{\delta x}\right)^{0.4} \tag{8.6-48}$$

This is the optimum effective number of samples integrated by the tracking loop. To relate this to the optimum gain crossover frequency of the loop, combine Eqs 8.6-20, -36 to obtain

$$N_i = \frac{1}{2f_bT} = \frac{2}{\omega_c(1 + \omega_i/\omega_c)} \tag{8.6-49}$$

Let us define the normalized time-constant parameter N_c by

$$N_c = \frac{\tau_c}{T} = \frac{1}{\omega_cT} \tag{8.6-50}$$

where $\tau_c = 1/\omega_c$. Combining Eqs 8.6-49, -50 gives

$$N_c = \left[\tfrac{1}{2}(1 + \omega_i/\omega_c)\right] N_i \tag{8.6-51}$$

For the optimum setting, $\omega_c = 1.8420\omega_i$, and so

$$N_c = 0.7714N_i = 0.409b^{0.4}(\sigma_n/\delta x)^{0.4} \tag{8.6-52}$$

Equations 8.6-48, -52 give the optimum values for N_i and N_c for steady-state conditions. However, immediately after the radar locks onto the target, the integration time of the tracking loop should be short, so that the loop can correct rapidly for initial errors in position and velocity. After lock-on, the integration time should be gradually increased to the desired steady-state value.

How rapidly should the integration time be increased? It is not useful to increase the integration time more rapidly than the elapsed time after lock-on because the filter would not have enough data to use the full integration time. Let us first consider the limiting case where the integration time is exactly equal to the elapsed time after lock-on. This characteristic is satisfied by the feedback filter shown in Fig 8.6-3a, which has a simple position loop with no integral network. The equations for the loop are

$$x = x_{(p)} + K_xx_e \tag{8.6-53}$$

$$x_e = x_i - x_{(p)} \tag{8.6-54}$$

where $x_{(p)}$ is the past value of x. Combining these gives

$$x = (1 - K_x)x_{(p)} + K_xx_i \tag{8.6-55}$$

The gain K_x is set equal to $1/N$, where N is the number of samples that have occurred after lock-on:

$$K_x = \frac{1}{N} \tag{8.6-56}$$

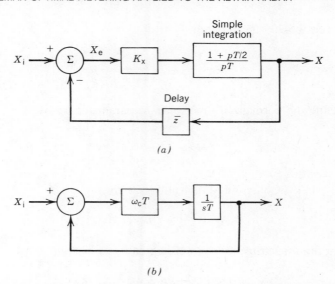

Figure 8.6-3 Simple feedback sampled-data tracking filter with no velocity memory: (a) complete sampled-data signal-flow diagram; (b) equivalent low-frequency signal-flow diagram.

Equation 8.6-55 becomes

$$x = \frac{1}{N}\left[(N-1)x_{(p)} + x_i\right] \tag{8.6-57}$$

Denote the input samples as $x_i[N]$. The values of x for the first three samples are

$$N = 1: \quad x = x_i[1] \tag{8.6-58}$$

$$N = 2: \quad x = \tfrac{1}{2}(x_{(p)} + x_i) = \tfrac{1}{2}(x_i[1] + x_i[2]) \tag{8.6-59}$$

$$N = 3: \quad x = \tfrac{1}{3}(2x_{(p)} + x_i) = \tfrac{1}{3}(x_i[1] + x_i[2] + x_i[3]) \tag{8.6-60}$$

Hence the value for the Nth sample is

$$x = \frac{1}{N} \sum_{k=1}^{N} x_i[k] \tag{8.6-61}$$

Thus the output x is the average of the N samples of the input x_i. This simple loop is an ideal integrator with an integration time equal to the elapsed time after lock-on. The number of samples integrated, which is N_i, is equal to the number of samples N that have been received.

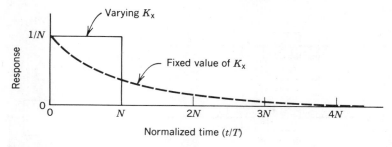

Figure 8.6-4   Impulse response of sampled-data tracking filter of Fig 8.6-3.

The impulse response of this filter is shown by the solid curve in Fig 8.6-4. If the variation of the gain constant is stopped and K_x is held fixed at a given value of $1/N$, the impulse response gradually changes from this rectangular pulse to the exponential pulse shown by the dashed curve. The loop can then be approximated by the steady-state form shown in Fig 8.6-3b, which has the gain crossover frequency

$$\omega_c = \frac{K_x}{T} = \frac{1}{NT} \tag{8.6-62}$$

The effective number of samples N_i integrated by this loop is obtained from Eq 8.6-49 by setting $\omega_i = 0$. This yields

$$N_i = \frac{T_i}{T} = \frac{2}{\omega_c T} = 2N \tag{8.6-63}$$

This value $2N$ is twice the effective number N of samples integrated while the constant K_x is being varied.

This shows that when the loop constants of the simple loop are being varied rapidly, one cannot use the normal steady-state expressions for noise bandwidth, integration time, or effective number of samples integrated. On the other hand, a simple loop such as this cannot be employed for tracking because it does not have velocity memory to compensate for target velocity. As will be shown, for a loop that does compensate in an effective manner for target velocity, the gain variation must be much slower. With this slow gain variation, the steady-state noise bandwidth approximately holds even when the loop gain is being varied.

Figure 8.6-5a shows the signal-flow diagram of a tracking loop having an integral network, which provides velocity memory. Both integrations of the loop are implemented by simple integration algorithms. The equations of the

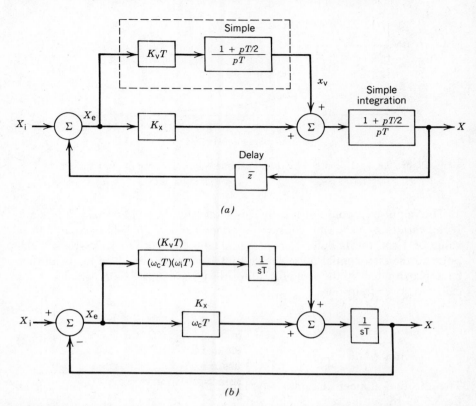

Figure 8.6-5 Sampled-data tracking filter, which compensates for input velocity: (*a*) complete signal-flow diagram; (*b*) equivalent low-frequency signal-flow diagram.

loop are

$$x = x_{(p)} + K'_x x_e + x_v \tag{8.6-64}$$

$$x_v = x_{v(p)} + K_v T x_e \tag{8.6-65}$$

$$x_e = x_i - x_{(p)} \tag{8.6-66}$$

The parameter K'_x has a prime because it corresponds to the effective position gain, not the parameter K_x of the Altair equations. For low frequencies, this loop reduces to the form shown in Fig 8.6-5*b*. The loop parameters are given by

$$\omega_c = \frac{K'_x}{T} \tag{8.6-67}$$

$$\omega_i \omega_c = \frac{K_v}{T} \tag{8.6-68}$$

By Eq 8.4-26 the optimum value of the ratio ω_c/ω_i is 1.842. However, for simplicity, let us round them off to 2:

$$\frac{\omega_c}{\omega_i} = 2 \qquad (8.6\text{-}69)$$

Combining Eqs 8.6-67 to -69 gives for $K_v T$:

$$K_v T = \frac{K_x'^2}{\omega_c/\omega_i} = \frac{K_x'^2}{2} \quad \text{(for } N > 1) \qquad (8.6\text{-}70)$$

The constant K_v is set to zero for the first sample because at that time the loop does not have sufficient information to compute velocity:

$$\text{For } N = 1: \quad K_v T = 0 \qquad (8.6\text{-}71)$$

The gain K_x' is set approximately equal to A/N, where the parameter A can be adjusted. To keep K_x' from exceeding unity at low values of N, the actual equation for K_x' is

$$K_x' = \frac{A}{N + A - 1} \qquad (8.6\text{-}72)$$

Figure 8.6-6 shows the normalized error responses of this loop to a ramp input for various values of the parameter A. (The error responses to a unit step are all unity at $N = 1$ and zero thereafter.) At $N = 2$, the normalized error responses to the ramp are all equal to unity, which is the amount the ramp input changes in one sample period. The plots show that the parameter A must be greater than 2.4 in order for the ramp error to reduce eventually to zero. A value of at least $A = 4$ is needed to achieve a reasonably rapid decrease of error. When A is as large as 4, the variation of loop parameters is so slow that time variations can be ignored in calculating the noise integration performance of the loop. Therefore, the normalized integration time N_i can be approximated by Eq 8.6-49 even when the gain is being varied. This gives

$$N_i = \frac{2}{\omega_c T(1 + \omega_i/\omega_c)} = \frac{2}{K_x'(1 + 0.5)} = \frac{1.333}{K_x'} \qquad (8.6\text{-}73)$$

The dashed curve in Fig 8.6-6 (which is negative) shows the normalized error response for $A = 4$ when the simple integration algorithm of the integral network is replaced by a trapezoidal integration algorithm. (The integration algorithm for the main loop is still simple.) With trapezoidal integration in the integral network, Eq 8.6-65 changes to

$$x_v = x_{v(p)} + \tfrac{1}{2} K_v T(x_e + x_{e(p)}) \qquad (8.6\text{-}74)$$

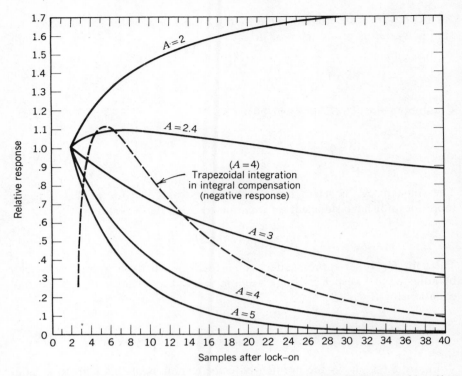

Figure 8.6-6 Error due to target velocity during acquisition for Kalman tracking filter with velocity memory.

where $x_{e(p)}$ is the past value of the error x_e. The other loop equations are the same. The dashed curve decays much more slowly than the solid curve for $A = 4$. This indicates that the error during lock-on is much greater when trapezoidal integration is used in the integral network. Thus, the tracking system should ideally use simple integration algorithms for both integrations, at least during the lock-on transient.

8.6.3 Responses of Optimal Algorithms for Setting Tracking Filter Gains

The simple optimal gain algorithm developed in the preceding section will now be used as a frame of reference for interpreting the responses of the Altair Kalman gain calculations. By Eq 8.2-2, the effective position gain constant for the Altair tracker is

$$K_x' = K_x + \tfrac{1}{2} K_v T \tag{8.6-75}$$

By Eq 8.2-7, this is equal to $\omega_c T$:

$$\omega_c T = K_x' = K_x + \tfrac{1}{2} K_v T \tag{8.6-76}$$

As in Eq 8.6-68, $\omega_i \omega_c$ is equal to

$$\omega_i \omega_c = \frac{K_v}{T} \tag{8.6-77}$$

Combining Eqs 8.6-76, -77 gives for the ratio ω_c / ω_i

$$\frac{\omega_c}{\omega_i} = \frac{\left(K_v + \tfrac{1}{2} K_v T \right)^2}{K_v T} \tag{8.6-78}$$

As defined in Eq 8.6-50, the normalized time constant of the loop is

$$N_c = \frac{\tau_c}{T} = \frac{1}{\omega_c T} = \frac{1}{K_x + \tfrac{1}{2} K_v T} \tag{8.6-79}$$

By Eq 8.6-51, the effective number of samples N_i that are integrated is related to this by

$$N_i = \frac{T_i}{T} = \frac{2}{(1 + \omega_i / \omega_c)} N_c \tag{8.6-80}$$

To simulate the Altair gain computations, the expressions in Eqs 8.6-1 to -9 were combined with Eqs 8.6-78 to -80 and solved iteratively for constant values of the ratio $\delta x / \sigma_n$. The initial values of σ_x and σ_v were set equal to σ_n, and the initial value of σ_{xv} was set equal to $\tfrac{1}{4} \sigma_n$. The results are shown in Figs 8.6-7 to -9, which give plots of N_c, N_i, and ω_c / ω_i versus normalized time N after lock-on. The variable N is the number of samples that have occurred since lock-on, which is equal to t/T, where t is the time and T is the sample period.

As shown in Fig 8.6-7, during the initial linear part of all transients, the normalized time constant follows the formula

$$N_c = \frac{\tau_c}{T} = 0.254N \tag{8.6-81}$$

Since N_c is equal to $1/K_x'$, the expression for K_x' during the linear part of the transient is

$$K_x' = \frac{1}{0.254N} = \frac{3.94}{N} \cong \frac{4}{N} \tag{8.6-82}$$

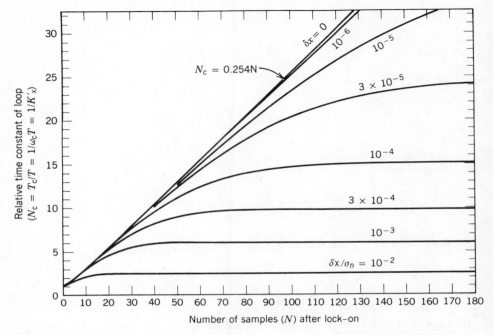

Figure 8.6-7   Variation of relative time constant $1/\omega_c T$ during acquisition transient, for Kalman-filter gain algorithm.

Hence the parameter A of Eq 8.6-72, which characterizes the ramp lock-on transient, is approximately equal to 4. The solid curve for $A = 4$ in Fig 8.6-6 gives a reasonably good lock-on transient. The curve for $A = 5$ settles much more rapidly, but the time to reach steady state with $A = 5$ exceeds that for $A = 4$ by the ratio $\frac{5}{4} = 1.25$.

On the other hand, as was shown in Section 8.2, Figs 8.2-9, -10, the velocity-memory integral network in the Altair tracker uses trapezoidal integration, rather than simple integration, in both the range tracker and the angle tracker. Hence, the actual Altair lock-on transient is the dashed curve in Fig 8.6-6. This response (which corresponds to $A = 4$ with a trapezoidal rate integration algorithm) settles much more slowly than the solid curve for $A = 4$ (which applies when both integrations use simple integration algorithms). This indicates that the lock-on performance of the Altair tracker could be improved by changing the integral-network algorithm from trapezoidal to simple integration during the acquisition transient.

Figure 8.6-9 shows plots of the ratio ω_c/ω_i. These are reasonably close to the theoretical optimum ratio 1.84 given in Eq 8.6-26. Thus, the Altair responses have reasonable damping during the lock-on transient.

Figure 8.6-8 shows the plots of N_i, the effective number of pulses integrated. Since the values of ω_c/ω_i are approximately equal to 2, these plots of

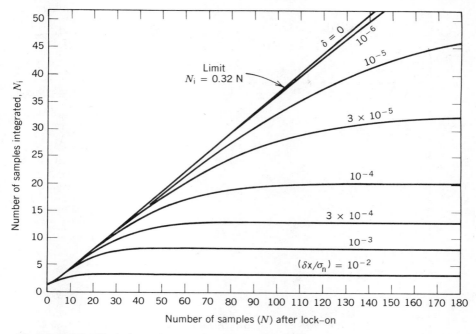

Figure 8.6-8 Variation of number of samples integrated, N_i, during acquisition transient, for Kalman-filter gain algorithm.

N_i are approximately equal to the plots of N_c in Fig 8.6-7 multiplied by the factor $2/(1 + \frac{1}{2})$, or 1.333.

Table 8.6-1 shows the steady-state values of N_i, N_c, and ω_c/ω_i derived from simulation for various values of the ratio $\delta x/\sigma_n$. These values for N_i are very close to those obtained from the theoretical expression of Eq 8.6-48, when the parameter b is set equal to unity. Table 8.6-2 shows the actual steady-state values of N_i and compares these with the theoretical values for $b = 1$. The errors between the actual values and the theoretical values are given in the last column and vary between $+2.3\%$ and -4.8%. Thus, the steady-state values provided by the Kalman algorithms agree closely with the theoretical equation for $b = 1.0$.

Remember that b is the ratio of the peak error Δx divided by the effective RMS error due to uncertainty of target motion. As was shown in the derivation of the Altair Kalman gain equations given in Section 8.4, the bound Δa on the uncertainty of target acceleration was treated as an RMS value, and so δx (which is equal to $\frac{1}{2} \Delta a T^2$) is also an RMS value. This shows that the factor b should be unity in the Altair equations, which agrees with the assumption used in Table 8.6-2. Thus, the theoretical expression for N_i in Eq 8.6-48 accurately predicts the steady-state value of N_i derived from the Altair gain algorithm.

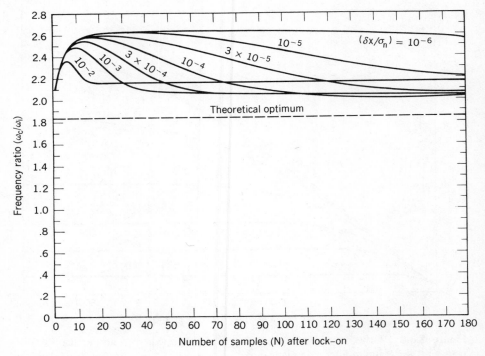

Figure 8.6-9 Variation of ω_c/ω_i ratio during acquisition transient, for Kalman-filter gain algorithm.

In the derivation of the Altair tracking equations, the parameter Δa was defined as a bound on the uncertainty of target acceleration. The preceding discussion indicates that Δa should be regarded as an "effective" RMS bound rather than a peak bound. Since this signal does not have a Gaussian distribution, it is not clear what the ratio of peak to effective RMS value should be. On the other hand, the peak bound should certainly be greater than the effective RMS bound.

TABLE 8.6-1 Steady-State Parameter Values Obtained by Simulating the Altair Gain Equations

$\delta x/\sigma_n$	N_i	N_c	ω_c/ω_i
10^{-2}	3.421	2.507	2.147
10^{-3}	8.221	6.107	2.059
10^{-4}	20.32	15.18	2.024
10^{-5}	50.70	37.96	2.010
10^{-6}	126.9	95.11	2.006

TABLE 8.6-2 Theoretical Number of Samples Integrated (n_i) Compared with Steady-State Values Obtained by Simulating Altair Gain Equations with $b = 1.0$

$\dfrac{\delta x}{\sigma_n}$	Number of Samples Integrated N_i		
	Actual	Theoretical	Error
10^{-2}	3.421	3.343	2.3%
10^{-3}	8.221	8.397	-2.1%
10^{-4}	20.32	21.09	-3.7%
10^{-5}	50.70	52.98	-4.4%
10^{-6}	126.9	133.1	-4.8%

In the Altair system, the bound Δa has apparently been set under the assumption that it is a peak bound rather than an effective RMS bound. This incorrect assumption would make the integration time lower than the theoretical optimal value. On the other hand, even though the setting of the integration time is theoretically nonoptimal, this setting is probably desirable in a practical sense because of uncertainty in the target model.

This point is illustrated in Fig 8.6-10, which gives normalized plots of the error components as functions of the normalized integration time. As shown in Eq 8.6-35, the effective RMS error due to target acceleration uncertainty is proportional to the square of the integration time T_i, while Eq 8.6-36 shows that the RMS error due to radar noise is inversely proportional to the square root of the integration time. In the figure, the integration time is normalized so that it is unity at the point where these two RMS error components are equal. The upper curve is the total RMS error, which is the square root of the sum of the squares of the two RMS error components. As shown, the total error is a minimum at a normalized integration time of 0.758.

The total error increases much more rapidly when the integration time exceeds the optimum value than when it is less than that value. For example, if the integration time T_i is a factor of 2 less than the optimum value, the total error is only 27.6% above the minimum, but if it is a factor of 2 greater than optimum, the total error is 89.8% above the minimum. Therefore, it is desirable to set the integration time lower than the theoretical optimum derived from the model because there is uncertainty in the model.

When the vehicle is in exoatmospheric flight, the value of K_x is appreciably less than unity except during the first few seconds after acquisition. At such low values of K_x, the parameter K_x' is closely approximated by K_x, and so the normalized time constant N_c is closely approximated by

$$N_c = \frac{1}{\omega_c t} = \frac{1}{K_x'} \cong \frac{1}{K_x} \qquad (8.6\text{-}83)$$

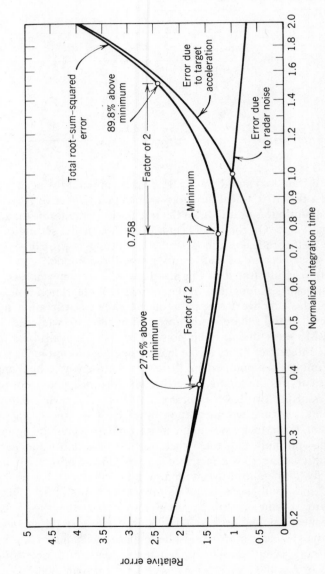

Figure 8.6-10 Plots of radar noise error, target acceleration error, and total root-sum-squared error, versus normalized integration time.

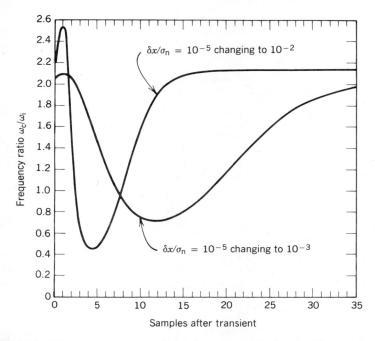

Figure 8.6-11 Transient response plots of ω_c/ω_i ratio for the Kalman-filter gain algorithm, which result from a large step increase in δ_x (reentry transient).

The ratio $\delta x/\sigma_n$ can be calculated from Eq 8.6-52 by setting $b = 1$, which gives

$$\frac{\delta x}{\sigma_n} = \left(\frac{0.409}{N_c}\right)^{2.5} = (0.409 K_x)^{2.5} \qquad (8.6\text{-}84)$$

After the lock-on transient has ended, the value of K_x during exoatmospheric flight typically may vary from about 0.05 to 0.005. By Eq 8.6-83, the corresponding range of variation of N_c would be 20–200, and by Eq 8.6-84 the ratio $\delta x/\sigma_n$ would vary from 6×10^{-5} to 2×10^{-7}.

The ratio $\delta x/\sigma_n$ varies slowly during exoatmospheric flight, but during reentry this ratio increases abruptly. The transients in Fig 8.6-11 show the effect of a large step increase in the ratio $\delta x/\sigma_n$. For both transients, the ratio $\delta x/\sigma_n$ was set equal to 10^{-5} until N_i reached its steady-state value (50.7). Then $\delta x/\sigma_n$ was abruptly increased to 10^{-3} and 10^{-2}. The plots show the variation of the ω_c/ω_i ratio during the resultant transient of the Kalman gain computation. As the plots show, the ratio ω_c/ω_i drops to very low values during the transients, which indicates poor stability.

The corresponding peak overshoot is calculated from Eq 8.6-27 and is plotted in Fig 8.6-12. As the plots show, during the transients there are

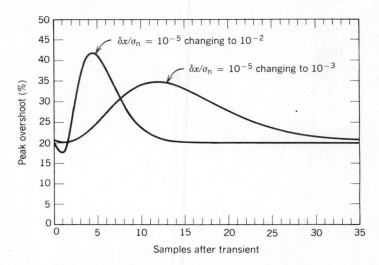

Figure 8.6-12 Plots of peak overshoot of Kalman-filter gain algorithm, which result from a large step increase of δ_x (reentry transient).

appreciable periods of high overshoot and hence poor stability. These plots suggest that the Altair tracker may experience poor stability at the moment of reentry.

8.6.4 Remainder Coefficient Analysis for Triple-Integration Loop

The remainder coefficient analysis that has been presented applies to the target estimator (angle tracker) feedback loops, which have a double integration at low frequency. As was shown in Section 8.2, Eq 8.2-6, the range tracker has the following loop transfer function, which has a triple integration at low frequency:

$$G_{(r)} = \frac{\omega_c(1 + \omega_{i1}/s)(1 + \omega_{i2}/s)}{s} \tag{8.6-85}$$

By Eq 8.2-5, the loop transfer function of the angle-tracking feedback loop is

$$G_{(a)} = \frac{\omega_c(1 + \omega_i/s)}{s} \tag{8.6-86}$$

The magnitude asymptote plots of $G_{(r)}$ and $G_{(a)}$ were shown in Fig 8.2-4. In the frequency region near the peak value of $|G_{ib}|$ and at higher frequencies, the loop transfer function $G_{(r)}$ of Eq 8.6-85 closely approximates that of $G_{(a)}$ of Eq 8.6-86, provided that

$$\omega_i = \omega_{i1} + \omega_{i2} \tag{8.6-87}$$

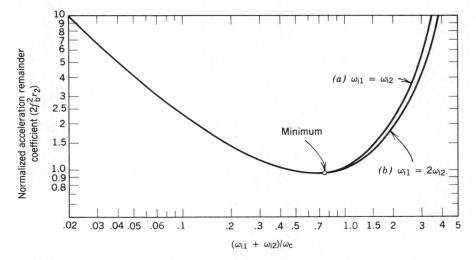

Figure 8.6-13   The acceleration remainder coefficient r_2 for a triple-integration loop, normalized in terms of the noise bandwidth f_b, expressed as a function of equivalent integral-network break frequency $(\omega_{i1} + \omega_{i2})$ divided by ω_c.

Hence, the noise bandwidth expression for $G_{(a)}$ given in Eq 8.6-20 applies approximately to $G_{(r)}$ when Eq 8.6-87 holds. By Eq 8.6-20, the noise bandwidth for the range tracker is approximately

$$f_b = \tfrac{1}{4}(\omega_c + \omega_i) = \tfrac{1}{4}(\omega_c t + \omega_{i1} + \omega_{i2}) \tag{8.6-88}$$

By simulation, the acceleration remainder coefficient r_2 for $G_{(r)}$ was calculated for various values of the loop parameters. This was achieved by plotting the error response to a unit step of acceleration, given by $x_i = \tfrac{1}{2}t^2$. The acceleration remainder coefficient r_2 was obtained by performing the calculations illustrated in Fig 8.6-1. Equation 8.6-88 was applied to express the remainder coefficient in terms of the noise bandwidth f_b. The resultant data is presented in Fig 8.6-13, which is a plot of the function $2f_b^2 r_2$. The horizontal axis gives the frequency ratio $(\omega_{i1} + \omega_{i2})/\omega_c$. Two curves are shown: for $\omega_{i1} = \omega_{i2}$ and for $\omega_{i1} = 2\omega_{i2}$.

Both curves of Fig 8.6-13 are a minimum at $(\omega_{i1} + \omega_{i2}) = 0.75\,\omega_c t$. The exact minimum values of r_2 are

$$\text{For } \omega_{i1} = \omega_{i2}: \quad r_{2(min)} = 0.4794/f_b^2 \tag{8.6-89}$$

$$\text{For } \omega_{i1} = 2\omega_{i2}: \quad r_{2(min)} = 0.4865/f_b^2 \tag{8.6-90}$$

Thus the minimum r_2 is slightly lower for $\omega_{i1} = \omega_{i2}$. These values should be compared with the value of $r_{2(min)}$ for the double-integration angle-tracking loop, which by Eq 8.6-25 is

$$r_{2(min)} = 0.3059/f_b^2 \qquad (8.6\text{-}91)$$

Comparing Eqs 8.6-89, -91 shows that the minimum acceleration remainder coefficient (at constant noise bandwidth) for the triple-integration loop exceeds that for the double-integration loop by the ratio $0.4794/0.3059$, which is a factor of 1.57.

For the triple-integration loop, the break frequency sum $(\omega_{i1} + \omega_{i2})$ is equivalent to the break frequency ω_i of the double-integration loop. As was shown in Eq 8.6-26, the optimum value of ω_i for the double-integration loop is $0.543\omega_c$. For the triple-integration loop, the optimum occurs at a somewhat higher value: $(\omega_{i1} + \omega_{i2}) = 0.75\omega_c$. However, the minimum in Fig 8.6-13 is very broad. At $(\omega_{i1} + \omega_{i2}) = 0.543\omega_c$, the value for r_2 in Fig 8.6-13 is $0.4925/f_b^2$, which is only 2.7% above the minimum.

In conclusion, the preceding analysis shows that a triple-integration loop achieves a minimum acceleration remainder coefficient (for constant noise bandwidth) at approximately the same loop parameters as a double-integration loop, provided that the integral-network break frequencies are set so that $(\omega_{i1} + \omega_{i2})$ for the triple-integration loop is equal to ω_i for the double-integration loop. However, at constant noise bandwidth, the minimum acceleration remainder coefficient for the triple-integration loop is 57% greater than that for the double-integration loop. Hence, for constant radar noise error, the error due to target motion is increased by 57% when an optimized double-integration loop is replaced by an optimized triple-integration loop.

8.7 SUMMARY OF KALMAN FILTERING

The Kalman filter can be separated into two parts: (1) the tracking filter itself and (2) the error covariance matrix calculations for computing the optimal gain parameters of the tracking filter. Section 8.5 demonstrated that the tracking filter portion of the Altair tracking system is equivalent to a conventional tracking filter, which filters the data in terms of the coordinates of the target. This same result should apply to any implementation of the Kalman filter. Therefore, it is only the second part of the Kalman filter that is unique to Kalman filter theory: the calculation of the optimal values of the tracking filter gains.

The Altair tracking system has demonstrated excellent tracking performance. The primary reason for the effectiveness of the Altair tracking system is that it performs tracking in terms of the coordinates of the target and compensates for the predictable portion of target motion. However, these aspects of the tracking system are independent of the fact that it uses a Kalman filter.

On the other hand, the Kalman filter algorithm, which adjusts the smoothing gains of the tracking loops, is also an important factor in the excellent tracking performance of Altair. It has provided very effective control of gains in the Altair tracking system, as has been demonstrated by experience with the Altair radar. Besides, the theoretical analysis of tracking response given in this chapter shows that the Kalman algorithm provides an excellent general method for optimal adjustment of tracking gains.

Despite the very good performance of the Altair Kalman gain algorithm, it has some weaknesses. As was shown in Fig 8.6-11, the tracking filters may experience poor damping during the reentry transient. However, one would have to simulate the tracking algorithms using actual tracking data during reentry to determine quantitatively how low the damping actually is. If poor damping were found to exist, it could be easily corrected by placing an upper limit on the K_v gain so that the ratio ω_c/ω_i cannot drop appreciably below the theoretical optimum value 1.84.

Another dynamic limitation of the Altair tracking equations is that the integral-network compensation, which corrects for velocity error, uses trapezoidal integration. Although trapezoidal integration may be desirable after the system has reached steady state, simple integration would be better during the acquisition transient because the tracking error would decrease faster.

The Kalman filter concept provides a remarkably effective theoretical tool for dealing with tracking applications. However, the abstract mathematical theory that has engulfed this concept weakens its effectiveness. The basic Kalman filter principles can be reduced to relatively simple engineering terms. When this is done, one can apply the Kalman filter concepts in a more effective manner than when the theory is encased in a cloud of esoteric mathematics.

Although the solution provided by the Kalman filter is optimal, it is not necessarily best even in a theoretical sense. The Kalman filter is only as good as the target model that it uses. Often the model is selected on the basis of mathematical convenience rather than on sound engineering grounds.

There are many subtle questions as to what is best, which can be obscured by the mathematics. Consider, for example, the plot of error versus integration time, which was given in Fig 8.6-10. The plot shows that the error is much greater when the integration time is too large by a given factor than when it is too small by the same factor. Since the target model is only an estimate, it is best to choose an integration time that is theoretically nonoptimal. The model should be biased so as to err on the side of reduced integration time.

The Altair Kalman filter is not a pure Kalman filter because

1. it is based on a deterministic bound placed on the uncertainty of target acceleration rather than on a true statistical forcing signal,
2. the separate range and angle tracking systems are not completely consistent with Kalman filter theory.

There have been long debates on these subjects by those who have worked on the Altair system. Nevertheless, despite its theoretical limitations, the Altair tracking system is very effective.

The evidence indicates that theoretical weakness 1 is a strong practical advantage. It is highly questionable that any pure statistical model of target motion could provide as effective a tracking system.

The issues relating to theoretical weakness 2 are more subtle and would require an extensive analysis to provide definitive comparisons. The reason for using two separate tracking systems is that range tracking is very much more accurate than angle tracking in terms of the equivalent errors of target position. It is possible to combine these two trackers into a single tracker by using coordinate axes that rotate with the antenna. However, complicated corrections are required to translate target acceleration information into these rotating coordinates. This approach, which is more compatible with basic Kalman filter theory, has been implemented in other systems. Whether or not the Altair system would be better or worse if this approach were used cannot be determined without a detailed study.

As has been shown, the Kalman filter theory yields very effective algorithms for providing optimum gain adjustments under different operating conditions. These algorithms are based on the covariance matrix of the target estimator error. In the Altair tracker, this error covariance matrix is used only for the purpose of optimal adjustment of tracking filter gains.

On the other hand, the Kalman error covariance matrix also has other uses. This point was explained to the author by R. Pike Green, who worked on a Kalman filter application in another classified tracking system, where many targets were tracked simultaneously in a multibeam phased-array radar. A time-shared tracking algorithm was used, and the sample period for tracking the individual targets varied from target to target. The parameters of the error covariance matrix were used to determine the setting of each sample period. The larger the error covariance parameters for a given target, the more often the target information was sampled.

Remember that the Kalman error covariance matrix is based on the estimate of target error rather than on the actual target tracking data. However, it is possible to modify the error covariance parameters to take into account tracking information. This approach can yield more accurate error covariance parameters for determining the optimum sampling period for a given target.

This illustrates the point that there are many complex engineering issues associated with an optimal real-time tracking system. The purpose of this chapter is to provide a frame of reference for explaining the basic concepts. With this insight, one should be better prepared to deal with the more subtle aspects of such a system.

Appendix A

Equations Relating True and Pseudo Complex Frequencies

This appendix derives the equations used in Chapter 5, Section 5.3, which express σ and ω, the real and imaginary parts of the complex frequency s, as functions of α and W, the real and imaginary parts of the sampled-data complex pseudo-frequency p. The following variables are defined:

$$s = \sigma + j\omega \tag{A-1}$$

$$\bar{z} = e^{-sT} \tag{A-2}$$

$$U = \frac{1 - \bar{z}}{1 + \bar{z}} \tag{A-3}$$

Substituting Eqs A-1, -2 into Eq A-3 gives

$$U = \frac{1 - e^{-(\sigma + j\omega)T}}{1 + e^{-(\sigma + j\omega)T}} \tag{A-4}$$

Rationalize this expression by multiplying the numerator and denominator by $(1 + e^{-(\sigma - j\omega)T})$:

$$
\begin{aligned}
U &= \frac{(1 - e^{-(\sigma + j\omega)T})(1 + e^{-(\sigma - j\omega)T})}{(1 + e^{-(\sigma + j\omega)T})(1 + e^{-(\sigma - j\omega)T})} \\
&= \frac{1 - e^{-2\sigma T} + e^{-\sigma T}(e^{j\omega T} - e^{-j\omega T})}{1 + e^{-2\sigma T} + e^{-\sigma T}(e^{j\omega T} + e^{-j\omega T})}
\end{aligned} \tag{A-5}
$$

Replace the complex exponentials by the following trigonometric expressions:

$$(e^{j\omega T} - e^{-j\omega T}) = j2 \sin[\omega T] \tag{A-6}$$

$$(e^{j\omega T} + e^{-j\omega T}) = 2 \cos[\omega T] \tag{A-7}$$

This gives for U

$$
\begin{aligned}
U &= \frac{1 - e^{-2\sigma T} + j2e^{-\sigma T} \sin[\omega T]}{1 + e^{-2\sigma T} + 2e^{-\sigma T} \cos[\omega T]} \\
&= \frac{(e^{\sigma T} - e^{-\sigma T}) + j2 \sin[\omega T]}{(e^{\sigma T} + e^{-\sigma T}) + 2 \cos[\omega T]}
\end{aligned} \tag{A-8}
$$

This becomes

$$U = \frac{\sinh[\sigma T] + j\sin[\omega T]}{\cosh[\sigma T] + \cos[\omega T]} \tag{A-9}$$

The complex pseudo-frequency p is defined as follows:

$$U = p\frac{T}{2} \tag{A-10}$$

The real and imaginary parts of p are defined as α and W, respectively. Thus,

$$p = \alpha + jW \tag{A-11}$$

$$U = p\frac{T}{2} = \frac{\alpha T}{2} + j\frac{WT}{2} \tag{A-12}$$

Comparing Eqs A-9 and -12 gives

$$\alpha\frac{T}{2} = \frac{\sinh[\sigma T]}{\cosh[\sigma T] + \cos[\omega T]} \tag{A-13}$$

$$W\frac{T}{2} = \frac{\sin[\omega T]}{\cosh[\sigma T] + \cos[\omega T]} \tag{A-14}$$

These equations can be expressed as

$$\cosh[\sigma T] + \cos[\omega T] = \frac{2}{\alpha T}\sinh[\sigma T] \tag{A-15}$$

$$\cosh[\sigma T] + \cos[\omega T] = \frac{2}{WT}\sin[\omega T] \tag{A-16}$$

Since the left-hand sides of the two equations are equal, the right-hand sides must be equal. Hence,

$$W\sinh[\sigma T] = \alpha\sin[\omega T] \tag{A-17}$$

Solve Eq A-15 for $\cos[\omega T]$ and Eq A-16 for $\cosh[\sigma T]$:

$$\cos[\omega T] = \frac{2}{\alpha T}\sinh[\sigma T] - \cosh[\sigma T] \tag{A-18}$$

$$\cosh[\sigma T] = \frac{2}{WT}\sin[\omega T] - \cos[\omega T] \tag{A-19}$$

Square both sides of Eqs A-18, -19:

$$\cos^2[\omega T] = \left(\frac{2}{\alpha T}\right)^2 \sinh^2[\sigma T] + \cosh^2[\sigma T]$$

$$- 2\left(\frac{2}{\alpha T}\right)\sinh[\sigma T]\cosh[\sigma T] \tag{A-20}$$

$$\cosh^2[\sigma T] = \left(\frac{2}{WT}\right)^2 \sin^2[\omega T] + \cos^2[\omega T]$$

$$- 2\left(\frac{2}{WT}\right)\sin[\omega T]\cos[\omega T] \tag{A-21}$$

Substitute into these the following identities:

$$\cos^2[\omega T] = 1 - \sin^2[\omega T] \tag{A-22}$$

$$\cosh^2[\sigma T] = 1 + \sinh^2[\sigma T] \tag{A-23}$$

Equations A-20, -21 become

$$1 - \sin^2[\omega T] = \left(\frac{2}{\alpha T}\right)^2 \sinh^2[\sigma T] + \left(1 + \sinh^2[\sigma T]\right)$$

$$- 2\left(\frac{2}{\alpha T}\right)\sinh[\sigma T]\cosh[\sigma T] \tag{A-24}$$

$$1 + \sinh^2[\sigma T] = \left(\frac{2}{WT}\right)^2 \sin^2[\omega T] + \left(1 - \sin^2[\omega T]\right)$$

$$- 2\left(\frac{2}{WT}\right)\sin[\omega T]\cos[\omega T] \tag{A-25}$$

Equation A-17 yields the following:

$$\sin[\omega T] = \frac{W}{\alpha} \sinh[\sigma T] \tag{A-26}$$

$$\sinh[\sigma T] = \frac{\alpha}{W} \sin[\omega T] \tag{A-27}$$

Substitute Eq A-26 into the left side of Eq A-24 and Eq A-27 into the left side of Eq A-25. Subtract unity from both sides. This gives

$$-\left(\frac{W}{\alpha}\right)^2 \sinh^2[\sigma T] = \left(\frac{2}{\alpha T}\right)^2 \sinh^2[\sigma T] + \sinh^2[\sigma T]$$

$$-\left(\frac{4}{\alpha T}\right)\sinh[\sigma T]\cosh[\sigma T] \qquad \text{(A-28)}$$

$$\left(\frac{\alpha}{W}\right)^2 \sin^2[\omega T] = \left(\frac{2}{WT}\right)^2 \sin^2[\omega T] - \sin^2[\omega T]$$

$$-\left(\frac{4}{WT}\right)\sin[\omega T]\cos[\omega T] \qquad \text{(A-29)}$$

Combining similar terms in Eqs A-28, -29 gives

$$\frac{4}{\alpha T}\sinh[\sigma T]\cosh[\sigma T] = \left[\left(\frac{2}{\alpha T}\right)^2 + 1 - \left(\frac{W}{\alpha}\right)^2\right]\sinh^2[\sigma T] \quad \text{(A-30)}$$

$$\frac{4}{WT}\sin[\omega T]\cos[\omega T] = \left[\left(\frac{2}{WT}\right)^2 - 1 - \left(\frac{\alpha}{W}\right)^2\right]\sin^2[\omega T] \quad \text{(A-31)}$$

Dividing the right-hand sides of the equations by the left-hand sides gives

$$1 = \left(\frac{1}{\alpha T} + \frac{\alpha T}{4} - \frac{W^2 T}{4\alpha}\right)\tanh[\sigma T]$$

$$= \frac{1}{\alpha T}\left(1 + \left(\frac{\alpha T}{2}\right)^2 + \left(\frac{WT}{2}\right)^2\right)\tanh[\sigma T] \qquad \text{(A-32)}$$

$$1 = \left(\frac{1}{WT} - \frac{WT}{4} - \frac{\alpha^2 T}{4W}\right)\tan[\omega T]$$

$$= \frac{1}{WT}\left(1 - \left(\frac{WT}{2}\right)^2 - \left(\frac{\alpha T}{2}\right)^2\right)\tan[\omega T] \qquad \text{(A-33)}$$

Solve Eq A-32 for tanh[σT] and Eq A-33 for tan[ωT]. This gives

$$\tanh[\sigma T] = \frac{\alpha T}{1 + \left[(\alpha T/2)^2 + (WT/2)^2\right]} \qquad \text{(A-34)}$$

$$\tan[\omega T] = \frac{WT}{1 - \left[(\alpha T/2)^2 + (WT/2)^2\right]} \qquad \text{(A-35)}$$

The magnitude of the complex pseudo-frequency p is equal to

$$|p| = \sqrt{\alpha^2 + W^2} \qquad \text{(A-36)}$$

Hence, Eqs A-34, -35 can be expressed as follows in terms of $|p|$:

$$\tanh[\sigma T] = \frac{\alpha T}{1 + (|p|T/2)^2} \qquad \text{(A-37)}$$

$$\tan[\omega T] = \frac{WT}{1 - (|p|T/2)^2} \qquad \text{(A-38)}$$

Appendix B

Integration of Runge-Kutta Polynomial

Equation 5.4-31 of Section 5.4 gave the following expression, used in the Runge–Kutta integration routine, for the integral of $\dot{y}[t]$ over the sampling period T:

$$\Delta y = \frac{T}{6}\left(\dot{y}_{(1)} + 4\dot{y}_{(2)} + \dot{y}_{(3)}\right) \tag{B-1}$$

This is the best estimate of the integral of $\dot{y}[t]$ over the time interval T, assuming one is given the values of $\dot{y}[t]$ only at the beginning, middle, and end of that time interval, which are designated $\dot{y}_{(1)}$, $\dot{y}_{(2)}$, $\dot{y}_{(3)}$, respectively. This expression will now be derived.

Three values of $\dot{y}[t]$ are specified at time t_1, t_2, t_3, where the two time intervals, $t_3 - t_2$ and $t_2 - t_1$, are equal. The best estimate of the continuous signal $\dot{y}[t]$ corresponding to these three values is obtained by fitting to them a polynomial of the following form:

$$\dot{y}[t] = a + bt + ct^2 \tag{B-2}$$

To simplify the fitting of this curve, the zero-time axis is chosen such that time t is zero at time t_2. Since the time interval between t_1 and t_3 is the sample period T, then $t_2 = 0$, $t_1 = -T/2$, and $t_3 = +T/2$. The values of $\dot{y}[t]$ at times t_1, t_2, and t_3 are therefore related as follows to the coefficients a, b, c:

$$\dot{y}_{(1)} = a + bt_1 + ct_1^2 = a + b(-T/2) + c(-T/2)^2 \tag{B-3}$$

$$\dot{y}_{(2)} = a + bt_2 + ct_2^2 = a \tag{B-4}$$

$$\dot{y}_{(3)} = a + bt_3 + ct_3^2 = a + b(T/2) + c(T/2)^2 \tag{B-5}$$

Solving these for the polynomial coefficients a, b, c gives

$$a = \dot{y}_{(2)} \tag{B-6}$$

$$b = \frac{1}{T}\left(\dot{y}_{(3)} - \dot{y}_{(1)}\right) \tag{B-7}$$

$$c = \frac{2}{T^2}\left(\dot{y}_{(3)} + \dot{y}_{(1)} - 2\dot{y}_{(2)}\right) \tag{B-8}$$

The integral of the polynomial expression for $\dot{y}[t]$ given in Eq B-2, over the interval from t_1 to t_3, is

$$\begin{aligned}
\Delta y &= \int_{-T/2}^{+T/2} (a + bt + ct^2)\, dt \\
&= \left\{ a(T/2) + (b/2)(T/2)^2 + (c/3)(T/2)^3 \right\} \\
&\quad - \left\{ a(-T/2) + (b/2)(-T/2)^2 + (c/3)(-T/2)^3 \right\} \\
&= aT + (c/12)T^3 \tag{B-9}
\end{aligned}$$

Substituting Eqs B-6, -8 into this gives the expression of Eq B-1, which was to be proved.

Appendix C

Summary of Matrix Formulas

C.1 SOME BASIC MATRIX DEFINITIONS

The elements of a matrix are labeled with a double subscript notation of the form a_{ij}. The first subscript i is the row of the matrix, and the second subscript j is the column. Thus, a 3-by-3 matrix has the form

$$\underset{\sim}{A} = \begin{bmatrix} a_{11} & a_{12} & a_{13} \\ a_{21} & a_{22} & a_{23} \\ a_{31} & a_{32} & a_{33} \end{bmatrix} \tag{C-1}$$

In this book, a matrix is designated by underlining its symbol with a tilde ($\sim$). If the number of rows equals the number of columns, the matrix is called *square*. The *diagonal* is defined only for a square matrix, and for Eq C-1 consists of the elements a_{11}, a_{22}, a_{33}. The *trace* of a square matrix, designated $\mathrm{tr}[\underset{\sim}{A}]$, is the sum of the diagonal elements, which for this matrix is

$$\mathrm{tr}\left[\underset{\sim}{A}\right] = \sum_{i=1}^{n} a_{ii} = a_{11} + a_{22} + a_{33} \tag{C-2}$$

The following are special types of square (n-by-n) matrices:

Diagonal matrix. A diagonal matrix, designated D or D_n, is a square matrix in which all of the off-diagonal elements are zero.

Unit matrix. A unit matrix, designated I or I_n, is a diagonal matrix in which all of the diagonal elements are unity.

Triangular matrix. An upper triangular matrix is a square matrix in which all of the upper off-diagonal elements are zero. A lower triangular matrix is a square matrix in which all of the lower off-diagonal elements are zero.

553

For a square 3-by-3 matrix, an upper triangular matrix, a diagonal matrix, and a unit matrix have the following forms:

Upper triangular matrix:

$$T_3 = \begin{bmatrix} a_{11} & 0 & 0 \\ a_{21} & a_{22} & 0 \\ a_{31} & a_{32} & a_{33} \end{bmatrix} \tag{C-3}$$

Diagonal matrix:

$$D_3 = \begin{bmatrix} a_{11} & 0 & 0 \\ 0 & a_{22} & 0 \\ 0 & 0 & a_{33} \end{bmatrix} \tag{C-4}$$

Unit matrix:

$$I_3 = \begin{bmatrix} 1 & 0 & 0 \\ 0 & 1 & 0 \\ 0 & 0 & 1 \end{bmatrix} \tag{C-5}$$

A *zero matrix* is a matrix in which all elements are zero.

C.2 MATRIX ADDITION

Matrix addition can be performed only on matrices having the same number of rows and the same number of columns. It is performed by adding the corresponding elements of the two matrices. For example:

$$\begin{bmatrix} a_{11} & a_{12} \\ a_{21} & a_{22} \\ a_{31} & a_{32} \end{bmatrix} + \begin{bmatrix} b_{11} & b_{12} \\ b_{21} & b_{22} \\ b_{31} & b_{32} \end{bmatrix} = \begin{bmatrix} a_{11} + b_{11} & a_{12} + b_{12} \\ a_{21} + b_{21} & a_{22} + b_{22} \\ a_{31} + b_{31} & a_{32} + b_{32} \end{bmatrix} \tag{C-6}$$

Matrix subtraction is defined in a similar manner.

C.3 MATRIX MULTIPLICATION

When a matrix is multiplied by a scalar, all elements of the matrix are multiplied by the scalar. For example,

$$\begin{bmatrix} a_{11} & a_{12} \\ a_{21} & a_{22} \end{bmatrix} \times K = \begin{bmatrix} a_{11}K & a_{12}K \\ a_{21}K & a_{22}K \end{bmatrix} \tag{C-7}$$

Two matrices can be multiplied together only if the number of columns of the first matrix is equal to the number of rows of the second matrix. One or both of these matrices can consist of a single column or row (i.e., it can be a vector). Consider the following matrix product:

$$C = AB \tag{C-8}$$

The elements of the product matrix C are given by the formula

$$c_{ij} = \sum_{k=1}^{n} a_{ik} b_{kj} \tag{C-9}$$

where n is the number of columns of the matrix $\underset{\sim}{A}$, which is equal to the number of rows of the matrix $\underset{\sim}{B}$. Consider, for example, the following matrix multiplication:

$$\underset{\sim}{C} = \begin{bmatrix} a_{11} & a_{12} \\ a_{21} & a_{22} \\ a_{31} & a_{32} \\ a_{41} & a_{42} \end{bmatrix} \begin{bmatrix} b_{11} & b_{12} & b_{13} \\ b_{21} & b_{22} & b_{23} \end{bmatrix} \tag{C-10}$$

The product matrix $\underset{\sim}{C}$ is

$$\underset{\sim}{C} = \begin{bmatrix} a_{11}b_{11} + a_{12}b_{21} & a_{11}b_{12} + a_{12}b_{22} & a_{11}b_{13} + a_{12}b_{23} \\ a_{21}b_{11} + a_{22}b_{21} & a_{21}b_{12} + a_{22}b_{22} & a_{21}b_{13} + a_{22}b_{23} \\ a_{31}b_{11} + a_{32}b_{21} & a_{31}b_{12} + a_{32}b_{22} & a_{31}b_{13} + a_{32}b_{23} \\ a_{41}b_{11} + a_{42}b_{21} & a_{41}b_{12} + a_{42}b_{22} & a_{41}b_{13} + a_{42}b_{23} \end{bmatrix} \tag{C-11}$$

The matrix product $\underset{\sim}{A}\underset{\sim}{B}$ is generally not equal to $\underset{\sim}{B}\underset{\sim}{A}$. Matrix division is not defined.

C.4 TRANSPOSE OF A MATRIX

The transpose of a matrix $\underset{\sim}{A}$ is designated A^T, and is obtained by interchanging the rows and columns of the matrix $\underset{\sim}{A}$. For example, the transpose of the 3-by-3 matrix $\underset{\sim}{A}$ of Eq C-1 is

$$\underset{\sim}{A}^T = \begin{bmatrix} a_{11} & a_{21} & a_{31} \\ a_{12} & a_{22} & a_{32} \\ a_{13} & a_{23} & a_{33} \end{bmatrix} \tag{C-12}$$

For a square matrix, the transpose can be formed by rotating the matrix about its diagonal. Consider also the following 3-by-2 matrix:

$$\underset{\sim}{F} = \begin{bmatrix} f_{11} & f_{21} \\ f_{12} & f_{22} \\ f_{13} & f_{23} \end{bmatrix} \tag{C-13}$$

The transpose of this matrix is

$$\underset{\sim}{F}^T = \begin{bmatrix} f_{11} & f_{12} & f_{13} \\ f_{21} & f_{22} & f_{23} \end{bmatrix} \tag{C-14}$$

The square of any matrix F^2 is obtained by multiplying the matrix by its transpose:

$$\underset{\sim}{F}^2 = \underset{\sim}{F}\underset{\sim}{F}^T \tag{C-15}$$

An important matrix equation involving the transpose is

$$(\underset{\sim}{A}\underset{\sim}{B})^T = \underset{\sim}{B}^T\underset{\sim}{A}^T \tag{C-16}$$

The *complex conjugate transpose* (or Hermitian conjugate) of a matrix $\underset{\sim}{A}$ is designated $\underset{\sim}{A}^\dagger$. Its elements are the complex conjugates of the elements of the matrix $\underset{\sim}{A}^T$, which is the transpose of the matrix $\underset{\sim}{A}$. An important relation concerning this matrix is

$$(\underset{\sim}{A}\underset{\sim}{B})^\dagger = \underset{\sim}{B}^\dagger\underset{\sim}{A}^\dagger \tag{C-17}$$

The following are definitions for special types of matrices:

$\underset{\sim}{A}$ is symmetric if	$\underset{\sim}{A} = \underset{\sim}{A}^T$	(C-18)
$\underset{\sim}{A}$ is skew-symmetric if	$\underset{\sim}{A} = -\underset{\sim}{A}^T$	(C-19)
$\underset{\sim}{A}$ is Hermitian if	$\underset{\sim}{A} = \underset{\sim}{A}^\dagger$	(C-20)
$\underset{\sim}{A}$ is skew-Hermitian if	$\underset{\sim}{A} = -\underset{\sim}{A}^\dagger$	(C-21)
A is normal if	$\underset{\sim}{A}^\dagger\underset{\sim}{A} = \underset{\sim}{A}\underset{\sim}{A}^\dagger$	(C-22)
$\underset{\sim}{A}$ is unitary if	$\underset{\sim}{A}^\dagger\underset{\sim}{A} = \underset{\sim}{I}$	(C-23)
$\underset{\sim}{A}$ is orthogonal if	$\underset{\sim}{A}^\dagger\underset{\sim}{A} = \underset{\sim}{D}$	(C-24)

C.5 DETERMINANT OF A MATRIX

The determinant is defined only for a square matrix $\underset{\sim}{A}$ and is designated $\det[\underset{\sim}{A}]$. The determinant of an n-by-n matrix is a sum of $n!$ terms, each of which is the product of n elements, multiplied by ± 1. Consider the following 2-by-2 matrix:

$$\underset{\sim}{A} = \begin{bmatrix} a_{11} & a_{12} \\ a_{21} & a_{22} \end{bmatrix} \tag{C-25}$$

The determinant of this matrix is

$$\det[\underset{\sim}{A}] = a_{11}a_{22} - a_{21}a_{12} \tag{C-26}$$

The determinants for higher-order matrices can be computed by the method of cofactors, which reduces a determinant to those of matrices of lower order. Let us apply this to the following 3-by-3 matrix $\underset{\sim}{A}$, which was given in Eq C-1:

$$\underset{\sim}{A} = \begin{bmatrix} - & a_{11} & - & a_{12} & - & a_{13} & - \\ & & & | & & & \\ & a_{21} & & a_{22} & & a_{23} & \\ & & & | & & & \\ & a_{31} & & a_{32} & & a_{33} & \end{bmatrix} \tag{C-27}$$

For each element of a matrix there is a submatrix called a *minor*, which is obtained by deleting the row and column of the element. The *cofactor* of an element is the determinant of its minor multiplied by $(-1)^{i+j}$, where i and j are the column and row indices of the element. Thus, the minor of the element a_{12} in the matrix $\underset{\sim}{A}$ of Eq C-27 is

$$\text{Minor}[a_{12}] = \underset{\sim}{A}_{12} = \begin{bmatrix} a_{21} & a_{23} \\ a_{31} & a_{33} \end{bmatrix} \tag{C-28}$$

This minor is formed by deleting the elements indicated by the lines in the matrix of Eq C-27, which are the elements in the row and column of element a_{12}. The cofactor of element a_{12} is the determinant of this minor matrix of Eq C-28 multiplied by $(-1)^{1+2}$, which is

$$\text{Cof}[a_{12}] = (-1)^{1+2} \det \begin{bmatrix} a_{21} & a_{23} \\ a_{31} & a_{33} \end{bmatrix}$$

$$= (-1)^3 (a_{21}a_{33} - a_{31}a_{23}) = a_{31}a_{23} - a_{21}a_{33} \tag{C-29}$$

The determinant of the matrix $\underset{\sim}{A}$ is the sum, over any row or column, of each element multiplied by its cofactor. Let us apply this rule to the elements a_{11}, a_{12}, a_{13} of the first row:

$$\det[\underset{\sim}{A}] = a_{11} \text{Cof}[a_{11}] + a_{12} \text{Cof}[a_{12}] + a_{13} \text{Cof}[a_{13}]$$

$$= a_{11}(-1)^{1+1} \det \begin{bmatrix} a_{22} & a_{23} \\ a_{32} & a_{33} \end{bmatrix} + a_{12}(-1)^{1+2} \det \begin{bmatrix} a_{21} & a_{23} \\ a_{31} & a_{33} \end{bmatrix}$$

$$+ a_{11}(-1)^{1+3} \det \begin{bmatrix} a_{21} & a_{22} \\ a_{31} & a_{32} \end{bmatrix} \tag{C-30}$$

Solving for the determinants of the minors gives

$$\det[\underset{\sim}{A}] = a_{11}(a_{22}a_{33} - a_{32}a_{23}) - a_{12}(a_{21}a_{33} - a_{31}a_{23})$$

$$+ a_{11}(a_{21}a_{32} - a_{31}a_{22}) \tag{C-31}$$

C.6   ADJOINT OF A MATRIX

The adjoint, designated adj[$\underset{\sim}{A}$], is defined for a square matrix. The adjoint of a square matrix $\underset{\sim}{A}$ is the transpose of a related matrix $\underset{\sim}{B}$ having elements that are the cofactors of the elements of matrix $\underset{\sim}{A}$. Thus

$$\text{adj}[\underset{\sim}{A}] = \underset{\sim}{B}^T \tag{C-32}$$

where each element b_{ij} of the matrix $\underset{\sim}{B}$ is the cofactor of the corresponding element a_{ij} of matrix $\underset{\sim}{A}$:

$$b_{ij} = \text{Cof}[a_{ij}] = (-1)^{i+j} \det[\underset{\sim}{A}_{ij}] \tag{C-33}$$

The matrix $\underset{\sim}{A}_{ij}$ is the minor matrix of the element a_{ij}, obtained by deleting the row and column of the element a_{ij}. Let us find the adjoint matrix for the 3-by-3 matrix $\underset{\sim}{A}$ given in Eq C-1. The corresponding matrix $\underset{\sim}{B}$ is defined as

$$\underset{\sim}{B} = \begin{bmatrix} b_{11} & b_{12} & b_{13} \\ b_{21} & b_{22} & b_{23} \\ b_{31} & b_{32} & b_{33} \end{bmatrix} \tag{C-34}$$

Consider, for example, the element b_{31} of this matrix: This is equal to

$$b_{31} = \text{Cof}[a_{31}] = (-1)^{3+1} \det \begin{bmatrix} a_{12} & a_{13} \\ a_{22} & a_{23} \end{bmatrix}$$

$$= (-1)^4 (a_{12}a_{23} - a_{22}a_{13}) = a_{12}a_{23} - a_{22}a_{13} \tag{C-35}$$

By repeating this procedure, all nine elements of the matrix $\underset{\sim}{B}$ can be calculated. By Eq C-32, the adjoint of $\underset{\sim}{A}$ is expressed as follows in terms of these nine elements:

$$\text{adj}[\underset{\sim}{A}] = \underset{\sim}{B}^T = \begin{bmatrix} b_{11} & b_{21} & b_{31} \\ b_{12} & b_{22} & b_{32} \\ b_{13} & b_{23} & b_{33} \end{bmatrix} \tag{C-36}$$

Let us calculate the adjoint of the 2-by-2 matrix $\underset{\sim}{A}$ given in Eq C-25. The corresponding matrix $\underset{\sim}{B}$ has the form

$$\underset{\sim}{B} = \begin{bmatrix} b_{11} & b_{12} \\ b_{21} & b_{22} \end{bmatrix} \tag{C-37}$$

The elements of this matrix $\underset{\sim}{B}$ are

$$b_{11} = \text{Cof}[a_{11}] = (-1)^{1+1} \det[a_{22}] = a_{22} \tag{C-38}$$

$$b_{12} = \text{Cof}[a_{12}] = (-1)^{1+2} \det[a_{21}] = -a_{21} \tag{C-39}$$

$$b_{21} = \text{Cof}[a_{21}] = (-1)^{2+1} \det[a_{12}] = a_{12} \tag{C-40}$$

$$b_{22} = \text{Cof}[a_{22}] = (-1)^{2+2} \det[a_{11}] = -a_{11} \tag{C-41}$$

Since $\underset{\sim}{A}$ is a 2-by-2 matrix, the minor of any element is a scalar, and the determinant of the minor is equal to that scalar. Substituting Eqs C-38 to -41 into Eq C-37 gives the matrix $\underset{\sim}{B}$:

$$\underset{\sim}{B} = \begin{bmatrix} a_{22} & -a_{21} \\ -a_{12} & a_{11} \end{bmatrix} \tag{C-42}$$

Applying Eq C-32 gives the following for the adjoint of $\underset{\sim}{A}$:

$$\text{adj}\big[\underset{\sim}{A}\big] = \underset{\sim}{B}^T = \begin{bmatrix} a_{22} & -a_{12} \\ -a_{21} & a_{11} \end{bmatrix} \tag{C-43}$$

This is the transpose of the matrix $\underset{\sim}{B}$ given in Eq C-42.

C.7   INVERSE OF A MATRIX

The inverse of a matrix $\underset{\sim}{A}$ is designated $\underset{\sim}{A}^{-1}$, and is defined by

$$\underset{\sim}{A}\underset{\sim}{A}^{-1} = \underset{\sim}{I} \tag{C-44}$$

where $\underset{\sim}{I}$ is an identity (unit) matrix, which was defined in Section C-1. If the matrix $\underset{\sim}{A}$ is square and its determinant is not zero, the inverse of the matrix $\underset{\sim}{A}$ is given as follows by Cramer's rule:

$$\underset{\sim}{A}^{-1} = \frac{\text{adj}\big[\underset{\sim}{A}\big]}{\det\big[\underset{\sim}{A}\big]} \tag{C-45}$$

Let us apply Eq C-45 to calculate the inverse of the 2-by-2 matrix $\underset{\sim}{A}$ of Eq C-25. The determinant of this matrix was given in Eq C-26, and the adjoint was given in Eq C-43. Substituting Eqs C-26, -43 into Eq C-45 yields the inverse matrix:

$$\underset{\sim}{A}^{-1} = \frac{\text{adj}\big[\underset{\sim}{A}\big]}{\det\big[\underset{\sim}{A}\big]}$$

$$= \frac{1}{a_{11}a_{22} - a_{21}a_{12}} \begin{bmatrix} a_{22} & -a_{12} \\ -a_{21} & a_{11} \end{bmatrix} \tag{C-46}$$

Appendix D

Values for Noise-Bandwidth Integrals

Frequency-response calculations of noise bandwidth and mean-square error can be greatly simplified by using tables of integrals of the general form

$$I_n = \frac{1}{2\pi j} \int_{-j\infty}^{+j\infty} H[s]H[-s]\, ds \tag{D-1}$$

The transfer function $H[s]$ is defined as follows as a ratio of two polynomials in s, where the order of the numerator is less than that of the denominator:

$$H[s] = \frac{c_{n-1}s^{n-1} + \cdots + c_1 s + c_0}{d_n s^n + d_{n-1}s^{n-1} + \cdots + d_1 s + d_0} \tag{D-2}$$

The parameter n is the order of s in the denominator of $H[s]$. This appendix gives equations for the integral I_n for values of n from 1 to 6, as functions of the coefficients c, d of the numerator and denominator of $H[s]$. These were obtained from Appendix E of Newton, Gould, and Kaiser [6.5]. That reference shows how these integrals are derived, and gives expressions for the integrals for values of n from 1 to 10. The integrals I_n, for $n = 1$ to 6, are as follows:

$$I_1 = \frac{c_0^2}{2 d_0 d_1} \tag{D-3}$$

$$I_2 = \frac{c_1^2 d_0 + c_0^2 d_2}{2 d_0 d_1 d_2} \tag{D-4}$$

$$I_3 = \frac{1}{2 D_3}\left[c_2^2 d_0 d_1 + \left(c_1^2 - 2 c_0 c_2 \right) d_0 d_3 + c_0^2 d_2 d_3 \right] \tag{D-5}$$

$$D_3 = d_0 d_3 (d_1 d_2 - d_0 d_3) \tag{D-6}$$

$$I_4 = \frac{1}{2 D_4}\left[c_3^2 m_{40} + \left(c_2^2 - 2 c_1 c_3 \right) m_{41} + \left(c_1^2 - 2 c_0 c_2 \right) m_{42} + c_0^2 m_{43} \right] \tag{D-7}$$

$$m_{40} = d_0 d_1 d_2 - d_0^2 d_3 \tag{D-8}$$

$$m_{41} = d_0 d_1 d_4 \tag{D-9}$$

$$m_{42} = d_0 d_3 d_4 \tag{D-10}$$

$$m_{43} = d_2 d_3 d_4 - d_1 d_4^2 \tag{D-11}$$

$$D_4 = d_0 d_4 \left(d_1 d_2 d_3 - d_0 d_3^2 - d_1^2 d_4 \right) \tag{D-12}$$

$$I_5 = \frac{1}{2D_5} \left[c_4^2 m_{50} + \left(c_3^2 - 2c_2 c_4 \right) m_{51} + \left(c_2^2 - 2c_1 c_3 + 2c_0 c_4 \right) m_{52} \right.$$

$$\left. + \left(c_1^2 - 2c_0 c_2 \right) m_{53} + c_0^2 m_{54} \right] \tag{D-13}$$

$$m_{51} = d_1 d_2 - d_0 d_3 \tag{D-14}$$

$$m_{52} = d_1 d_4 - d_0 d_5 \tag{D-15}$$

$$m_{53} = \frac{d_2 m_{52} - d_4 m_{51}}{d_0} \tag{D-16}$$

$$m_{54} = \frac{d_2 m_{53} - d_4 m_{52}}{d_0} \tag{D-17}$$

$$m_{50} = \frac{d_3 m_{51} - d_1 m_{52}}{d_5} \tag{D-18}$$

$$D_5 = d_0 \left(d_1 m_{54} - d_3 m_{53} + d_5 m_{52} \right) \tag{D-19}$$

$$I_6 = \frac{1}{2D_6} \left[c_5^2 m_{60} + \left(c_4^2 - 2c_3 c_5 \right) m_{61} + \left(c_3^2 - 2c_2 c_4 + 2c_1 c_5 \right) m_{62} \right.$$

$$\left. + \left(c_2^2 - 2c_1 c_3 + 2c_0 c_4 \right) m_{63} + \left(c_1^2 - 2c_0 c_2 \right) m_{64} + c_0^2 m_{65} \right] \tag{D-20}$$

$$m_{61} = d_0 d_3^2 + d_1^2 d_4 - d_0 d_1 d_5 - d_1 d_2 d_3 \tag{D-21}$$

$$m_{62} = d_0 d_3 d_5 + d_1^2 d_6 - d_1 d_2 d_5 \tag{D-22}$$

$$m_{63} = d_0 d_5^2 + d_1 d_3 d_6 - d_1 d_4 d_5 \tag{D-23}$$

$$m_{64} = \frac{d_2 m_{63} - d_4 m_{62} + d_6 m_{61}}{d_0} \tag{D-24}$$

$$m_{65} = \frac{d_2 m_{64} - d_4 m_{63} + d_6 m_{62}}{d_0} \tag{D-25}$$

$$m_{60} = \frac{d_4 m_{61} - d_2 m_{62} + d_0 m_{63}}{d_6} \tag{D-26}$$

$$D_6 = d_0 \left(d_1 m_{65} - d_3 m_{64} + d_5 m_{63} \right) \tag{D-27}$$

Appendix E

Acceleration Remainder Coefficient for Integral-Compensated Loop

This appendix calculates the acceleration remainder coefficient for a loop with the following loop transfer function:

$$G = \frac{\omega_c(1 + \omega_i/s)}{s} = \frac{\omega_c(s + \omega_i)}{s^2} \tag{E-1}$$

The error transfer function of this loop is

$$G_{ie} = \frac{1}{1 + G} = \frac{s^2}{s^2 + \omega_c s + \omega_c \omega_i}$$

$$= \frac{s^2}{s^2 + 2\zeta\omega_n s + \omega_n^2} \tag{E-2}$$

where

$$\omega_n = \sqrt{\omega_c \omega_i} \tag{E-3}$$

$$2\zeta = \frac{\omega_c}{\omega_n} = \sqrt{\frac{\omega_c}{\omega_i}} \tag{E-4}$$

The acceleration error coefficient for this loop is

$$c_2 = \frac{1}{\omega_c \omega_i} = \frac{1}{\omega_n^2} \tag{E-5}$$

The acceleration remainder coefficient is derived from the transient response to a unit step of acceleration, which is characterized by

$$\frac{d^2 x_i}{dt^2} = u[t] \tag{E-6}$$

The transform of this acceleration is

$$\mathscr{L}\left[\frac{d^2 x_i}{dt^2}\right] = \frac{1}{s} \tag{E-7}$$

Hence, the transform of the input is

$$X_i = \frac{1}{s^2} \mathscr{L}\left[\frac{d^2 x_i}{dt^2}\right] = \frac{1}{s^3} \qquad \text{(E-8)}$$

Applying this to Eq E-2 gives the following for the transform of the error for a unit step of acceleration:

$$X_e = G_{ie} X_i = \frac{1}{s\left(s^2 + 2\zeta\omega_n s + \omega_n^2\right)} \qquad \text{(E-9)}$$

For $\zeta \geq 1$, the inverse transform of Eq E-9 has no overshoot. Hence, for $\zeta \geq 1$ the acceleration remainder coefficient r_2 is equal to the acceleration error coefficient c_2:

$$\text{For } \zeta \geq 1: \quad r_2 = c_2 \qquad \text{(E-10)}$$

for $\zeta < 1$, the inverse transform of Eq E-10 is

$$x_e = \frac{1}{\omega_n^2}\left\{1 - \exp[-\zeta\omega_n t]\left(\cos[\omega_o t] + \frac{\zeta}{\sqrt{1-\zeta^2}}\sin[\omega_o t]\right)\right\} \qquad \text{(E-11)}$$

where ω_o is equal to

$$\omega_o = \omega_n\sqrt{1-\zeta^2} \qquad \text{(E-12)}$$

The remainder coefficient is derived by considering the values of error x_e at minima and maxima points. This error response of Eq E-10 has minima and maxima at values of time t for which $\omega_o t = N\pi$, where N is a positive integer. At these maxima and minima points, $\sin[\omega_o t] = 0$ and $\cos[\omega_o t] = (-1)^N$. Hence the values of x_e of Eq E-10 at the maxima and minima points are

$$x_e[N] = \frac{1}{\omega_n^2}\left\{1 - (-1)^N \exp[-\zeta\omega_n t]\right\}$$

$$= c_2\left\{1 - (-1)^N \exp\left[-\frac{\zeta\omega_n N\pi}{\omega_o}\right]\right\}$$

$$= c_2\left\{1 - (-1)^N \exp\left[-\frac{N\pi\zeta}{\sqrt{1-\zeta^2}}\right]\right\} \qquad \text{(E-13)}$$

In accordance with Eq E-5, $1/\omega_n^2$ was replaced by the acceleration error coefficient c_2. Equation E-13 can be expressed in the form

$$x_e[N] = c_2\left\{1 - (-A)^N\right\} \qquad \text{(E-14)}$$

where the constant A is equal to

$$A = \exp\left[-\frac{\zeta\pi}{\sqrt{1 - \zeta^2}}\right] \tag{E-15}$$

In Chapter 8, Fig 8.6-1 shows how the acceleration remainder coefficient is calculated. In that figure, point (1) is $x_e[1]$, point (2) is $x_e[2]$, etc., where $x_e[1], x_e[2], \ldots$ are obtained from Eq E-14 by setting $N = 1, 2, 3, \ldots$. The figure shows that the acceleration remainder coefficient is equal to

$$r_2 = x_e[1] + (x_e[1] - x_e[2]) + (x_e[3] - x_e[2])$$
$$+ (x_e[3] - x_e[4]) + \cdots \tag{E-16}$$

Combining Eqs E-14, -16 gives

$$r_2 = c_2\{(1 + A) + (A + A^2) + (A^3 + A^2) + (A^3 + A^4) + \cdots\}$$
$$= c_2\{1 + 2A + 2A^2 + 2A^3 + 2A^4 + \cdots\} \tag{E-17}$$

Using long division, divide 1 by $(1 - A)$. This gives the following infinite series:

$$\frac{1}{1 - A} = 1 + A + A^2 + A^3 + A^4 + \cdots \tag{E-18}$$

Hence,

$$\frac{1 + A}{1 - A} = \frac{2}{1 - A} - 1$$
$$= 1 + 2A + 2A^2 + 2A^3 + 2A^4 + \cdots \tag{E-19}$$

Comparing Eqs E-17, -19 allows r_2 to be expressed in closed form:

$$r_2 = \frac{c_2(1 + A)}{1 - A} \tag{E-20}$$

Combining Eqs E-15, -20 gives the following for the acceleration remainder coefficient for $\zeta < 1$:

$$r_2 = \frac{c_2\left\{1 + \exp\left[-\zeta\pi/\sqrt{1 - \zeta^2}\right]\right\}}{1 - \exp\left[-\zeta\pi/\sqrt{1 - \zeta^2}\right]} \tag{E-20}$$

as shown previously, $r_2 = c_2$ for $\zeta \geq 1$.

Computation of Steady-State and Initial-Value Coefficients

The long-division method for calculating the steady-state and initial-value coefficients will now be described by applying it to the following feedback transfer function:

$$G_{ib} = \frac{5 + 5.5s}{(s + 1)(s + 5)} = \frac{5 + 5.5s}{5 + 6s + s^2} \tag{F-1}$$

The steady-state coefficients are obtained by expressing the numerator and denominator polynomials in ascending powers of s, and dividing numerator by denominator in a long-division process as follows:

$$
\begin{array}{r}
1 - 0.1s - 0.08s^2 + 0.116s^3 \\
5 + 6s + s^2 \overline{\smash{)}\ 5 + 5.5s } \\
5 + 6.0s + 1.0s^2 \\
\hline
-0.5s - 1.0s^2 \\
-0.5s - 0.6s^2 - 0.10s^3 \\
\hline
-0.4s^2 + 0.10s^3 \\
-0.4s^2 - 0.48s^3 - 0.08s^4 \\
\hline
0.58s^3 + 0.08s^4
\end{array}
\tag{F-2}
$$

Thus, G_{ib} can be expanded in the series

$$
\begin{aligned}
G_{ib} &= 1 - 0.1s - 0.08s^2 + 0.116s^3 + \cdots \\
&= c_0' + c_1's + c_2's^2 + c_3's^3 + \cdots
\end{aligned}
\tag{F-3}
$$

TABLE F-1 Feedback-Signal Steady-State and Initial-Value Coefficients for G_{ib} of Eq. F-1

Steady State	Initial value
$c_0' = 1$	$a_0 = 0$
$c_1' = -0.1 \text{ sec}$	$a_1 = 5.5 \text{ sec}^{-1}$
$c_2' = -0.08 \text{ sec}^2$	$a_2 = -28 \text{ sec}^{-2}$
$c_3' = 0.116 \text{ sec}^3$	$a_3 = 140.5 \text{ sec}^{-3}$

The steady-state coefficients c'_n for the feedback signal are shown in the first column of Table F-1. When the terms of the polynomials are reversed, so they are expressed in descending powers of s, the long division process yields the initial-value coefficients, as follows:

$$
\begin{array}{r}
5.5/s - 28/s^2 + 140.5/s^3 \\
\hline
s^2 + 6s + 5 \overline{)\,5.5s \quad\quad +5} \\
5.5s \quad\quad +33 \quad +27.5/s \\
\hline
-28 \quad -27.5/s \\
-28 \quad -168/s \; -140/s^2 \\
\hline
140.5/s \; +140/s^2
\end{array}
$$

$$(\text{F-4})$$

Hence G_{ib} can be expanded as

$$
\begin{aligned}
G_{ib} &= \frac{5.5}{s} - \frac{28}{s^2} + \frac{140.5}{s^3} + \cdots \\
&= a_0 + \frac{a_1}{2} + \frac{a_2}{s^2} + \frac{a_3}{s^3} + \cdots
\end{aligned}
$$

$$(\text{F-5})$$

The initial-value coefficients are shown in the second column of Table F-1.

References

CHAPTER 1

1.1. G. Biernson, *Principles of Feedback Control*, Vol. 1, *Feedback System Design*, Wiley, New York, 1988.

1.2. G. Biernson, *Principles of Feedback Control*, Vol. 2, *Advanced Control Topics*, Wiley, New York, 1988.

1.3. E. Brookner, *Radar Technology*, Artech House, Dedham, MA, 1977.

CHAPTER 2

2.1. E. I. Green, "The Decilog, a Unit for Logarithmic Measurement," *Electrical Engineering*, Vol. 73, No. 7, July 1954, pp. 597–599.

2.2. H. Chestnut and R. W. Mayer, *Servomechanism and Regulating System Design*, Vol. 1, Wiley, New York, 1950.

CHAPTER 3

3.1. G. Biernson, "A Simple Method for Calculating the Time Response of a System to an Arbitrary Input," *AIEE Transactions*, Vol. 74, Pt. II, 1955, pp. 227–245.

CHAPTER 5

5.1. J. R. Ragazzini and G. F. Franklin, *Sampled Data Control Systems*, McGraw-Hill, New York, 1958.

5.2. A. V. Oppenheim, A. S. Willsky, and I. T. Young, *Signals and Systems*, Prentice-Hall, Englewood Cliffs, NJ, 1983, p. 527, Fig. 8.15.

5.3. G. W. Johnson, D. P. Lindorff, and G. A. Nordling, "Extension of Continuous Data Systems Design Techniques to Sample-Data Control Systems," *Trans. AIEE*, Vol. 74, Pt. II, Sept. 1955, pp. 252–263.

5.4. David P. Lindorff, *Theory of Sample Data Control Systems*, Wiley, New York, 1965.

5.5. K. P. Phillips, "Current-source Inverter for AC Motor Drives," *IEEE Trans. Industry Applications*, Vol. IA-8, No. 6, Nov./Dec. 1972, pp. 679–683.

5.6. E. D. Rainville and P. E. Bedient, *Elementary Differential Equations*, 5th ed., Macmillan, New York, 1974, pp. 389–392.

CHAPTER 6

6.1. William Feller, *An Introduction to Probability Theory and Its Applications*, Vol. 1, 2nd ed., Wiley, New York, 1957, p. 166.

6.2. Al Ryan and Tim Scranton, "DC Amplifier Noise Revisited," *Analog Dialog*, Vol. 18, No. 1, 1984. (Analog Devices, Norwood, MA.)

6.3. John G. Truxal, *Automatic Feedback Control System Synthesis*, McGraw-Hill, New York, 1955.

6.4. R. S. Phillips, "Statistical Properties of Time-Varying Data" and "RMS Error Criterion in Servomechanisms Design," Chapters 6 and 7 of H. M. James, N. B. Nichols, and R. S. Phillips, *Theory of Servomechanisms*, McGraw-Hill, New York, 1947.

6.5. G. C. Newton, L. A. Gould, and J. F. Kaiser, *Analytical Design of Linear Feedback Controls*, Wiley, New York, 1957.

6.6. G. Biernson, "Fundamental Equations for the Application of Statistical Techniques to Feedback Control Systems," *IRE Trans. Automatic Control*, Vol. PGAC-2, Feb. 1957, pp. 56–78.

6.7. James G. Titus, "Wind-induced Torques Measured on a Large Antenna," U.S. Naval Research Laboratory Report 5549, Dec. 27, 1960.

6.8. Robert S. Briggs, Jr., "Control System for a Large Steerable Tracking Antenna," MSEE Thesis, University of Texas, Jan. 1967.

CHAPTER 7

7.1. Merrill I. Skolnick, *Introduction to Radar Systems*, McGraw-Hill, New York, 1962.

7.2. Merrill I. Skolnick, ed., *Radar Handbook*, McGraw-Hill, New York, 1970.

7.3. M. Schwartz, W. Bennett, and S. Stein, *Communication Systems and Techniques*, McGraw-Hill, New York, 1966.

7.4. David K. Barton, *Radar System Analysis*, Prentice-Hall, Englewood Cliffs, NJ, 1964; revised 1977 (Artech House, Dedham, MA).

7.5. David K. Barton and Harold R. Ward, *Handbook of Radar Measurement*, Prentice-Hall, Englewood Cliffs, NJ, 1969.

7.6. Peter W. Hannan, "Optimum Feeds for All Three Modes of a Monopulse Antenna," *IRE Trans. Antennas and Propagation*, Vol. AP-9, No. 5, Sept. 1961, pp. 444–461.

7.7. Denny D. Pidhayny, "Effect of Error Sensor Crosstalk on System Operation for Linear and Nonlinear Cases," Aerospace Report No. TOR-0088(3491-05)-1, Aerospace Corporation, Los Angeles, CA, April 9, 1988.

CHAPTER 8

8.1. R. E. Kalman, "A New Approach to Linear Filtering and Prediction Problems," *Trans. ASME* (*Journal of Basic Engineering*), Vol. 82, 1960, pp. 35–45.

8.2. R. E. Kalman and R. S. Bucy, "New Results in Linear Filtering and Prediction Theory," *Trans. ASME* (*Journal of Basic Engineering*), March 1961, pp. 95–107.

8.3. D. E. Gray, Ed., *American Institute of Physics Handbook*, 2nd ed., McGraw-Hill, Englewood Cliffs, NJ, 1963.

8.4. Sighard F. Hoerner, *Fluid Dynamic Drag*, 1958 (published by the author), Library of Congress Catalog No. 57-13009.

8.5. Arthur Gelb, Ed., *Applied Optimal Estimation*, MIT Press, Cambridge, MA, 1974.

8.6. D. E. Johansen, "Solution of a Linear Mean-Square Estimator Problem when Process Statistics Are Undefined," *IEEE Trans. Automatic Control*, Vol. AC-11, Jan. 1966, pp. 20–30.

Problems

The following problems are numbered in terms of the section or chapter of the book to which they apply.

CHAPTER 2

2-1. For the *RLC* lowpass filter shown in Fig 2.2-1 of Chapter 2, the circuit parameters are: $R = 3000 \ \Omega$, $L = 25$ H, $C = 1.0 \ \mu\text{f}$. Compute the following parameters:

(a) The natural frequency ω_n.

(b) The damping ratio ζ.

(c) The oscillation frequency ω_o.

Using the general transient response and frequency response plots of Figs 2.2-3, -4, find the following parameters:

(d) The rise time T_R of the step response.

(e) The delay time T_d of the step response.

(f) The peak overshoot of the step response.

(g) The half-power frequency ω_{hp} of the frequency response.

(h) The frequency ω_ϕ of one radian of phase lag.

(i) The maximum magnitude of the frequency response.

From the values of ω_{hp}, ω_ϕ, estimate the values of the rise time T_R and delay time T_d. What is the percent difference between these estimates of T_R, T_d and the values obtained in parts (d), (e) from the transient response plot. Using the median curve of Fig 2.3-8, estimate, from the peak overshoot of the step response, the maximum magnitude of the frequency response. What is the ratio of this estimate to the actual value of the maximum magnitude obtained in part (i)?

2-2.   Consider three feedback loops with the following loop transfer functions, where $\omega_c = 10 \ \text{sec}^{-1}$, $\omega_i = 2.5 \ \text{sec}^{-1}$, and $\tau = 0.1 \ \text{sec}$:

$$G_A = \frac{\omega_c}{s}$$

$$G_B = \frac{\omega_c}{s(1 + \tau s)}$$

$$G_C = \frac{\omega_c(1 + \omega_i/s)}{s} = \frac{\omega_c(s + \omega_i)}{s^2}$$

For all three loops, do the following:

(a) Calculate the feedback and error transfer functions G_{ib} and G_{ie}.

(b) Plot the error and feedback responses to a unit step input at increments of 0.05 sec up to 0.2 sec, and at increments of 0.1 sec up to 1.0 sec.

(c) Plot the error and feedback responses to a unit ramp input $(x_i = t)$ using the same values of time.

Show the step and ramp input signals as dashed curves on the plots. Pick scales to show the feedback or error signal responses clearly.

2-3.   For the feedback-signal step responses in Fig P2-1, estimate the following characteristics of the G_{ib} and G frequency responses:

(a) The frequency where $|G_{ib}| = 0.7$.

(b) The frequency where $\text{Ang}[G_{ib}] = -57°$.

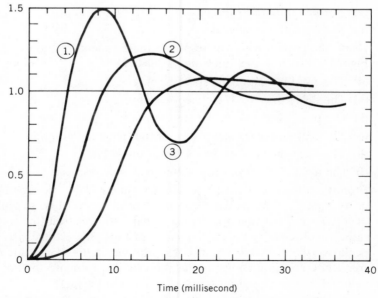

Figure P2-1   Step responses of feedback signal x_b for three feedback loops.

(c) The frequency where $|G| = 1$.

(d) The maximum value of $|G_{ib}|$.

CHAPTER 3

3.2-1. As in Fig 3.2-9 of Chapter 3, consider a target moving in a straight line at the constant velocity: $V = 480$ ft/sec. A tracking radar is located at a point 400 ft from the target trajectory, so that the minimum range is $R_{min} = 400$ ft. Consider five servos with the following loop transfer functions:

$$G_1 = \frac{\omega_c}{s}$$

$$G_2 = \frac{\omega_c}{s(1 + \omega_1/s)}$$

$$G_3 = \frac{\omega_c(1 + \omega_3/s)}{s}$$

$$G_4 = \frac{\omega_c(1 + \omega_3/s)}{s(1 + \omega_2/s)}$$

$$G_5 = \frac{\omega_c(1 + \omega_3/s)}{s(1 + \omega_1/s)(1 + \omega_2/s)}$$

The frequency parameters in sec^{-1} are $\omega_c = 20$, $\omega_1 = 0.1$, $\omega_2 = 0.32$, $\omega_3 = 8$. Do the following:

(a) Draw magnitude asymptote *sketches* of G and approximate G_{ie} for the five cases. What are the approximate equations for G_{ie}?

(b) What are the values for the applicable position, velocity, and acceleration error coefficients for the five servos and the values for the time-delay parameters τ_c or τ_{ic}?

(c) Plot the waveforms of angular velocity and angular acceleration of the target as observed by the servo for values of time t from -2 to $+2$ sec, in steps of 0.2 sec. Also plot the points of maximum acceleration on the angular-acceleration plot. Label the axes of these plots in rad/sec and rad/sec^2.

(d) Plot the approximate angular error responses for loops G_1, G_3, G_4, ignoring the effects of time delays in the filters H_c, H_{ic}. Multiply the error in radians by 1000 to express it in milliradians.

(e) Replot the curves in part (d), to include the time delays produced by the filter responses H_c, H_{ic}. You can trace and time-shift the curves. Plot these on the same graph as the corresponding plots of part (d).

(f) Calculate by means of error coefficients, for all five servos, the approximate error due to the following input command angle:

$$\Theta_k = 8 \sin[\omega t] \text{ deg}, \quad \text{where} \quad \omega = 3 \text{ rad/sec}$$

CHAPTER 4

4-1. A bandpass filter operating at a carrier frequency of 400 Hz has the following equivalent signal-frequency transfer function:

$$H = \frac{1}{1 + s/\omega_f}$$

The break frequency is $\omega_f/2\pi = 25$ Hz. Determine the parameters for implementing this filter with the RLC circuit of Fig P4-1. The inductor L has an inductance of 0.10 H and a Q of 40 at 400 Hz. The resistor R_p is the equivalent parallel circuit resistance corresponding to this Q, which is

$$R_p = Q\omega_r L$$

where $\omega_r/2\pi = 400$ Hz. Find the capacitance C and resistance R_1. What is the gain E_o/E_i at the carrier frequency?

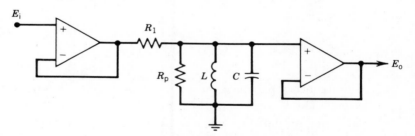

Figure P4-1 Circuit for RLC bandpass filter.

4-2. Assume that the bandpass filter in Problem 4-1 is implemented by the feedback RC circuit of Fig P4-2. The capacitance values are

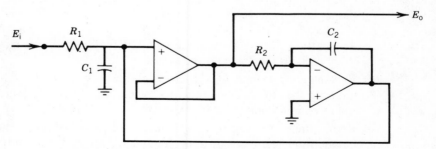

Figure P4-2 Circuit of feedback RC bandpass filter.

$C_1 = 1\ \mu\mathrm{F}$, $C_2 = 0.01\ \mu\mathrm{F}$. Find the values for resistors R_1 and R_2. What is the gain E_o/E_i at the carrier frequency.

4-3. A bandpass filter operating at a carrier frequency of 400 Hz has the following equivalent signal-frequency transfer function:

$$H = \frac{1}{1 + 2\zeta(s/\omega_n) + (s/\omega_n)^2}$$

where $\omega_n/2\pi = 50$ Hz and $\zeta = 0.707 = 1\sqrt{2}$. Assume that two circuits of the form of Fig P4-2 are cascaded to implement this transfer function, where $C_1 = 1\ \mu\mathrm{F}$, $C_2 = 0.01\ \mu\mathrm{F}$. Find the values for resistors R_1 and R_2 for both circuits. What is the total gain E_o/E_i for the two circuits at the carrier frequency?

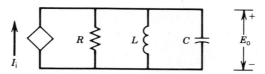

Figure P4-3   Circuit of tuned RLC IF amplifier stage.

4-4. As a simplified radar-system model, assume that the IF amplifier of a radar receiver operates at a carrier frequency of 60 MHz and contains the simple bandpass filter stage shown in Fig P4-3. The filter input signal is the current source I_i, and the output signal is the voltage E_o. The circuit parameters are $L = 1.0\ \mu\mathrm{H}$, $C = 7.036$ pF, and $R = 2.5\ \mathrm{k}\Omega$. This is a bandpass filter tuned to 60 MHz. The input to the IF amplifier is a 60-MHz pulse with a rectangular pulse modulation of pulse width 0.1 μsec. Plot the envelope of the 60-MHz wave at the output of the filter.

4-5. Assume a 60-MHz amplifier, as described in Problem 4-4, which uses a two-stage stagger-tuned filter, where each stage has the form of Fig P4-3. The equivalent signal-frequency transfer function of this bandpass filter has the same form shown in Problem 4-3, where $\omega_n/2\pi = 6.37$ MHz, $\zeta = 0.5$. Find the values of R and C for the two stages, assuming $L = 1.0\ \mu\mathrm{H}$. Plot the envelope of the 60-MHz IF signal at the output of the filter, assuming the input signal is the same as for Problem 4-4.

CHAPTER 5

5.3-1. Calculate the sampled-data algorithm to simulate the following lead-network transfer function:

$$\frac{Y}{X} = \frac{1 + s/\omega_L}{1 + s/\alpha\omega_L}$$

To simplify the computations, define the factor $2/\omega_L T$ as B. Give the numerical values of the algorithm constants for $T = 0.0001$ sec, $\alpha = 10$, $\omega_L = 200$ sec^{-1}.

5.3-2. Assume that the stage-positioning servo of Fig 5.3-1 in Chapter 5 has the following parameters, expressed in sec^{-1}:

$$\omega_e = 800, \quad \omega_{cc} = 500, \quad \omega_{cv} = 270, \quad \omega_{cp} = 80, \quad \omega_{ip} = 16$$

Note that $\tau_e = 1/\omega_e$. Use serial simulation to simulate the response to a unit step of command position x_k. Plot the controlled position, velocity, and acceleration. Ignore the effect of back EMF of the motor. Use a sample period T of 0.5×10^{-3} sec, and plot for 200 msec.

5.3-3. Using the servo of Problem 5.3-2, plot the error response x_e to the following normalized ramp input: $x_k = \omega_{cp}t$. The velocity ω_{cp} of the input ramp is chosen so that the peak ramp error is approximately unity. Perform this simulation for the following two values of the integral-network break frequency: $\omega_{ip} = 16$ sec^{-1}, 10 sec^{-1}.

5.5-1. For the ELF transmitter circuit of Fig 5.5-2 in Chapter 5, approximate the source current I_s with the following waveform: $I_s = I_{s0} \sin[\omega_o t]$. The waveform starts at time $t = 0$; I_s is zero prior to that point. Assume the frequency $f_o = \omega_o/2\pi = 100$ Hz, $I_{s0} = 100$ A, $R_a = 10$ Ω. The antenna is tuned to the frequency $\omega_o = 2\pi f_o$ and has a Q of 5 at this frequency. The inductor has a Q of 20 at the frequency f_o. Hence the circuit parameters are $L_a = 79.5$ mH, $C_a = 31.8$ μf, $L = 15.9$ mH, $R = 0.5$ Ω, $C_1 = C_2 = 159$ μF. Simulate the circuit and plot the antenna current for 4 cycles of I_s (to 40 msec). The sample period is $T = 0.1$ msec. Plot points every 0.4 msec. This response approximates the startup transient of the transmitter.

5.5-2. Another possible circuit for driving the ELF antenna described in Section 5.5.1 is shown in diagram *a* of Fig P5-1. The circuit uses a thyristor voltage-source inverter, which generates a voltage waveform equivalent to the current waveform I_s of Fig 5.5-1. (The voltage-source thyristor inverter is commonly used at megawatt power levels in uninterruptable AC power supplies to protect critical facilities, such as computers and broadcast stations, from loss of prime AC power. The 60-Hz prime power is rectified and stored in batteries. A voltage-source inverter converts the DC battery power back to AC power, which drives the critical equipment.) The values of L_1, L_2 in Fig P5-1 are the same as L in Problem 5.5-1; R_1, R_2 are the same as R; C is the same as C_1 or C_2; and the values for R_a, L_a, and C_a are the same. To develop the signal-flow diagram of the circuit, it is convenient to combine the series-connected inductors L_a and L_2 into a single inductor L_{a2} equal to $(L_a + L_2)$, as shown in diagram *b*. Do the following:

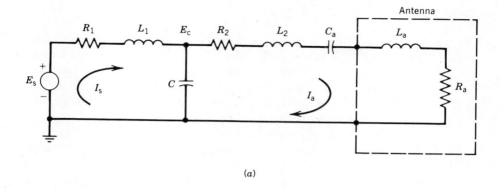

(a)

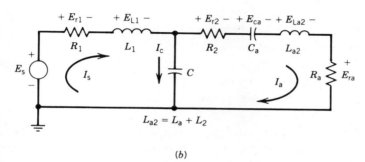

$$L_{a2} = L_a + L_2$$

(b)

Figure P5-1   ELF transmitter circuit with voltage-source inverter, used for serial simulation: (a) complete circuit; (b) simplified form used in simulation.

(a) Draw a signal-flow diagram for the circuit.

(b) Write the state equations for the circuit.

(c) Express these state equations in matrix format.

(d) Convert the signal-flow diagram for the circuit to the form for serial simulation.

(e) Perform the serial simulation described in Problem 5.5-1, using this circuit in response to the signal $E_s = E_{s0} \sin[\omega_o t]$ from the inverter, for $E_{s0} = 1000$ V.

CHAPTER 6

6.3-1.   Find the noise bandwidth in hertz of the following transfer function:

$$H = \frac{1 + s/20}{(1 + s/10)(1 + s/100)}$$

CHAPTER 7

7-1. A radar operating at a frequency of 6 GHz has a parabolic steerable antenna with an antenna efficiency of 55%. The radar is tracking a target having a radar cross section of 2 m² at a range of 80 km. The transmitter generates 300-kW peak power with rectangular pulses 0.07 μsec in duration. Assume that the total waveguide circuit loss (transmition plus reception) is 2.5 dB. The noise temperature at the input of the receiver (including noise from the antenna and the atmosphere) is 300°C. The matching loss of the receiver is 0.7 dB. Find the following:

(a) The wavelength.

(b) The peak antenna gain, expressed in decibels and as a power ratio.

(c) The approximate half-power beamwidth. (Assume this is the exact value in the following.)

(d) The beamwidth between -1.5-dB points.

(e) The range resolution of the radar.

In steps (f) to (k), assume that the target is located a the peak of the antenna beam, and find:

(f) The peak power received from the target at the input to the receiver.

(g) The energy received from one pulse at that point.

(h) The noise power density referenced to the input of the receiver.

(i) The signal/noise power ratio for an ideal matched receiver.

(j) The actual signal/noise power ratio at the output of the receiver.

(k) The output signal from the receiver is fed through a peak detector. Find the (approximate) noncoherent detection loss of the detector, and the effective detected signal/noise ratio.

(l) Repeat steps (j) and (k), assuming that the target is 0.5° from the center of the beam.

7-2. A radar transmits rectangular pulses 0.15 μsec wide, with a pulse-repetition frequency of 600 Hz. The signal/noise ratio at the output of a radar receiver (prior to the envelope detector) is 2.0 (or 3 dB). The matching loss of the receiver is 1.1 dB. Find the following:

(a) The approximate loss in the noncoherent detector.

(b) The signal/noise ratio E_s/p_n of an ideal matched receiver.

(c) The ratio E_s/p_{nd} of signal energy to detected noise power density.

(d) Assuming the range-gate expression of Eq 7.2-19, find the RMS range-gate timing error per pulse, and the RMS range error per pulse.

(e) The noise bandwidth of the IF amplifier, and the gate width τ_g corresponding to the range-gate expression used in part (d).

(f) The noise bandwidth of the range-tracking loop in hertz, assuming that the range-tracking loop has the following loop transfer function:

$$G = \frac{\omega_{cr}(1 + \omega_{ir}/s)}{s}$$

where $\omega_{cr} = 80$ sec^{-1}, $\omega_{ir} = 40$ sec^{-1}.

(g) The effective integration time of the range-tracking loop.

(h) The effective number of pulses integrated by the range-tracking loop.

(i) The RMS noise tracking error of the range-tracking loop.

(j) The peak (3-sigma) range-tracking error due to noise.

(k) The peak error at lock-on due to target velocity, for a maximum target velocity of 400 m/sec. Compare this with the linear region of the range gate.

(l) The maximum tracking error due to target motion after the lock-on transient has settled, assuming a maximum target acceleration of $4g$ (39.2 m/sec^2).

7-3. A monopulse tracking radar transmits 0.12-μsec rectangular pulses at a pulse-repetition frequency of 500 Hz. The signal/noise ratio at the output of the sum-channel receiver, prior to the detector, is 2.5 (or 4 dB). The matching loss of the receiver is 0.9 dB. The normalized monopulse slope k_m is 1.8, and the antenna beam has a 3-dB half-power beamwidth of 1.5°. The monopulse detector has a range-gate width τ_{ga} of 0.2 μsec. Find the following:

(a) The signal/noise ratio E_s/p_n of an ideal receiver.

(b) The RMS angular monopulse error due to noise for a single pulse.

(c) The noise bandwidth of the angle-tracking loop in hertz, and the effective integration time, assuming that the loop transfer function of the angle-tracking loop is approximated by the expression of Eq 7.2-52, where $\omega_i = 3$ sec^{-1}, $\omega_c = 9$ sec^{-1}, and $\omega_f = 27$ sec^{-1}.

(d) The effective number of pulses integrated by the angle-tracking loop.

(e) The RMS angle-tracking error per axis due to noise, expressed in milliradians.

(f) The peak (3-sigma) two-axis angle-tracking error due to noise, expressed in milliradians.

(g) The maximum angle-tracking error due to target motion, expressed in milliradians, for a target following a straight-line course at a velocity of 400 m/sec, reaching a minimum range of 500 m.

Author Index

Subject Index